CRUSTACEAN EGG PRODUCTION

CRUSTACEAN ISSUES

7

General editor:

FREDERICK R. SCHRAM
Natural History Museum of Los Angeles County, California

A.A.BALKEMA / ROTTERDAM / BROOKFIELD / 1991

CRUSTACEAN EGG PRODUCTION

Edited by
ADRIAN WENNER & ARMAND KURIS
University of California, Santa Barbara

CRC Press
Taylor & Francis Group
Boca Raton London New York

CRC Press is an imprint of the
Taylor & Francis Group, an **informa** business
A BALKEMA BOOK

CRC Press
Taylor & Francis Group
6000 Broken Sound Parkway NW, Suite 300
Boca Raton, FL 33487-2742

Published by:
CRCPress/Balkema
P.O. Box 447, 2300 AK Leiden, The Netherlands
e-mail: Pub.NL@taylorandfrancis.com
www.crcpress.com – www.taylorandfrancis.com

CIP-DATA KONINKLIJKE BIBLIOTHEEK, DEN HAAG

Crustacean

Crustacean egg production / ed. by Adrian Wenner & Armand Kuris. – Rotterdam [etc.]: Balkema. – Ill. – (Crustacean issues, ISSN 0168-6356; 7)
With index.
ISBN 90 6191 098 6 bound
Subject heading: crustaceans; reproduction.

Published by
A.A. Balkema, P.O. Box 1675, 3000 BR Rotterdam, Netherlands
A.A. Balkema Publishers, Old Post Road, Brookfield, VT 05036, USA

ISBN 13: 978-90-6191-098-5 (pbk)
ISBN 13: 978-1-138-44075-3 (hbk)

Visit the Taylor & Francis Web site at
http://www.taylorandfrancis.com

and the CRC Press Web site at
http://www.crcpress.com

Table of contents

2 *Variations in life history patterns*

3 *Population and fisheries aspects*

Preface

This volume is a sequel to two earlier *Crustacean Issues*, Volumes 2 and 3, which dealt specifically with crustacean growth. Thus, the related topics of growth and egg production are now covered more thoroughly than considered possible only a few years ago. Just as in the earlier volumes on growth, this volume on egg production arose in conjunction with a symposium held at the annual meetings of The Crustacean Society and the American Society of Zoologists (ASZ). That symposium, in 1987, included two sessions of invited papers on crustacean egg production. We also arranged an additional day-long series of contributed papers as a supplement to the symposium.

Our thanks go to the officers of those societies for their support, an acknowledgement which includes both the Ecology Division and the Invertebrate Zoology Division of the ASZ. Mary Adams-Wiley of the ASZ was particularly helpful, also assisting with financial arrangements through the ASZ. The Department of Biological Sciences of the University of California, Santa Barbara assisted greatly these past two years with telephone costs, mailing expenses, and the cost of duplication services. We also thank the authors of individual chapters for providing the material, for being so patient during the seemingly long time required for the editing and re-editing, and for being so cooperative in dealing with the changes suggested during that process. In particular, we feel privileged to have had the opportunity to get to know the authors better, both during the 1987 meetings and subsequent to that time, and to have been able to work closely once again with Frederick Schram. His patience, understanding, and cooperation have been most welcome. We also thank A.T. Balkema for his continued interest in invertebrate marine biology and for his recognition of the need for this type of volume.

Adrian M. Wenner
Armand M. Kuris

Introduction

As with the issue of crustacean growth, we found the contributions on egg production to be far more diverse and to have more depth than anything imagined before we started requesting authors to submit manuscripts. Fortunately, the papers herein fall rather equally into three categories: 1) egg formation, release and development, 2) variations in life history patterns, and 3) population and fisheries aspects. In order to keep the volume within manageable bounds, both embryogenesis and growth were excluded from consideration unless the material related directly to one of the three categories mentioned above. Eventually, manuscripts were submitted on virtually all relevant aspects of egg production. Penaeid shrimp provide excellent material for studies of egg production, maturation and jelly synthesis (Clark), especially since spawning can be induced at will under the proper circumstances. Knowledge of extrusion, attachment and retention of eggs (Talbot) is extended in studies herein on lobsters. Within the egg itself, our understanding of yolk utilization (Subramoniam) and the role of hemocyanins (Terwilliger) are further clarified from what was previously known.

Once eggs complete development, they may either enter a resting stage (Onbé) or not hatch immediately unless conditions are appropriate (Crisp, Hill & Holland); and cladocerans and barnacles, respectively, provide the research material for the elucidation of those interruptive phenomena. Egg loss from the pleopods in some genera can be considerable, but such loss varies according to species and/or locality (Corey). Last, but certainly not least, egg production does not operate independently of growth and molting (Nelson); those phenomena can be quite interdependent.

The environment, in its broadest definition, can have a marked effect on life history patterns. Interest is focused on brood mortality in commercial crabs as a consequence of parasitism (Kuris). Egg size varies considerably, as revealed in studies of squat lobsters (Van Dover & Williams); and both morphological and environmental restraints on egg production are found in marine amphipods (Steele & Steele). Wider treatments of reproductive patterns are described in *Armadillidium* (Brody), in a caridean shrimp guild (Bauer), and in *Cancer* crabs (Shields).

Increasing attention of late has been focused on egg production at the population level, especially in those forms living in specialized habitats and/or in commerically valuable species. Fiddler crabs dependent on sporadic rainfall for successful spawning (Rabalais) and freshwater crayfish (Huner & Lindqvist) have unusual developmental and egg production patterns in those relatively non-typical crustacean habitats. Special attention has also been paid to crayfish and lobsters because of their obvious economic importance. Studies of crayfish and lobsters in South Africa (Pollock) indicate that alterations in mean growth rates and fecundity of females can effect changes in egg and larval production in populations. Other summaries of studies of egg production in the American lobser in Canada (Waddy & Aiken, Ennis) and of the spiny lobster, *Jasus*, in New Zealand (Annala) round out our current understanding on lobsters. These contributions are followed by a similar analysis of season and spatial dynamics in penaeid prawns in Australia (Crocos). A few contributions deal with more theoretical aspects of fecundity at the population level. Egg production may vary during the year in those species which produce more than one brood per season; studies of *Emerita* (Wenner, Dugan & Wells) reveal how total season and lifetime egg production can be estimated despite that variation. Theoretical considerations along that same line (Somers), illustrate the various ways in which data on egg production as a function of size can be analyzed. Finally, the application of our extensive knowledge of crustacean egg production can be applied to fisheries management (Botsford) whenever size limits and stock-recruitment considerations need re-examination.

All too often, publication of a volume such as this may depress research in an area, despite the usual admonition provided in the introduction; that is, hope is usually expressed that publication will stimulate research. We feel the ultimate answers are nowhere near at hand on issues of crustacean egg production, despite the material we have been able to amass. The complexity of the crustacean life cycle has actually permitted only the beginning of investigations of this field of research.

Adrian M. Wenner
Armand M. Kuris

1. Egg formation, release and development

Egg production, release and activation in the marine shrimp, *Sicyonia ingentis*

Wallis H. Clark, Jr. & Muralidharan C. Pillai
Department of Animal Science & Bodega Marine Laboratory, University of California, Bodega Bay, USA

ABSTRACT

The decapod *Sicyonia ingentis* reproduces from the middle of June to the middle of October, during which the females undergo ovarian development and spawning. Multiple spawning without an intervening molt commonly occurs among both field and laboratory held populations. While the details on the vitellogenic stages of oögenesis are largely unknown, post-vitellogenic stages of oöcyte development have been characterized. During the late stages of oögenesis, large jelly-like inclusions (jelly precursor) develop in the oöplasm and eventually become externalized in deep extra-oöcytic crypts around the periphery of the oöcyte. This stage of the oögenesis (cortical specialization phase) occurs prior to germinal vesicle break down (GVBD). GVBD occurs before ovulation and spawning and is an asynchronous event. Prior to spawning, the females exhibit a characteristic prespawning behavior which is correlated with ovulation. The ovulated females can be induced to spawn on demand under laboratory conditions; upon contact with sea water, spawned ova undergo a series of activational events that include: 1) release of the jelly precursor from the extra-oöcytic crypts, 2) formation of a jelly layer, 3) resumption of meiosis, and 4) formation of an extracellular hatching envelope. These events are Mg^{++} dependent in fertilized ova; however, in unfertilized ova both Mg^{++} and Ca^{++} are required.

1 INTRODUCTION

The order Decapoda is divided into two suborders, the Pleocyemata, animals that brood their eggs, and the Dendrobranchiata, whose members are broadcast spawners. Although much is known concerning the reproductive biology of the Pleocyemata, a basic understanding of the reproductive biology of the Dendrobranchiata is still needed. This is somewhat curious, since members of this suborder, especially the Penaeoidea, represent one of the largest wild caught and cultured fisheries in the world. With the exception of older treatises by Hudinaga (1942) and King (1948), most of the available literature deals with techniques for the induction of ovarian growth (i.e. eyestalk ablation as well as nutritional and environmental requirements). While these studies have been important to culturists, for the most part, they have done little towards furthering our understanding of the basic process associated with egg production and early development.

The most thoroughly examined Penaeoidea is the marine shrimp, *Sicyonia ingentis* (Clark et al. 1984). Extensive knowledge is available on the reproductive biology of males, including studies on male gamete formation (Shigekawa & Clark 1986), sperm structure and sperm activation (for review see: Clark & Griffin 1988). Although less is known concerning the reproductive biology of the female, the reproductive cycle, postvitellogenic ovarian development (Anderson et al. 1985) and egg activation (Pillai & Clark 1987) have been studied. This paper reviews our knowledge of female reproductive biology in this species.

2 BIOGEOGRAPHY

Sicyonia ingentis is the only member of its genus recorded from the west coast of the United States. Its geographic range extends from Monterey Bay, California in the north to Isla Maria Madre, Mexico in the south; in addition it is also reported to be found in the Gulf of California (Perez Farfante 1985). *Sicyonia ingentis* is commonly found at depths of 61 to 183 m (Carlisle 1969). Females of this species attain a maximum length of 180 mm, measured from the telson to the base of the antenna. Males, on the other hand, are generally smaller and attain a maximum length of 157 mm (Herkelrath 1977). This species is commercially fished in the Santa Barbara Channel off California.

3 FIELD BIOLOGY

The reproductive season of *S. ingentis* females, sampled off Santa Barbara, California, between 1979-1981, extended from the middle of June to the middle of October. During this period the

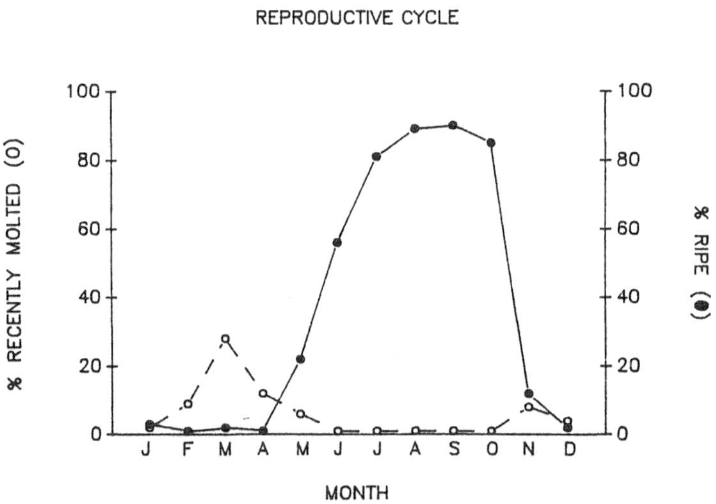

Figure 1. Molt activity (based on the frequency of animals with soft exoskeletons) and reproductive activity (based on the frequency of females with advanced stages of oöcytic development) for female *S. ingentis*. Solid line represents reproductive season and broken line represents the molting season. (Modified from Anderson et al. 1985.)

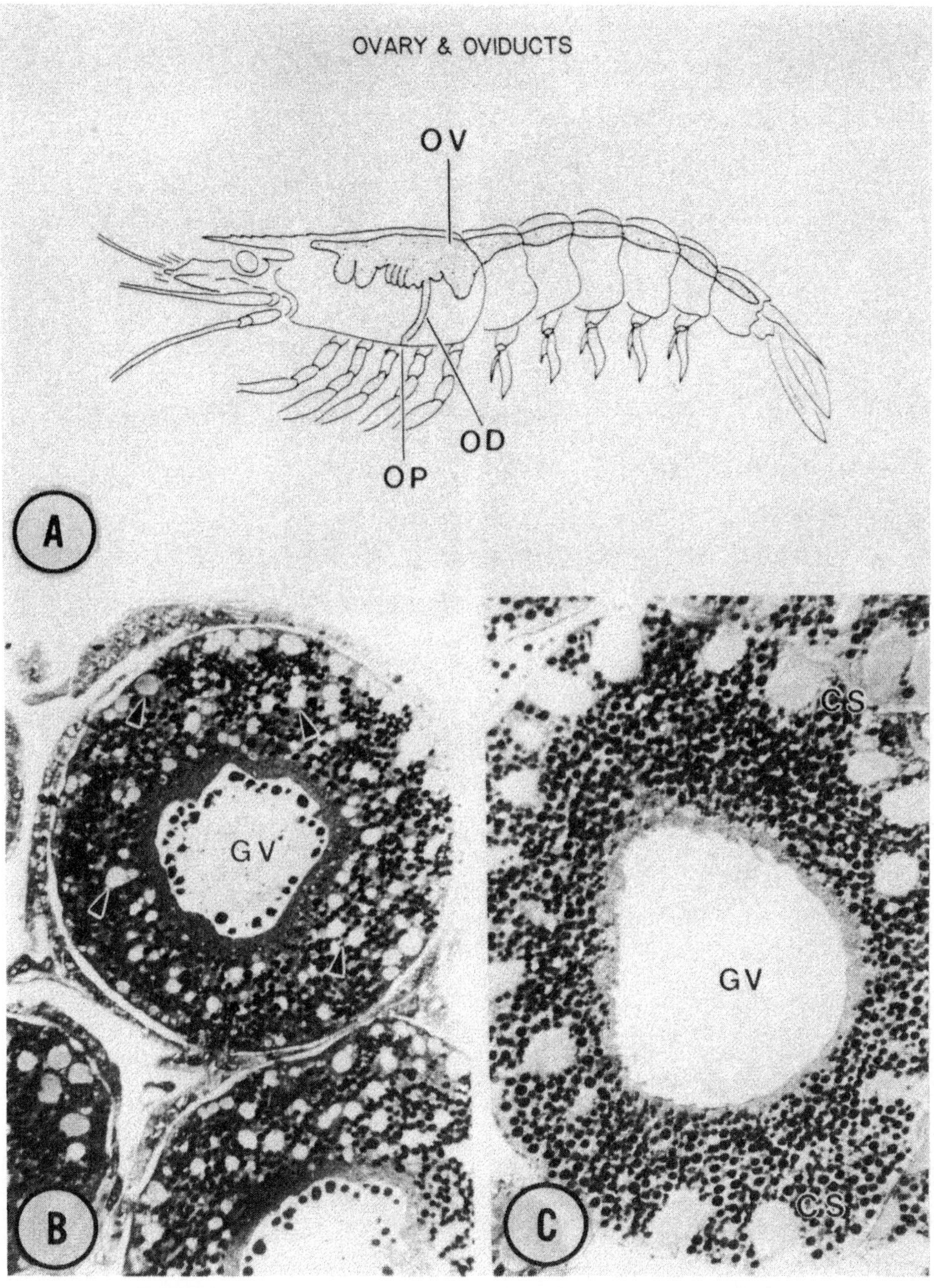

Plate 1. A; Diagrammatic representation of a gravid female *S. ingentis* illustrating organization of ovary (OV). The oviducts (OD) pass from lateral lobes of the ovary to ovipores (OP) situated at the base of each third pereiopod. B; Post vitellogenic oöcytes showing jelly-like inclusions (arrows) in the oöplasm. Germinal vesicle (GV) is very prominent. ×260. C; Oöcyte in cortical specialization (CS) phase. Germinal vesicle is still present. ×260.

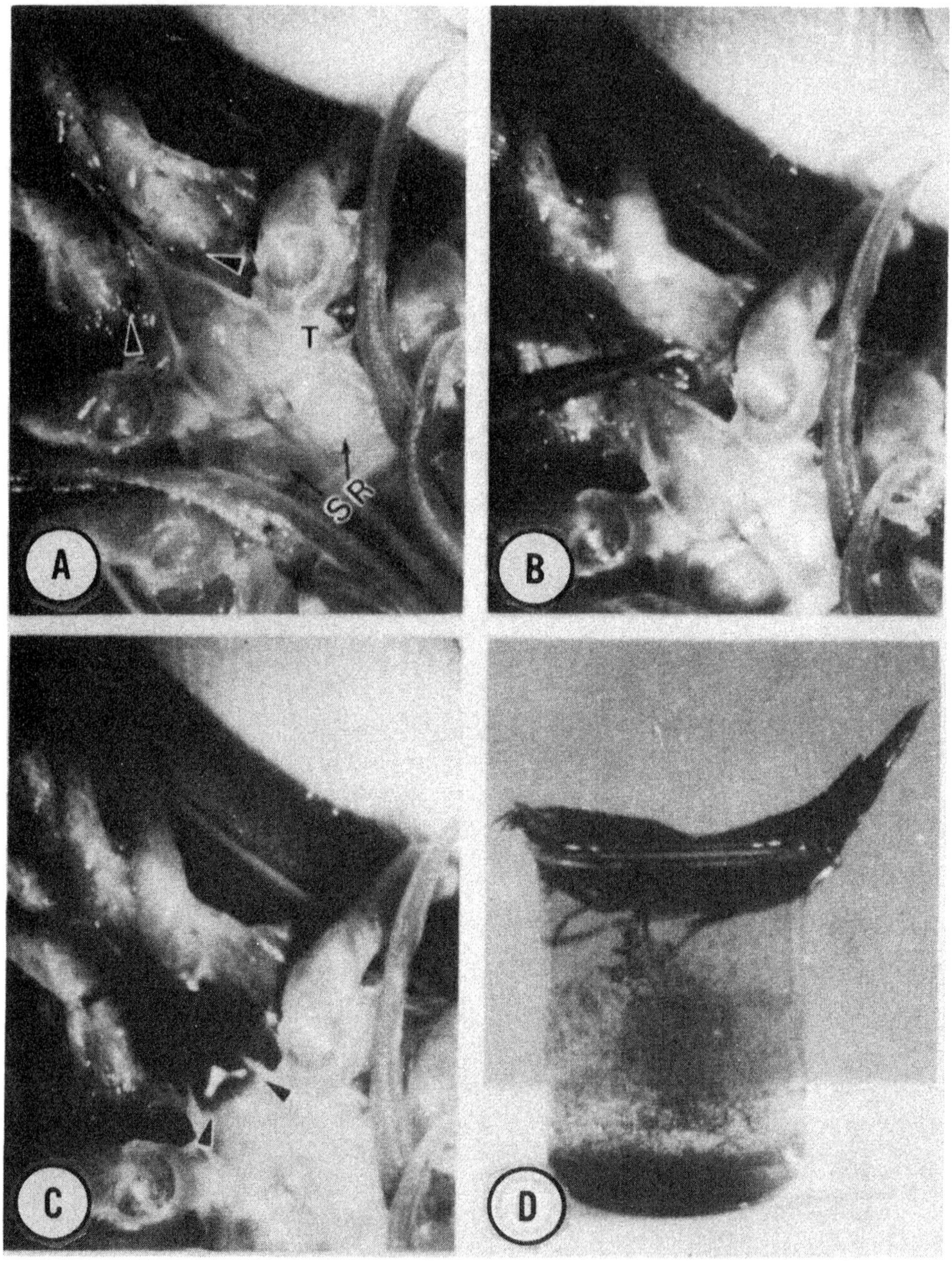

Plate 2. A; Ventral view of female. Ovipores (arrows) located at bases of the third pereiopods. Seminal receptacles (SR) are located in posterior lobes of the thelycum (T). B; The tip of the steel probe being inserted into the right ovipore. C; Ova have been extruded (arrows) from the ovipore as a result of probing. ×3.5. D; The animal after having the right ovipore probed was placed on a beaker containing sea water. Note ova are being spawned from both ovipores. ×0.5 (modified from Pillai et al. 1988).

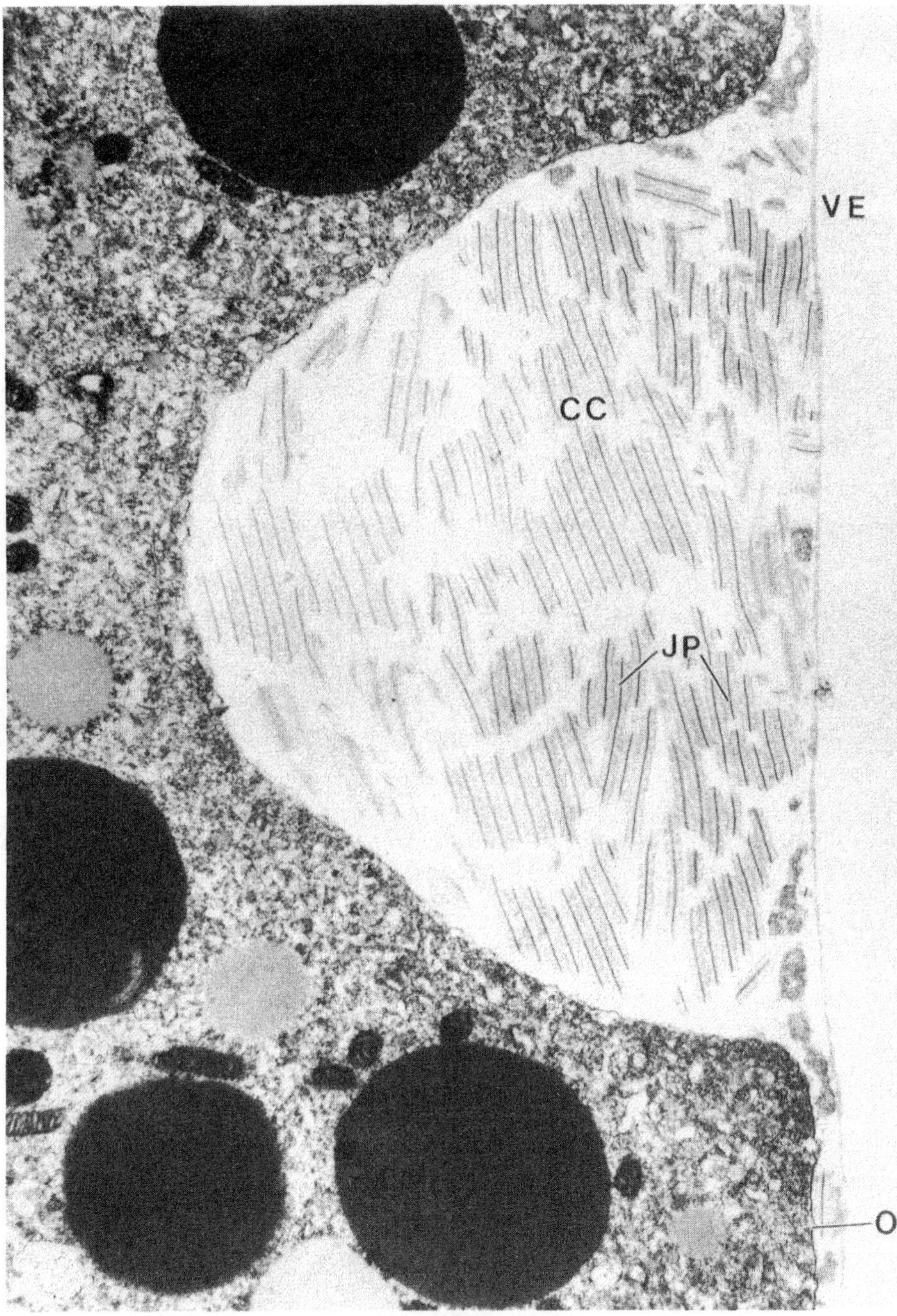

Plate 3. Transmission electron micrograph of an ovulated ovum (prior to activation). CC–cortical crypt; JP–jelly precursor elements; O–oölemma; VE–vitelline envelope. × 10 000.

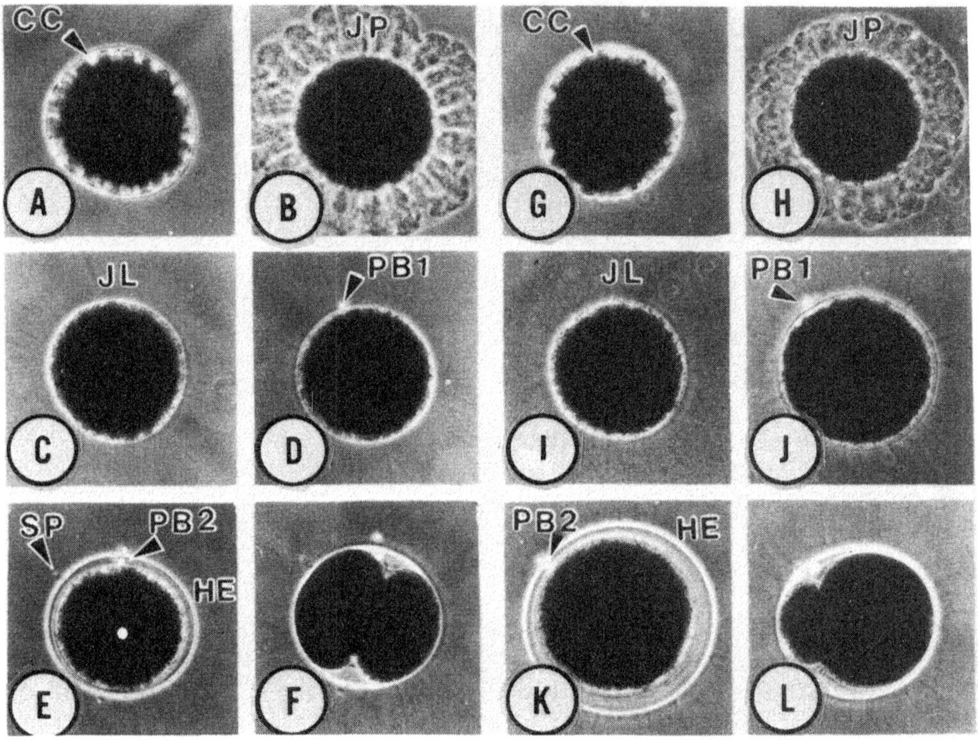

Plate 4. A-F; Phase micrographs showing sequence of activational events that a fertilized ovum undergoes upon contacting sea water. A; An ovulated ovum prior to initiation of activation. B-C; The activational events include extrusion of jelly precursor and formation of the jelly layer. D-E; Formation of the polar bodies. E; Hatching envelope formation. F; Normal cleavage. CC– cortical crypts; JL– jelly layer; JP– jelly precursor elements, HE– hatching envelope; PB– polar body; SP– sperm. × 90 (from Pillai & Clark 1987). G-L; Phase micrographs showing the sequence of events (paralleling A-F; above) that an unfertilized ovum of *S. ingentis* undergoes in sea water. Note the unequal cleavage pattern (L). × 90 (from Pillai & Clark 1987).

female population entered anecdysis; ecdysis resumed at the end of the reproductive season. Females exhibited two periods of molt activity, one between late February and early May and a second, more extended period, between late October and early November (Fig. 1). Males did not show the dramatic peaks in molt activity demonstrated by the females, though they did undergo anecdysis during the summer months (Anderson et al. 1985).

Anderson et al. (1985) suggested that females undergo multiple spawns within a single molt cycle. This was based on the high frequency of females in wild population in advanced stages of oöcytic development throughout the reproductive season. This hypothesis was supported by multiple spawning, without intervening molts of laboratory held *S. ingentis*. The average time between spawns for a single female was 20 days. Multiple spawns without an intervening molt have also been reported for laboratory held *Penaeus indicus* (Emmerson 1980).

4 EGG DEVELOPMENT

The highly fecund *S. ingentis* possesses bilateral ovaries which extend from the anterior region of the cephalothorax to the telson. Lateral extensions of the anterior lobes in the cephalothorax course in a ventrolateral direction. Oviducts pass from each anterior lobe to an ovipore situated at the base of each third pereiopod (Pl. 1A). With the exception of the gross morphology, however, the structural organization of the ovary is largely unknown. The vitellogenic stages of oögenesis also remain an enigma. For example, it is not known whether yolk production is autosynthetic, sequestered from extra oöcytic sources (e.g. pinocytotic uptake from hemolymph), or if both mechanisms occur. In the oöcytes of *Penaeus* sp. both mechanisms may in fact be used (Duronslet et al. 1975). The vitellogenic stages in the oöcytes of *S. ingentis* appear similar to the vitellogenic stages of *Penaeus* sp.; however, preliminary data for *S. ingentis* suggests that, what has been described as autosynthetic yolk production in *Penaeus* sp., may be the production of a jelly precursor (unpubl. data).

During the late stages of oögenesis, large inclusions appear within the oöplasm (Pl. 1B); these structures fuse with each other, forming rod-shaped bodies which become externalized in deep oöplasmic crypts around the periphery of the oöcytes (Pl. 1C).

The rod-shaped bodies in *S. ingentis* possess feather-like substructural elements (Pl. 3) similar to that described in *Penaeus* sp. (Duronslet et al. 1975, Clark et al. 1980, Lynn & Clark 1987). These elements appear to be of oöcytic origin; this is interesting in view of recent evidence suggesting that similar substructural material, of follicular origin, is responsible for extracellular coats of lobster ova (Talbot & Goudeau 1988).

The appearance of jelly-like inclusions around the cortex of penaeoidean oöcytes [corticle specialization phase (CS)] apparently represents one of the last postvitellogenic stages of oögenesis; however, their appearance in relation to the time of germinal vesicle breakdown (GVBD), ovulation and spawning remained unclear (Hudinaga 1942, Duronslet et al. 1975, Clark et al. 1980, Clark et al. 1984). Anderson et al. (1984), using a biopsy technique, examined the timing of the postvitellogenic events in *S. ingentis* ovaries and found that the CS phase occurred approximately 90 hours prior to spawning. In addition, GVBD occurred asynchronously, in CS phase oöcytes, and, in some instances, was completed 24 hours prior to ovulation. GVBD, however, was always completed prior to ovulation. After ovulation the longest documented time to spawning was 4 hours (Anderson et al. 1984), a time period in agreement with spawning induction data discussed in this paper.

5 SPAWNING

5.1 *Natural spawning*

Gravid *S. ingentis* females held in the laboratory spawn at night, an observation in accord with observations of other Penaeoidea (reviewed in Pillai et al. 1988). During the day animals remain on the bottom of the aquarium and rarely enter the water column, except during feeding. At night some of the females begin swimming and repeatedly rise to the top of the water column, slowly sinking back to the bottom of the aquaria. A similar behavior noted in *Penaeus monodon* was termed prespawning behavior (Motoh 1981). Animals that have initiated this behavior eventually remain at the surface of the aquaria and spawn after a variable period of time; the time from initiation of prespawning behavior to spawning varies from several minutes to two or three hours. During spawning, ova released through paired ovipores are mixed externally with sperm ejected from seminal receptacles (Pl. 2A) of the female thelycum (Clark et al. 1984). Females hold their pereiopods closely together and project them forward while vigorously beating their pleopods, a behavior similar to that described for *P. japonicus* (Hudinaga 1942) and *P. monodon* (Motoh 1981). The beating of the pleopods apparently assures sperm-egg mixing and the dispersal of the ova.

5.2 *Induced spawning*

Anderson et al. (1985), working with *S. ingentis*, suggested that prespawning behavior might be correlated with ovulation. This correlation has now been substantiated, and a technique to circumvent the time lag between ovulation and spawning has been developed (Pillai et al. 1988). Gravid females maintained under constant light are inhibited from spawning (Griffin et al. 1987). As soon as a dark environment was provided, regardless of the time of day, animals initiated prespawning behavior (Pillai et al. 1988). Once animals exhibited prespawning behavior, they were removed from the aquaria, and each animal had one ovipore momentarily opened using a steel probe (Pl. 2B). As a result of probing, ova were extruded from the manipulated ovipores (Pl. 2C), and the females were placed on beakers containing sea water or an experimental solution. Within seconds the probed females initiated spawning behavior, as described above, and ova were released from both ovipores, probed as well an non-probed (Pl. 3D).

6 EGG ACTIVATION

6.1 *Morphology*

The ova are isolecithal and approximately 225 μm in diameter. Ovulated ova, prior to contacting sea water, are in meiotic arrest (first meiotic metaphase) and possess deep extracellular cortical crypts which contain a jelly precursor. Surrounding these crypts, and closely opposed to the oölemma in surface regions between crypts, is an extracellular vitelline envelope (Pl. 3). Upon contact with sea water the ova become 'activated', resulting in the resumption of meiosis, extrusion of the jelly precursor and jelly layer formation, elaboration of a hatching envelope and cleavage (Pillai & Clark 1987).

When the ova contact sea water, the jelly precursor elements are expelled from the crypts, pushing the vitelline envelope off the surface of the ova. Within 3 minutes these elements have completely left the crypts, resulting in the formation of a corona of jelly precursor material surrounding an ovum. Ten to fifteen minutes after expulsion, this material undergoes a transition, resulting in the formation of an optically homogeneous jelly layer that encompasses the ovum; within minutes of this transition the resumption of meiotic maturation is manifested by the appearance of the first polar body. Approximately 20 minutes after the formation of the first polar body, a hatching envelope elevates from the surface of the zygote; 4-5 minutes later the second polar body is formed (Pl. 4A-F).

The foregoing events are independent of an ovum's interaction with a sperm, although unfertilized eggs undergo abnormal cleavage (Pl. 4G-L) and seldom survive past the 4 cell stage (Pillai & Clark 1987). The activational events through second polar body formation take approximately 40 minutes in fertilized ova (zygotes) and 50 minutes in unfertilized ova (Pillai & Clark 1987). A time differential is also noted for cleavage; unfertilized ova cleave in approximately 135 minutes while fertilized ova do so in 90 minutes (unpubl. data).

6.2 *Physiology*

The mechanisms of oöcyte activation in members of the Penaeoidea have not been exhaustively investigated; however, in the ova of *P. aztecus* and *P. setiferus* both jelly precursor extrusion and the transition of the precursor to the homogeneous state have been shown to be dependent on external Mg^{++} (Clark & Lynn 1977). In *S. ingentis* ova, external Mg^{++} is required for this transition, but its requirement for precursor extrusion is in question. External Mg^{++} has also been shown to be a requirement for the resumption of meiosis and hatching envelope formation in the ova of *S. ingentis* (Pillai & Clark 1987). External Ca^{++} is not required for activation if the ova have been fertilized; however, Ca^{++} is required for resumption of meiosis and hatching envelope elevation in unfertilized ova (Pillai & Clark 1987). The activity of a protease also appears to be required for the jelly transition stage in *S. ingentis* (unpubl. data), as has been suggested for the ova of other Penaeoidea (Lynn & Clark 1987). Lynn & Clark (1987) suggested that the protease(s) implicated in this transition may be Mg^{++} dependent.

7 SUMMARY AND CONCLUSIONS

Egg production in members of the Dendrobranchiata is largely unstudied, with the exception of *S. ingentis*. Reproduction in *S. ingentis* is seasonal. Unlike other closed thelyca Penaeoidea found on the Atlantic and Gulf coasts of the United States gravid, *S. ingentis* are not found outside of the reproductive season. During the spawning season ecdysis ceases in females, and individuals undergo multiple spawns; the latter phenomenon has been substantiated with laboratory held animals. In addition to undergoing ovarian maturation, *S. ingentis* is a very hardy laboratory organism and is amenable to numerous manipulations. Ovarian stages can be determined via a simple biopsy technique, the onset of spawning can be manipulated by light cycling, ovulation can be determined, and the actual spawning event can be initiated by probing an ovipore. As such, the system is available for further examination of many interesting phenomena. For example the physiological/endocrinological controls of GVBD, ovulation, and oöcyte activation have received limited attention. These topics are of general biological interest, as well as being critical to individuals concerned with the husbandry of this group of animals.

ACKNOWLEDGEMENTS

We thank Dr Fred Griffin for his suggestions and Elisabeth Clark for her editorial assistance. We are also grateful to Kevin Uhlinger for his technical expertise. This work was supported in part by NOAA, National Sea Grant College Program, Department of Commerce, grant #NA85AA-D-SG140 project #R/A-61, through the California Sea Grant Program and USDA Competitive Grant #87CRCR-1-2514.

REFERENCES

Anderson, S.L., E.S. Chang & W.H. Clark, Jr 1984. Timing of postvitellogenic ovarian changes in the ridgeback prawn *Sicyonia ingentis* (Penaeidae) determined by ovarian biopsy. *Aquaculture* 42: 257-271.

Anderson, S.L., W.H. Clark, Jr & E.S. Chang 1985. Multiple spawning and molt synchrony in a free spawning shrimp (*Sicyonia ingentis*: Penaeidae). *Biol. Bull.* 168: 377-394.

Carlisle, J.G., Jr 1969. Invertebrates taken in six year trawl study in Santa Monica Bay. *Veliger* 11: 237-242.

Clark, W.H., Jr & F.J. Griffin 1988. The morphology and physiology of the acrosome reaction in the sperm of the decapod *Sicyonia ingentis*. *Develop. Growth and Differ.* 30: 451-462.

Clark, W.H., Jr & J.W. Lynn 1977. A Mg^{++} dependent cortical reaction in the eggs of penaeid shrimp. *J. Exp. Zool.* 200: 177-183.

Clark W.H., Jr, J.W. Lynn, A.I. Yudin & H.O. Persyn 1980. Morphology of the cortical reaction in the eggs of *Penaeus aztecus*. *Biol. Bull.* 158: 175-186.

Clark, W.H., Jr, A.I. Yudin, F.J. Griffin & K. Shigekawa 1984. The control of gamete activation and fertilization in the marine Penaeidae, *Sicyonia ingentis*. *Adv. Invert. Repro.* 3: 459-471.

Duronslet, M.J., A.I. Yudin, R.S. Wheeler & W.H. Clark, Jr 1975. Light and fine structural studies on naturally and artificially induced egg growth of penaeid shrimp. *Proc. World Mariculture Soc.* 6: 105-132.

Emmerson, W.D. 1980. Induced maturation of prawn *Penaeus indicus*. *Mar. Ecol. Progr. Ser.* 2: 121-131.

Griffin, F.J., W.H. Clark, Jr, J.H. Crowe & L.M. Crowe 1987. Intracellular pH decreases during the *in vitro* induction of the acrosome reaction in the sperm of *Sicyonia ingentis*. *Biol. Bull.* 173: 311-323.

Herkelrath, J.M. 1977. Some aspects of the growth and temperature tolerance of the ridgeback prawn, *Sicyonia ingentis* Burkenroad, in southern California waters. M.S. thesis, Whittier College, 67 pp.

Hudinaga, M. 1942. Reproduction, development and rearing of *Penaeus japonicus* Bate. *Jpn. J. Zool.* 10: 305-393.

King, J.E. 1948. A study of the reproductive organs of the common marine shrimp *Penaeus setiferus* (Linnaeus). *Biol. Bull.* 94: 244-262.

Lynn, J.W. & W.H. Clark, Jr 1987. Physiological and biochemical investigations of the egg jelly release in *Penaeus aztecus*. *Biol. Bull.* 173: 451-460.

Motoh, H. 1981. Studies on the fisheries biology of the giant tiger prawn, *Penaeus monodon* in the Phillipines. Tech. Rep. No. 7, SEAFDC, Phillipines.

Perez Farfante, I. 1985. The rock shrimp genus *Sicyonia* (Crustacea: Decapoda: Penaeidea) in the eastern Pacific. *Fish. Bull.* 83: 1-80.

Pillai, M.C. & W.H. Clark, Jr 1987. Oöcyte activation in the marine shrimp, *Sicyonia ingentis*. *J. Exp. Zool.* 244: 325-330.

Pillai, M.C., F.J. Griffin & W.H. Clark, Jr 1988. Induced spawning of the decapod crustacean *Sicyonia ingentis*. *Biol. Bull.* 174: 181-185.

Shigekawa, K. & W.H. Clark, Jr 1986. Spermiogenesis in the marine shrimp, *Sicyonia ingentis*. *Develop. Growth and Differ.* 28: 95-112.

Talbot, P. & M. Goudeau 1988. A complex cortical reaction leads to the formation of fertilization envelope in the lobster *Homarus*. *Gamete Res.* 19: 1-18.

Ovulation, attachment and retention of lobster eggs

P. Talbot
Department of Biology, University of California, Riverside, USA

ABSTRACT

The mechanisms involved in ovulation and attachment of lobster (*Homarus*) eggs and the factors which affect the retention of eggs following attachment are reviewed. Ovulation has been examined *in vitro* using ovaries that were ripe and close to natural ovulation. Small strips of ovary were subjected to collagenase treatment, and oöcytes were ovulated within 4-5 hours. The morphological changes occurring during ovulation suggest that a contractile process is involved in extrusion of the oöcyte from the follicle. At spawning, oöcytes are expelled from the ovary by contraction of muscle cells surrounding the ovary. These muscle cells are rich in microtubules as well as actin and myosin. Ultrastructural examination of mature ovarian oöcytes has revealed that a complex coat surrounds oöcytes prior to ovulation. This coat is secreted at least in part by the follicle cells. No additional coats are added to the oöcyte during passage through the oviduct. Following spawning this coat is deflected off the oöcyte's surface and wraps around the ovigerous setae, forming the attachment stalk. An additional inner coat is secreted by the oöcyte in response to fertilization. The two coats adhere to each other and form the fertilization envelope.

In captivity, many female lobsters lose a significant portion of their clutch prior to hatching. This has multiple causes. One factor contributing to poor egg retention is the formation of faulty attachment stalks. In addition, some females produce stunted ovigerous setae following a critical molt at 60-70 mm carapace length. This clearly precludes proper attachment of eggs to the pleopods.

1 GENERAL CHARACTERISTICS OF LOBSTER EGGS

There are four major families and hundreds of species of lobsters in existence (Phillips et al. 1980). Most of our knowledge of lobster egg biology has come from studies on relatively few of these species. The focus of this review will be on the genus *Homarus*.

1.1 *Egg development*

In *Homarus*, eggs develop in an H-shaped ovary on a biennial cycle (Herrick 1909), although yearly cycles have been reported, particularly for females greater than 120 mm in carapace length (Waddy & Aiken 1986). In captivity the cycle may be accelerated (D'Abramo & Conklin 1985, P. Wilson pers. comm.). When a female is close to ovulating, her ovaries extend from the eyestalk to the fourth or fifth abdominal segment (Herrick 1891, Talbot 1981a, & Pl. 1). The ovaries are dark green to black and fill much of the dorsal cavity. It is possible to estimate ovarian maturity of females by 'candling' and examining the degree of darkening visible through the intersegmental membranes of the abdomen (Hedgecock et al. 1978).

1.2 *Mature eggs*

Ovarian eggs at maturity measure 1.5 to 1.7 mm in diameter and weigh 321 m̃g (Herrick 1909). Their dark green color is due to the presence of ovoverdin (Stern & Salomon 1937), which is comprised of a green pigment astaxanthin (Goodwin 1951) bound to a protein (Wallace et al. 1967). The subunit structure of ovoverdin in the eggs of *H. gammarus* and *H. americanus* apparently differs (Zagalsky 1985). The function of the carotenoid astaxanthin is not certain, but it is not metabolized during embryonic development (Goodwin 1951). Donahue (1948, 1957) and later Lisk (1961) reported that estrogenic activity, specifically estradiol-17β, is present in lobster eggs. Recently Couch et al. (1987) confirmed these results and further demonstrated the presence of progesterone in lobster ovaries. The biological significance of these findings remains uncertain, although a preliminary study suggests that estrogenic activity inhibits molting (Donahue 1955).

The mature lobster egg is surrounded by a single investing coat, originally called a 'chorion' (Herrick 1909, Talbot 1981a, b), but more recently described as 'envelope 1' (Talbot & Goudeau 1988) to be consistent with the literature on *Carcinus* (e.g. Goudeau & LaChaise 1983). This coat can be subdivided into a thin granular region (layer 1A) adjacent to the follicle cells and a thicker layer (1B) adjacent to the oöcyte (Pl. 2). Layer 1B is comprised of structures resembling 'bottlebrushes' (Talbot 1981a, b). Because of their distinctive structure, bottlebrush synthesis and assembly has been identified in the follicle cells immediately outside envelope 1. The origin of the remainder of envelope 1 has not been determined, but it is probable that the coat is produced entirely by the follicle cells. This is the only coat present around lobster eggs at the time they are spawned (Cheung 1966, Talbot & Goudeau 1988).

2 OVULATION

Ovulation specifically refers to the escape of a mature oöcyte from the follicle in which it developed. It is a process common to all sexually reproducing animals with the possible exception of coelenterates which lack well defined follicles (Campbell 1974). The process of ovulation has been studied extensively in mammals (reviewed by Schroeder & Talbot 1985, Yoshimura & Wallach 1987), but it has rarely been examined as a process separate from spawning in marine invertebrates.

2.1 *Structure of the ovarian follicle*

In lobsters, the follicle is solid and simple in structure (Talbot 1981b, & Pl. 2). The oöcyte surrounded by envelope 1 is in turn surrounded by a single layer of follicle cells. A thick multilamellar basal lamina is present outside the follicle cells. Blood vessels and muscle cells insert into this basal lamina, and fibroblasts form a discontinuous layer outside of the basal lamina. Prior to ovulation, the follicle cells are closely applied to the surface of envelope 1. Processes extend from the follicle cells into the pores of envelope 1.

For successful ovulation in the lobster at least three events must occur. The follicle cell processes must be retracted from envelope 1, a rupture site must form in the apical follicle wall, and the oöcyte must pass through the rupture site into the lumen of the ovary. *In vivo* ovulation is very difficult to study in lobsters because most females are on a biennial ovarian cycle. We also lack a precise method to predict when ovulation will occur (although the stage of development of the pleopod tegumental glands can provide some clues, Aiken & Waddy 1982).

2.2 *In vitro ovulation*

Experiments done *in vitro* have increased our understanding of lobster ovulation (Talbot 1981c). Ovarian strips were prepared from females that were judged to be in stage 6 of maturation (i.e. close to the time of natural ovulation) using the classification scheme of Aiken & Waddy (1980). These strips were washed and then incubated in sea water containing various enzyme solutions (Talbot 1981c). Of eleven enzymes tested, only collagenase produced *in vitro* ovulation of lobster oöcytes. The collagenase inhibitor, cysteine, blocked collagenase-induced ovulation in a dose dependent manner, indicating that collagenase was required for the *in vitro* event. However, crude preparations of collagenase worked better than highly purified preparations, suggesting that other factors in the collagenase solutions were also involved. Transmission electron microscopic examination of ovarian strips treated with collagenase revealed that collagenase effectively removed the thick basal lamina of the follicle and thus allowed the rupture site to begin forming (Talbot, unpubl. observ.). Other contaminating enzymes may have facilitated this process or may also have degraded some of the follicle cells near the rupture site. Collagenase has been implicated in ovulation in mammals, where its role has been more widely studied. It appears in rat follicular fluid as ovulation approaches (Curry et al. 1985), and ovulation can be blocked in mammals with collagenase inhibitors (Ischikawa et al. 1983a, b). It may degrade the rich collagen matrix of the theca externa in the mammalian follicle and also disrupt the basal lamina near the follicle apex.

2.3 *Follicle wall contraction during ovulation*

When follicles treated with collagenase were viewed with a dissecting microscope during *in vitro* ovulation, the thin wall of the follicle noticeably contracted (that is, became smaller) and squeezed the oöcyte out of the follicle (Pl. 3). Contractile events are a common feature of ovulation throughout the animal kingdom (reviewed by Schroeder & Talbot 1985). In animals having solid follicles, the oöcyte is squeezed into a 'dumbell' shape as it passes through the rupture site. Such was observed in lobsters. In lobsters, the solid follicle probably would not readily allow escape of the oöcyte following the formation of a rupture site, unless the rupture site were unusually large. Contraction of the follicle wall thus guarantees extrusion of the oöcyte from its follicle. In oöcytes that are ovulated *in vitro* with collagenase, the follicle cell extensions into envelope 1 are withdrawn. The collagenase solutions must contain factors capable not only of dissolving the basal lamina but also initiating withdrawal of these follicle cell processes and contraction of the follicle wall. It is not yet known what factors regulate these events *in vivo* in the lobster or in any other decapod.

Delicate muscle cells extend like 'guy wires' from the muscular wall surrounding the lobster ovary into the basal lamina surrounding mature follicles (Talbot 1981a). Perhaps contraction of these muscle cells during ovulation pulls the contracting follicle wall away from the apex or rupture site.

2.4 *Relationship between ovulation and spawning*

In lobsters, ovulation may precede spawning by several weeks or longer. Mature ovaries may contain numerous eggs in the lumen (Herrick 1891, 1909, Aiken & Waddy 1980). It is not known how long lobsters can hold ovulated eggs prior to spawning. Spawning is a complicated process and requires precise positioning of the female on her back in an undisturbed state (Templeman 1937). Behavioral and environmental conditions may have to be satisfied before spawning occurs. It is thus advantageous to have eggs available in the lumen of the ovary and ready when a ripe female assumes the precarious inverted spawning position.

2.5 *Spawning and ovarian muscle*

Spawning appears to be facilitated by the muscle coat surrounding the ovary. These muscles have been described as smooth muscles (Herrick 1909), and indeed they are not striated. However, they differ significantly from other smooth muscles in that they are packed with microtubules in addition to actin and myosin (Talbot 1981a, Howard et al. 1985a, b). The muscles around the crayfish ovary (Pl. 3B) are similar in this respect (Howard et al. 1985a). These muscles are one of the richest souces of tubulin known. The gel in Plate 3C compares the relative abundance of tubulin to actin and myosin in the lobster ovarian muscle. Following polymerization-depolymerization cycles, the tubulin band shows significant enrichment (Pl. 3C, lane 6). Immunofluorescent microscopy of lobster ovarian muscle labelled with a monoclonal antibody to tubulin indicates that microtubules are distributed throughout this cell. It is not yet known if and how the various cytoskeletal components in these cells interact to bring about cycles of contraction and relaxation. Nor is it known how contraction of these cells is regulated *in vivo*. This cell nevertheless provides an excellent system for the study of microtubule, actin and myosin interactions.

3 ATTACHMENT

During spawning, female lobsters lie in an inverted spawning position. Oöcytes pass single file through the gonopores at the base of the third walking legs (Scott 1902, Herrick 1909, Templeman 1937). As an egg is extruded through the gonopore, it pushes the eggs preceding it along the female's ventral surface. This displacement enables the eggs to pass over the thelycum, which is the site of sperm storage. Although fertilization in lobsters has not been observed, most evidence suggests that it occurs as the eggs pass over the thelycum. It is not known how the sperm are released from the spermatophore and transferred to the oöcytes. However, only a small fraction of the total number of sperm in the spermatophore is released at spawning (Templeman 1936). If molting does not occur prior to the next spawn, sperm from the same spermatophore are used for fertilization (Waddy & Aiken 1986). Following fertilization, eggs continue to move along the ventral surface of the famale to a 'basket' formed by flexure of the abdomen. The beating of the pleopods may draw eggs into this chamber where they attach to the ovigerous setae.

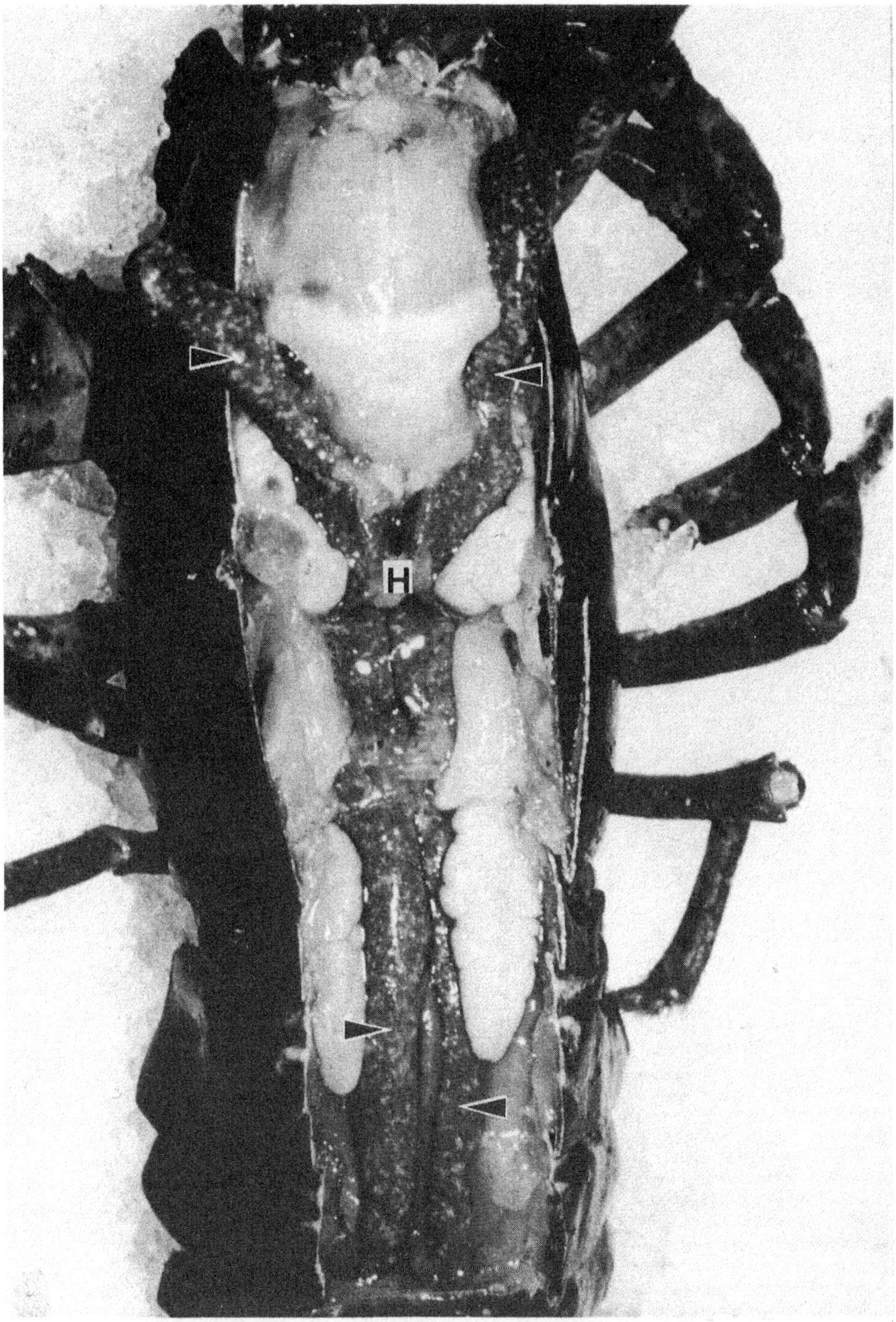

Plate 1. Dorsal view of a dissected female (*Homarus americanus*) showing organization and position of the ovary (arrows), which is H-shaped. The cross bar of the H is located beneath the heart (H). The small white spheres in the ovary are atretic follicles.

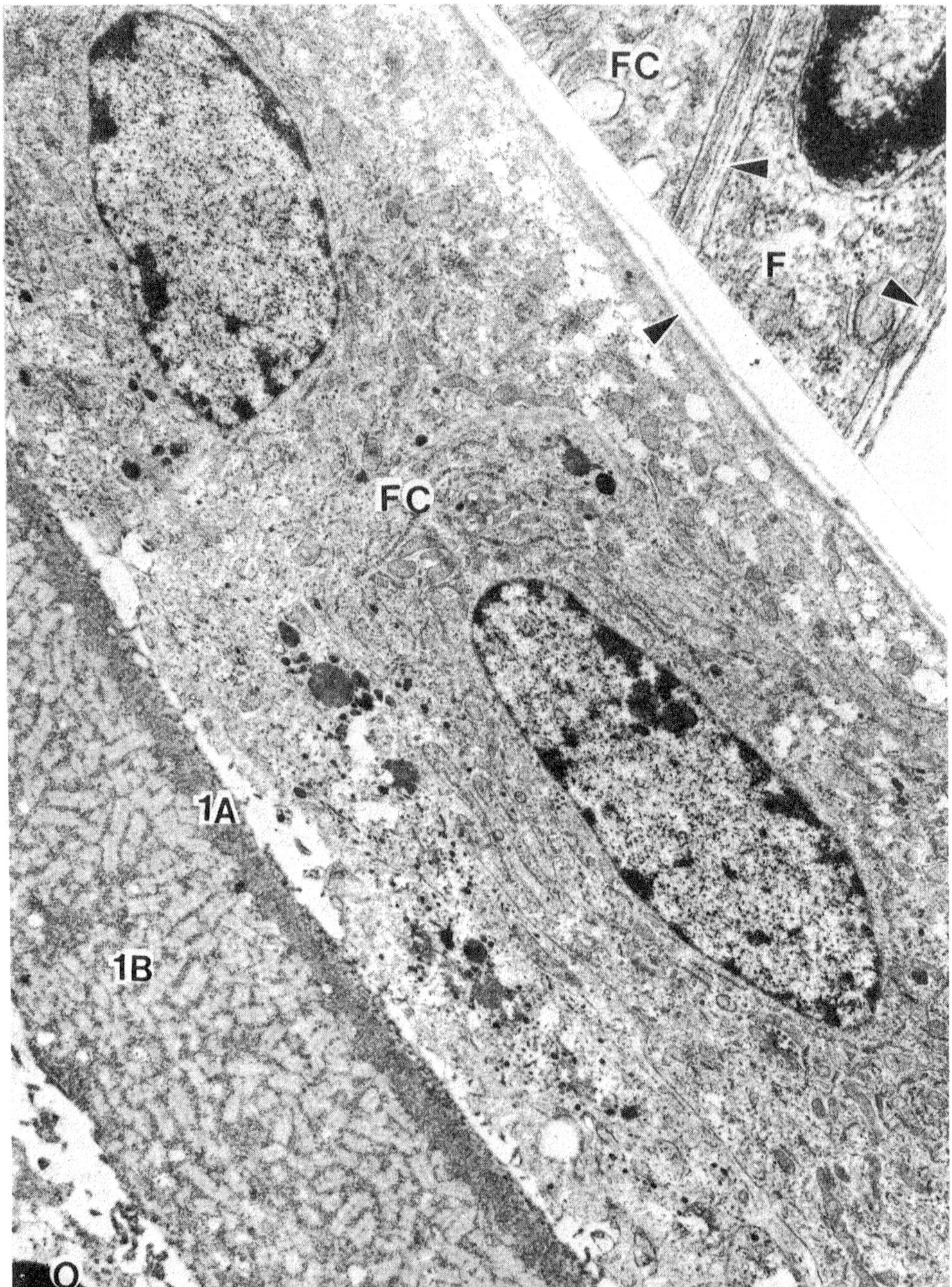

Plate 2. Transmission electron micrograph of a section through an ovarian follicle that was close to natural ovulation. A portion of the oöcyte's cortex (O) is shown surrounded by the perivitelline space and envelope 1, which can be subdivided into layer 1A and 1B. Layer 1B contains structures resembling bottlebrushes. A single layer of follicle cells (FC) surrounds envelope 1. In most sections the outermost barrier of the follicle is a thick basal lamina (arrow); however in some regions (upper right inset) the follicle cells are surrounded by a layer of fibroblasts (F). The thick lamellae of the basal lamina are apparent in the inset (arrows). ×5500, ×25 200 (inset).

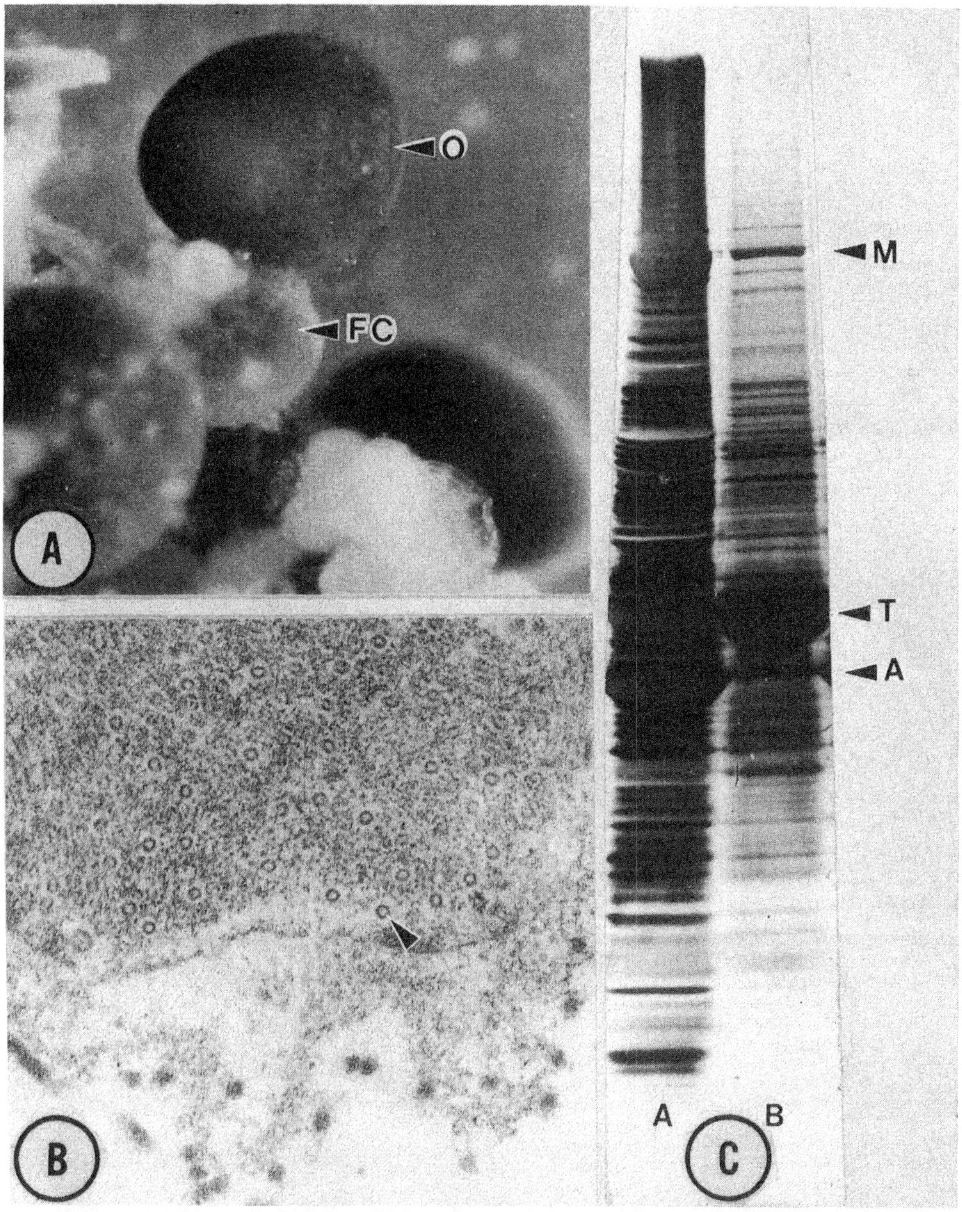

Plate 3. A. Micrograph of a lobster oöcyte undergoing *in vitro* ovulation induced by incubating ovarian strips in collagenase. A portion of the oöcyte (O) has emerged from the thin follicle wall; however, the wall of the follicle cell (FC) still remains around the lower aspect of the oöcyte. The oöcyte is being squeezed into a 'dumbell' shape characteristic of ovulating lobster eggs. × 20. B. Transmission electron micrograph through the *Procambarus clarkii* ovarian wall musculature showing the abundance of microtubules (arrows) in these muscle cells. A similar arrangement of microtubules is found in lobster ovarian muscle. Micrograph provided by courtesy of David Howard. × 62 200. C. SDS polyacrylamide gel showing the bands obtained with whole homogenates of lobster ovarian muscle (lane A). Significant levels of actin (A), myosin (M), and tubulin (T) are present. Lane B shows a muscle preparation cycled to enrich the tubulin band. Data provided by David Howard.

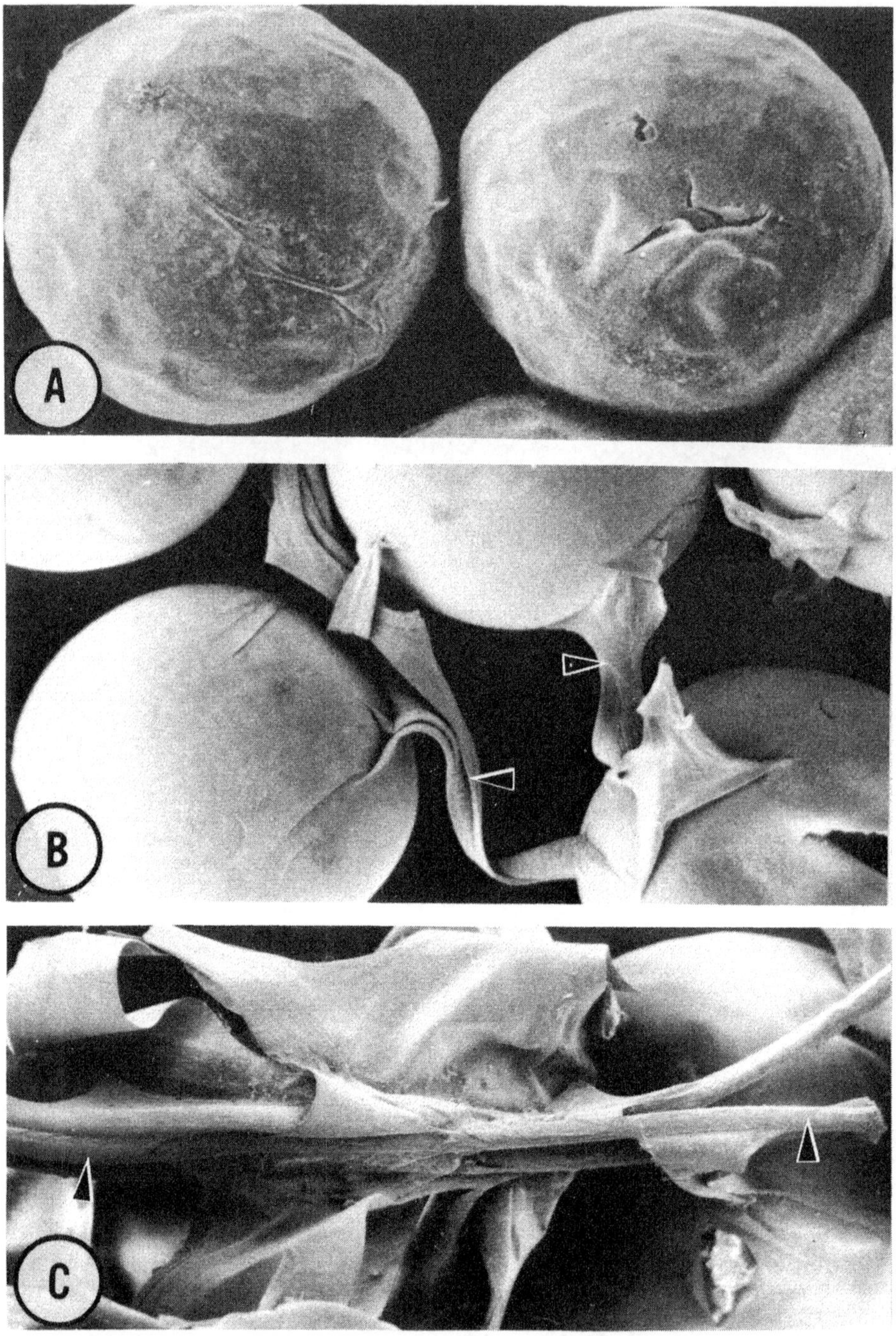

Plate 4. Scanning electron micrographs of lobster eggs at various times during spawning. A. The eggs have just emerged from the gonopore before fixation; they do not have attachment stalks. B, C. The eggs have attached to the pleopods prior to fixation. Attachment stalks interconnect eggs (B, arrows) and also wrap around ovigerous setae (C, arrows). ×29, ×32, ×35.

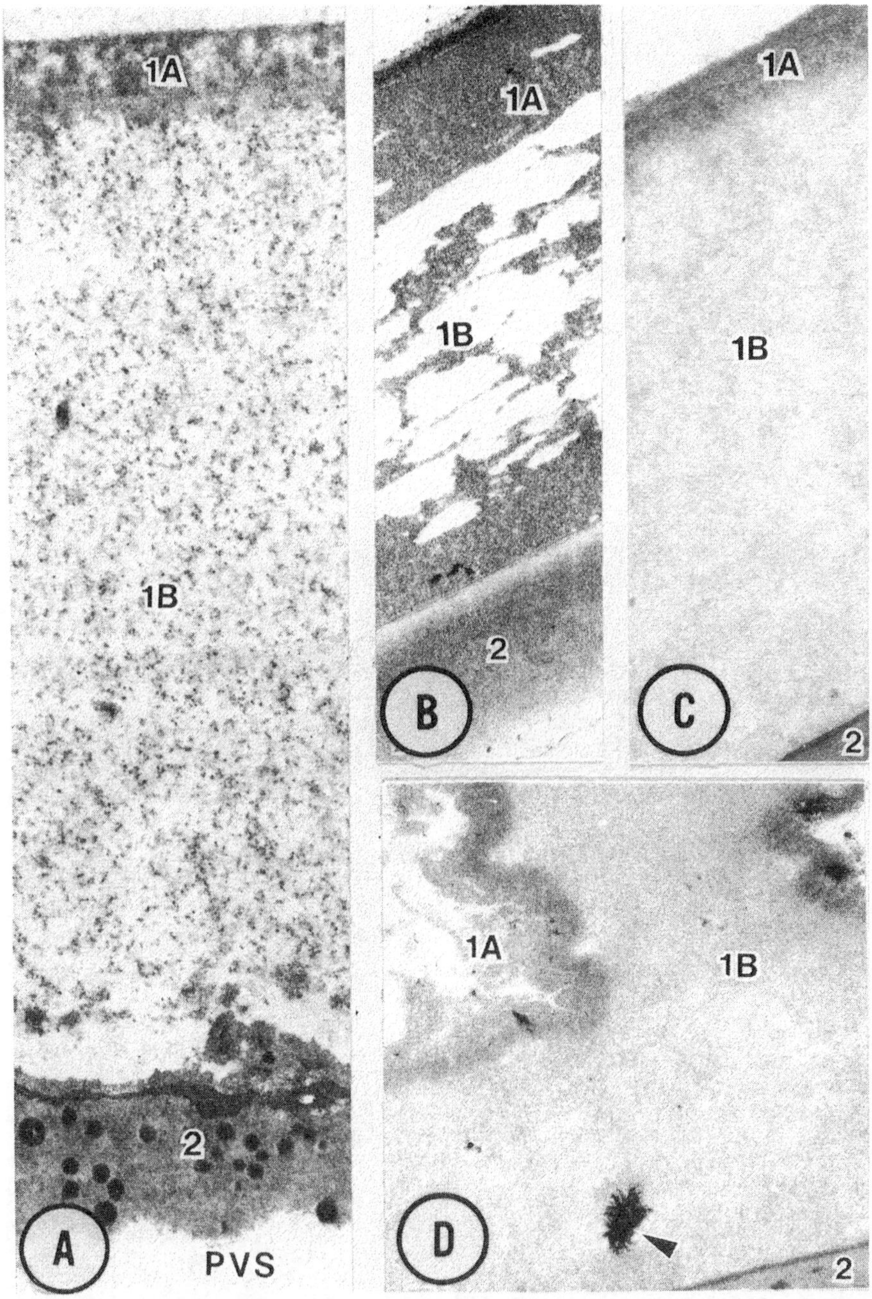

Plate 5. Transmission electron micrographs showing various stages in attachment stalk formation. A. Shows a recently spawned fertilized egg. Layers 1A and 1B have expanded, and envelope 2 is forming from cortical granule contents. B, C. In later stages, envelope 1 condenses and envelope 2 becomes homogenous. D. Shows a portion of envelope 1 which has been deflected off the egg's surface and forms the attachment stalk. A crystal (arrow) is present in layer 1B; these are frequently observed in *H. americanus.* (PVS = perivitelline space;). × 9200, × 16 400, × 19 500, × 21 000.

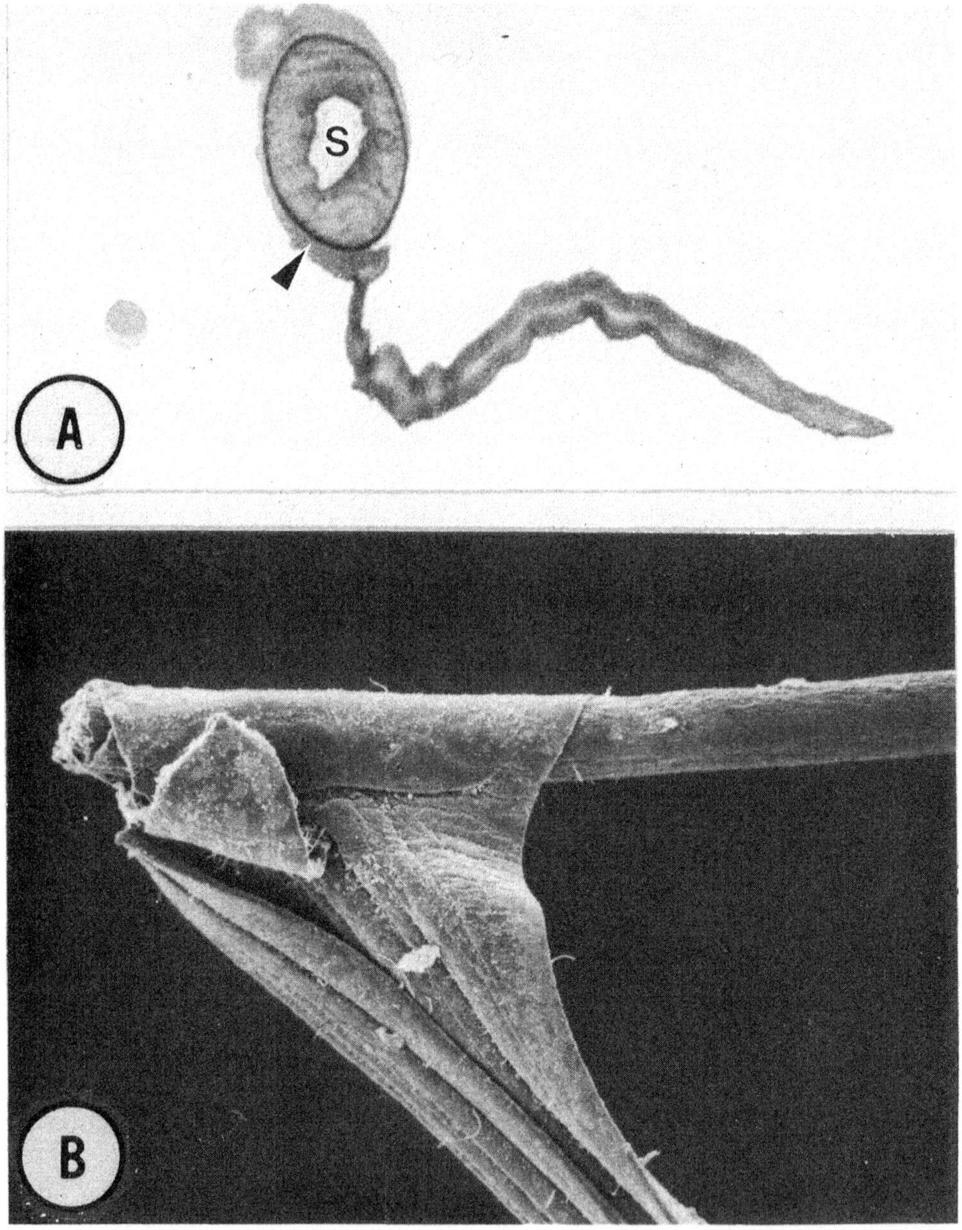

Plate 6. A, B. Micrographs depicting ovigerous setae with attachment stalks wrapped around them. S = seta; arrow points to attachment stalk. × 1000, × 180.

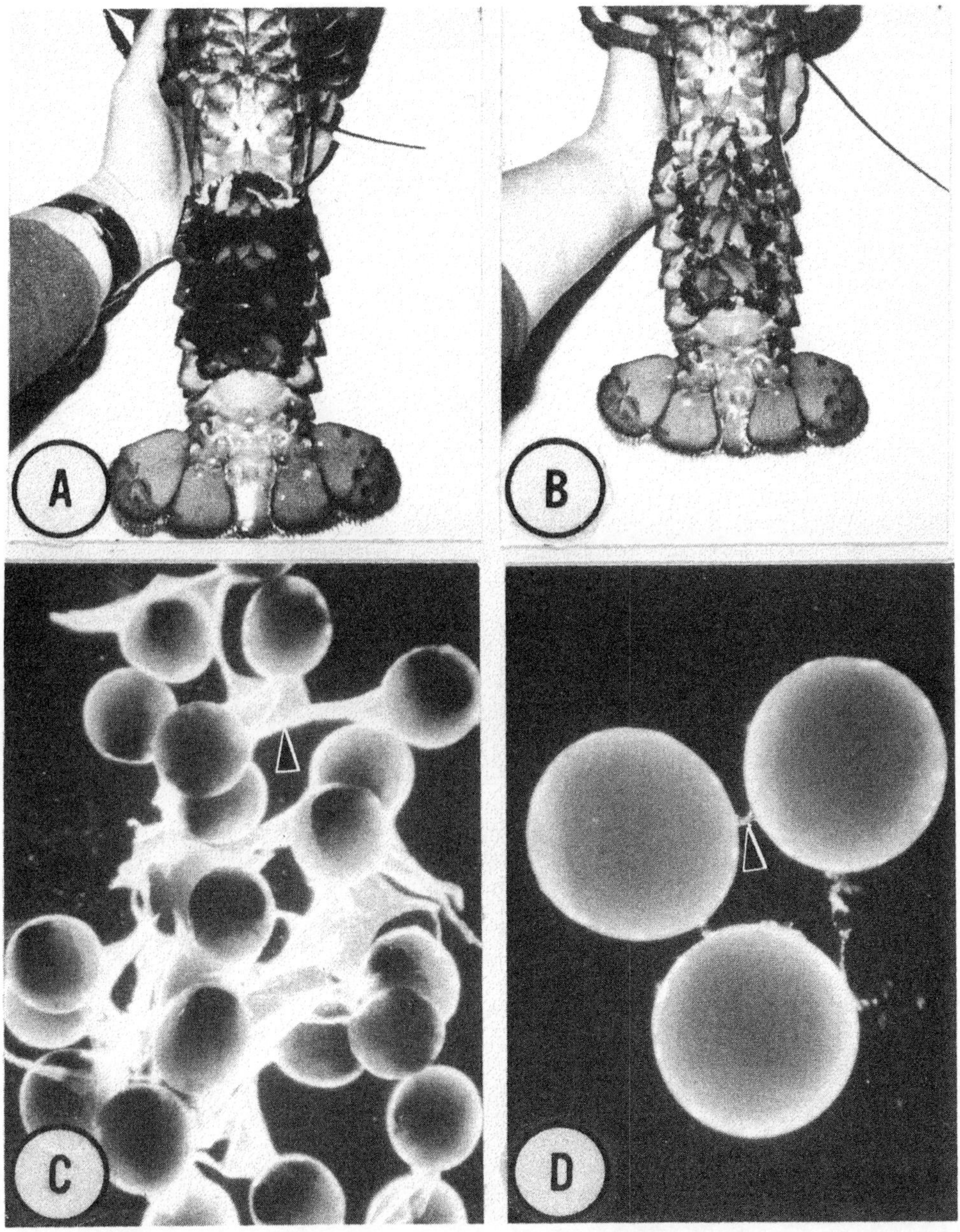

Plate 7. A, B. Photographs showing a complete (A) and incomplete (B) egg clutch. In A, the female has attached a full clutch of eggs numbering 13 815. In B, 1982 eggs remain in the clutch. Such losses are often observed in captive females. C. This micrograph shows the condition of attachment stalks found on females that spawned and attached eggs in the wild. The stalks are thick and well formed. × 10. D. These eggs came from a female that was born and matured in the laboratory. Such females often produce attachment stalks that are small and thin, as depicted here (arrows). × 10.

3.1 *Mechanism of egg attachment to ovigerous setae*

The mechanism of egg attachment has been studied in lobsters for more than 100 years. The earlier work and controversy in this area has recently been reviewed (Aiken & Waddy 1980, Talbot & Goudeau 1988). Studies on *Homarus americanus* have revealed that envelope 1 is the only coat present around eggs as they pass through the gonopore (Talbot & Goudeau 1988). Moreover, this coat appears identical ultrastructurally both in the ovary and immediately after exiting the gonopore. However, if oviposited eggs are placed in sea water, and then fixed for electron microscopy, both layers of envelope 1 swell and bottlebrushes are no longer evident. Eggs with swollen coats are sticky and will adhere to each other if placed in sea water. Such swelling is presumed to occur *in situ* as the eggs are passing along the female's ventral surface. Prior to reaching the abdomen, lobster eggs lack attachment stalks (Pl. 4A). When eggs arrive at the abdominal region they encounter the beating pleopods, which may move them into the abdominal basket and aid in attachment to the ovigerous setae.

Attachment is accomplished by elaboration of a stalk or funiculus. These stalks interconnect adjacent eggs and wrap around the ovigerous setae (Plates 4B, C). The stalk is formed from and is directly continuous with envelope 1 (Goudeau et al. 1987). The swelling of envelope 1 is probably important in the formation of the attachment stalks. The swollen coat is not only sticky but also more supple and thus easily extruded from the egg's surface and wound around setae. On the basis of existing evidence, it appears that eggs, which are sticky, adhere to each other and to ovigerous setae upon entering the abdominal basket. The beating pleopods may promote the elaboration of attachment stalks from the adherent eggs. These stalks are often highly twisted perhaps due to the beating action of the pleopods. The beating of the pleopods probably also causes the stalks to wrap around the ovigerous setae, securing the egg mass to the female. The entwined stalks give the egg mass a very cohesive property.

It is unclear why the egg coat sticks only to other egg coats and to the ovigerous setae. Eggs do not attach to plumose or marginal setae on the pleopods; nor do they stick on the female's ventral surface. Layer 1A of envelope 1 must be the adhesive surface, as it is the outermost coat. Its adhesive properties appear to be very specific.

The coat remains swollen for a relatively short time (Pl. 5A). Within 24 hours of spawning, envelope 1 of the egg coat and attachment stalk have condensed to about their original size (Plates 5B-D). Following condensation, envelope 1 appears homogeneous, although 1A and 1B can still be distinguished (Pl. 5C). The coat with its continuous stalk is now a more rigid, tough structure. The stalks secure eggs to the ovigerous setae of the pleopods for the long duration of embryogenesis. Although this interval may vary depending on water temperature, it can be as long as 16 months (Perkins 1972, Branford 1978). Plates 6A and 6B show attachment stalks wrap around ovigerous setae.

3.2 *Role of pleopod tegumental glands in egg attachment*

The role of the pleopod tegumental glands in the egg attachment process of the lobster is unclear (reviewed by Aiken & Waddy 1982, Johnson & Talbot 1987). These glands cycle with the ovary and become extremely engorged just prior to spawning (Herrick 1909, Aiken & Waddy 1982, Johnson & Talbot 1987). Yonge (1937) assumed that the pleopod tegumental glands formed the attachment stalk material, since their contents are released at spawning and they seemed to be in the appropriate location. However, Cheung (1966) correctly challenged this idea; as is evident from electron microscopy, the stalk forms from envelope 1 (Goudeau et al. 1987). The pleopod tegumental glands may play an as yet unidentified role in this process. Tegumental glands in other regions of the arthropod body may produce polyphenoloxidase, which functions in tanning or hardening reactions (Stevenson 1961). Moreover, in crayfish, Stevenson & Schneider (1962) found evidence that polyphenoloxidase is present in pleopod tegumental glands *per se*. It is plausible the pleopod tegumental glands release an enzyme(s), such as polyphenoloxidase, that causes hardening or condensation of envelope 1. However, this has not been demonstrated experimentally. Tegumental gland secretions might also create an environment favorable for fertilization or early development, as suggested by Cheung (1966). Clearly more work needs to be done on the role of these glands during egg attachment.

3.3 *Formation of envelope 2*

A second coat, envelope 2, forms inside envelope 1, after fertilization has occurred (Plates 5A-D). Envelope 2 originates from a complex cortical reaction in which four different types of secretory vesicle release their contents into the perivitelline space (Talbot & Goudeau 1988). The contents of at least two of these vesicles coalesce to form envelope 2. This envelope does not participate in formation of the attachment stalk.

4 RETENTION

4.1 *Egg retention in natural populations*

Lobsters normally carry eggs on their pleopods for 9-11 months, although longer times are also possible (Perkins 1972, Branford 1978). In nature, a significant percentage (30-50%) of the clutch is usually lost during this interval (Perkins 1971, Campbell & Brattey 1986). Because the brooding interval is long, some attrition would be expected to occur. However, the exact causes of natural egg loss are not well understood. Waddy & Aiken (1985) found that the nemertean, *Pseudocarcinonemertes homari* can feed on lobster eggs and potentially damage a clutch. However, in natural populations, this is usually not a major contributor to egg loss unless a high intensity of nemerteans becomes established on eggs during the early stages of incubation (Cambell & Brattey 1986). Other extrinsic factors that could affect egg retention include removal or eating of eggs by the female (Knight 1918), severe fouling by particular epibionts (Miller & Fleming 19183, Harper & Talbot 1984), and attemped cleaning of infested egg masses (Aiken et al. 1985). Capture, handling and release of gravid females in lobster traps may contribute to clutch attrition (Herrick 1909).

4.2 *Egg retention in captive populations*

In natural populations, the initial size of the clutch varies from 3000 to 100 000 depending on the size of the female (Herrick 1909). A 25 cm female typically spawns and attaches an average of 10 550 eggs (Herrick 1909). In captivity, the number of eggs spawned per clutch is similar to that predicted from the values in nature (Talbot et al. 1984). In the relatively controlled environment of a laboratory, egg retention could be equivalent to or better than retention observed in nature. While this is sometimes the case (Waddy & Aiken 1984), significant egg loss may occur in captivity (Plates 7A, B; Hedgecock et al. 1978, Talbot et al. 1984). Several factors are responsible for this. Some females actually spawn the normal number of eggs, but fail to attach them (Talbot et al. 1984). This can occur if the female is disturbed during spawning or if she does not have proper balance during spawning, e.g. because of the absence of 1 or more walking legs (Wilson pers. comm.). Eggs may roll off her ventral surface before arriving at the pleopods for attachment. Also, captive females may fail to produce normal ovigerous setae and plumose setae after passing through a critical molt that occurs at about 65-70 mm in carapace length (Gerrero & Talbot 1984, Talbot & Wilson unpubl. data), clearly precluding proper attachment of a full clutch of eggs.

4.2.1 *Egg stalk formation*

Some captive females attach a full complement of eggs, but lose them rapidly, i.e. within the first 2-3 weeks (Talbot et al. 1984). One cause of this is the formation of poor quality egg stalks (Talbot & Harper 1984, Plates 7C, D). Females that produce small, thin eggs stalks (Pl. 7D) do not carry their eggs very long. Some females produce a stalk of intermediate quality, not as thick as the wild type nor as thin as the example in Plate 7D. These females tend to lose their eggs at a somewhat slower rate and carry few if any eggs through to hatching (Talbot et al. 1984). The formation of poor attachment stalks has not yet been studied directly, but it is probably caused by incomplete formation of envelope 1 during oöcyte maturation. If this envelope were incompletely developed at the time of spawning, the attachment stalk probably would be incompletely formed. In laboratories, females are often placed on accelerated ovarian cycles (Nelson et al. 1983, D'Ambramo & Conklin 1985, Wilson unpubl. data); eggs and their coats then may not be comparable to eggs matured in the wild. In fact, eggs in captive females are often pale green (Hedgecock & Wilson pers. comm.), suggesting incomplete deposition of ovoverdin. Females in laboratories where water temperatures mimic those found in their natural environment do not experience massive egg losses (Waddy & Aiken 1984). Thus a simple solution to egg loss in captive females may be to return broodstock females to conditions characteristic of their natural habitat. When accelerated growth rates are preferred to reproduction, animals could be maintained at higher water temperatures to encourage molting (Hughes et al. 1972).

4.2.2 *Bacteria on egg surfaces*

Lobster eggs have numerous bacteria on their surfaces (Fisher et al. 1978, Harper & Talbot 1984). Although they are apparently able to tolerate high bacterial loads without egg loss (Harper & Talbot 1984), lower numbers of bacteria are probably desirable for several reasons. First large numbers of filamentous bacteria or fungi on egg surfaces can interfere with proper hatching of larvae which become entangled in the filamentous web (Nilson et al. 1975). Also some bacteria may be harmful to the developing lobsters or may compete with them metabolically. Certain bacteria on the exoskeleton of lobsters apparently regulate growth of other epibionts on lobster eggs by releasing a natural fungicide (Gils-Turnes unpubl. data). Encouragement of this microorganism may be a simple and safe way to improve bacterial populations on the lobster egg surface.

4.3 *Hatching*

At the completion of successful brooding, first stage larvae hatch from the egg coats during a several day period. The egg cases and attachment stalks remain on the female and are presumably not lost until the next molt (Goudeau et al. 1987).

5 SUMMARY

We now have a general knowledge of how lobster eggs are ovulated and attached to the ovigerous setae, and many of the controversies regarding attachment have been clarified in the past few years. However, we lack information on the specific details and regulation of these events, and much work remains to be done before these processes are adequately understood. Ovulation is an essential reproductive event that has not been extensively studied in marine invertebrates. Attachment is important from both a biological and economic perspective, as lobster aquaculture requires satisfactory techniques for broodstock management and production of larvae for grow out. A complete knowledge of the attachment mechanism and how to control it should be a goal for future research with lobsters. The problem of egg retention on the ovigerous setae is important for both natural and cultured populations of lobsters. Data clearly show that egg loss from both these populations can be quite high. We are just beginnning to understand the factors that contribute to poor egg retention in this commerically important group of decapods.

ACKNOWLEDGEMENTS

I am indebted to Li Wen-Yu for her help in preparing the plates and to my colleagues who have participated in work on lobster reproduction with me during the past 10 years. I would also like to thank Georgia Lovely and Debbie Drake for their help with word processing. Work done in my laboratory was supported by R/A-29 and R/A-59 from the California College Sea Grant Program and by funds from the University of California, Riverside, Academic Senate.

REFERENCES

Aiken, D.E. & S. Waddy 1980. Reproductive biology. In S.C. Cobb & B. Phillips (eds), *The Biology and Management of Lobsters* vol. 1: pp. 215-275. New York: Academic Press.

Aiken, D.E. & S.L. Waddy 1982. Cement gland development, ovary maturation, and reproductive cycles in the American lobster *Homarus americanus*. *J. Crust. Biol.* 2: 315-327.

Aiken, D.E., S.L. Waddy & L. Uhazy 1985. Aspects of the biology of *Pseudocarcinonemertes homari* and its association with the American lobster, *Homarus americanus*. *Can. J. Fish. Aquat. Sci.* 42: 351-356.

Branford, J.R. 1978. Incubation period for the lobster *Homarus gammarus* at various temperatures. *Mar. Biol.* 47: 363-368.

Campbell, A. & J. Brattey 1986. Egg loss from the American lobster, *Homarus americanus*, in relation to nemertean, *Pseudocarcinonemertes homari*, infestation. *Can. J. Fish. Aquat. Sci.* 43: 772-780.

Campbell, R.D. 1974. Cnidaria. In A.C. Giese & J. Pearse (eds), *Reproduction in Marine Invertebrates* vol 1: pp. 133-139. New York: Academic Press.

Cheung, T.S. 1966. The development of egg membranes and egg attachment in the shore crab, *Carcinus maenas*, and some related decapods. *J. Mar. Biol. Assc. UK*. 46: 373-400.

Couch, E.F., N. Hagino & J.W. Lee 1987. Changes in estradiol and progesterone immunoreactivity in tissues of the lobster, *Homarus americanus*, with developing and immature ovaries. *Comp. Biochem. Physiol.* 87A: 765-770.

Curry T., D. Dean, J. Woessner & W. LeMaire 1985. The extraction of a tissue collagenase associated with ovulation in the rat. *Biol. Reprod.* 33: 981-991.

D'Abramo, I.R. & D.E. Conklin 1985. Lobster aquaculture. In J.V. Huner & E.E. Brown (eds), *Crustacean and Mollusk Aquaculture in the United States*. Westport, CT: AVI Publishing Co.

Donahue, J.K. 1948. Fluorimetric and biological determination of estrogens in the eggs of the American lobster (*Homarus americanus*). *Proc. Soc. Exp. Biol. Med.* 69: 179-181.

Donahue, J.K. 1955. Studies on ecdysis in the American lobster. 4. Estrogenic hormone as a possible moult inhibitor in the egg bearing female. *Maine Dept. Sea & Shore Fisheries Bull.* No. 24.

Donahue, J.K. 1957. Chromatographic identification of lobster estrogen. *Maine Dept. Sea & Shore Fisheries Bull.* No. 28.

Fisher, W.S., E.H. Nilson, J.F. Steinbergen & D.V. Lightner 1978. Microbial diseases of cultured lobsters: a review. *Aquaculture* 14: 115-140.

Gerrero, R. & P. Talbot 1984. A microscopic study of the setae on the necrotic pleopods of the American lobster. *Abst. World Mariculture Soc.*

Goodwin, T.W. 1951. Carotenoid metabolism during development of lobster eggs. *Nature* 7: 559.

Goudeau, M. & F. LaChaise 1983. Structure of the egg funiculus and deposition of embryonic envelopes in a crab. *Tissue & Cell* 15: 47-62.

Goudeau, M., P. Talbot & R. Harper 1987. Mechanism of egg attachment stalk formation in the lobster, *Homarus*. *Gamete Res.* 18: 279-289.

Harper, R. & P. Talbot 1984. Analysis of the epibiotic bacteria of lobster (*Homarus*) eggs and their influence on the loss of eggs from the pleopods. *Aquaculture* 36: 9-26.

Hedgecock, D., W.L. Moffett, W. Borgeson & K. Nelson 1978. Progress and problems in lobster broodstock development. *Proc. World Maricul. Soc.* 9: 497-506.

Herrick, F.H. 1891. The reproductive organs and early stages of development of the American lobster. *John Hopkins Univ. Circ.* 88: 98-101.

Herrick, F.H. 1909. Natural history of the American lobster. *Bull. US Bur. Comm. Fish.* 29: 149-408.

Howard, D., L. Haimo & P. Talbot 1985a. A demonstration of actin and microtubules in non-striated muscle cells of the lobster ovary. *J. Cell Biol.* 101: 166a.

Howard, D., L. Haimo & P. Talbot 1985b. Microtubules in non-striated muscle cells. *Proc. West Coast Regional Devel. Biol. Conf.*

Hughes, J.T., J.J. Sullivan & R.A. Shleser 1972. Enhancement of lobster growth. *Science* 177: 1110-1111.

Ichikawa, S., M. Ohta, H. Morioka & S. Murao 1983a. Blockage of ovulation in the explanted hamster ovary by a collagenase inhibitor. *J. Reprod. Fert.* 68: 17-19.

Ichikawa, S., H. Morioka, M. Ohta, K. Oada & S. Murao 1983b. Effect of various proteinase inhibitors on ovulation of explanted hamster ovaries. *J. Reprod. Fert.* 68: 407-412.

Johnson, B. & P. Talbot 1987. Ultrastructural analysis of the pleopod tegumental glands in male and female lobsters, *Homarus americanus*. *J. Crust. Biol.* 7: 288-301.

Knight, A.P. 1918. Lobster investigation at Long Beach Pond, New Jersey. *Contrib. Can. Biol.* 1918: 53-71.

Lisk, R.D. 1961. Estadiol-17b in the eggs of the American lobster, *Homarus americanus*. *Can. J. Biochem. Physiol.* 39: 659-663.

Miller, J. & L. Fleming 1983. Fungi associated with an infestation of *Pseudocarcinonemertes homari* in *Homarus americanus*. *Trans. Brit. Mycol. Soc.* 80: 9-12.

Nelson, K., D. Hedgecock & W. Borgenson 1983. Photoperiodic and ecdysial control of vitellogenesis in lobsters (*Homarus*) (Decapoda, Nephropidae). *Can. J. Fish. Aquat. Sci.* 40: 940-947.

Nilson, E.H., W.S. Fisher & R.A. Shleser 1975. Filamentous infestations observed on eggs and larvae of cultured crustaceans. *Proc. World Maricult. Soc.* 6: 367-375.

Perkins, H.C. 1971. Egg loss during incubation from offshore northern lobsters (Decapoda: Homaridae). *Fish. Bull.* 69: 451-453.

Perkins, H.C. 1972. Developmental rates at various temperatures of embryos of the northern lobster (*Homarus americanus*). *Fish. Bull.* 70: 95-99.

Phillips, B.F., J.S. Cobb & R.W. George 1980. General Biology. In J.S. Cobb & B. Phillips (eds), *The Biology and Management of Lobsters*. pp. 1-82. New York: Academic Press.

Schroeder, P. & P. Talbot 1985. Ovulation in the animal kingdom: a review with emphasis on the role of contractile processes. *Gamete Res.* 11; 191-221.

Scott, A. 1902. On the spawning of the common lobster. *Trans. Liverpool Biol. Soc.* 17: 106-113.

Stern, K. & K. Salomon 1937. Ovoverdin, a pigment chemically related to visual purple. *Science* 86: 310-311.

Stevenson, J.R. 1961. Polyphenol oxidase in the tegumental glands in relation to the molting cycle of the isopod crustacean *Armadillidium vulgare*. *Biol. Bull.* 121: 554-560.

Stevenson, J.R. & R.P. Schneider 1962. Tyrosinase activity of organs containing tegumental glands in the crayfish. *J. Exp. Zool.* 150: 17-25.

Talbot, P. 1981a. The ovary of the lobster, *Homarus americanus*. I. Architecture of the mature ovary. *J. Ultrastruct. Res.* 76: 235-248.

Talbot, P. 1981b. The ovary of the lobster, *Homarus americanus*. II. Structure of the mature folicle and origin of the chorion. *J. Ultrastruct. Res.* 76: 249-262.

Talbot, P. 1981c. Collagenase solutions induce *in vitro* ovulation in lobsters (*Homarus americanus*). *J. Exp. Zool.* 216: 181-185.

Talbot, P. & M. Goudeau 1988. A complex cortical reaction leads to formation of the fertilization envelope in the lobster (*Homarus*). *Gamete Res.* 19: 1-18.

Talbot, P. & R. Harper 1984. Abnormal egg stalk morphology is correlated with clutch attrition in laboratory-maintained lobsters (*Homarus*). *Biol. Bull.* 166: 349-356.

Talbot, P., C. Thaler & P. Wilson 1984. Spawning, egg attachment, and egg retention in captive lobsters (*Homarus americanus*). *Aquaculture* 37: 239-249.

Templeman, W. 1936. Further contributions to mating in the American lobster. *J. Biol. Board. Can.* 2: 223-226.

Templeman, W. 1937. Egg laying and hatching postures and habits of the American lobster (*Homarus americanus*). *J. Biol. Board Can.* 3: 339-342.

Waddy, S. & D. Aiken 1984. Broodstock managements for year-round production of larvae for culture of the American lobster. *Can. Tech. Rep. Fish. Aquat. Sci.* 1272: iii+14 p.

Waddy, S. & D. Aiken 1985. Immunofluorescent localization of American lobster egg yolk protein in the alimentary tract of the nemertean *Pseudocarcinonemertes homari*. *Can. J. Fish. Aquat. Sci.* 42: 357-359.

Waddy, S. & D. Aiken 1986. Multiple fertilization and consecutive spawning in large American lobsters, *Homarus americanus*. *Can. J. Fish. Aquat. Sci.* 43: 2291-2294.

Wallace, R.A., S.L. Walker & P.V. Hauschka 1967. Crustacean lipovitellin. Isolation and characterization of the high molecular weight lipoprotein from the eggs of decapods. *Biochem.* 6: 1582-1590.

Yonge, C.M. 1937. The nature and significance of the membranes surrounding the developing eggs of *Homarus vulgaris* and other Decapoda. *Proc. Roy. Soc. London* (A)120: 15-41.

Yoshimura, Y & E. Wallach 1987. Studies of the mechanism(s) of mammalian ovulation. *Fert. Steril.* 47: 22-34.

Zagalsky, P.F. 1985. A study of astaxanthin lipovitellin ovoverdin isolated from ovaries of the lobster *Homarus gammarus*. *Comp. Biochem. Physiol.* 80B: 589-598.

Yolk utilization and esterase activity in the mole crab *Emerita asiatica* (Milne Edwards)

T. Subramoniam
Unit of Invertebrate Reproduction, Department of Zoology, University of Madras, Guindy Campus, India

ABSTRACT

The mole crab, *Emerita asiatica* lays a large number of yolky eggs, and incubates them attached to pleopodal hairs. The present paper reports on the biochemical alterations in the major organic substrates, as well as the activity of non-specific esterases during egg development leading to the release of the zoea larva. Egg development has been divided into nine stages based on color change and other concomittant morphological features of the embryo. The water content steadily increased during embryogenesis; conversely, protein content showed a steady decline. Although the relative concentration of free carbohydrates and glycogen exhibited an increasing trend, the protein-bound carbohydrates declined during embryogenesis. Lipid level remained unaltered almost up to stage V; thereupon, the value fell precipitously, reaching the minimum in stage IX. Polyacrylamide gel electrophoresis revealed that the yolk protein is mainly concentrated in the form of a thick, slow-moving glycolipoprotein, lipovitellin. This protein band becomes diffused during the course of embryogenesis, giving rise to several smaller units; concomittantly new protein fractions also appear towards the end of egg development. Non-specific esterase activity appears from stage IV onwards, reaching a peak in stage V, and declines thereafter. Slab gel electrophoresis of esterases revealed up to five fractions, all of them appearing to be carboxyl esterases. This stage-specific appearance of esterase isozymes may indicate changes in cellular differentiation and metabolic pathways signifying differential gene action during embryogenesis.

1 INTRODUCTION

Many decapod crustaceans lay a large number of yolky eggs and incubate them on pleopodal setae. The yolk serves as a source of energy and biochemicals during development. In Crustacea, several species have been investigated with respect to yolk utilization and energy transformation during embryogenesis (Pandian 1970a, b). However, the main emphasis of those studies was on ecophysiology of the organism; the biochemical intricacies of the preferential substrate utilization were given only scanty attention. During vitellogenesis, besides the accumulation of the main yolk components, other metabolically important substances such as RNA, as well as a host of hydrolytic enzymes, are also synthesized and stored (Adiyodi & Subramoniam 1983). In a prosobranch mollusc, *Lymnea stagnalis*, Blumink (1967) demonstrated several hydrolytic enzymes packaged in lysosomes of the multivesicular yolk-like bodies. Recently, Ezhilarasi & Subramoniam (1984) reported the accumulation of carboxyl esterases in the ovary during vitellogenesis of mud crab, *Scylla serrata*. However, enzyme activity, as related to substrate utilization during embryogenesis, has not been adequately understood in decapod crustaceans.

The eggs of *Emerita asiatica* offer several advantages for such a biochemical investigation. Berried females with morphologically distinguishable egg stages are obtainable all through the year (Subramoniam 1977, 1979). The present study reports on the biochemical alterations in the organic constituents during the progressive egg development leading to the release of the zoea larva. As lipids form the main energy-yielding substrate of several decapod crustaceans, the pattern of non-specific esterase activity in the developing eggs of *E. asiatica* has also been investigated.

2 PROCEDURES

Specimens of *Emerita asiatica* were collected from Marina Beach, opposite the University Buildings of the Chepauk Campus in Madras. Different stages of egg development were recognized using criteria in Table 1. The time taken for each stage during egg development was estimated by rearing the freshly ovulated females in the laboratory (Temp. 26° C; salinity 34‰).

Eggs were stripped off the pleopodal setae and used for biochemical estimation in the wet condition. To determine water content, the eggs were blot-dried with filter paper and weighed

Table 1. Classification of egg development in *Emerita asiatica*. Adapted from Subramoniam (1982).

Stages	Approximate number of days	Description
I	5	Yellow yolk granules seen; egg mass bright orange in color
II	5	Cleavage has taken place and blastomeres are seen; egg mass bright orange in color
III	1	A yolk-free white streak makes its appearance at the animal pole
IV	1	One quarter of the yolk cleared; the white band encircles the yolk material which is now in the center; at the animal pole a periodic twitching is recognized; red pigments are seen at the edge of the yolk; color of the egg mass is dull orange
V	2	One third of the yolk is cleared; two eye spots appear; red spots prominent and seen at the end of the animal pole; color of the egg mass dull orange
VI	2	Egg mass brownish orange in color; eyes well developed; yolk is found in the vegetal pole; two thirds of the yolk is cleared; red pigments seen all over the white space
VII	1	Egg mass greyish orange in color; yellow yolk is found as two clusters in the center; appendages of the embryo are developed; heart beat seen; eye spots very well developed
VIII	1	Egg mass pale grey in color; heart beat more prominent; embryo almost developed
IX	1	Embryo fully formed; egg mass white in color; no yolk globules seen; colorless yolk in the form of oil globules seen just below the eyes; about to hatch

in a monopan electrical balance and dried in a desiccator with calcium chloride to standardize weight. Water content was calculated as the difference between wet weight and dry weight of the eggs and expressed as a percentage. Total protein was determined using the Folin-ciocalteu method (Lowry et al. 1951); total free sugar, occurring as mono-, di- and oligosaccharides, was determined by the anthrone method (Roe 1955); protein-bound sugar (PBS) was determined after the method of Carroll et al. (1956), as adopted by Parvathy (1970). Glycogen was determined by the method of Carroll et al. (1956). The gravimetric method of Folch et al. (1957) was employed to quantify total lipid using chloroform and methanol as solvents. Separation of yolk proteins was carried out using polyacrylamide gel electrophoresis (Davis 1964). The gels (7%) were stained in Coomassi brilliant blue.

Enzyme assay of non-specific esterase was carried out using 1-naphthyl acetate as the substrate, following the method outlined by Kapin & Ahmed (1980) and Ezhilarasi & Subramoniam (1984). Fractionation of esterase isozymes was made on 7% polyacrylamide slab gel, according to Smith (1968). Non-specific esterase fractions on the gel were stained by the method of Ruddle & Harrington (1967), in which the complete reaction products of naphthyl substrates were stained with a diazonium salt. The multiple forms of esterases were characterized histochemically on polyacrylamide gel after fractionation, using specific inorganic inhibitors such as 1) p-chloromercuric benzoic acid (CMB), 10^{-2}M, 2) ethylene diaminetetra acetic acid (EDTA), 10^{-2}M, 3) malathione (organophosphate), 10^{-3}M, 4) silver nitrate, 10^{-3}M, 5) eserine sulphate, 10^{-3}M.

The observations made on the esterase activity in different stages of egg development were subjected to statistical analyses, including analysis of variance (ANOVA) as described by Zar (1974). The significant differences in the enzyme activity among various stages were further analyzed using the student Newman-Keul procedure.

3 BIOCHEMICAL CHANGES DURING EGG DEVELOPMENT

Water content increased steadily from an initial value of 59% in stage I to 82% in stage IX. However, there was a striking fall in stage II (51%) from stage I; the water content was high in stages VIII & IX (Fig. 1). Changes in percentage values of protein, lipid and carbohydrates, calculated on wet weight basis are summarized in Table 2. Protein value steadily declined from stage I to IX, corresponding to increasing water content. On the other hand, lipid content remained almost unaltered up to stage V; thereupon, the value fell precipitously, reaching a minimum of 0.49 mg/10 mg in stage IX. Compared to protein and lipid, the total carbohydrate content was low, but the different carbohydrate components exhibited an interesting pattern of fluctuation during egg development. The total free carbohydrates increased from a low value of 0.12 mg/10 mg in stage I to 0.72 mg/10 mg in stage IX. The free glycogen content also exhibited a similar increase during egg development. Conversely, the protein bound polysaccharide increased and decreased from an initial high value of 0.072 mg/10 mg in stage I to 0.021 mg/10 mg in stage IX. The above biochemical changes during egg development are statistically significant ($p < .01$).

The increasing water content of eggs in later developmental stages is expected to reduce correspondingly the concentration of other biochemical substances. To standardize for this, the weight of water in each stage was subtracted. Constituents such as protein, carbohydrate and lipid were expressed as proportions to the total weight of the remaining components (Table 3).

The percentage value of protein among the non-aqueous components steadily declined from

Table 2. Major organic composition of eggs during different stages of development in *E. asiatica*. Values expressed as mg/10 mg wet weight (mean ± SD). Esterase activity expressed as nM 1naphthol/mg protein/min.

Biochemical constituents	Stages of eggs I	II	III	IV	V	VI	VII	VIII	IX
Protein	1.32±0.150	1.09±0.143	0.949±0.107	0.90±0.03	0.77±0.09	0.66±0.043	0.44±0.08	0.28±0.021	0.16±0.029
Free carbohydrates	0.12±0.013	0.24±0.009	0.28±0.012	0.32±0.012	0.36±0.010	0.48±0.012	0.52±0.008	0.68±0.009	0.72±0.015
Glycogen	0.005±0.001	0.0057±0.0001	0.0119±0.0002	0.0168±0.003	0.0109±0.0002	0.026±0.0022	0.0231±0.0013	0.0197±0.002	0.0207±0.0017
Protein bound sugar	0.0729±0.0031	0.0849±0.0024	0.0467±0.0032	0.0457±0.0014	0.0205±0.0071	0.0339±0.0021	0.0261±0.0018	0.0250±0.007	0.0210±0.003
Lipid	1.89±0.014	1.88±0.014	1.87±0.010	1.85±0.014	1.80±0.012	0.64±0.009	0.52±0.014	0.51±0.015	0.49±0.009
Non specific esterases	-	-	-	0.670±0.123	1.08±0.214	1.73±0.530	0.89±0.090	0.62±0.035	0.29±0.055

Table 3. Major organic composition of eggs during different stages of development in *E. asiatica* after correction for water content (expressed as mg/10 mg of dry tissue, mean values only).

Biochemical constituents	Stages of eggs I	II	III	IV	V	VI	VII	VIII	IX
Protein	3.219	2.224	2.433	2.727	2.484	2.357	1.692	1.272	0.889
Free carbohydrates	0.293	0.489	0.718	0.969	1.161	1.714	2.000	3.090	4.000
Glygogen	0.012	0.012	0.031	0.051	0.036	0.093	0.089	0.089	0.115
Protein bound sugars	0.178	0.173	0.119	0.139	0.066	0.121	0.100	0.114	0.117
Lipid	4.609	3.837	4.795	5.606	5.806	2.286	2.000	2.318	2.722

Table 4. Esterase activity in different stages of egg development in *E. asiatica* at various pH (values expressed as nM 1naphthol/mg protein/min., mean ± SD).

Egg stages	pH 6.0	6.5	7.0	7.5	8.0	8.5	9.0
III	0.068±0.015	0.112±0.014	0.119±0.022	0.077±0.013	0.072±0.018	0.017±0.005	0.010±0.001
IV	0.117±0.016	0.171±0.020	0.198±0.038	0.117±0.016	0.099±0.017	0.057±0.005	-
V	0.203±0.065	0.229±0.070	0.309±0.095	0.205±0.043	0.119±0.013	0.098±0.015	0.065±0.011
VI	0.099±0.016	0.134±0.015	0.159±0.016	0.119±0.019	0.068±0.008	0.059±0.013	0.009±0.002
VII	0.076±0.015	0.093±0.007	0.112±0.006	0.075±0.009	0.032±0.008	0.004±0.001	-
VIII	0.019±0.002	0.025±0.003	0.053±0.009	0.017±0.003	0.003±0.001	-	-

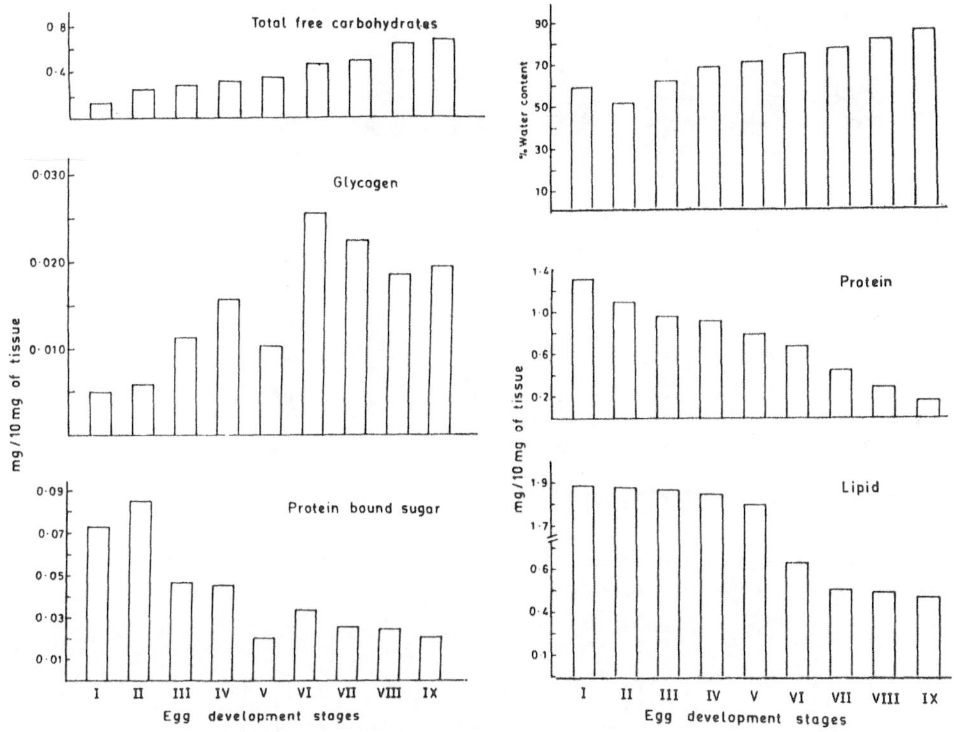

Figure 1. Bar diagrams showing the fluctuation in the major biochemical components during egg development in *E. asiatica*. That in the upper right is for total biomass.

stage I onwards, excepting stages III and IV when there was a slight increase from the preceding stage. The lipid level exhibited an initial increase, but declined sharply from stage VI onwards. On the contrary, the total free carbohydrate and glycogen increased linearly, although the protein bound polysaccharide showed an inverse relationship to other carbohydrate values.

Electrophoretic separation of the yolk proteins was noted in different stages of the egg development. In stage I eggs, the lipovitellin (LV) resolved into a thin fraction LV1 and a thick fraction LV2 with a slightly higher mobility in the slow-moving region of the electropherogram (Fig. 2). Both these fractions were colored bright yellow even in the unstained gels. They stained avidly for amido black, Oil Red O, and PAS suggesting the glycolipoprotein nature. The thin LV1 cleaved apart into two fractions in stage VII, before it totally disappeared in stage VIII/IX. Similarly, the thick LV2 also gradually gave rise to three fast-migrating fractions that resolved below the LV2 and were designated as LV2', LV2" and LV2"' to indicate their origin. From stage VII onwards, the staining intensities of LV2 and its subunits declined and disappeared in stage VIII/IX.

4 NON-SPECIFIC ESTERASE ACTIVITY

Progress curves for 1-naphthyl acetate hydrolysis by eggs in different stages of development showed that the optimum pH was 7.0 (Fig. 4a, Table 4). Enzyme assay revealed no activity in

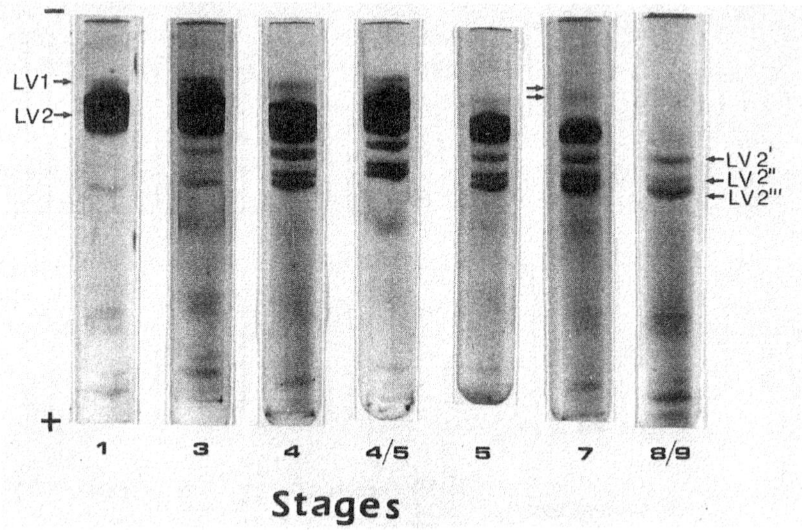

Figure 2. Electropherograms of protein pattern in different stages of egg development. LV1 and LV2 – Lipovitellins. Double arrows indicate the subunits of LV1; LV2', LV2" and LV2''' are the subunits of LV2.

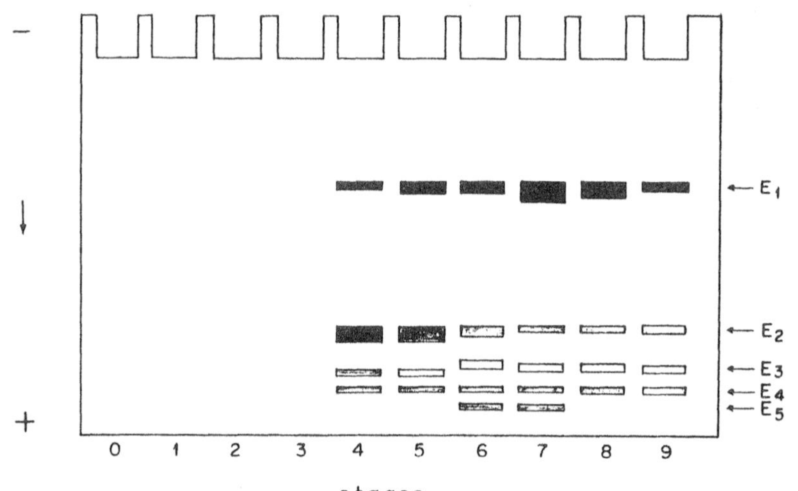

Figure 3. Zymogram pattern of esterases from the ovary (O) and different egg developmental stages (1-9) in *E. asiatica*. (E_1-E_5 represent different isozymes).

the first three stages of egg development. The enzyme activity appeared from stage IV onwards, reaching a peak in stage V and thereupon declined gradually in the subsequent stages (Fig. 4B). The analysis of variance (ANOVA) test indicated a significant change in the esterase activity in egg developmental stages ($p < .05$). Significant preponderance of esterase activity was observed in the egg stage V and low activity in stage VIII. Moderate enzyme activity was also seen

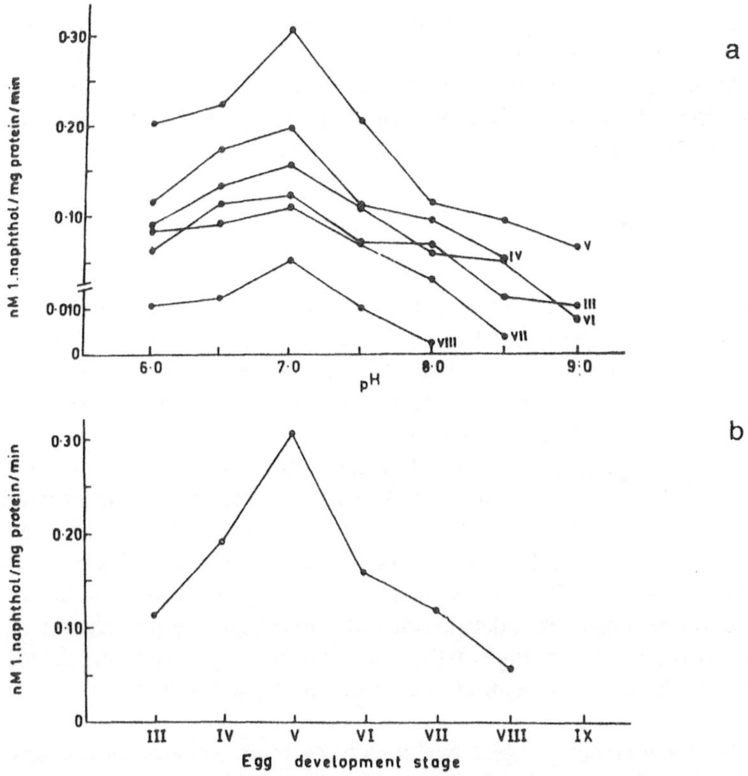

Figure 4. A. Esterase activity of different egg stages related to different pH. B. Differences in enzyme activity at pH7 in different egg stages.

Table 5. Esterase activity in different stages of egg development in *E. asiatica*. Lines connect homogeneous sets; student Newman-Keul Test.

Egg stages	Esterase activity
VIII	0.0523±0.009
VII	0.1115±0.006
III	0.1198±0.022
VI	0.1585±0.016
IV	0.1983±0.038
V	0.3086±0.094

in stages III, IV, VI & VII and the difference among these stages was insignificant (Table 5).

The zymogram pattern of esterase as revealed by slab gel electrophoresis, indicated up to five fractions (Fig. 3). In the mature ovary and the egg stages up to three, no fractions could be visualized. In stage IV, the zymogram consisted of a major fraction (E1), a moderately staining E2 fraction and two other thin fractions (E3 and E4). In stage V and VI, E2 fraction declined in intensity, but yet another fraction E5 appeared in the fast moving zone in stage VI. E5 persisted up to the VII stage but was absent in stages VIII and IX. In general, the E1 fraction did not significantly change in intensity, but others decreased in intensity and disappeared in the last stages of egg development (Fig. 3). Further characterization of the esterases indicated that all of the five fractions are isozymes of the carboxylesterase, as they were inhibited by silver nitrate and malathione and unaffected by pCMB, EDTA and eserine sulphate.

5 DISCUSSION

Earlier workers on crustacean egg development have discussed yolk utilization from the energetic aspect and environmental adaptations. Many crustacean eggs are large and contain substantial reserves of organic substrates for their independent embryonic development. Yolk utilization necessitates extensive reshuffling of substrates, especially during the early stages of embryonic development. Analysis of the interconversion among organic substrates during egg development has not been attemped previously mainly because the classification of egg developmental stages was arbitrary. For instance, many workers have divided the egg stages based on the appearance of distinctive morphological features such as the eye, heart beat, and appendage formation. However, such morphological characters only begin to appear mid-way during embryonic development. Cellular differentiation starts soon after gastrulation and requires enormous energy expenditure. Therefore, a detailed classification of early egg development is a necessary prerequisite to understand changes in the metabolic pathways involving interconversion of already stored substrates within the closed system of egg development of decapod crustaceans.

The biochemical data reported here suggest several such possibilities of interconversion of metabolic substrates as well as changes in the metabolic pathways during the course of egg development. In *Emerita asiatica*, protein and lipid are the major organic substrates, although both free and bound carbohydrates are substantial also. Protein declines monotonically from stage I up to stage IX. The electrophoretic data suggest that the storage protein in the beginning of egg development is mainly in the form of lipovitellins, typical of many decapod crustaceans (Adiyodi & Subramoniam 1983). This study provides evidence that lipovitellins are gradually dismantled during embryogenesis, as indicated by the electrophoretic transformation that these proteins undergo during egg development. Concurrently, several subunits of this main yolk protein appear in the electropherogram. In addition to this change in the main yolk protein, there are also many new proteins that appear as a result of cellular differentiation and histogenesis during embryogenesis.

Although carbohydrates constitute a minor proportion of the organic substances present in developing eggs, they show interesting changes during egg development. Total free carbohydrates, including glycogen, show a steady increase during egg development. Concurrently, there is a decline of protein-bound carbohydrates. Such a possibility is implicit in the gradual utilization of protein, which is mainly found in the form of glycolipoprotein. The presence of increased free sugars, in spite of their possible use as an energy source during development, may also be due to synthesis of carbohydrates from other sources such as proteins. Carbohydrates are needed in large quantities during later phases of embryonic development for the synthesis of chitin. The participation of free and protein bound sugars in chitin synthesis of adult crustaceans has been reported by many authors (e.g. McWhinnie & Sallar 1960, Meenakshi & Scheer 1961, Parvathy 1970).

The initial high content of lipid is characteristic of lecithotropic eggs. In *E. asiatica*, initial high lipid value does not change significantly up to stage V. Tardiness in the lipid utilization of early embryogenesis has also been shown for two gammarid amphipods, where there is an initial rise in the triglycerides during embryonic development (Clarke et al. 1985). In *E. asiatica*, lipid utilization starts only from stage V onwards, suggesting that protein may be the chief source of energy for initial embryonic development. However, the decline in lipid in later stages is so sharp that lipid use increases, along with a steady utilization of proteins. In many decapod crustacean eggs, lipid forms the major component that is used up, right from the initial stage of development. As an extreme example, the eggs of a freshwater crab, *Parathelphusa hydrodromus* expend their enormous reserves of lipid continuously during embryogenesis with a concomittant increase in the protein level (Pillai & Subramoniam 1985). Suppression of protein utilization and enhanced lipid metabolism is characteristic of cleidoic eggs, a feature found also in many crustacean species (see Pandian 1970a, Pillai & Subramoniam 1985). Conversely, in the bony fishes, protein is preferentially used during the entire course of embryogenesis (Lasker 1962). Egg development of *E. asiatica* represents a condition intermediate between cleidoic and non-cleidoic developmental extremes. Protein is continuously used up, although lipid forms the major reserve that gets depleted in the second half of embryogenesis.

Although lipid forms the major energy source during the second half of embryogenesis, when tissue synthesis and organ formation occur rapidly, a substantial quantity of lipid is still retained within the embryo at stage IX. These lipid reserves are in the form of colorless oil globules, which may increase the buoyancy of the pelagic larvae on their release. Such lipid reserve may also be of advantage in delaying starvation during the rather protracted larval life of *E. asiatica*. In general, the protein/lipid ratio is high in typically planktotrophic larvae such as cirripedes (Architur & Wortzlavski 1983), whereas lipids form the major reserves in lecithotrophic eggs. Although *E. asiatica* releases planktotrophic zoea with a long pelagic life, it produces many yolky eggs in which the lipid/protein ratio is much higher.

Lecithotrophic eggs are fewer in numbers per brood and their larvae spend only a very short time as plankton, but use yolk reserve and do not feed upon release. A short period of planktonic existence reduces the mortality rate and increases the survival rate. On the other hand, planktotrophic larvae have a high mortality rate and low survival in view of high exposure to predation and starvation. The condition found in *Emerita* with respect to egg development (present study) and larval development (studies of *E. analoga* by Efford 1970) present an intermediate condition to lecithotrophy and planktotrophy. *Emerita* eggs possess rather heavy yolk materials. However, the larvae live a long pelagic life (ranging from 30-130 days: Efford 1970) before settling on the beach as megalopa.

Lipid is the chief organic reserve of *Emerita* embryogenesis. Therefore, aspects of enzyme activity related to lipid utilization merit discussion. Esterases are involved in the breakdown of various lipids. In crustaceans, they function as hydrolytic enzymes which dismantle the complex lipid yolk during embryogenesis. Hence, the initial tardiness in lipid utilization may be attributed to esterase inactivity. This inactivity might be due to the proenzyme nature of esterases or there may be specific inhibitors. Alternatively, the gene encoding esterase synthesis is 'turned on' at stage IV of egg development.

In *Emerita* eggs, the esterases resolve into two groups; one is a homogeneously thick proximal fraction occurring in stage IV through IX. This common esterase probably possesses a very generalized function such as the breakdown of lipids to shorter fatty acid chains which may be used as energy sources. This fraction appears to overlap the main lipovitellin in its relative mobility. Interestingly, Ezhilarasi & Subramoniam (1984) reported in the ovary of the brachyuran crab, *Scylla serrata*, that an intense slow moving esterase (E1), corresponding to the main lipovitellin in its relative mobility, appeared during peak vitellogenesis. These authors, however, did not observe any esterase fraction in the freshly laid eggs of *Scylla serrata*.

While the main esterase fraction (E1) may be synthesized during vitellogenesis, kept in an inactive condition, and activated during embryogenesis, the other esterases may be synthesized by the embryonic cells. Their stage-specific appearance or disappearance may further indicate changes in cellular differentiation and metabolic pathways, signifying differential gene action essential for the synthesis of a specific ester by a particular organ system. In an isopod, Doyle et al. (1959) found that freshly laid eggs contained a significant quantity of esterases that gradually increased during embryogenesis, suggesting embryonic synthesis of this enzyme in addition to the stored esterases found in the eggs.

ACKNOWLEDGEMENTS

This work was financially supported by the International Foundation of Sciences (IFS), Sweden. I am thankful to Drs Ezhilarasi, Balasubramanian, N. Munuswamy and K. Nelliappan for help. Mr Kalimur Rahman helped in maintaining *Emerita* in the laboratory. My thanks are also due to Prof. G. Durairaj for encouragement.

REFERENCES

Achitur, Y. & A. Wortzlavski 1983. Studies in the biochemistry of cirripede eggs VII. Changes in the general biochemical composition during development of *Chthamalus dentatus* Krauss and *Octomeris angulosa* Sowerby. *J. Exp. Mar. Biol. Ecol.* 69: 137-144.

Adiyodi, R.G. & T. Subramoniam 1983. Oogenesis, oviposition and oosorption, Arthropoda – Crustacea. In K.G. Adiyodi & R.G. Adiyodi (eds), *Reproductive Biology of Invertebrates*, Vol. 1, p. 443-495. J. Wiley and Sons Ltd., Chichester.

Blumink, J.G. 1967. The subcellular structure of the blastula of *Limnaea stagnalis* L. (Mollusca) and the mobilization of the nutrient reserve. Thesis. Utrecht.

Carroll, W.W., R.W. Zongley & J.H. Roe 1956. The determination of glycogen in the liver and muscle by the use of anthrone reagent. *J. Biol. Chem.* 220: 586-593.

Clarke, A., A. Skadsheim & L.J. Holmes 1985. Lipid biochemistry and reproductive biology in two species of Gammaridae (Crustacea, Amphipoda). *Mar. Biol.* 88: 247-263.

Davis, B.J. 1964. Disc electrophoresis. II. Method and application to human serum proteins. *Ann. Nat. Acad. Sci.* 121: 404-427.

Doyle, W.L., R. Rappaprort & M.E. Doyle 1959. Formation and distribution of esterase in gammarids. *Physiol. Zool.* 32: 246-255.

Efford, I.E. 1970. Recruitment to sedentary marine populations as exemplified by the sand crab, *Emerita analoga* (Decapoda, Hippidae). *Crustaceana* 18: 293-308.

Ezhilarasi, S. & T. Subramoniam 1984. Esterase activity in *Scylla serrata* (Forskal) during ovarian development. *J. Exp. Biol. Ecol.* 83: 1-12.

Folch, J., M. Lees & G.H. Bloane-Stanley 1957. Purification of total lipids from animal tissues. *J. Biol. Chem.* 266: 497-509.

Herring, P.J. & R.J. Morris 1975. Embryonic metabolism of carotenoid pigments and lipid in species of *Acanthephyra* (Crustacea : Decapoda). *Proc. Europ. Mar. Biol. Symp.* 95: 299-310.

Kapin, M.A. & S. Ahmed 1980. Esterases in larval tissues of gypsy moth, *Lymantria dispar* (L), optimum assay conditions, quantification and characterization. *Insect Biochem.* 10: 331-337.

Lasker, R. 1962. Efficiency and rate of yolk utilization by developing embryos and larvae of the Pacific sardine *Sardinops caerula* (Girard). *J. Fish. Res. Bd. Canada* 19: 867-875.

Lowry, D.H., N.J. Rosenbrough, F.L. Farr & R.J. Randell 1951. Protein measurement with the folin phenol reagent. *J. Biol. Chem.* 193: 265-275.

McWhinnie, M.A. & P.N. Sallar 1960. Analysis of blood sugars in the crayfish, *Orconectes virilis*. *Comp. Biochem. Physiol.* 1: 110-122.

Meenakshi, V.R. & B.T. Scheer 1961. Metabolism of glucose in the crabs, *Cancer magister* and *Hemigrapsus nudus*. *Comp. Biochem. Physiol.* 3: 30-41.

Pandian, T.J. 1970a. Ecophysiological studies on the developing eggs and embryos of the European lobster *Homarus gammarus*. *Mar. Biol.* 5: 154-167.

Pandian, T.J. 1970b. Yolk utilization and hatching time in the Canadian lobster *Homarus americanus*. *Mar. Biol.* 7: 249-254.

Parvathy, K. 1970. Blood sugars in relation to chitin synthesis during cuticle formation in *Emerita asiatica*. *Mar. Biol.* 5: 108-112.

Pillai, C.K. & T. Subramoniam 1985. Yolk utilization as an adaptive strategy of terrestrialization in the freshwater crab *Parathelphusa hydrodromus* (Herbst). *Physiol. Zool.* 58: 445-457.

Roe, J.H. 1955. The determination of sugar in the blood and spinal fluid with anthrone reagent. *J. Biol. Chem.* 20: 335-343.

Ruddle, F.H. & L. Harrington 1967. Tissue specific esterase isozymes of the mouse. *J. Exp. Zool.* 166: 51-69.

Schwab, W.E. 1974. Sex-specific proteins in *Artemia salina* and their possible function in cryptobiosis. *Comp. Biochem. Physiol.* 48B: 175-180.

Smith, I. 1968. *Chromatographic and Electrophoretic Techniques* Vol. 2. Heineman, London.

Subramoniam, T. 1977. Continuous breeding in the tropical anomuran crab, *Emerita asiatica* Milne Edwards from Madras Coast. In K.G. Adiyodi & R.G. Adiyodi (eds), *Advances in Invertebrate Reproduction* 1: pp. 166-174.

Subramoniam, T. 1979. Some aspects of reproductive ecology of a mole crab *Emerita asiatica* (Milne Edwards). *Exp. Mar. Biol. Ecol.* 65: 259-268.

Subramoniam, T. 1982. Determination of reproductive periodicity in the intertidal mole crab, *Emerita asiatica*. In T. Subramoniam (ed.), *Manual of Research Methods for Marine Invertebrate Reproduction*, pp. 163-167. CMFRI no. 9. Madras: Diocesan Press.

Zar, J.H. 1974. *Biostatistical Analysis*. Prentice Hall, Englewood Cliffs.

Hemocyanins in crustacean oöcytes and embryos

Nora Barclay Terwilliger
Oregon Institute of Marine Biology, University of Oregon, Charleston & Department of Biology, University of Oregon, Eugene, USA

ABSTRACT

Hemocyanin, the copper-containing respiratory protein, is present in brachyuran crustacean oöcytes and embryos. Oöcyte hemocyanin increases in concentration during vitellogenesis, resembles maternal hemocyanin, and is probably derived via endocytosis from maternal hemolymph. Embryo hemocyanin differs in subunit composition from hemocyanin of oöcyte and adult crab and may represent the onset of hemocyanin ontogenic expression.

1 INTRODUCTION

Arthropod oöcytes, like other oöcytes, contain materials which initiate and maintain growth and development of the oöcyte and the embryo. The developing oöcyte synthesizes many of its own RNAs and proteins; it also actively accumulates extraoöcytic materials which may act as food reservoirs for the developing embryo or which may have other functions. Hemoglobin, the iron-containing respiratory protein, is one of these molecules found in oöcytes, embryos and larvae of some arthropods. For example, certain blood-sucking insects such as *Rhodnius* produce hemoglobin-containing oöcytes (Wigglesworth 1943); in these species, the hemoglobin is not synthesized by the female but is produced by the host. In another insect, *Chironomus thummi*, maternal hemoglobin is deposited in the yolk granules of developing oöcytes (Travis & Shin 1976, Trewitt et al. 1986). Later during development, the larva synthesizes sequentially several different hemoglobins (cf., Manwell 1966, Laufer et al. 1976). In crustaceans, the red color of *Daphnia* parthenogenic eggs is due to a hemoglobin (Teissier 1932) whose presence accelerates embryonic development (Fox 1948, Hoshi 1957). During the last few hours preceding egg-laying, extracellular hemoglobin concentration in the maternal hemolymph decreases by one third as hemoglobin in the *Daphnia* oöcytes increases (Dresel 1948).

While the above examples show that some arthropods may sequester hemoglobin into the oöcyte, most arthropods contain hemocyanin rather than hemoglobin as their circulating respiratory protein (Van Holde & Miller 1982, Mangum 1983). Arthropod hemocyanin is a copper-containing protein composed of polypeptide chains or subunits with molecular weights of about 75 000. Each subunit contains two copper atoms and combines reversibly with one molecule of oxygen. Six hemocyanin subunits self-assemble into a 450 000 molecular weight 16S hexamer. The hexamers further aggregate into 25S dimers, 36S tetramers and 48S octomers of the basic hexameric unit. The aggregation state is species specific and sometimes class specific. Brachyuran crabs, for example, usually contain hexamer and two-hexamer assemblages of hemocyanin in their hemolymph, whereas *Limulus* has a hemocyanin composed of eight-hexamer polymers.

Since polymeric hemoglobins and other hemolymph proteins, including high molecular weight vitellogenins, are actively endocytosed by arthropod oöcytes (see Meusy 1980, Adiyodi & Subramoniam 1983, Charniaux-Cotton 1985 for reviews), it is reasonable to ask whether hemocyanin might be endocytosed by the oöcytes. If hemocyanin is present in the oöcyte, how might this respiratory protein be involved in crustacean egg production? Furthermore, it has recently been shown in brachyuran crabs that different stages of development, megalopae and early instar juveniles, contain hemocyanin whose subunit composition and functional properties differ from that of the adult crab (Terwilliger & Terwilliger 1982, Terwilliger & Terwilliger 1983, Terwilliger et al. 1986, Markl et al. 1986, Markl 1986). When does the ontogenic expression of hemocyanin begin, and what role does it play in crustacean development? In this chapter, we examine the presence, structure and possible functions of oöcyte and embryo hemocyanin in some decapod crustaceans.

2 HEMOCYANIN IN CRUSTACEAN OÖCYTES

Hemocyanin has been identified in ovarian homogenates of several crustaceans, including *Uca pugilator* (Fielder et al. 1971), *Emerita analoga* (Gilchrist & Lee 1972) and *Astacus leptodactylus* (Durliat 1984). However, because crustacean ovaries are highly vascularized (Pearson 1908), it is possible that the putative ovarian hemocyanin was not in the oöcytes but rather in the circulating maternal hemolymph. We have looked for hemocyanin in oöcytes of three species of cancrid crabs, *Cancer magister, C. productus* and *C. gracilis*, under conditions designed to obviate contamination of the sample by maternal hemolymph. The procedures we used included extraction of isolated oöcytes (ripe oöcytes are easily shaken loose from the ovary) instead of ovarian tissue and extensive washing of the isolated oöcytes in either filtered sea water or a buffer solution which matched the species' hemolymph ionic composition. We also employed limited trypsin digestion of isolated, washed oöcytes to digest or remove extra-oöcytic proteins which might be adhering to the surface of the oöcyte and centrifugation of the homogenized isolated oöcyte preparation to separate oöcyte contents from cell membranes to which extraneous hemocyanin might be bound.

When ripe *Cancer magister* oöcytes are homogenized and purified by chromatography on a BioGel A-5m column, the elution pattern seen in Plate 1a is obtained. Further analysis of the fractions by pH 7.4 PAGE (Pl. 1b) indicates the presence of oöcyte 25S and 16S hemocyanin in the chromatography elution positions corresponding to those of adult 25S and 16S hemocyanin.

Hemolymph proteins from female *C. magister* are compared with oöcyte homogenates by pH 7.4 polyacrylamide gel electrophoresis (PAGE) in Plate 2. This non-denaturing, non-hemocyanin-dissociating buffer system permits electrophoresis of crustacean hemocyanin in its native aggregation states (Terwilliger & Terwilliger 1982). The major adult hemolymph protein components are 25S two-hexamer hemocyanin, 16S hexamer hemocyanin, and often an unidentified protein (X) which migrates to an intermediate position between the two hemocyanins. While protein X could be related to oögenesis, its presence in hemolymph of both non-vitellogenic (Pl. 2A) and vitellogenic (Pl. 2B) females is variable; furthermore, it is sometimes found in male and juvenile crabs. Identification of the hemocyanin components was ascertained by indication of the presence of copper in the putative hemocyanin bands due to quenching of fluorescence (Bruyninckx et al. 1978) and by comparative electrophoresis with purified hemocyanin of known molecular weight from other decapods. Two step electrophoresis, wherein the protein bands from a pH 7.4 gel are excised and subsequently electrophoresed on an SDS gel, showed the typical subunit SDS pattern of *C. magister* hemocyanin (Larson et al. 1981), further confirming the identity of 25S and 16S hemocyanin on the pH 7.4 gels. The slowly migrating bands near the top of the pH 7.4 gel increase in concentration as vitellogenesis progresses (Pl. 2A, B); they have a yellow-orange color and probably correspond to lipoglycocarotenoproteins, the vitellogenins (Zagalsky et al. 1967, Kerr 1969).

The oöcyte homogenates (Pl. 2C, D) contain many bands in common with adult hemolymph proteins. This is not surprising, since endocytosis of maternal vitellogenins from crustacean hemolymph into the oöcyte has been documented (Wolin et al. 1973, Zerbib & Mustel 1984). The relative concentrations of the various proteins differ between maternal *C. magister* hemolymph and oöcyte. For example, the hemocyanin concentration is lower in the oöcyte than in the hemolymph, while the carotenoprotein concentration is much higher. (Oöcyte lipoglycocarotenoproteins are usually referred to as vitellins to contrast them with the hemolymph vitellogenins, although the molecules are not always distinguishable from one another). Furthermore, some hemolymph proteins are absent from the oöcyte. The differences between hemolymph and oöcyte proteins suggest the oöcyte is capable of selective uptake of proteins from the hemolymph. The oöcyte also contains several proteins not present in the maternal hemolymph which are presumably synthesized by the oöcyte.

The pattern of oöcyte protein bands changes during vitellogenesis. In Plate 2C, white, previtellogenic oöcytes measuring 114 μm in diameter contain 25S hemocyanin but no detectable 16S hemocyanin. Bright orange, ripe oöcytes measuring 380 μm in diameter contain both 25S and 16S hemocyanin (Pl. 2D). Parallel with the increase in concentration and number of aggregation states of hemocyanin is the appearance of a large amount of slowly migrating orange proteins. Thus, hemocyanin accumulates in the oöcyte during the active vitellogenic stage of development, the time when brachyuran crab oöcytes are know to be engaged in vigorous endocytosis (Charniaux-Cotton 1985).

When we examined the oöcytes of *Cancer gracilis*, a species exhibiting a ripe ovary in mid-summer in contrast to the early winter ripening of *Cancer magister*'s oöcytes, we found the same phenomenon. Early, cream colored oöcytes did not contain any detectable 16S hemocyanin, although some 25S hemocyanin was present; during vitellogenesis, an increase in 25S hemocyanin and the appearance of 16S hemocyanin occurred.

Many crustacean hemocyanins, including those of *C. magister*, dissociate into subunits but do not denature when electrophoresed at pH 8.9. Plate 3A shows the five subunits, a-e, of 25S hemocyanin from adult *C. magister* hemolymph which are resolved on a pH 8.9 gel (Larson et al. 1981). In Plate 3B, an electropherogram of oöcyte homogenate, a very similar hemocyanin subunit pattern is seen with the possible exception of subunit e, which migrates slightly faster than adult subunit e. In addition, there are slowly migrating orange bands, probably vitellogenins, present in female (Pl. 3A top) but absent in male and juvenile hemolymph (not shown). Corresponding orange bands, putative vitellins, are present in high concentration in the oöcyte (Pl. 3B top).

Cancer magister hemolymph and oöcyte hemocyanin subunits can also be compared by sodium dodecyl sulfate (SDS) PAGE in the presence of reducing agents as shown in Plate 4. The six subunits of both adult (Pl. 4A) and oöcyte (Pl. 4D) 25S hemocyanin are indistinguishable on an SDS gel. There may be slightly more subunit VI in the oöcyte preparation. Oöcyte 16S hemocyanin is shown in Plate 4E. Like adult 16S hemocyanin, it lacks band III, presumably a linker subunit necessary for association of the two hexamers of 25S hemocyanin. The two major orange carotenoproteins of the oöcyte homogenate (from Pl. 2D, arrows) are shown in Plate 4B and 4C; their apparent subunit molecular weights are slightly larger than or similar to the hemocyanin subunits.

The same analyses were carried out on oöcytes of *C. productus* (Wache et al. 1986) and *C. gracilis*. Biogel A-5m chromatography, pH 7.4 PAGE, pH 8.9 PAGE and SDS PAGE all reveal the presence of oöcyte hemocyanin that is indistinguishable in quaternary structure and subunit composition from that of the maternal crab. Negative staining and transmission electron microscopy (TEM) of *C. productus* oöcyte homogenate show the two-hexagon and single hexagon molecules characteristic of crustacean 25S and 16S hemocyanin (Pl. 5A). Clearly the oöcytes of these crabs also contain hemocyanin.

Is crustacean oöcyte hemocyanin derived from maternal hemolymph via endocytosis or is it synthesized by the oöcyte? In a study of chelicerate oöcytes, hemocyanin was found inside the oöcytes of *Eurypelma*, but there was no evidence of hemocyanin synthesis by the oöcytes (Kempter 1986). The increase in *Cancer* oöcyte 25S hemocyanin and 16S hemocyanin during the stages of crustacean oögenesis when active endocytosis is occurring (Schade & Shivers 1980, Zerbib & Mustel 1984) strongly suggest that oöcyte hemocyanin could be obtained from the maternal hemolymph. The close structural resemblance between oöcyte and maternal hemocyanin is also compatible with this idea. In order to determine whether crab oöcytes were capable of endocytosis of hemocyanin, we incubated *C. magister* oöcytes with *C. productus* hemolymph and vice versa. Hemocyanin from the two species can be differentiated easily from one another by SDS PAGE; the former has six subunits and the latter, four, and the subunits from each species have slightly different molecular weights. We found that when isolated oöcytes from one species were incubated in hemolymph from the other species, the oöcytes took up the foreign hemocyanin. *In vivo* studies have not been done, but the potential for hemocyanin endocytosis by oöcytes from these *Cancer* species certainly exists. Transmission electron micrographs of *C. productus* oöcytes (Pl. 6) show many coated pits and vesicles indicative of the internalization of extracellular molecules (Moore et al. 1987). Similar coated pits have been identified in other crustacean oöcytes (cf. Schade & Shivers 1980, Zerbib 1980). Although it is very likely that hemocyanin is taken up by the oöcytes, the possibility of synthesis of some hemocyanin by the oöcyte cannot yet be excluded.

3 HEMOCYANIN IN CRAB EMBRYOS

Hemocyanin is present in the developing crab embryo as well as in the oöcyte. It is known from other studies, especially those of sea urchin development, that one of the effects of fertilization seems to be an immediate burst in the rate of protein synthesis. As cleavage and further embryonic development continue, there is a sequential expression of certain proteins such as actins and calcium binding proteins (cf. Cox et al. 1986). Another protein that changes its expression during crab embryogenesis is hemocyanin. The quaternary structure of the hemocyanin is apparently the same in both oöcyte and embryo; i.e. 25S and 16S hemocyanin molecules of negatively stained *C. magister* embryo homogenate in Plate 5B have the same appearance by TEM as do oöcyte hemocyanin molecules (Pl. 5A) and adult hemocyanin. *Cancer magister* embryo hemocyanin is also indistinguishable from oöcyte and adult hemocyanin on pH 7.4

PAGE. In addition, both 25S and 16S hemocyanin from the embryo give a positive reaction for the presence of copper, and the pattern of pH 8.9 PAGE (Pl. 3D) of dissociated subunits of embryo 25S hemocyanin purified by chromatography on BioGel A-5m is similar to that of adult 25S hemocyanin (Pl. 3C), with the exception of the slightly faster moving band e. However, when the embryo 25S hemocyanin band is cut out from a pH 7.4 gel and re-electrophoresed on SDS PAGE, three lower molecular weight bands which are absent in adult and oöcyte hemocyanin are now present in the embryo sample. Biogel A-5m purified embryo 25S hemocyanin shows the same additional three bands on SDS PAGE (Pl. 7).

Studies of *C. productus* embryos gave results similar to those for *C. magister* embryos (Wache et al. 1986). The major soluble proteins of *C. productus* oöcytes and embryos are compared with adult hemocyanin on a pH 7.4 gel in Plate 8. Several proteins are present only in the embryo homogenate; other proteins, including 25S and 16S hemocyanin, are present at both stages of development. Embryo hemocyanin resembles adult and oöcyte hemocyanin on pH 7.4 and 8.9 gels, but analysis by SDS PAGE reveals the presence of two extra subunits with lower apparent molecular weights. Thus in both crab species, embryo hemocyanin can be differentiated from adult and oöcyte hemocyanin.

Are these additional subunits seen in *C. magister* and *C. productus* embryos an artifact of analysis generated by proteolysis? A novel protease in the embryo might cause proteolysis during preparation of the embryo sample. To address this concern, we incubated *C. productus* maternal hemolymph with a *C. productus* embryo homogenate to see whether the maternal hemocyanin would be digested. If a protease were present and had produced the embryo subunits, this experiment should generate an increase in concentration of the lower molecular weight components. However, no increase was detected. Furthermore, experiments were done to enhance putative proteolysis by incubation of embryo homogenate at warm temperatures in the absence of protease inhibitors prior to SDS analysis. We also sought to minimize proteolysis by immediate preparation of SDS samples in the presence of a mixture of protease inhibitors as soon as the embryos were removed from the female crab. In neither case was there a change in concentration of the lower molecular weight components (Wache et al. 1988). Thus these lower molecular weight subunits do not appear to be proteolytic artifacts induced during preparation of embryo samples. A more likely explanation for the extra subunits observed in both *C. magister* and *C. productus* embryo hemocyanin is that these subunits are synthesized *de novo* in the embryo as hemocyanin subunits with similar charges, but with either different molecular weights than the subunits with which they co-electrophorese on pH 8.9 PAGE or with different binding affinities for SDS. That hemocyanin subunits could have similar chain lengths but different mobilities on SDS gels has been suggested for other arthropod hemocyanin subunits (Linzen et al. 1985). The novel embryo hemocyanin chains could also be due to post-translational modifications of the polypeptide chains such as glycosylation or *in vivo* proteolysis. Whether embryo hemocyanin represents a direct change in hemocyanin genome expression or an indirect effect at another point in the processing of the protein is not known; however, embryo hemocyanin is different from oöcyte hemocyanin.

Cancer gracilis embryos were compared at three different stages of development: early, orange colored embryos about 310 μm in diameter; older, brown colored embryos about 340 μm in diameter, and prezoeae about 340 μm in egg capsule diameter and just about to hatch (siblings were already swimming about in the tank). Equal numbers of embryos in each aliquot were homogenized in the same volume of buffer and protease inhibitor; equal volumes of supernatant were applied to a pH 7.4 gel. No differences could be detected in the concentration of hemocyanin between early and late embryos. However, concentrations of both hemocyanin and the slowly migrating orange carotenoproteins were greatly reduced in the prezoeae. Further work on zoea hemocyanin is in progress.

Hemocyanin may commonly occur in arthropod embryos. Embryos of other crustaceans, *Carcinus maenas, Eriocheir sinensis* and *Portunus holsatus* (Busselen 1971) and *Emerita analoga* (Gilchrist & Lee 1972) also contain hemocyanin. As in the *Cancer* species described above, the embryo hemocyanin of *Carcinus maenas* differs electrophoretically from that of the adult crab. Embryos of *Hemigrapsus nudus* also contain a hemocyanin with a subunit composition different from adult hemocyanin (Terwilliger & Terwilliger 1983, Terwilliger et al. 1983), while hemocyanin in embryos of the crayfish, *Astacus leptodactylus*, appears to be immunologically identical to that of the female crayfish (Durliat 1984). Hemocyanin is also present in the pre-hatching stages of the isopod, *Ligia pallasii*, which undergoes direct development (Terwilliger, N., pers. observ.). In the Chelicerata, hemocyanin appears in the perivitelline fluid of horseshoe crab, *Tachypleus tridentata*, embryos after the first embryonic molt (Sugita & Sekiguchi 1979). They found no difference between embryonic and adult hemocyanin using immunological and electrophoretic techniques. Embryo hemocyanin thus may represent a stage specific expression of protein structure and possibly protein function in some but not all crustaceans.

4 FUNCTION OF OÖCYTE AND EMBRYO HEMOCYANIN

What is the function of oöcyte hemocyanin? One possibility is that the oöcyte uses the hemocyanin as a source of protein. However, the hemocyanin molecules are present in their native quaternary structures in the oöcyte; there is no evidence of dissociated subunits or degraded fragments of subunits, which might be expected if the oöcyte were digesting the protein. Another function for hemocyanin might be as a copper source for the oöcyte. However, the protein bands on the pH 7.4 gel corresponding to 25S protein show a positive reaction for copper, suggesting that copper is not removed from the protein. A possibility remains that hemocyanin is deposited in the oöcyte and used after fertilization, as are so many components of the oöcyte.

Is oöcyte hemocyanin involved in oxygen transfer or storage? At first glance, this would seem unnecessary because of the extensive ovarian vascularization. Oöcyte hemocyanin might facilitate diffusion throughout the metabolizing oöcyte much as myoglobin may function in muscle tissue (Wittenberg 1970). If so, perhaps we have in crustacean oöcytes the elusive 'myohemocyanin' or tissue hemocyanin. We do not know at present whether oöcyte hemocyanin binds oxygen reversibly and whether its binding properties are similar to those of the adult. Even if its structure and inherent oxygen binding properties were identical to maternal hemocyanin, its *in vivo* affinity could be different due to the striking sensitivity of *C. magister* hemocyanin to allosteric regulators (Graham et al. 1982). Thus, an oxygen transfer system between maternal blood and developing oöcyte could exist.

What is the function of embryonic hemocyanin in crustaceans? Fertilization in brachyuran crabs occurs just prior to oviposition, as the oöcytes pass through the oviduct and out of the ovipores on the abdominal surface. The newly fertilized embryos are then attached onto the hairlike setae of the pleopods, where they remain until hatching. The egg mass can be quite large and may contain up to 2.5 million embryos. In a crab such as *C. magister*, development from fertilization to hatching is at least three months. Although the female crab cares for the egg mass, grooming it with her legs and aerating it by moving her abdominal flap back and forth, embryos

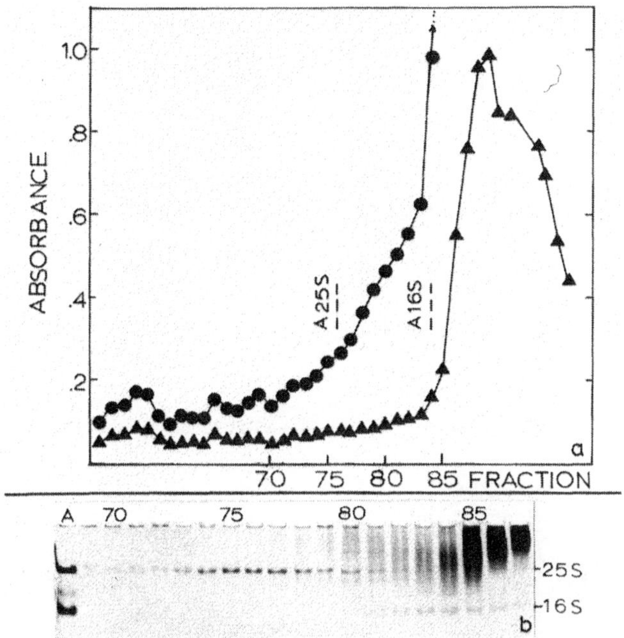

Plate 1. a. BioGel A-5m chromatography of *C. magister* oöcyte supernatant. Buffer, 0.1 ionic strength Tris-HC1, 0.1 M in NaC1, 10 mM in $CaCl_2$, 50 mM in $MgCl_2$, pH 7.5. Column, 100 cm × 1.5 cm. Absorbance at 280 nm (●) and 340 nm (▲5). b. pH 7.4 PAGE, 5% polyacrylamide of fractions from a. Calibrants: A 25S, adult *C. magister* 25S hemocyanin; A 16S, adult *C. magister* 16S hemocyanin.

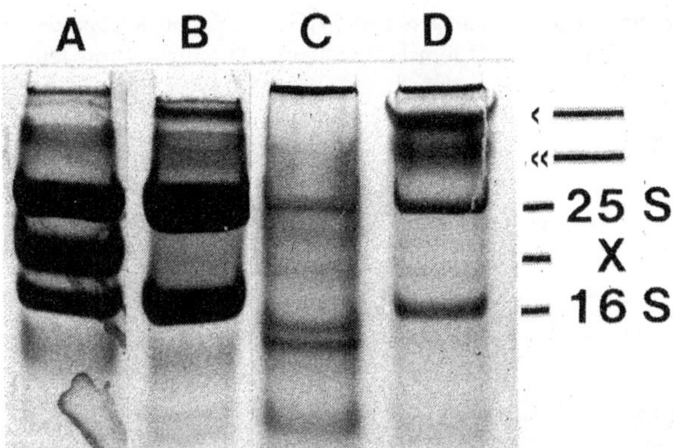

Plate 2. *C. magister* hemolymph and oöcyte supernatants, pH 7.4 PAGE, 5% polyacrylamide. A, previtellogenic female hemolymph; B, vitellogenic female hemolymph; C, previtellogenic oöcytes; D, vitellogenic oöcytes.

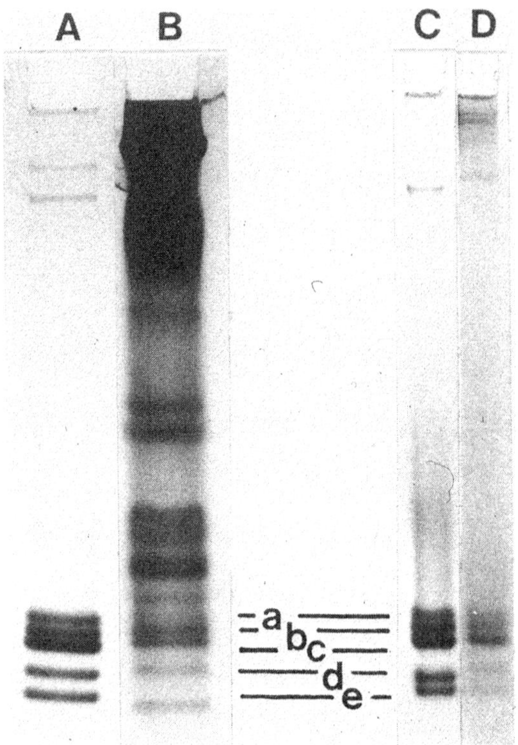

Plate 3. *C. magister* hemolymph, oöcyte and embryo supernatants, pH 8.9 PAGE, 7.5% polyacrylamide. A, vitellogenic female hemolymph; B, oöcytes from female in A; C, female hemolymph; D, embryos from female in C.

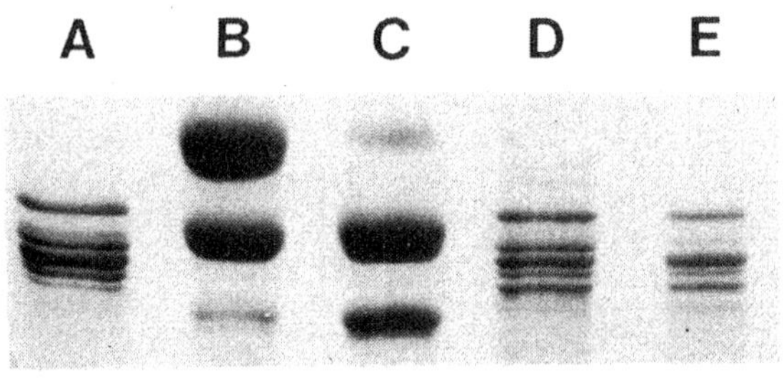

Plate 4. SDS PAGE, dithiothreitol, 7.5% polyacrylamide, of *C. magister* hemolymph and oöcyte protein bands from pH 7.4 PAGE. A, Adult female, 25S hemocyanin; B, upper vitellin band, oöcyte homogenate (arrow, Fig. 2); C, lower vitellin band, oöcyte homogenate (double arrow, Fig. 2); D, 25S hemocyanin, oöcyte homogenate; E, 16S hemocyanin, oöcyte homogenate.

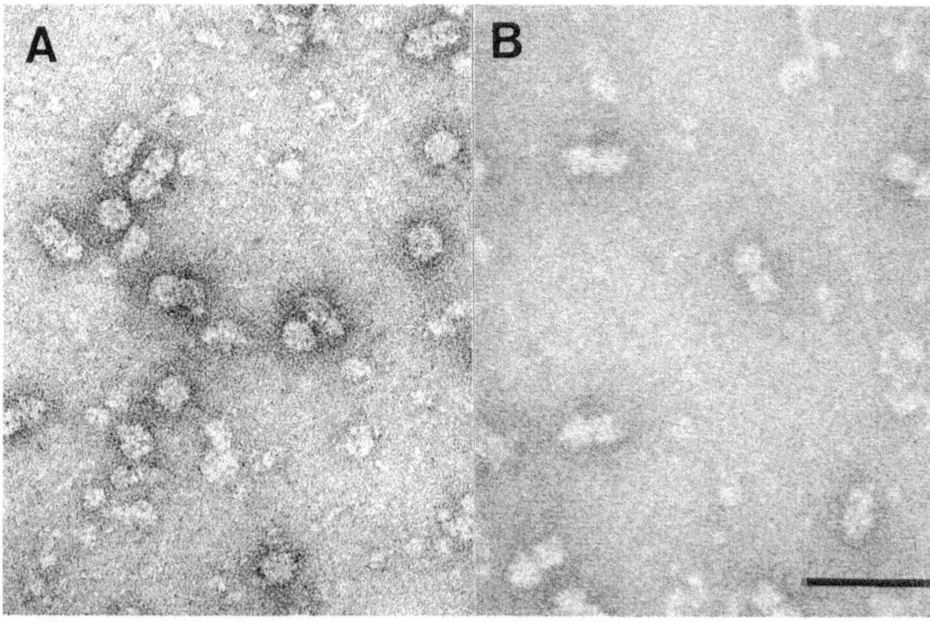

Plate 5. A, TEM of *C. productus* oöcyte homogenate negatively stained with 0.5% uranyl acetate; B, TEM of *C. magister* embryo 25S fraction from BioGel A-5m chromatography negatively stained with 2% phosphotungstate. Bar = 50 nm.

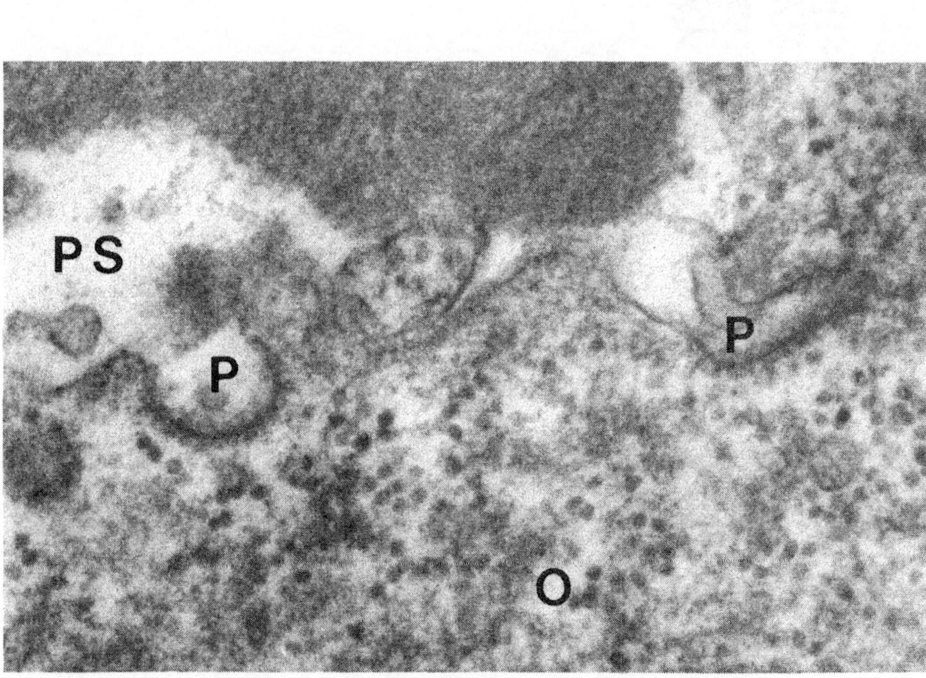

Plate 6. TEM, *Cancer productus* oöcyte. O, oöcyte; PS, perivitelline space; P, coated pit.

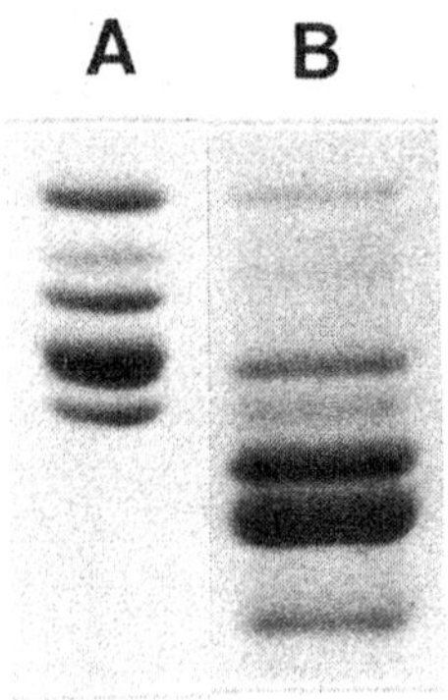

Plate 7. SDS PAGE, dithiothreitol, 7.5% polyacrylamide. A, *C. magister* female 25S hemocyanin; B, *C. magister* embryo 25S hemocyanin from BioGel A-5m chromatography.

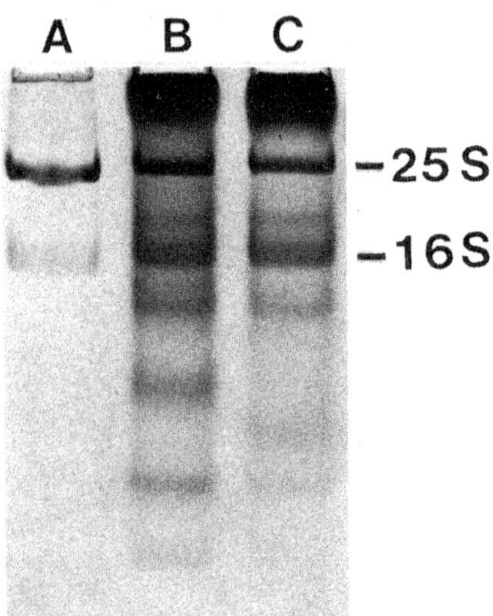

Plate 8. *C. productus*. A, adult female hemolymph; B, oöcyte homogenate; C, embryo homogenate. pH 7.4 PAGE, 5% polyacrylamide.

in the center of the egg mass may experience a lower partial pressure of oxygen than do embryos on the outer surface. Since it is likely that there is an oxygen gradient from the outside to the center of the egg mass, embryo hemocyanin may help in oxygen transport or diffusion.

Sequential changes in structure of hemoglobin during vertebrate development are associated with changes in hemoglobin functional properties (Bunn & Forget 1986). Similar ontogenic changes occur in crustacean hemocyanin structure and function during the transition from larval stages to adult (Terwilliger et al. 1986). Much remains to be learned about the functional role of hemocyanin in the physiology of both oöcyte and embryo.

ACKNOWLEDGEMENTS

This chapter is dedicated to Dr Robert C. Terwilliger (1940-1989), my colleague and husband, for his contributions to this manuscript and to the field of comparative physiology and biochemistry of invertebrate respiratory proteins. His encouragement and energy are greatly missed, but his creativity and appreciation for the beauty and wonder of biology continue to inspire fellow scientists, students and friends.

The assistance of Susanne Wache, Brett Larson and Kristen Larson in these studies are greatly appreciated. I thank Eric Schabtach, Department of Biology, University of Oregon, for the electron microscopy and Dr A.O.D. Willows for use of facilities at Friday Harbor Laboratories, University of Washington, where some of this research was done. The study was supported by NSF grant DMB-8811150.

REFERENCES

Adiyodi, R.G. & T. Subramoniam 1983. Arthropoda-Crustacea. In K.G. Adiyodi & R.G. Adiyodi (eds), *Reproductive Biology of Invertebrates*. Vol. 1 pp. 443-495. Chichester: Wiley.

Bruyninckx, W.J., S. Gutteridge & H.S. Mason 1978. Detection of copper on polyacryamide gels. *Anal. Biochem.* 89: 174-177.

Bunn, H.F. & B.G. Forget 1986. *Hemoglobins: Molecular, Genetic and Clinical Aspects*. pp. 61-91. Philadelphia: Saunders.

Busselen, P. 1971. The presence of haemocyanins and of serum proteins in the eggs of *Carcinus maenas, Eriocheir sinensis,* and *Portunus holsatus*. *Comp. Biochem. Physiol.* 38A: 317-328.

Charniaux-Cotton, H. 1985. Vitellogenesis and its control in malacostracan Crustacea. *Amer. Zool.* 25: 197-206.

Cox, K.H., L.M. Angerer, J.J. Lee, E.H. Davidson & R.C. Angerer 1986. Cell lineage-specific programs of expression of multiple action genes during sea urchin embryogenesis. *J. Mol. Biol.* 188: 159-172.

Dresel, E.I.B. 1948. Passage of haemoglobin from blood into eggs of *Daphnia*. *Nature* 162: 736-737.

Durliat, M. 1984. Occurrence of plasma proteins in ovary and egg extracts from *Astacus leptodactylus*. *Comp. Biochem. Physiol.* 78B: 745-753.

Fielder, D.R., K.R. Rao & M. Fingerman 1971. A female-limited lipoprotein and the diversity of hemocyanin components in the dimorphic variant of the fiddler crab *Uca pugilator*, as revealed by disc electrophoresis. *Comp. Biochem. Physiol.* 39B: 291-297.

Fox, H.M. 1948. The haemoglobin of *Daphnia*. *Proc. Roy. Soc.* (B) 135: 195-211.

Gilchrist, B.M. & W.L. Lee 1972. Carotenoid pigments and their possible role in reproduction in the sand crab *Emerita analoga*. *Comp. Biochem. Physiol.* 42B: 263-294.

Graham, R.A., C.P. Mangum, R.C. Terwilliger & N.B. Terwilliger 1982. The effect of organic acids on oxygen binding of hemocyanin from the crab *Cancer magister*. *Comp. Biochem. Physiol.* 74A: 45-50.

Hoshi, T. 1957. Studies on physiology and ecology of plankton. XII. Changes in oxygen consumption of the daphnid, *Simocephalus vetulus*, with the decrease of oxygen concentration. *Sci. Rep. Tohoku Univ. Fourth Ser. Biol.* 23: 27-33.

Kempter, B. 1986. Intracellular hemocyanin and site of biosynthesis in the spider *Eurypelma californicum*. In B. Linzen (ed.), *Invertebrate Oxygen Carriers*. pp. 489-494. Berlin: Springer-Verlag.

Kerr, M.S. 1969. The hemolymph proteins of the blue crab, *Callinectes sapidus*. II. A lipoprotein serologically identical to oöcyte lipovitellin. *Devel. Biol.* 20: 1-17.

Larson, B., N.B. Terwilliger & R.C. Terwilliger 1981. Subunit heterogeneity of *Cancer magister* hemocyanin. *Biochim. Biophys. Acta* 667: 294-302.

Laufer, H., G. Bergtrom & R. Rogers 1976. Stage and tissue-specific hemoglobin synthesis in an invertebrate. In E. Kurstak & K. Maramorosch (eds), *Invertebrate Tissue Culture Applications in Medicine, Biology and Agriculture*. pp. 227-240. New York: Academic Press.

Linzen, B., N.M. Soeter, A.F. Riggs, H.J. Schneider, W. Schartau, M.D. Moore, E. Yokota, P.Q. Behrens, H. Nakashima, T. Takagi, T. Nemoto, J.M. Vereijken, H.J. Bak, J.J. Beintema, A. Volbeda, W.P.J. Gaykema & W.G.J. Hol 1985. The structure of arthropod hemocyanins. *Science* 229: 519-524.

Mangum, C.P. 1983. Oxygen transport in the blood. In L.H. Mantel (ed.), *The Biology of Crustacea*. Vol. 5. pp. 373-429. New York: Academic Press.

Manwell, C. 1966. Starch gel electrophoresis of the multiple haemoglobins of small and large larval *Chironomus* – a developmental haemoglobin sequence in an invertebrate. *J. Embryol. Exp. Morphol.* 16: 259-270.

Markl, J. 1986. Evolution and function of structurally diverse subunits in the respiratory protein hemocyanin from arthropods. *Biol. Bull.* 171: 90-115.

Markl, J., W. Stocker, R. Runzler & E. Precht 1986. Immunological correspondences between the hemocyanin subunits of 86 arthropods: evolution of a multigene family. In B. Linzen (ed.), *Invertebrate Oxygen Carriers*. pp. 281-292. Berlin: Springer-Verlag.

Meusy, J.J. 1980. Vitellogenin, the extraovarian precursor of the protein yolk in Crustacea: a review. *Reprod. Nutr. Develop.* 20: 1-21.

Moore, M.S., D.T. Mahaffey, F.M. Bradley & R.G.W. Anderson 1987. Assembly of clathrin-coated pits onto purified plasma membranes. *Science* 236: 558-563.

Pearson, J. 1908. *Cancer* (the edible crab). *Proc. Trans. Biol. Soc. Liverpool* 22: 291-499.

Schade, M.L. & R.R. Shivers 1980. Structural modulation of the surface and cytoplasm of oöcytes during vitellogenesis in the lobster, *Homarus americanus*. An electron microscope-protein tracer study. *J. Morph.* 163: 13-26.

Sugita, H. & K. Sekiguchi 1979. Protein components in the perivitelline fluid of the embryo of the horseshoe crab, *Tachypleus tridentatus*. *Dev. Biol.* 73: 183-192.

Teissier, G. 1932. Les pigments des oeufs de *Daphnia pulex*. *C.R. Soc. Biol. Paris* 109: 813-816.

Terwilliger, N.B. & R. Terwilliger 1982. Changes in the subunit structure of *Cancer magister* hemocyanin during larval development. *J. Exp. Zool.* 221: 181-191.

Terwilliger, N.B. & R.C. Terwilliger 1983. Changes in crustacean hemocyanin structure during development. In E. Wood (ed.), *Structure and Function of Invertebrate Respiratory Proteins*, pp. 357-360. Life Chem Reports Suppl. Harwood Acad. Publ.

Terwilliger, N.B., R. Terwilliger & R. Graham 1986. Crab hemocyanin function changes during development. In B. Linzen (ed.), *Invertebrate Oxygen Carriers*, pp. 333-335. Berlin: Springer-Verlag.

Terwilliger, R.C., N. Terwilliger & K. Larson 1983. Developmental changes in structure of crustacean hemocyanin. *Fed. Proc.* 42: 2051.

Travis, J.L. & K. Shin 1976. Evidence for haemoglobin uptake by oöcytes of *Chironomus thummi*. *J. Insect Physiol.* 22: 1601-1608.

Trewitt, P.M., D.R. Boyer & G. Bergtrom 1986. Characterization of maternal haemoglobins in the eggs and embryos of *Chironomus thummi*. *J. Insect Physiol.* 32: 963-969.

Van Holde, K. & K. Miller 1982. Haemocyanins. *Quart. Rev. Biophys.* 15: 1-129.

Wache, S.C., N.B. Terwilliger & R.C. Terwilliger 1986. Oöcyte and embryo hemocyanins of *Cancer productus*, a brachyuran crab. *Amer. Zool.* 26: 49A.

Wache, S.C., N.B. Terwilliger & R.C. Terwilliger 1988. Hemocyanin structure changes during early development of the crab *Cancer productus*. *J. Exp. Zool.* 247: 23-32.

Wigglesworth, V.B. 1943. The fate of haemoglobin in *Rhodnius prolixus* (Hemiptera) and other blood-sucking arthropods. *Proc. R. Soc. London* (B)131: 313-339.

Wittenberg, J.B. 1970. Myoglobin facilitated oxygen diffusion: role of myoglobin in oxygen entry into muscle. *Physiol. Rev.* 50: 559-636.

Wolin, E.M., H. Laufer, & D.F. Albertini 1973. Uptake of the yolk protein, lipovitellin, by developing crustacean oöcytes. *Devel. Biol.* 35: 160-170.

Zagalsky, P.F., D.F. Cheesman & H.J. Ceccaldi 1967. Studies on carotenoid-containing lipoglycoproteins and a glycoprotein from the eggs and ovaries of certain aquatic invertebrates. *Comp. Biochem. Physiol.* 22: 851-871.

Zerbib, C. 1980. Ultrastructural observation of oogenesis in the Crustacea Amphipoda *Orchestia gammarellus* (Pallas). *Tissue and Cell* 12: 47-62.

Zerbib, C. & J.J. Mustel 1984. Incorporation de la vitellogenine tritiee dans les ovocytes du Crustaces Amphipode *Orchestia gammarellus* (Pallas). *Int. J. Invert. Reprod.* 7: 63-68.

Some aspects of the biology of resting eggs of marine cladocerans

Takashi Onbé
Faculty of Applied Biological Science, Hiroshima University, Saijo, Higashi-Hiroshima, Japan

ABSTRACT

Marine cladocerans, comprising only 8 species in the world oceans, have a mode of gamogenetic (bisexual) reproduction, although in most cases parthenogenetic generations dominate in their populations. Resting eggs are produced as a result of gamogenesis and, after release from mother animals, sink in the water column onto the sea-bottom sediment. Apart from those of the brackish-water cladoceran *Bosmina maritima* in the Baltic Sea, resting eggs are described and illustrated for 6 marine species: *Penilia avirostris, Evadne nordmanni, E. tergestina, Podon intermedius, P. leuckarti* and *P. polyphemoides*. Development and hatching processes of eggs are described for *P. avirostris, E. tergestina* and *P. polyphemoides*. Distribution and abundance of resting eggs in sediments have been studied in southern Japan, in particular, in the Inland Sea of Japan; Buzzards Bay, Massachusetts, USA; and the Gulf of Marconi, northern Ligurian Sea, Italy. In the majority of cases, the eggs of *Penilia avirostris* were most abundant. Eggs, while sinking, tend to be transported to areas with sediment predominantly composed of silty-clayish mud at the mercy of currents, and more eggs are accumulated there than in areas with sandy bottoms. Evidence indicates that organic sediment pollution reduces the survival of benthic resting eggs. The density of the resting eggs in sediments fluctuates seasonally, reaching highest levels in late summer shortly before the disappearance of the planktonic populations, and decreasing to the lowest level in spring just before the resurgence of population development in the water column. Adaptive significance of resting eggs is discussed in view of the life history patterns of marine cladocerans.

1 INTRODUCTION

Cladocerans are a group of small, mostly planktonic, crustaceans that thrive in limnetic environments, where over 600 species have been recorded (Schram 1986). By contrast, marine cladocerans include only 8 species belonging to 3 genera in 2 superfamilies: 1 species in Sidoidea, *Penilia avirostris;* and 7 species in Polyphemoidea, *Evadne nordmanni, E. spinifera, E. tergestina, Podon intermedius, P. leuckarti, P. polyphemoides* and *P. schmackeri* (Della Croce 1974, Onbé 1978a). In addition to these, a species belonging to the superfamily Daphnoidea, *Bosmina maritima* (formerly *B. coregoni maritima*), is known to inhabit the Baltic Sea, the largest brackish-water system in the world.

Like freshwater counterparts, each of these species of marine cladocerans has a heterogonous life cycle, characterized by the alternation of parthenogenetic and gamogenetic generations.

2 GAMOGENESIS

The transition from parthenogenesis to gamogenesis, typified by the appearance of males and resting-egg bearing females, is usually observed just before or at the time the population reaches its peak density (Onbé 1968 for *Penilia avirostris*, Gieskes 1970, 1971 for *Evadne nordmanni*). However, little is known about the mechanism triggering the initiation of gamogenesis for marine cladocerans. As in freshwater *Daphnia*, temperature and possibly photoperiod might control the process (Stross 1966).

Prior to the transition to gamogenesis, parthenogenetic egg production is reported to decrease progressively to a low level in *P. avirostris* (Pavlova 1959, Onbé 1968, 1974, Della Croce & Venugopal 1973), *E. nordmanni* (Onbé 1974) and *Podon polyphemoides* (Onbé 1974). Earlier, Berg (1931) demonstrated such a phenomenon in freshwater *Daphnia*.

In marine species, environmental stresses, such as scarcity of food, inadequate food quality, oxygen deficiency, temperature changes and overcrowding, act singly or in combinations as contributing factors causing the decrease in parthenogenetic egg production, as shown experimentally for some freshwater species (cf. Green 1966, Hutchinson 1967).

Gamogenesis becomes most intense just before the population is about to disappear from the planktonic community, thereby accelerating the rate of resting egg production. In *E. nordmanni*, gamogenetic individuals, i.e. males and females with resting eggs, sometimes comprise up to 80% of the total population in the Clyde Sea (Bainbridge 1958), 70% in the North Sea (Gieskes 1970) and 30% in the Inland Sea of Japan (Onbé 1974). For other species, 23% has been reported for *E. tergestina* in the Inland Sea of Japan (Onbé 1978d), 60-80% for *P. leuckarti* in the North Sea (Gieskes 1970), 50-70% for *P. intermedius* in the North Sea (Gieskes 1970), 10% for *P. polyphemoides* in the Chesapeake Bay (Bosch & Taylor 1973) and 20% in the Inland Sea of Japan (Onbé 1974), and approximately 15% for *P. avirostris* in the Black Sea (Paulova 1959) and 50% in the Inland Sea of Japan (Onbé 1974).

3 RESTING EGGS

Fertilized eggs are produced by females mated with males through gamogenesis. These eggs are large, having considerable yolk, unlike smaller, yolk-poor, parthenogenetic eggs. In parthenogenetic eggs, development proceeds to hatching within the brood chamber of the females. In the fertilized eggs (usually only a single egg is formed in marine species), development is suspended at an early stage, and eggs are released from the females into the water column during molting and fall to the bottom where they endure adverse conditions until the situation becomes favorable for their development. Such physiological characteristics of the fertilized eggs permit them to be called resting, dormant, or resisting eggs.

Resting eggs of marine cladocerans remain on the bottom of the sea-bed after release from females (Caroli 1924, Lochhead 1954, Bainbridge 1958, Wickstead 1963). Purasjoki (1945, 1958) first discovered the resting eggs of a brackish-water *Bosmina maritima*, a freshwater species, *Daphnia*, and marine species, *Evadne* and *Podon*, in the Gulf of Finland. The discovery of resting eggs of marine cladocerans in truly marine environments was not until 1970, when the eggs of *Penilia avirostris* and *Evadne* spp. were found in the central Inland Sea of Japan (Onbé 1972, 1973). The resting eggs of 6 of the 8 species of marine cladocerans have now been recovered from sediments of the sea-bottom: i.e. *P. avirostris* (Onbé 1972), *Evadne tergestina* (Onbé 1973), *E nordmanni* (Onbé 1979), *Podon polyphemoides* (Onbé et al. 1977), *P. leuckarti* (Onbé & Hayashida unpubl.) and *P. intermedius* (Onbé & Grice unpubl.).

4 MORPHOLOGY OF RESTING EGGS

4.1 *Penilia avirostris*

The resting egg of *Penilia avirostris* (Onbé 1972, 1973, 1974, 1985) is ovoid, 210 to 290 μm (mean: 250 μm) long, 140 to 200 μm (mean: 180 μm) wide, and approximately 100 μm thick (Pl. 1A). In most cases, one side of the egg is characteristically concave. Observation by SEM revealed that the whole surface of the outer membrane is covered by many minute pits, each pit being bordered by a flap (Pl. 2A). The egg looks dark in transmitted light, and grayish in reflected light.

4.2 *Evadne tergestina*

The resting egg of this species (Onbé 1973, 1974) is spherical, measuring 190 to 240 μm (mean: 200 μm) in diameter (Pl. 1C). The thick outer membrane appears to be provided with no ornamentations under a compound microscope, but is covered by numerous villiform structures as seen with SEM (Pl. 2C). The egg has compact, dense yolk substances, which are light-brown in color.

4.3 *Evadne nordmanni*

The resting egg of *Evadne nordmanni* measures almost the same as the former species, being also light-brown in color. However, the yolk is more transparent and globular in appearance (Cheng 1947, Onbé 1985) (Pl. 1B). SEM observation shows that the whole surface of the outer membrane is smooth, without any apparent ornamentation (Pl. 2B).

4.4 *Podon polyphemoides*

The resting egg of *Podon polyphemoides* (Onbé et al. 1977, Onbé 1985) is also spherical as is the egg of *Evadne*, being 150 to 210 μm (mean: 180 μm) in diameter (Pl. 1D). Numerous crater-like pits (2-3 μm in diameter) are scattered over the entire surface of the outer membrane as seen by SEM (Pl. 2D). The space in between the pits appears smooth and flat.

4.5 *Podon leuckarti*

The resting egg of this species (Onbé 1985) is much larger that that of *P. polyphemoides*, measuring 210 to 230 μm in diameter. The ornamentations on the outer membrane differ in structure from those in *P. polyphemoides*, in that those pits are more densely scattered over the surface, thus making the space in between narrower and somewhat bulged (Pl. 2E).

4.6 *Podon intermedius*

A resting egg of *Podon intermedius*, recovered from the bottom sediment of Vineyard Sound off Woods Hole, Massachusetts, USA, was 270 μm in diameter (Onbé & Grice unpubl.). The egg surface appears to have a complex array of irregularly-shaped groove-like structures, apparently different from those of the congeneric species, as seen by compound microscope. However, SEM examination is needed for a more precise description of these eggs.

5 DEVELOPMENT AND HATCHING PROCESSES OF RESTING EGGS

At present, the development and hatching of resting eggs are known for 3 species (Onbé 1974, Onbé et al. 1977).

5.1 *Penilia avirostris*

5.1.1 *Development*

The overall pattern of resting egg development is similar to that of parthenogenetic eggs within the brood chamber as described and illustrated by Della Croce & Bettanin (1965).

An undeveloped egg (Pl. 3A) has a finely granuled yolk, and the cytoplasm looks compact in appearance. The first sign of development is evident when a narrow perimeter (perivitelline space) appears; yolk granules increase in size, and the peripheral part of cytoplasm becomes increasingly transparent (Pl. 3B). Next, the anlage of the 2nd antenna appears, and its tip eventually becomes biramous. A thin inner membrane enclosing the embryo now becomes visible (Pl. 3C). The 1st antenna then appears in front of the 2nd, followed by the appearance of the mandible, the labrum and the anlagen of thoracic legs. At both stages shown in Plates 3D &

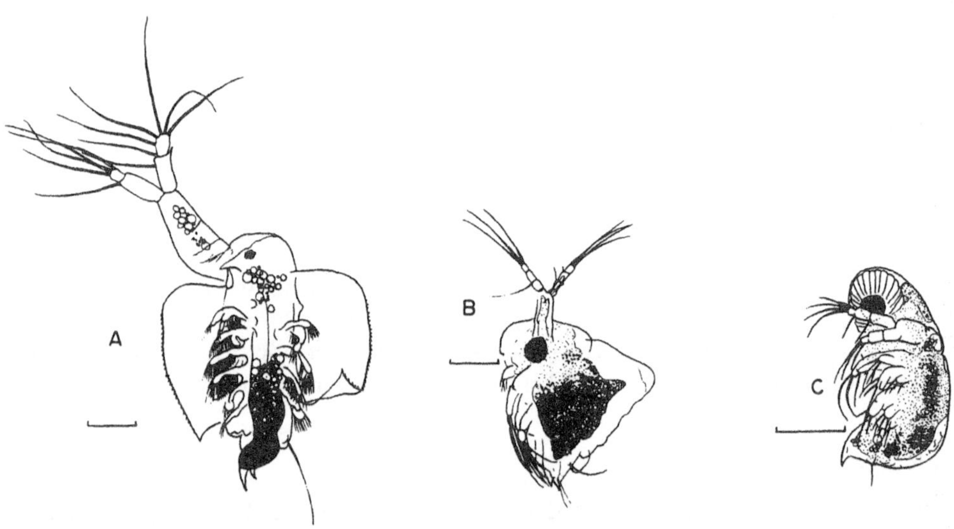

Figure 1. Young of 3 species of marine cladocerans newly hatched from the resting egg. A: *Penilia avirostris*, B: *Evadne tergestina*, C: *Podon polyphemoides* (A, B: from Onbé 1974, C: from Onbé et al. 1977). Scale bar: 100 μm.

E, the embryo has respectively 4 and 5 pairs of thoracic legs. All 6 pairs of thoracic legs are finally completed, the abdomen becomes bifurcated, the eye is now present (but without any pigmentation), and the 2nd antennae move spasmodically at times (Plates 3F, G). Then, the eye becomes pigmented in black, the 2nd antennae further elongate, and morphogenesis is completed (Pl. 3H).

Approximately 90 hours were needed for the embryo to develop from the stage shown in Plate 3B to the last stage (Pl. 3H) at a temperature of 19.4°C.

5.1.2 *Hatching*

Soon after the embryo reaches the stage shown in Plate 3H, it turns at right angles, which is the characteristic posture always seen in the embryo as it nears hatching. Now the embryo moves vigorously, movements of the 2nd antennae and the postabdomen are especially marked. Along the center of the outer membrane, perpendicular to the major axis, a dehiscent line becomes distinct. The outer membrane then ruptures at the line, and a thin inner membrane swells out of the slit. The inner membrane distends rapidly, with concurrent swelling of the embryo (Pl. 3I). At this stage, the embryo restlessly moves its 2nd antennae, postabdomen, and caudal furca. The inner membrane ruptures instantaneously, and the embryo hatches. The time required from the appearance of the first sign of development (Pl. 3B) to hatching was ca. 100 hours at a mean temperature of 19.4°C.

Osmotic hatching, as described for the resting eggs of freshwater cladocerans (Sebestyén 1949) and the subitaneous eggs of marine calanoid copepods (Marshall & Orr 1954, Davis 1968), may provide the hatching mechanism for *P. avirostris* resting eggs.

5.1.3 *Newly hatched young*

Newly hatched young of *Penilia avirostris* range between 410 and 510 μm (mean 457 μm) in standard length (the distance between the anterior edge of the head and the tip of caudal furca) (Fig. 1A). The following characters are different from those of the adult female. The basal segments of the 2nd antennae are greatly expanded, and exopods of the 2nd antennae bear only 7 setae, in contrast to 8 setae in the adult. The carapace is short, and the caudal furca is short but stout. The postabdomen, head, and basal segments of the 2nd antennae each have a patch of dense yolky substance.

Without exception, all the young hatched from resting eggs were females with a pointed rostrum and short 1st antennae.

5.2 *Evadne tergestina*

5.2.1 *Development and hatching*

Compared with the resting egg of the former species, the resting egg of this species has a thicker outer membrane and denser cytoplasm, which makes it difficult to observe its developmental processes. At an advanced stage of development, a part of the egg covering the head portion of embryo becomes transparent. At the stage shown in Plate 3J, formation of all the appendages and the eye has been completed; a clear dehiscent line on the outer membrane cuts the egg at one-third of its diameter from the anterior end of the embryo. Just before hatching, the egg contains a black-eyed embryo (Pl. 3K). By lifting up the smaller half of the outer membrane, the embryo hatches out.

5.2.2 *Newly hatched young*

The hatchling measures ca. 380 µm in standard length. It differs from the adult in that its dorsal part is much reduced and dense yolk substances still remain within the body (Fig. 1B).

5.3 *Podon polyphemoides*

5.3.1 *Development and hatching*

Embryonic development is difficult to trace in detail due to the presence of a thick, ornamented outer membrane and the possession of dense residual yolk. Just before hatching, the egg is quite similar in appearance to that of *Evadne tergestina* (Pl. 3L). Hatching is also similar to that in the former species.

5.3.2 *Newly hatched young*

The young hatched from the resting egg measures ca. 300 µm in standard length, having a large head and low depth of body owing to a reduced brood chamber (Fig. 1C).

6 DISTRIBUTION AND ABUNDANCE OF RESTING EGGS IN SEA-BED SEDIMENTS

6.1 *Collection, sorting, and identification*

6.1.1 *Collection*

Behavior of resting eggs after settling to the bottom can only be surmised at present. The eggs of *Penilia avirostris* are found in the upper 6 cm of sediments (Onbé 1974, 1978c), with 60-100% of eggs being present in the uppermost 0-2 cm layer. However, as demonstrated in copepod eggs by Marcus & Schmidt-Gengenbach (1986), they can be transported to deeper layers by, for example, the action of current or of benthic animals (bioturbation), which can also bring the eggs from deeper layers back to the surface and resuspend them in the water column. In areas where the vertical flux of organic matter is great, due to pollution coupled with stagnant flow systems, eggs are buried in the sediments at a greater rate than expected in clearer, or less stagnant, water.

For quantitative use, undisturbed sediments of at least 6-10 cm deep should be carefully collected. For this purpose, a gravity corer can be most effectively employed, although multiplicate samples should be taken for an accurate estimation of egg density.

6.1.2 *Sorting*

Sediment samples are sieved through a 100 µm mesh screen in the case of cladoceran resting eggs. The residue remaining on the screen must be observed under a dissecting microscope by reflected light; the eggs are then sorted by a micropipet.

A sugar flotation method (Onbé 1978c) greatly lessens the time required for sorting eggs. This technique, originally adopted for extracting the meiobenthos from mud samples (Heip et al. 1974), involves the following procedure: the sieved residue on the screen is washed into centrifuging tubes with a dense solution of sugar (1 kg sucrose dissolved in 1 liter of water) and centrifuged at 3000 rpm for 5 min. The eggs floated up to the surface of the solution are collected.

Examinations showed that almost all the eggs of *Penilia, Evadne* and *Podon* can be extracted after two centrifugations without any apparent loss of egg viability. The percent hatching is reported to be nearly equal between the eggs extracted by this procedure and the hand-sorted controls.

6.1.3 *Identification*

Size, shape, and color are primarily important for species identification of resting eggs. Surface structure of the egg membrane, whether it looks transparent or not, is also useful in discriminating the spherical eggs of *Evadne* and *Podon*. In *Evadne*, the membrane is more transparent than in *Podon*, so that clear perimeter (perivitelline space) is always seen by transmitted light in the egg of *Evadne* (Pl. 1B, C). In contrast, the perimeter of *Podon*'s egg is obscure due to the characteristic surface structures of the egg membrane (Pl. 1D).

Identification of eggs to species level can be finally attained by hatching them. Because the embryonic development of cladocerans is direct, the newly-hatched young already has species-specific characteristics for identification (Fig. 1).

6.2 *Distribution and abundance in sediments*

6.2.1 *Previous information*

According to Purasjoki's (1945) pioneering work done at nearshore stations in the Gulf of Finland, Baltic Sea, the resting eggs of *Bosmina maritima* were most abundant, ranging from 240 to 31 280 eggs/m^2, followed by those of *Evadne* and *Podon* (range: 20-10 200 /m^2) and *Daphnia* (range: 20-4420/m^2). Purasjoki (1958) revealed that the abundance of *Bosmina* eggs fluctuated with season.

In 1970, the resting eggs of *Penilia avirostris* were first recovered from sediments of the central part of the Inland Sea of Japan (Onbé 1972, 1973, 1974), together with the eggs of *Evadne* and *Podon*. Since then, marine cladoceran eggs have been collected from various parts of the Inland Sea (Onbé unpubl.) and from Ise Bay and Uragami Inlet in central Japan (Onbé 1978b). Outside Japan, they have been recorded from Buzzards Bay, Massachusetts, USA. (Onbé & Grice unpubl.) and the Gulf of Marconi, northern Ligurian Sea, Italy (Onbé & Della Croce unpubl.).

The standing stocks of the resting eggs of marine cladocerans in terms of the number of eggs per square meter of sea floor vary with both locality and season. In most cases, the eggs of *Penilia avirostris* were most abundant, followed by those of *Podon polyphemoides* and *Evadne tergestina*. All these 3 species are warm-water forms with abundant planktonic populations. The eggs of cold-water, less abundant species, *Evadne nordmanni, Podon intermedius,* and *P. leuckarti*, were much lower in abundance than those of the former 3 species.

The maximum number of *Penilia avirostris* hitherto recorded was 142 000 eggs/m^2 recovered from a station located in the eastern part of Ise Bay. 64 980 eggs of *Podon polyphemoides* were concurrently collected, which also marks the highest record for this particular species (Onbé 1978b).

6.2.2 *Distribution in Bingo-Hiuchi Nada, the Inland Sea of Japan*

Figure 2 shows the results of a survey of the distribution and abundance of the resting eggs of 4 species of marine cladocerans at 19 stations in Bingo-Hiuchi Nada, central Inland Sea of Japan, in December 1978. The egg density was variable, ranging from 4.0-40.5×10^3 eggs/m^2 of sea

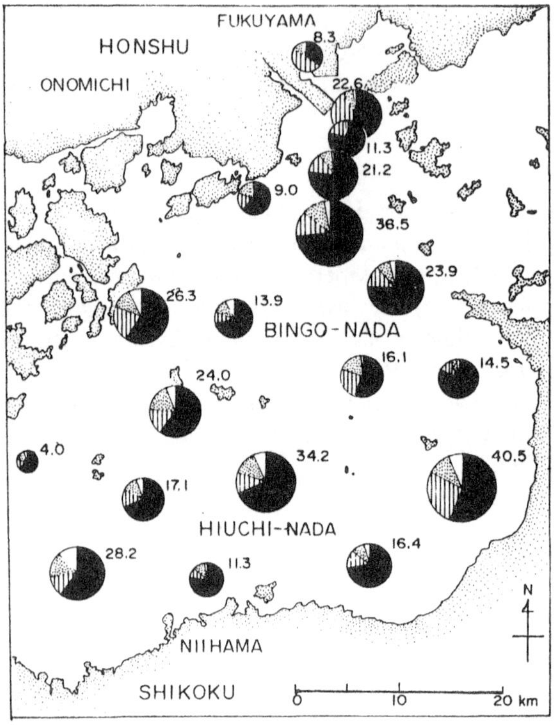

Figure 2. Distribution of resting eggs of 4 species of marine cladocerans in Bingo- and Hiuchi-Nada, central Inland Sea of Japan, in December 1978 (Onbé & Ariyoshi unpubl.). Solid part: *Penilia avirostris*, striped part: *Podon polyphemoides*, stippled part: *Evadne tergestina*, open part: *Evadne nordmanni*. A few eggs of *Podon leuckarti* were found but not counted. Figures indicate the total number of eggs ($\times 10^3$)/m^2 on the sea floor.

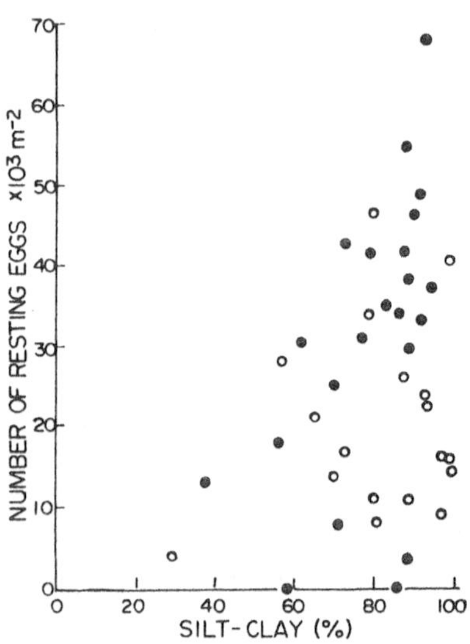

Figure 3. Abundance of resting eggs of marine cladocerans (number $\times 10^3$ m^{-2}) in Bingo- and Hiuchi-Nada, central Inland Sea of Japan, in relation to sediment particle composition (silty-clay, %). Solid circle: Nov. 1977, open circle: Dec. 1978.

floor, of which eggs of *Penilia avirostris* were generally the highest. *Podon polyphemoides* eggs were next in abundance, followed by those of *Evadne tergestina*.

The resting eggs of cladocerans tended to be distributed more abundantly at stations with sediments predominating in fine particles of less than 53 μm in diameter, as seen clearly in Figure 3. This suggests that the resting eggs after being released from females are, while descending in the water column, gradually transported by current. Resting eggs accumulate more densely in areas with muddy bottoms than in areas having sandy sediments. Eggs could also be redistributed after being on the bottom by resuspension and transport (Marcus 1984, Marcus & Schmidt-Gengenbach 1986). This pattern of egg distribution might be influenced largely by local hydrodynamics.

The survival rate of eggs appears to be reduced by the organic pollution of sediments. Generally, the resting eggs of cladocerans have a thick outer membrane and may be able to endure a certain degree of environmental deterioration. The eggs of *Podon polyphemoides* recovered from the sediments of the innermost station of Fukuyama Harbor (the northernmost station in Fig. 2) did not hatch at all, suggesting the death of these eggs by the highly polluted sediments there, which created strongly reduced, H_2S-rich, anoxic conditions.

6.2.3 *Distributions in Buzzards Bay and the Gulf of Marconi*

A survey was made at 6 stations in Buzzards Bay, Massachusetts, USA, in November 1980 (Onbé & Grice unpubl.). Most numerous again were the eggs of *Penilia avirostris* (6600 eggs/m^2), followed in order by those of *Podon polyphemoides, Evadne tergestine, E. nordmanni*, and *P. intermedius*. The tendency was consistent at a station near Nonamesett Island off Woods Hole, where a series of collections were made from early July through mid- November 1980. Because this station was in a Thalassia bed with sediment composed of sandy mud, egg density was significantly less than at the muddy bottom stations in Buzzards Bay. At this station, however, a maximum density of 14 000 eggs/m^2 of *P. avirostris* occurred in early October.

In early March 1981, sediments were collected at 15 stations in the Gulf of Marconi, northern Ligurian Sea, Italy (Onbé & Della Croce unpubl.). Except for a few stations in the Harbors of Rapallo and Chiavari, the stations visited were of sandy bottoms. Resting eggs of 4 species were recovered, but were scarce compared to densities at other localities. However, the relative abundance of *Podon polyphemoides* was greater; this species being more abundant than *Penilia avirostris* and *Evadne tergestina*. A few *Evadne*-like type of eggs were also collected. These eggs are here tentatively attributed to *E. spinifera*, which is known to abound in the plankton community of the area (cf. Bogliolo et al. 1979, Basso et al. 1980).

6.2.4 *Seasonal fluctuations in abundance in sediments*

In association with their planktonic populations, seasonal fluctuations in the abundance of resting eggs of 4 species: *Penilia avirostris, Evadne nordmanni, E. tergestina*, and *Podon polyphemoides*, were investigated at a station off Sensui-jima near Fukuyama, central part of the Inland Sea of Japan, for a period of over 2 consecutive years (Onbé 1985). As seen in Figure 4, the density of resting eggs changed with seasons so that highest numbers were reached in late summer shortly before the population disappeared from the plankton community. Density of resting eggs decreased to its lowest level in the spring, just before the initiation of population development in the water column. This pattern of fluctuation was consistent in both years for each of the four species studied. A high abundance of the benthic resting eggs undoubtedly reflects the sequence of events resulting in intense production of eggs by gamogenetic individuals in declining populations. Drastic decreases in abundance of eggs during colder seasons were apparent in both years, presumably due to hatching, predation by deposit-feeding invertebrates, and/or death from environmental deterioration (Marcus 1984 for benthic copepod eggs, Onbé 1985).

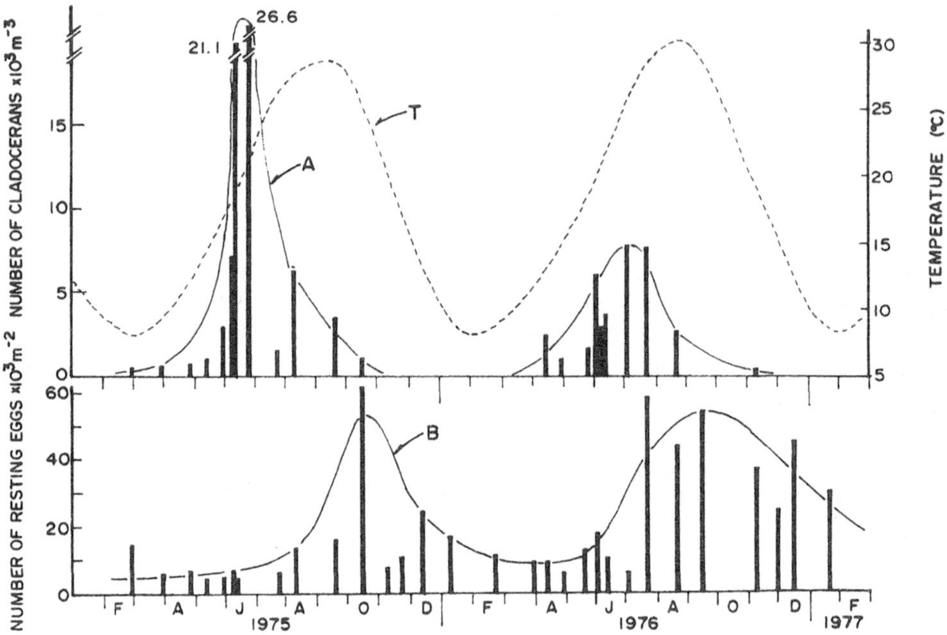

Figure 4. Seasonal fluctuations in abundance of 5 species of marine cladocerans (combined) in the water column (A), and their resting eggs (combined) in the sediments (B), with changes in water temperature (T) at Station C, Sensui-jima, Fukuyama, central Inland Sea of Japan (Data taken from Onbé 1985).

Considerable numbers of eggs remained in the sediments during fall and mid-winter when the water column was devoid of planktonic cladocerans. These overwintering eggs may serve as the 'seed population' for the resurgence of the next planktonic population.

7 HATCHING OF RESTING EGGS

7.1 *Effect of salinity and temperature on hatching*

The range of salinity that favors the hatching of resting eggs of *Penilia avirostris* was found to be 24-30 ‰ (Onbé 1974, Iwasaki et al. 1977, Onbé & Ariyoshi unpubl.). Under the 5 experimental temperatures between 12 and 23°C, *Penilia*'s eggs hatched most successfully at 17-20° (Onbé 1974).

Hatching of resting eggs of *Podon polyphemoides* has been studied in more detail than in the former species (Onbé & Ariyoshi unpubl.). In the eggs of *P. polyphemoides,* hatching occurred at a wide range of salinities between 3.0-33.7‰ at 20°C. At salinities higher than 40‰ hatching never took place. Hatching occurred at temperatures ranging from 5 to 26°C, with an optimum temperature lying between 18 and 23°C. At 26°C, hatching success drastically reduced and no eggs hatched at 30°C. When the unhatched eggs kept at 26°C were transferred to 18°C, much additional hatching occurred. Time required for 50% hatch changed with temperatures: at 5°C, it took as long as 76.5 days, but shortened with increasing temperatures to 15.4-17.2 days at

18-23°C; a further confirmation of the optimal temperatures for this species. Hatching tended to be suppressed in deoxygenated seawater. In addition, hatching was apparently inhibited at temperatures higher than 26°C, whereas hatching did occur at a temperature as low as 5°C.

7.2 *Seasonal changes in hatchability of resting eggs recovered from sediments*

Hatchability was tested for the resting eggs of *Penilia avirostris* and *Podon polyphemoides* recovered from sediments once a month for 2 consecutive years at a station near Fukuyama, central Inland Sea of Japan (Onbé & Ariyoshi unpubl.).

P. avirostris eggs showed a clear seasonal cycle in hatching performance, being low during a period from September to December, and gradually becoming higher until May to August, when maximum hatching success was reached. On the contrary, there were no apparent seasonal changes in hatchability in the resting eggs of *P. polyphemoides*. Hatching was consistently high throughout the year. The eggs of the former species appear to require a diapause (or refractory) period before development occurs, whereas those of the latter may need only a short refractory period.

8 DISCUSSION

In temperate and boreal coastal waters, many zooplankters occur seasonally for some period of time year after year in the local plankton community and disappear from the water column for the rest of the year. Traditional explanation for the resurgence of the next planktonic populations had been the recruitment of individuals from other surviving populations by transportation, although some authors speculated the existence of resting stages from which new populations should derive. The discovery of resting cysts of tintinnids (Reid & John 1978, Paranjape 1980), and resting eggs of marine cladocerans (Onbé 1972, 1973, 1974) and calanoid copepods (Grice & Marcus 1981, Uye 1985, for review) from coastal sea-bed sediments has made it possible to furnish a reasonable explanation for their seasonal distribution. Such resting stages can insure the survival and maintenance of populations during adverse condtions which lead to the extinction of pelagic populations.

Overwintering populations of dormant stages will, therefore, primarily be ecologically important as determinants of the initial population size in a water column devoid of parent populations. It is possible that young individuals hatched from the dormant cells may have some biological properties to promote better survivorship in such an environment.

The young marine cladocerans hatched from the resting eggs are significantly larger than the parthenogenetic young released from females. Newly hatched young of *Penilia avirostris*, for example, are 8-34% (mean: 20%) longer than the smallest parthenogenetic individuals in the plankton samples (Murakami & Onbé 1967) or fully developed young within the brood chamber just before the liberation from a female (Onbé 1974). Similarly, in *Podon polyphemoides*, the former (300 μm in standard length, Onbé et al. 1977) is 25% longer than the latter (240 μm in s. l., Onbé 1974). Larger size may increase the range of swimming and feeding activities of these young and thus enhance their survival. Another feature of the newly hatched young of marine cladocerans is the retention of a large quantity of residual yolk, which would also be advantageous for survival of young in an environment of uncertain nutritional conditions.

On the other hand, the resting eggs produced by gamogenesis are thought to enable each species to enhance adaptation and to survive in an ever-changing coastal environment, because the gamogenesis theoretically accelerates the rate of gene flow and promotes greater genetic variation. However, according to Della Croce & Angelino (1968-69), *Penilia avirostris* occurs year round in Agulhas Bank off South Africa, where temperature is stable (annual temperature range: 15.0-21.8°C) and seemingly favorable for this species. Nevertheless, the population exhibits both parthenogenesis and gamogenesis. Perhaps the populations cannot persist by parthonogenesis alone even in such a stable environment. The resting eggs of *Bosmina maritima* are reported to have a resting period of 5 to 6 months before development is initiated (Purasjoki 1958). In *Penilia avirostris*, 4-5 months appear to be necessary for the eggs to terminate the refractory period in the laboratory, whereas *Podon polyphemoides* eggs have no apparent diapause. Benthic eggs of *P. avirostris* remain dormant in the Inland Sea of Japan for 8-9 months, until the middle of May, the following year. In contrast, *P. polyphemoides* usually appears in the plankton community during the temperature-increasing period from April through July. On the basis of our experimental results, the resting eggs of *P. polyphemoides* could hatch at any time of the year except for August and September when the bottom water temperatures are higher than 26°C, above which egg hatching might be suppressed. This species reappeared in the plankton in November and December 1974 (Onbé 1985), apparently derived from resting eggs produced mostly in July. Except for the peak summer season, temperature alone does not appear to be the primary factor controlling the development and maintenance of the planktonic population of this species.

Kankaala (1983) calculated the production of resting eggs (Pr) of the brackish-water *Bosmina maritima* based on the equation of Winberg (1971):

$$Pr = B't^{1/2}(No + Nt)$$

where B' is equal to E, the number of resting eggs per female, divided by D, the instar duration related to temperature, $B' = E/D$. No and Nt are the initial and final numbers of sexual females present during the period t. However, the resting egg production cannot be estimated for marine cladocerans because parameters for these equations are not yet available. Only *Penilia avirostris* has been successfully reared in the laboratory (Takami et al. 1978, Gore 1980, Paffenhöfer & Orcutt 1986, Turner et al. 1988).

The durability of the resting eggs of marine cladocerans to various environmental stresses has not been examined in detail. However, some of the eggs of *Penilia avirostris* recovered from a sediment sample, which had been kept in a plastic bag under ambient laboratory temperatures for 13 months, hatched successfully (Onbé 1974). Development of effective preservation methods for live eggs may lead to the possible use of resting eggs for cultivation or bioassay materials for pollution.

The life cycle and population dynamics of marine cladocerans previously have been studied by examining only pelagic populations without giving much attention to their benthic resting eggs. If we try to gain a deeper insight into the mechanisms controlling the seasonal succession of each species, then production of dormant eggs, and their distribution, abundance and survival in sea-floor sediments should be investigated concurrent with studies on pelagic populations. Such an approach has already been taken for calanoid copepods (Uye 1985, for review).

ACKNOWLEDGEMENTS

I would like to thank Dr George D. Grice, Woods Hole Oceanographic Institution and Professor Norberto Della Croce, University of Genoa, for hosting my stay and supporting my studies at

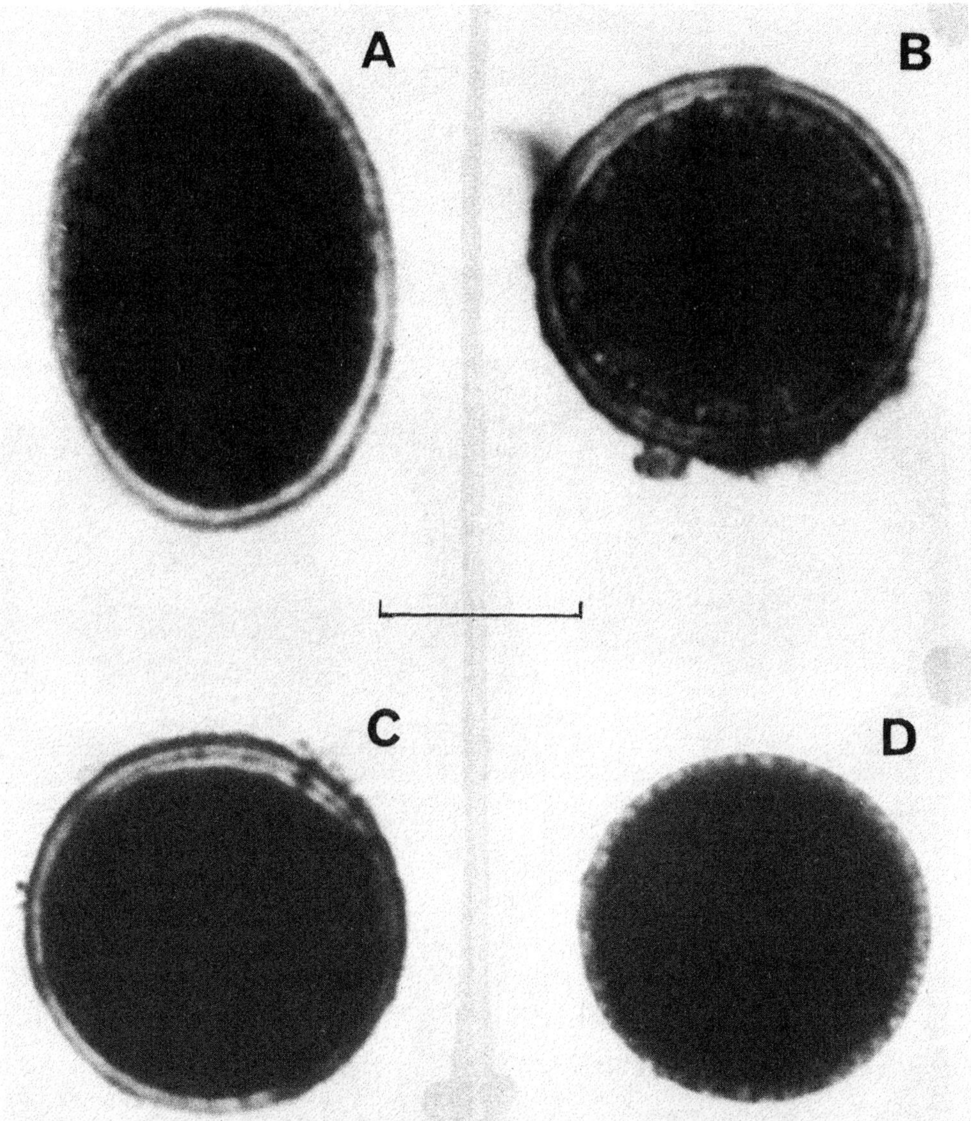

Plate 1. Resting eggs of 4 species of marine cladocerans recovered from sea-bed sediments. A: *Penilia avirostris*, B: *Evadne nordmanni*, C: *Evadne tergestina*, D: *Podon polyphemoides*. Scale bar: 100 μm.

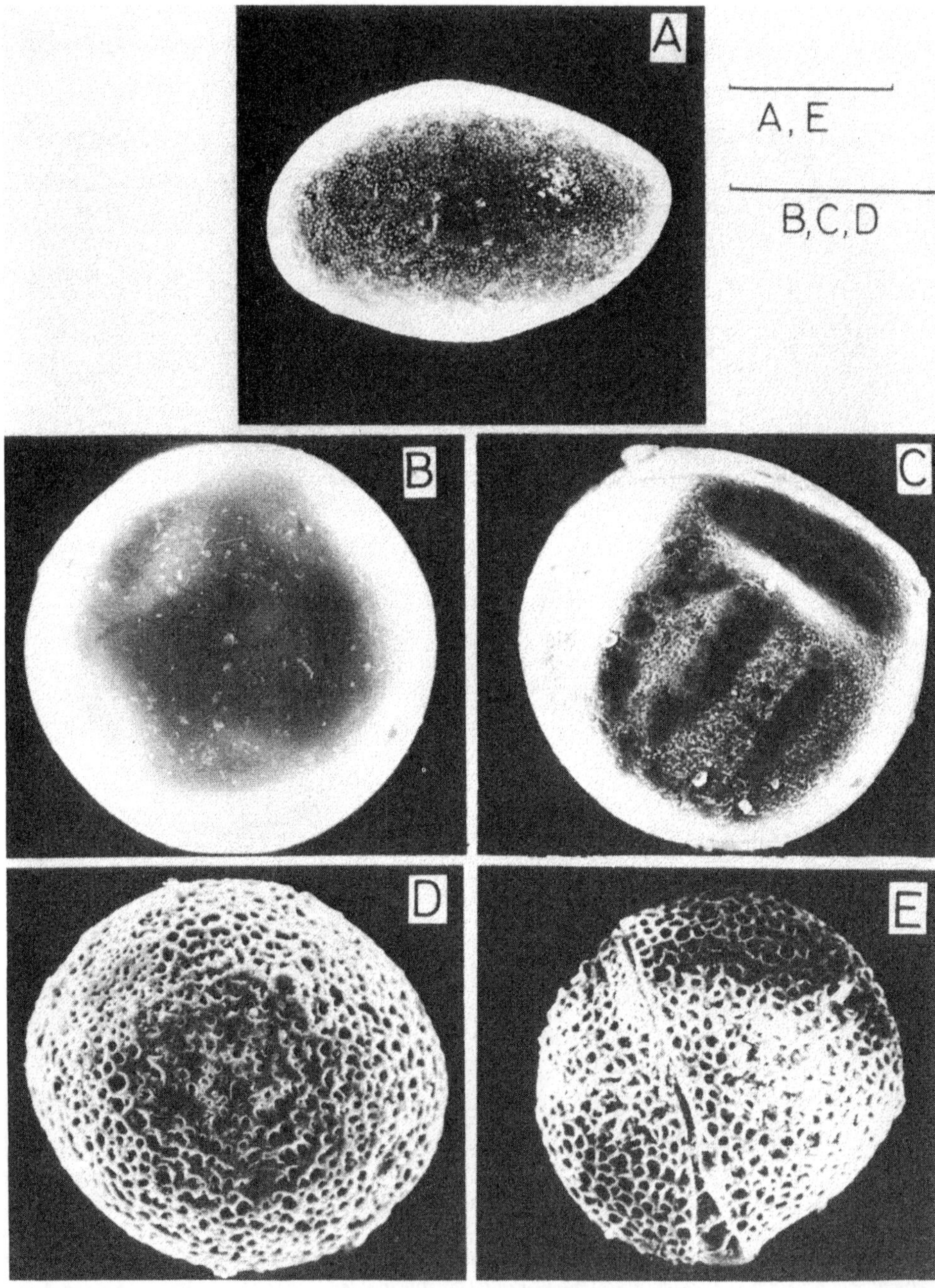

Plate 2. Scanning electron micrographs of resting eggs of 5 species of marine cladocerans (Onbé & Hayashida unpubl.). A: *Penilia avirostris*, B: *Evadne nordmanni*, C: *Evadne tergestina*, D: *Podon polyphemoides*, E: *Podon leuckarti*. Scale bar: 100 μm.

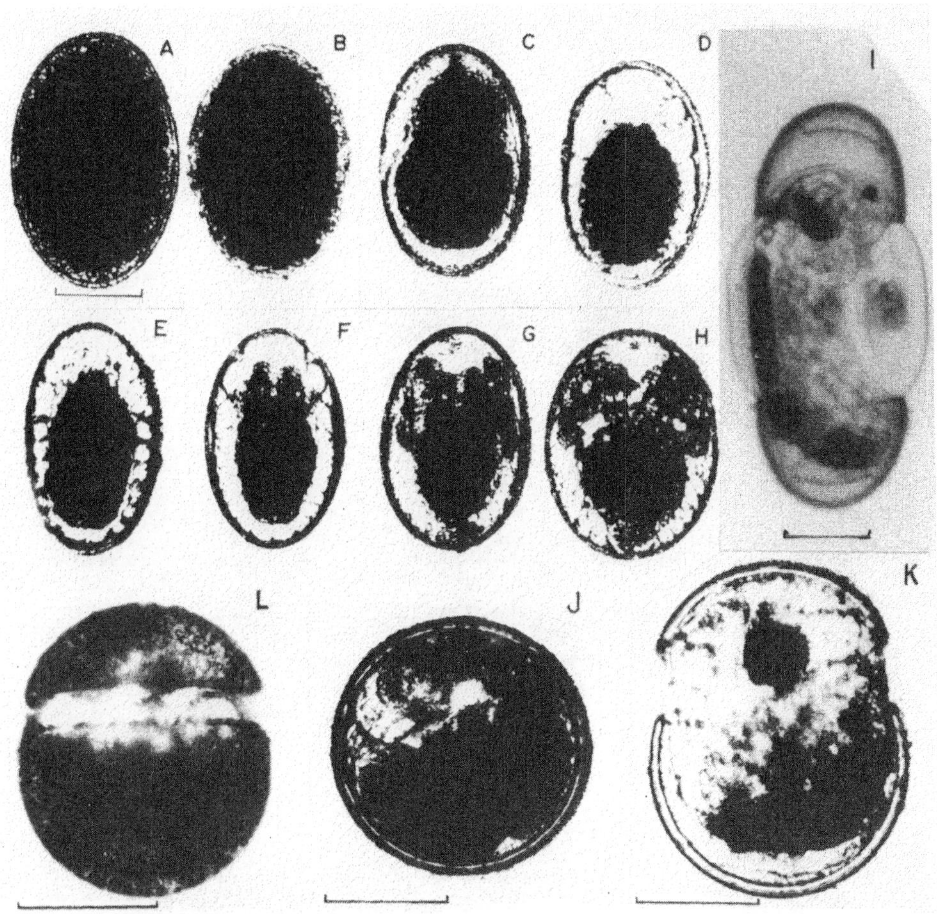

Plate 3. Development of the resting eggs of 3 species of marine cladocerans. A-I: *Penilia avirostris*, J-K: *Evadne tergestina*, L: *Podon polyphemoides* (Modified from Onbé 1974, Onbé et al. 1977). Scale bar: 100 μm.

their laboratories, and the late Professor Yutaka Murakami, Hiroshima University, for his constant encouragement. Sincere thanks are also extended to Dr Nancy H. Marcus, Florida State University, for her kind help at sea in the Woods Hole region and valuable comments on the manuscript, and Professors Adrian M. Wenner and Armand Kuris, University of California, Santa Barbara, for their critical review and careful editing of the final manuscript. Messrs Toshikazu Ariyoshi and Hidekazu Hayashida respectively helped me with some of the experiments described here, and provided me with the SEM micrographs used in Plate 2. Financial support was given by grants-in-aid from the Ministry of Education, Science and Culture, and the Ministry of Agriculture, Forestry and Fisheries of Japan.

REFERENCES

Bainbridge, V. 1958. Some observations on *Evadne nordmanni* Lovén. *J. Mar. Biol. Ass. UK* 37: 349-370.

Basso, M.P., M.G. Cevasco, N. Della Croce & P. Picone 1980. Caratteristice ecologice e popolamento zooplanctonico in ambienti portuali del Mar Ligure e Alto Tirreno. *Univ. Genova Catt. Idrobiol. Pescic., Fac. Sci. Fis. Mat. Nat., Rap. Tec.* 12: 1-50.

Berg, K. 1931. Studies on the genus *Daphnia* O.F. Müller, with special reference to the mode of reproduction. *Vidensk Meddr. Dansk. Naturh. Foren* 92: 1-222.

Bogliolo, A., G. Cevasco, N. Della Croce & T. Zunini Sertorio 1979. Composizione qualitativa e quantitativa dello zooplancton (Chiavari): Ottobre 1977 – Ottobre 1978. *Univ. Genova, Fac. Sci. Fis. Mat. Nat., Rap. Tec.* 9: 1-14.

Bosch, H.F. & W.R. Taylor 1973. Distribution of the cladoceran *Podon polyphemoides* in the Chesapeake Bay. *Mar. Biol.* 19: 161-171.

Caroli, E. 1924. Sulla presenza di *Penilia schmackeri* Rich. nel Golfo di Napoli. *Boll. Soc. Natur. Napoli* 35: 96-99.

Cheng, C. 1947. On the fertility of marine Cladocera with a note on the formation of the resting eggs in *Evadne nordmanni* Lovén and *Podon intermedius* Lilljeborg. *J. Mar. Biol. Ass. UK* 26: 551-561.

Davis, C.C. 1968. Mechanisms of hatching in aquatic invertebrate eggs. *Oceanogr. Mar. Biol. Ann. Rev.* 6: 325-326.

Della Croce, N. 1974. Cladocera. *Cons. Internat. Explor. Mer, Zoopl. Sheet* 143: 1-4.

Della Croce, N. & M.I. Angelino 1968-69. *Penilia avirostris* Dana nelle acque del Banco Agulhas. *Boll. Mus. Ist. Biol. Univ. Genova* 36: 5-14.

Della Croce, N. & S. Bettanin 1965. Sviluppo embrionale della forma partenogenetica di *Penilia avirostris* Dana. *Cah. Biol. Mar.* 6: 269-275.

Della Croce, N. & P. Venugopal 1973. *Penilia avirostris* Dana in the Indian Ocean (Cladocera). *Int. Rev. Ges. Hydrobiol.* 58: 713-721.

Gieskes, W.W.C. 1970. The Cladocera of the North Atlantic and the North Sea – Biological and Ecological Studies. Ph.D. thesis, McGill Univ., Montreal, 204 p.

Gieskes, W.W.C. 1971. Ecology of the Caldocera of the North Atlantic and the North Sea. *Neth. J. Sea Res.* 5: 342-376.

Gore, M.A. 1980. Feeding experiments on *Penilia avirostris* Dana (Cladocera: Crustacea). *J. Exp. Mar. Biol. Ecol.* 44: 253-260.

Green, J. 1966. Seasonal variation in egg production by Cladocera. *J. Anim. Ecol.* 35: 77-104.

Grice, G.D. & N.H. Marcus 1981. Dormant eggs of marine copepods. *Oceanogr. Mar. Biol. Ann. Rev.* 19: 125-140.

Heip, C., N. Smol & W. Hautekiet 1974. A rapid method of extracting meiobenthic nematodes and copepods from mud and detritus. *Mar. Biol.* 28: 79-81.

Hutchinson, G.E. 1967. *A Treatise on Limnology. Vol. II. Introduction to Lake Biology and the Limnoplankton.* New York, London, Sydney: John Wiley & Sons.

Iwasaki, H., A. Takami & T. Onbé 1977. Studies on the cultivation of marine Cladocera – I. Factors affecting the hatch of resting eggs. *Bull. Plankton Soc. Japan* 24: 62-65. (In Japanese with English abstract).

Kankaala, P. 1983. Resting eggs, seasonal dynamics, and production of *Bosmina longispina maritima* (P.E. Müller) (Cladocera) in the northern Baltic proper. *J. Plankton Res.* 5: 53-69.
Lochhead, J.H. 1954. On the distribution of a marine cladoceran, *Penilia avirostrs* Dana (Crustacea, Branchiopoda), with a note on its reported bioluminescence. *Biol. Bull.* 107: 92-105.
Marcus, N.H. 1984. Recruitment of copepod nauplii into the plankton: Importance of diapause eggs and benthic processes. *Mar. Ecol. Prog. Ser.* 15: 47-54.
Marcus, N.H. & J. Schmidt-Gengenbach 1986. Recruitment of individuals into the plankton: The importance of bioturbation. *Limnol. Oceanogr.* 31: 206-210.
Marshall, S.M. & A.P. Orr 1954. Hatching of *Calanus finmarchicus* and some other copepods. *J. Mar. Biol. Ass. UK.* 33: 393-410.
Murakami, Y. & T. Onbé 1967. Notes on the marine cladocerans observed at a definite station near Tomo Marine Biological Station. *Inform. Bull. Planktol. Japan, Comm. No. Dr. Matsue:* 123-127. (In Japanese with English abstract).
Onbé, T. 1968. Studies on the marine cladocerans – I. A biological note of *Penilia. J. Fac. Fish. Anim. Husb. Hiroshima Univ.* 7: 269-279. (In Japanese with English abstract).
Onbé, T. 1972. Occurrence of the resting eggs of a marine cladoceran, *Penilia avirostris* Dana, on the sea-bottom. *Bull. Japan. Soc. Sci. Fish.* 38: 305.
Onbé, T. 1973. Preliminary notes on the biology of the resting eggs of marine cladocerans. *Bull. Plankton Soc. Japan* 20: 74-77. (In Japanese with English abstract).
Onbé, T. 1974. Studies on the ecology of marine cladocerans. *J. Fac. Fish. Anim. Husb. Hiroshima Univ.* 13: 83-179. (In Japanese with English abstract).
Onbé, T. 1978a. The life cycle of marine cladocerans. *Bull. Plankton Soc. Japan* 25: 41-54. (In Japanese with English abstract).
Onbé, T. 1978b. Distribution of the resting eggs of marine cladocerans in the bottom sediment of Ise Bay and Uragami Inlet, central Japan. *Bull. Japan. Soc. Sci. Fish.* 44: 1053.
Onbé, T. 1978c. Sugar flotation method for sorting the resting eggs of marine cladocerans and copepods from sea-bottom sediment. *Bull. Japan. Soc. Sci. Fish.* 44: 1411.
Onbé, T. 1978d. Gamogenetic forms of *Evadne tergestina* Claus (Branchiopoda: Cladocera) of the Inland Sea of Japan. *J. Fac. Fish. Anim. Husb. Hiroshima Univ.* 17: 43-52.
Onbé, T. 1979. Resting eggs of marine cladocerans. In, *Mass Culture of Zooplankton as Food for Fish Larvae:* pp. 22-33. Tokyo: Japan Fish. Res. Protection Assoc. (In Japanese).
Onbé, T. 1985. Seasonal fluctuations in the abundance of populations of marine cladocerans and their resting eggs in the Inland Sea of Japan. *Mar. Biol.* 87: 83-88.
Onbé, T., T. Mitsuda & Y. Murakami 1977. Some notes on the resting eggs of the marine cladoceran *Podon polyphemoides. Bull. Plankton Soc. Japan* 24: 85-93.
Paffenhöfer, G.-A. & J.D. Orcutt, Jr 1986. Feeding, growth and food conversion of the marine cladoceran *Penilia avirostris. J. Plankton Res.* 8: 741-754.
Paranjape, M. 1980. Occurrence, and significance of resting cysts in a hyaline tintinnid, *Helicostomella subulata* (Ehre.) Jorgensen. *J. Exp. Mar. Biol. Ecol.* 48: 23-33.
Pavlova, E.V. 1959. Development cycle and some data on the growth of *Penilia avirostris* Dana in the Sevastopol Bay. *Trudy Sevastopol. Biol. Sta. Akad. Nauk SSSR* 2: 54-62. (Fish. Res. Bd Canada, Translation Series 967, 1968).
Purasjoki, K.J. 1945. Quantitative Untersuchungen über die Mikrofauna des Meeresbodens in der Umgebung der Zoologischen Station Tvärminne an der Südküste Finlands. *Soc. Scient. Fenn., Comm. Biol.* 9(14): 1-24.
Purasjoki, K.J. 1958. Zür Biologie der Brackwasserkladozere *Bosmina coregoni maritima* (P.E. Müller). *Ann. Zool. Soc. Zool. Bot. 'Vanamo'* 19(2): 1-117.
Reid, P.C. & A.W.G. John 1978. Tintinnid cysts. *J. Mar. Biol. Ass. UK.* 58: 551-557.
Schram, F. 1986. *Crustacea.* New York: Oxford Univ. Press.
Sebestyén, O. 1949. On the life method of the larva of *Leptodora kindtii* (Focke), (Cladocera, Crustacera). *Hung. Acta Biol.* 1: 71-81.
Stross, R.G. 1966. Light and temperature requirements for diapause development and release in *Daphnia. Ecology* 47: 368-374.

Takami, A., H. Iwasaki & M. Nagoshi 1978. Studies on the cultivation of marine Cladocera – II. Cultivation of *Penilia avirostris* Dana. *Bull. Fac. Fish. Univ. Mie* 5: 47-63. (In Japanese with English abstract).

Turner, J.T., P.A. Tester & R.L. Ferguson 1988. The marine cladoceran *Penilia avirostris* and the 'microbial loop' of pelagic food webs. *Limnol. Oceanogr.* 33: 245-255.

Uye, S. 1985. Resting egg production as a life history strategy of marine copepods. *Bull. Mar. Sci.* 37: 440-449.

Wickstead, J.H. 1963. The Cladocera of the Zanzibar area of the Indian Ocean, with a note on the comparative catches of two plankton nets. *East Afr. Agric. For. J.* 29: 164-172.

Winberg, G.D. (ed.) 1971. *Methods for the Estimation of Production of Aquatic Animals*. London, New York: Academic Press.

A review of the hatching process in barnacles

D.J. Crisp,[†] E.M. Hill & D.L. Holland
School of Animal Biology, UCNW, Bangor, Gwynedd, UK

ABSTRACT

Though the barnacle embryo hatches by mechanically disrupting its eggshell, this does not happen automatically when it reaches full term. A stimulating substance secreted by the parent incubating the eggs in its mantle cavity is required. The parent possesses a precursor of this 'hatching subtance', throughout the year, namely eicosapentaenoic acid. When this substrate is acted upon by a non-specific lipoxygenase, it produces the hatching substance which can penetrate the eggshell and awake the embryos from dormancy. Hatching substance is produced artificially when the tissues that contain enzyme and substrate are macerated or naturally when the adult begins to resume feeding after a period of starvation.

The hatching substance bears a resemblance to some prostaglandins in being acid labile, in having similar Rf mobilities in TLC solvent systems designed to separate polar lipids and in its production being inhibited by aspirin. However, gas-liquid chromatography combined with mass spectrometry has shown it to be 10, 11, 12-trihydroxy-5, 8, 14, 17-eicosatetraenoic acid formed from the putative precursor 20:5w3 via a linear pathway.

The hatching substance is the first of this type of eicosanoid to have been shown to have a specific physiological function. As yet, it has been found in no organism other than thoracican barnacles, though the gorgonian coral *Plexaura homomalla* possesses barnacle hatching activity. It appears to be regularly present in barnacles and probably co-evolved with the development of a dormant condition in the fully developed embryo. The advantage of this mechanism is likely to be the control by the parent of the time of hatching, ensuring that the planktonic environment is suitable for the liberation of the young.

1 INTRODUCTION

Crisp (1954) described in *Balanus balanus*, (L) (=*B. porcatus* da Costa) how the fertilized eggs are contained in two egg masses, placed on either side of the mantle space, each egg mass held within a single envelope. Walley (1965) subsequently described exactly how the enveloping membrane was formed within the oviducal gland, and how it expanded when unfertilized rounded eggs were discharged into it. Her account is in line with that of Krohne, who wrote to Darwin correcting his original misconception (see Crisp 1983). At fertilization the sperm reach the eggs via minute pores in the envelope (Walley 1969, Walley et al. 1971, Walker 1977). In response to sperm penetration the fertilized egg throws off a membrane that hardens into the individual egg case within which development takes place. The egg cases are bound together by small amounts of proteinaceous cement (Crisp 1956). The supposition that development to the cyprid took place not in the sea but in the mantle space (Barnes 1953) was incorrect for the great majority of barnacles (Crisp & Knight-Jones 1953). Normally eggs or hatched larvae are discharged with an exhalent pulse of water from the parent (Crisp & Southward 1961) and are thenceforth planktotrophic.

Of the mechanisms leading to hatching the first to be investigated were purely physical. Crisp (1954) and Barnes (1955) showed oxygen to be necessary, but eggs could hatch at tensions as low as 0.08 atm (Crisp & Spencer 1958). Evidently there is sufficient oxygen in the mantle cavity fluid to allow hatching to take place there, since Barnes (1955) carefully watched the violent liberation of hatched first stage nauplius larvae together with groups of unhatched cemented eggs and larval egg cases from *Semibalanus balanoides*. Hatching could also take place in water containing up to 5.7% atm of carbon dioxide at pH 6.0. The deterioration of the glue binding the eggs causes the mature egg mass to become soft and friable (Crisp 1954, Barnes & Barnes 1977), but this does not of itself lead to hatching. Barnes (1955) examined also the possibility that osmotic pressure changes might influence hatching as in some copepods (Marshall & Orr 1954). However, the fluid inside the egg case was found to be isoosmotic with sea water, and the egg case itself was water permeable but elastic. Eggs therefore swelled in hypotonic sea water and shrunk in hypertonic; the latter condition inhibited hatching because it required extra pressure from the embryo struggling to get free. Barnes (1955) therefore concluded that hatching was a mechanical process due to the embryos' movements. Thus there remained the outstanding problem; why do the eggs not hatch at full term but often remain in the egg cases for indefinite periods after full development.

2 THE HATCHING SUBSTANCE

2.1 *Early experiments*

The first to point out that the critical determinant in hatching was chemical was Crisp (1956), who showed that a substance (called 'hatching substance') could be extracted from the tissues of the adult barnacle. If kept in clean running sea water, gravid *S. balanoides* did not liberate the eggs or nauplii. However, if the parents were fed or if they, or the carefully removed egg masses, were treated with hatching substance, hatching invariably took place. The view that eggs that were removed from the mantle cavity and placed in sea water hatched in response to well oxygenated water was thus shown to be a misinterpretation. It was the damage to the parent in removing it from the rock, thus releasing hatching substance, that was the cause of hatching. Eggs carefully and quickly removed and immediately washed in sea water did not hatch. Thus, although hatching resulted from muscular movements of the embryo as Barnes (1955) stated, this was only the proximate cause. It was the hatching substance, acting directly on the embryo in the egg case, that stimulated embryo movement. Ten to fifteen minutes after treatment with hatching substance, almost 90% of the ripe eggs hatched whether *in vivo* or *in vitro*. Without being fed or treated with hatching substance, the eggs would remain in the parental mantle cavity and eventually perish there. However, from these experiments one could not conclude that the active principle extracted from barnacle tissue was the same as that which acted naturally on the egg masses when barnacles resumed feeding in the spring. Extracts of hatching substance certainly stimulated release from live barnacles retaining mature eggs (Crisp 1956), but such a result for the reasons given above might have been an artefact. All that could be rigorously deduced was that the mechanism, by which the eggs remained dormant until planktonic food became plentiful, would prevent release until the hatched larvae were reasonably sure of a food source. Such timing, militating against premature hatching, would be especially valuable for high latitude species. The spring bloom in the Arctic is unpredictable and in the continued presence of ice a temperature change does not always signal the change in biological seasons (Crisp 1987a).

2.2 *Synchrony between release and the spring phytoplankton outburst*

The liberation of nauplii from *S. balanoides* would be expected to follow closely on the advent of the spring bloom of phytoplankton if Crisp's (1956) experiments were applicable under natural conditions.

Barnes (1957, 1962) summarized the evidence from the existing literature. The information from Norway, the Faroes and Iceland was sketchy, and the investigators did not identify the larvae as those of *S. Balanoides*. From the American coast and from Scotland the spring bloom and the liberation of *S. balanoides* were roughly in synchrony and generally occurred later at higher latitudes. However small numbers of *S. balanoides* nauplii may appear very early, even when the main release occurs many weeks later (Evans & Burrows 1987). A strictly quantitative measure, both of the algal population and of the density of larvae, is still needed.

Evidence of synchrony from an experimental approach was obtained from transplant experiments (Crisp 1964). Having shown that the population of *S. balanoides* on the east coast of North America was genetically different, (see also Flowerdew & Crisp 1976, Flowerdew 1983a) had larger eggs and experienced faster embryogenesis than European *S. balanoides*, samples of the American population were transported to the Menai Strait. The local population reached full development several weeks later than their American counterparts, but both liberated nauplii at the same time. This observation shows that hatching does not follow immediately on full development, but is triggered by external conditions, presumably the availability of food to the parent.

Barnes (1957) also described experiments designed to elucidate how liberation was effected at the time of the spring outburst. He suggested several possibilities. He attributed some importance to the breakdown of the egg masses caused by increased feeding and ventilation when the barnacle began to feed and showed that feeding barnacles artificially on diatom cultures during the winter accelerated development and release. He suggested that the molt that occurs at release (Patel & Crisp 1961) might also break down the egg masses. Although Barnes prepared an extract of the prosoma and found that it stimulated hatching, he did not find the extract essential and stated that ripe egg masses will hatch in sea water without it (see also Barnes & Barnes 1963). Unfortunately, nothing was known, at the time of these experiments, of the importance of removing and washing the ripe egg masses quickly with least injury to them or to the parent, lest they become contaminated with hatching substance released from the damaged tissues. For these reasons the importance of the hatching substance was not given the emphasis it deserved. Barnes (1957) suggested, but later withdrew (Barnes & Barnes 1963), that the hatching substance might be derived directly from diatoms and might be present in concentrated media in which *Skeletonema costatum* had been growing.

2.3 *Detection of natural produced hatching substance*

Crisp & Spencer (1958) described a bioassay by means of which they demonstrated that the hatching substance could not only be formed when the barnacle tissue was artificially macerated, but also that it actually played a part in the natural release of nauplii. It was not therefore an artefact, but was the link by which the information acquired by the parent's feeding was transmitted to the offspring and stimulated it to hatch.

The assay method adopted by Crisp & Spencer (1958) relied on the fact that the ability to induce hatching in ripe eggs of *S. balanoides* was concentration dependent. Pairs of carefully removed and washed egg masses were set out in watch glasses. One of each was treated with twice serially diluted fractions of the sample, the other one formed a control with only sea water. The point of extinction of the hatching affect was noted.

Three groups of *S. balanoides* were collected when naupliar liberation was taking place in nature. Two groups were starved for one or two weeks. One group, however, immediately liberated clouds of nauplii. The water was filtered and extracted with isobutanol and the extract was dried and redissolved in a few ml of sea water. Hatching substance was then detected by bioassay. Subsamples of the two starved groups were eventually fed on *Artemia* nauplii, barnacle nauplii, *Phaeodactylum* or raw plankton. The two groups then spawned and hatching substance was again detected by isobutanol extraction of the ambient sea water, followed by bioassay. No detectable hatching substance was found either in the laboratory sea water or in the food controls. Thus the hatching substance must have been synthesized by the barnacle itself and then secreted into the mantle cavity after feeding, thereby awakening the fully developed embryos from a state of dormancy. A recent repeat of this experiment (Song pers. comm.) demonstrated that the hatching substance released by the parent into the sea co-ran on thin layer chromatography (TLC) plates with hatching substance derived from the macerated adult tissue.

Thus, the only significant influences on the time of hatching were: 1) the presence of hatching substance and 2) the uptake of food by the parent. The action of hatching substance *in vitro* was certain and rapid (10-15 min), whereas feeding the adults required varying lengths of time for larval release (i.e. production of hatching factor independent of the kind of food provided).

2.4 *Enzymatic production*

Using the assay developed by Crisp & Spencer (1958), Crisp (1969) measured the amount of hatching substance in barnacle tissue macerates as a function of time. Macerates of standard tissue/sea water ratio were held at room temperature for a given time and quenched in water at 80°C, which stopped all further production of hatching substance. It is important to note that the amount of hatching substance present initially was undetectable, showing that a precursor (not the hatching substance itself) is stored in the barnacle tissue. The yield of hatching substance increased rapidly during the first 15 minutes, with the maximum being formed in less than three hours. Thus when the precursor present in the tissues was made accessible, presumably to an enzyme system, the hatching substance was formed. However, the amount of hatching substance that was extractable depended on the ratio of tissue macerate to sea water. There was a partition coefficient to approximately two in favour of tissue against sea water, with the result that a high dilution or a series of multiple extractions was needed to yield the maximum, amounting to some 200 units per g of tissue. The unit was defined as the amount present in 1 ml of sea water just sufficient to promote hatching. Attempts to generate hatching substance from preboiled tissue by adding very small quantities of living macerate, presumed to contain active enzyme, were unsuccessful.

2.5 *Properties and chemical structure*

Crisp & Spencer (1958) investigated the properties of the hatching substance and found it to be a heat stable, moderately polar lipid that partitioned strongly from water into such solvents as isobutanol and was freely soluble in 50% aqueous acetone, ethanol and diethyl ether. It did not dissolve readily in less polar solvents such as petroleum ether. It passed freely through dialysis tubing and cationic exchange resins, but disappeared on charcoal or anionic exchangers. It was stable to mild oxidizing agents (e.g. potassium ferricyanide) and mild alkali (pH 11), but was destroyed by acid in the region of pH 3 or below. This acid liability was irreversible in alkali and appeared to be a key property in resolving the nature of the substance. Crisp & Spencer (1958) suggested a possible lactone or lactam structure.

Two decades later, interest in the biochemistry of hatching substance was revived. Clare et al. (1982, 1985) provided evidence that the hatching substance was either a prostaglandin or a prostaglandin-related compound. Barnacle extracts were macerated, extracted, separated and purified by TLC. Using a solvent system designed to separate polar lipids, a readily available commercial source of prostaglandins was found that shared the same Rf value of the hatching substance region. This material was a fraction of the gorgonian coral *Plexaura homomalla.* Like the hatching substance it was acid labile, had hatching activity and ran close to prostaglandin standards. However, later studies showed that the prostaglandin extract from the coral and the barnacle hatching substance were not identical.

Clare et al. (1985) investigated the effect of the inhibitors of cyclo-oxygenase activity: aspirin is known to inhibit the conversion of the polyunsaturated fatty acids into prostaglandins. They showed that aspirin inhibited the production of hatching substance in barnacle tissue and this was considered to be strong evidence of its prostaglandin nature. However, aspirin inhibited fully only at a very high concentration (10^{-2}M), whereas indomethacin was not inhibitory. Thus the situation in hatching substance synthesis contrasted strongly with prostaglandin synthesis in vertebrates, though such a difference has been found in other invertebrates (Freas & Grollman 1980). Clare et al. (1985) concluded that hatching substance might be a prostaglandin or a prostaglandin-like substance. Clare (1987) claimed that the hatching substance was a prostaglandin and its precursor in barnacle tissue was arachidonic acid (20:4w6). However, in view of the structure of hatching substance and its more probable formation from the precursor eicosapentaenoic acid 20:5w3 (Holland et al. 1985) this can be refuted. Clare's persuasive argument was based on his experiments and similar experiments reported by Holland et al. (1985) and Holland (1987) in which barnacle macerates were shown to be capable of incorporating ^{14}C-labelled arachidonic acid into a product that co-ran with hatching substance on TLC. Clare (1987) implied that it was identical with hatching substance, but identification solely by TLC must be regarded only as tentative; nor are the postulated enzymes involved in the synthesis of hatching substance specific to given polyunsaturated fatty acids (Holland et al. 1985).

Subsequent work was published by Holland et al. (1985) using gas-liquid chromatography-mass spectroscopy (GC-MS) of the active region from the TLC plates. That study revealed a single major GC peak which was identified as the eicosanoid, 10, 11, 12-trihydroxy-5, 8, 14, 17-eicosatetraenoic acid (Fig. 1). The term eicosanoid includes the prostaglandins and other similar biologically active compounds. Eicosanoid formation involves a cyclic pathway for the

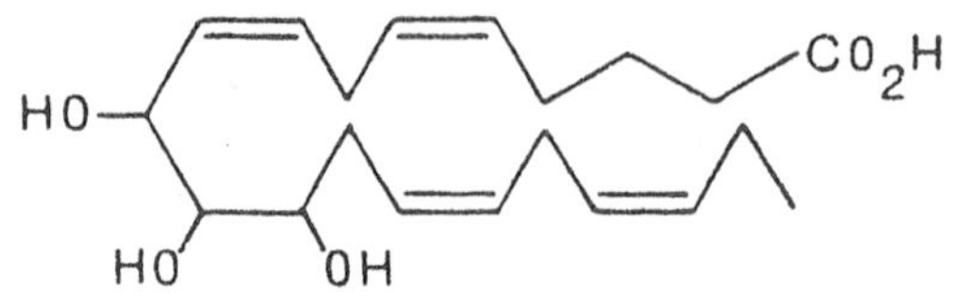

10,11,12-trihydroxy-5,8,14,17-eicosatetraenoic acid

Figure 1. The structure of barnacle hatching substance (from Holland et al. 1985).

formation of prostaglandins and a linear pathway via lipoxygenases for the formation of leukotrienes and other hydroxy fatty acids. Hatching substance is a hydroxy fatty acid produced from the putative precursor 20:5w3 via a linear pathway (Holland et al. 1985). Hatching activity is the first specific physio-logical function to be established for this type of eicosanoid. Hatching substance is a w3 fatty acid. In animal tissues investigated to date the interconversion of w3 and w6 fatty acids does not take place and so hatching substance cannot be formed from the w6 fatty acid, arachidonic acid.

3 HATCHING SUBSTANCE AND RELATED EICOSANOIDS IN OTHER CIRRIPEDES

Tests with ripe egg masses of *S. balanoides* failed to detect the presence of the hatching substance at similar levels in any organism tested other than barnacles (Crisp & Spencer 1958). The sole published exception is that of the gorgonian, *Plexaura homomalla*, where the same or a similar substance may be present (Clare et al. 1985). However, the hatching substance of *S. balanoides* was capable of hatching the eggs of six other sessile barnacles, and in most cases the effect was reciprocal. Only two of eight donor species that were tested failed to provide detectable amounts of hatching substance on maceration and extraction; these were *Balanus perforatus* and *Verruca stroemia*. However Patel & Crisp (1960) were able to induce hatching *in vivo* by applying *S. balanoides* extract to *V. stroemia*, and W.C. Song (pers. comm.) recently found hatching substance in *B. perforatus*.

The tissues in *S. balanoides* that provided hatching substance were found to be primarily those lining the shell. There was less substance found in the prosoma extract and little from the egg masses themselves. The macerates from both well-fed and starved barnacles contained hatching substance, but more in extracts from starved barnacles according to Crisp (1956).

Clare & Walker (1986) investigated the quantity of hatching substance precursor present in *S. balanoides* at different seasons by measuring hatching activity of extracts – the two are not necessarily identical. Although some activity was present all year round (Crisp 1969), the amounts varied widely, being maximal in April at the time of release and minimal in midsummer. Clare & Walker measured the percentage hatch from a standard dilution, but did not provide a quantitative dilution assay (Crisp & Spencer, 1958). As can be seen from their Figure 3 (Clare & Walker 1986), a four times dilution reduced the percentage hatch from 100 to 10% and two fold dilution at the critical region from 98 to 30%. Percentage hatch, therefore, does not bear a linear relationship with concentration and may exaggerate a small reduction in the titre of the hatching substance extract and underestimate it in the high range. The experiment needs to be repeated before any quantitative conclusions can be accepted other than that some degree of seasonal increase occurs in spring and a reduction in summer.

In an analogous survey (Clare & Walker 1986), using barnacles starved since August, a similar increase in hatching activity in standard macerates was observed, but the level was higher than in animals growing and feeding in the sea at that time. A measurement of eicosapentaenoic acid, the putative precursor of hatching substance, as the percentage of the total fatty acids present in barnacles, was obtained by analysis. Eicosapentaenoic acid was found to be present throughout the year, but the results given as a percentage of the variable amounts are difficult to interpret quantitatively. Nevertheless the results suggested that the seasonal variation in total eicosapentaenoic acid was much smaller than the changes based on hatching activity of the macerates. We consider that failure to detect hatching activity in macerates of a given species on isolated occasions should not be regarded as conclusive of its absence.

Clare & Walker (1986) found hatching activity in *Chirona hameri* macerates and complete reciprocity of hatching activity between the European and Pacific races of *S. balanoides* (cf. Flowerdew 1983b). However, extracts of *Lepas anatifera* were not active on its own ripe egg masses, but *Pollicipes* and *Lepas* extracts are active on *S. balanoides* and *Eliminius modestus* eggs (W.C. Song pers. comm.). Thus on the basis of current evidence, it seems that all thoracican barnacles contain hatching substance.

4 IS HATCHING SUBSTANCE NECESSARY?

There is some debate whether hatching substance is necessary to stimulate hatching in all species. It appears to be necessary in *S. balanoides* and in *E. modestus* since egg masses either kept in frequently changed sea water, or not released from unfed adults, eventually die. In nature, one sometimes finds individuals that have oviposited a new batch of eggs overlying the brown leathery remains of the previous brood, implying that insufficient hatching substance had been released to liberate the earlier clutch.

Hatching substance indubitably stimulates hatching of mature embryos in virtually every species that has been adequately investigated. Equally certainly, as the eggs develop up to and beyond maturity, they become increasingly easy to hatch; this fact makes the question of spontaneous hatching difficult to resolve.

Table 1 includes published and unpublished observations that illustrate the disagreement regarding spontaneous hatching in many species. Any mishandling of adults that causes damage when removing the eggs or any failure to wash the egg mass thoroughly and immediately after removal could expose them to hatching substance and cause hatching in the short term.

The eggs themselves yield on maceration small amounts of a hatching substance leaving open the possibility that autolysis, bacterial or fungal attack might release hatching substance. However, Crisp (1959) found fungal attack prevented hatching of eggs *in vitro*.

Although column I (Table 1) provides good evidence of delayed hatching when the parent is starved, only in *S. balanoides* and *E. modestus* have decayed remains of unhatched broods been repeatedly found beneath a new brood – but perhaps only because of lack of sufficient investigation of other species.

Column II (Table 1) lists studies that claim that eggs *in vitro* do not eventually hatch without added hatching substance. Eventually is a long time, but it can be accepted if, as in *Eliminius*, decayed eggs have been observed as the end product. More long term work on this question is clearly needed.

Two possibilities emerge: 1) hatching substance is needed not only in *S. balanoides* and *E. modestus* but in all species, or 2) hatching substance in some species is not required, except at

Table 1. Evidence for delayed and spontaneous hatching of mature eggs in barnacles. D = decayed eggs at end of experiment.

Species	Multiple or annual broods	Delayed hatching in starved adults I	Delayed hatching *in vitro* II	Spontaneous hatching *in vitro* III
Semibalanus balanoides	A	Crisp (1956), Crisp & Spencer (1958) D	Crisp (1959) D	Barnes (1957), Barnes & Barnes (1959), Clare (1984)
Chirona hameri	A		Crisp (1987, pers. comm.), if washed frequently Song (1987 pers. comm.), if washed frequently	Clare (1984)
Balanus balanus	A		Patel & Crisp (1960), reared below 3°C	Patel & Crisp (1960), reared above 3°C
Balanus crenatus	M	Crisp & Spencer (1958), Patel & Crisp (1961)	Patel & Crisp (1960), reared below 6°C	Patel & Crisp (1960), reared above 6°C
Balanus perforatus	M	Patel & Crisp (1960)		Patel & Crisp (1960)
Balanus amphitrite	M	Patel & Crisp (1960)		Patel & Crisp (1960)
Elminius modestus	M	Patel & Crisp (1961) D	Patel & Crisp (1960), reared below 3°C, Biagi (1984, pers. comm.), if washed frequently, D	Clegg (1959, pers. comm.), Crisp (1987, pers. comm.), Patel & Crisp (1960), above 3°C
Chthamalus montagui	M	Patel & Crisp (1960), Hill (1987, pers. comm)	Hill (1987, pers. comm.), if well washed	Patel & Crisp (1960)
Verruca stroemia	M	Patel & Crisp (1960)	Patel & Crisp (1960), if reared below 3°C	Patel & Crisp (1960), above 3°C

normal maturity. In this case hatching may be brought about at first only by hatching substance, but if the threshold required gradually falls, the eggs may eventually hatch spontaneously, a phenomenon known to ethologists as 'overflow activity'. Hatching may be genuinely spontaneous, for even normally mature eggs show spontaneous weak muscular activity in the egg case. Alternatively, the eggs may mobilize their own stimulus from precursors after remaining unhatched for a long period.

In summary, there seems to be no doubt that embryos of cirripedes uniquely have a block to hatching. Of necessity the hatching substance must have co-evolved with the development of this block. All the classical work on delayed hatching was carried out on *S. balanoides*, in which the process is more readily demonstrated without adopting any special techniques. The same process may apply to the other species from high latitudes where the biological seasons are irregular and the timing of larval release more critical. In species from low latitudes, where plankton production extends over a longer period and where species usually have multiple broods, timing is less important. Hence, in such species there would be a tendency for naupliar release to occur more readily, perhaps even spontaneously. Furthermore, retaining mature eggs for too long can result in low viability of offspring because of loss of reserves due to basal respiration (Lucas & Crisp 1987). In warmer waters, such losses would be greater and constitute a greater hazard. Consequently, the hatching substance in temperate or tropical species, or, alternatively, the development of a spontaneous hatch might be valuable in maintaining a more regular succession of young, whatever the amounts of food available, and avoiding the eventual decomposition of the brood in the mantle cavity.

5 MODE OF ACTION AND ROLE OF NEUROTRANSMITTERS

Clare et al. (1985) provided a detailed account of hatching as observed in eggs of *S. balanoides* and recorded it cinematographically. Stage 12 and 13 (fully developed) embryos had already swollen to fill the enlarged egg case (Crisp 1987b). Gut movement preceded spasmodic beats of the limbs against the anterior of the egg case until the case ruptured and the nauplius swam off.

Crisp & Spencer (1958) artificially removed dormant embryos from their egg cases. They remained more or less inactive until treated with hatching substance. If narcotics were applied to the ripe eggs, the hatching substance was ineffective, but became so when the anaesthetic was washed out. Thus the hatching substance stimulated the embryo directly rather than altering the egg case or material binding the eggs together (cf. Barnes 1957).

The association between molting and naupliar liberation (Crisp & Patel 1958, Patel & Crisp 1961), together with an isolated observation implicating tyrosine (Crisp & Spencer 1958) suggested the possible involvement of substances associated with the molting cycle in the hatching process. Neuro-transmitters may also be involved in the hatching process because the embryonic muscles are stimulated. Barnes & Barnes (1959) used as test subjects the ripe eggs of *Balanus nubilus* and investigated several candidates, namely histamine, tryptamine, tyramine and 5-hydroxytryptamine. They claimed that only the last stimulated hatching. Crisp (1969) pointed out that none of these amines could comply with the acid liability of the hatching substance, so that the effect, if any, could only be mimetic. In any case, Barnes' assay was not conclusive since the controls as well as the experimental eggs hatched in significant amounts in a very short time, indicating a spontaneous hatch. Comparative physiological tests (Crisp 1969) proved that 5-hydroxytryptamine was quite different from hatching substance or its action.

Further bioassays on *S. balanoides* eggs have also eliminated the following neurotransmitters as mimics; acetylcholine, 5-hydroxytryptamine and noradrenaline (J. Louttit pers. comm.).

Subsequently, Clare et al. (1985) found dopamine caused hatching. It produced a full hatch when applied at concentrations down to 10^{-4} M and partial hatch down to 10^{-5}M. However, the realted compounds n-acetyl dopamine and DOPA (3,4 dihydroxy-phenylalanine) did not produce a hatch in the usual assay, but DOPA had some effect after one hour. Clare & Walker (1986) suggested the DOPA was converted into dopamine by the embryo and proceeded to investigate a possible role for dopamine. They found that epinine, which acts similarly to dopamine with comparable potency on crustacean (Lingle 1981) and mammalian receptors (Goldberg & Kohli 1981), induced hatching over approximately the same concentration range (10^{-3}-10^{-5}M).

Benserazide is an inhibitor of DOPA decarboxylase that will block the conversion of DOPA into dopamine, a capacity which the embryos were believed to possess. The embryos were thought to synthesize dopamine in response to the hatching substance but continued treatment with 2×10^{-3}M benserazide in no way inhibited the effect of hatching substance. Hence, if dopamine were involved it would have to be released from storage and not newly synthesized from DOPA. To test whether stored dopamine might be utilized in the hatching process, Clare & Walker (1986) employed solutions of tetrabenazine at 10^{-4} and 10^{-6}M. This drug in mammals depletes central nervous system stores of dopamine and other neurotransmitters. Ripe eggs treated with 10^{-4}M tetrabenazine solution completely failed to hatch under the influence of hatching substance, although the sea water control egg masses from the same animal hatched normally. No test of the toxicity of tetrabenazine was reported nor was added dopamine tested. Finally, the nonspecific dopamine antagonist, chlorpromazine, prevented hatching entirely when applied at concentrations down to 5×10^{-4}M and partially down to 10^{-6}M (Clare & Walker 1986). Although most of these tests were carried out using dopamine specific drugs they were used at high, non-physiological, concentrations and the tests are therefore not conclusive (contra Clare 1987).

REFERENCES

Barnes, H. 1953. Orientation and aggregation in *Balanus balanus* (L) Da Costa. *J. Anim. Ecol.* 22: 141-148.

Barnes, H. 1955. The hatching process in some barnacles. *Oikos* 6: 114-123.

Barnes, H. 1957. Processes of restoration and synchronisation in marine ecology. the Spring diatom increase and the 'spawning' of the common barnacle *Balanus balanoides* (L). *Année Biol.* 33: 67-85.

Barnes, H. 1962. Note on the variations in the release of nauplii of *Balanus balanoides* with special reference to the spring diatom outburst. *Crustaceana* 4: 118-122.

Barnes, H. & M. Barnes 1959. Note on the stimulation of cirripede nauplii. *Oikos* 10: 19-23.

Barnes, H. & M. Barnes 1963. *In vitro* development of cirripede eggs. *Vidensk. Medd. fra Dansk naturh. Foren.* 125: 93-100.

Barnes, H. & M. Barnes 1977. Studies on the reproduction of cirripedes ll. Setting of the lamellae; action of protease and disintegration of the oviducal sac. *J. Exp. Mar. Ecol. Biol.* 27: 219-231.

Clare, A.S. 1984. Hormones in cirripedes. Ph.D. Thesis. University of Wales, UK.

Clare, A.S. 1987. Endocrinology of Cirripedes. In A.J. Southwood (ed.), *Barnacle Biology Crustacean Issues* 5: 249-266. Rotterdam: A.A. Balkema.

Clare, A.S. & G. Walker 1986. Further studies on the control of the hatching process in *Balanus balanoides* (L). *J. Exp. Mar. biol. Ecol.* 97: 295-304.

Clare, A.S., G. Walker, D.L. Holland & D.J. Crisp 1982. Barnacle egg hatching: a novel role for a prostaglandin-like compound. *Mar. Biol. Letters* 3: 113-120.

Clare, A.S., G. Walker, D.L. Holland & D.J. Crisp 1985. The hatching substance of the barnacle *Balanus balanoides*. *Proc. Roy. Soc. Lond.* (B)224: 131-147.

Crisp, D.J. 1954. The breeding of *Balanus porcatus* (Da Costa) in the Irish Sea. J. Mar. Biol. Assc. UK. 33: 473-496.

Crisp, D.J. 1956. A substance promoting hatching and liberation of young in cirripedes. *Nature* 178: 263.

Crisp, D.J. 1959. The rate of development of *Balanus balanoides* (L) embryos in vitro. *J. Anim. Ecol.* 28: 119-132.

Crisp, D.J. 1964. Racial differences between North American and European forms of *Balanus balanoides. J. Mar. Biol. Assc. UK.* 44: 33-45.

Crisp, D.J. 1969. Studies of the barnacle hatching substance. *Comp. Biochem. Physiol.* 30: 1037-1048.

Crisp, D.J. 1983. Extending Darwin's investigations on the barnacle life-history. *Biol. J. Linn. soc.* 20: 73-83.

Crisp, D.J. 1987a. A comparison between the reproduction of high and low latitude barnacles, including *Balanus balanoides* and *Tetraclita (Tesseropora) pacifica.* In M.F. Thompson, R. Sarojini & R. Nagabhushanam (eds), *Biology of Benthic Marine Organisms* pp. 69-84. New Dehli: Oxford and IBH Publ. Co.

Crisp, D.J. 1987b. On the sizes and shapes of barnacle eggs. In, *Contributions in Marine Sciences. Dr. S.Z. Qasim Sixtienth Birthday Felicitation*, vol. no. 10, pp. 1-26. Goa, India: Dona Paula.

Crisp, D.J. & P.A. Davies 1955. Observations *in vivo* on the breeding of *Elminius modestus* grown on glass slides. *J. Mar. Biol. Assc. UK.* 34: 357-380.

Crisp, D.J. & E.W. Knight-Jones 1953. The mechanism of aggregation in barnacle populations: A note on recent contribution by Dr H. Barnes. *J. Animal Ecol.* 22: 360-362.

Crisp, D.J. & B.S. Patel 1958. The relationship between breeding and ecdysis in Cirripedes. *Nature* 181: 1078-1079.

Crisp, D.J. & A.J. Southward 1961. Different types of cirral activity in barnacles. *Phil. Trans. Roy. Soc. Lond.* (B)243: 271-308.

Crisp, D.J. & C.P. Spencer 1958. The control of the hatching process in barnacles. *Proc. Roy. Soc. Lond.* (B)148: 278-299.

Evans, F. & M.T. Burrows 1987. The nauplii of *Balanus balanoides* (Cirripedia: Thoracica) taken in January and February in Northumberland waters. *J. Mar. Biol. Assc. UK.* 67: 815-818.

Flowerdew, M.W. 1983a. Electrophoretic investigations of populations of the cirripede *Balanus balanoides* (L) around the North Atlantic seaboard. *Crustaceana* 45: 260-278.

Flowerdew, M.W. 1983b. The circumboreal barnacle *Balanus balanoides* (L) and its subpopulations. In G.S. Oxford & D. Rollinson (eds), *Protein Polymorphism 'Adaptive and Taxonomic Significance'*. pp. 75-84. London: Academic Press.

Flowerdew, M.W. & D.J. Crisp 1976. Esterase heterogeneity and an investigation into racial differences in the Cirripede *Balanus balanoides* using acrylamide gel electrophoresis. *Mar. Biol.* 35: 319-325.

Freas, W. & S. Grollman 1980. Ionic and osmotic influences on prostaglandin release from the gill tissue of a marine bivalve, *Modiolus demissus. J. Exp. Biol.* 84: 169-185.

Goldberg, L.I. & J.D. Kohli 1981. Agonists and antagonists of peripheral and pre- and post- synaptic dopamine receptors: clinical implications. In G.L. Gessa & G.D. Gorsini (eds), *Ampomorphine and Other Dopamino Mimetics. vol. 1. Basic Pharmacology.* pp. 273-284. New York: Raven Press.

Holland, D.L. 1987. Lipid biochemistry of barnacles. In A.J. Southward (ed.), *Barnacle Biology Crustacean Issues* 5: 227-248. Rotterdam: A.A. Balkema.

Holland, D.L., J. East, K.H. Gibson, E. Clayton & A. Oldfield 1985. Identification of the hatching factor of the barnacle *Balanus balanoides* as the novel eicosanoid 10, 11, 12 trihydroxy-5, 8, 14, 17-eicosatetraenoic acid. *Prostaglandins* 29: 1021-1029.

Lingle, C. 1981. The modulatory action of dopamine on crustacean neuromuscular preparations. *J. Exp. Biol.* 14: 285-299.

Lucas, M.I. & D.J. Crisp 1987. Energy metabolism of eggs during embryogenesis in *Balanus balanoides. J. Mar Biol. Assc. UK* 67: 27-54.

Marshall, S.M. & A.B. Orr 1954. Hatching in *Calanus finmarchicus* and some other copepods. *J. Mar. Biol. Assc. UK.* 33: 393-401.

Patel, B. & D.J. Crisp 1960. Rates of development of embryos of several species of barnacles. *Physiol. Zool.* 33: 104-119.

Patel, B. & D.J. Crisp 1961. Relation between the breeding and moulting cycles in cirripedes. *Crustaceana* 2: 89-107.

Walker, G. 1977. Observations by scanning electron microscope (SEM) on the oviducal gland sacs of *Balanus balanoides* at egg laying. *J. Mar. Biol. Assc. UK.* 57: 969-972.

Walley, L.J. 1965. The development and function of the oviducal glands in *Balanus balanoides. J. Mar. Biol. Assc. UK* 45: 115-128.

Walley, L.J. 1969. Studies on the larval structure and metamorphosis of *Balanus balanoides* (L). *Phil. Trans. Roy. Soc. Lond.* (B)256: 237-280.

Walley, L.J., F. White & K.M. Brander 1971. Sperm activitation and fertilisation in *Balanus balanoides. J. Mar. Biol. Assc. UK.* 51: 489-494.

Comparative potential reproduction and actual production in several species of North American crayfish

Susan Corey
Department of Zoology, College of Biological Sciences, University of Guelph, Ontario, Canada

ABSTRACT

The reproductive potential (ovarian counts), realized reproduction (abdominal egg counts) actual production (stage III larval counts) and the reproductive efficiency (percent survival to stage III) vary interspecifically and intraspecifically in many species of North American crayfish. The emphasis here is on *Orconectes propinquus*, *O. rustucus* and *Cambarus robustus*.

The causes of losses in the mean counts at the various stages of development are frequently a combination of biotic and abiotic factors such as: presence of the internal parasite *Crepidostomum cornutum*, size of female, capacity of the abdomen to protect the developmental stages, infertile eggs, lack of proper attachment of eggs to the pleopods, and growth of larvae and river current speed.

1 INTRODUCTION

Freshwater crayfish occur in large numbers, propagate rapidly, and are among the most energetically important benthic invertebrates. Crayfish feed on detritus and in turn are prey for many vertebrates (Crocker & Barr 1968).

The life cycles of most crayfish in the northern United States and Southern Canada consist of females being fertilized in the fall and again in the spring. Soon after the spring mating the females release eggs, which are attached to the pleopods. After 4 to 6 weeks the eggs hatch into Stage I larvae. Stage I, stage II and early stage III larvae remain attached to the pleopods. After a few days the stage III become completely independent from the female (Corey 1987b, Momot 1984).

Reproduction in crayfish is usually expressed as the reproductive potential (the number of ova in the ovaries) and/or realized reproduction (the number of eggs carried on the abdomen) (Smart 1962, Abrahamsson 1971). Realized reproduction related to carapace length has been quantified for *Pacifastacus leniusculus*, *Orconectes limosus*, *O. propinquus* and *Cambarus robustus* (Abrahamsson & Goldman 1970, Mason 1977, Smith 1981, Corey 1987a). The difference between the number of ova in the ovary and the number of eggs carried on the abdomen has been described for *Procambarus clarkii*, *P. hayi*, *Cambarus longulus longulus*, *Orconectes causeyi*, *O. kentuckiensis*, *O. rusticus* and *O. virillis* (Penn 1943, Payne 1971, Smart 1962, Dean 1969, Boyd & Page 1978, Prins 1968, Momot 1967, Weagle & Ozburn 1972,

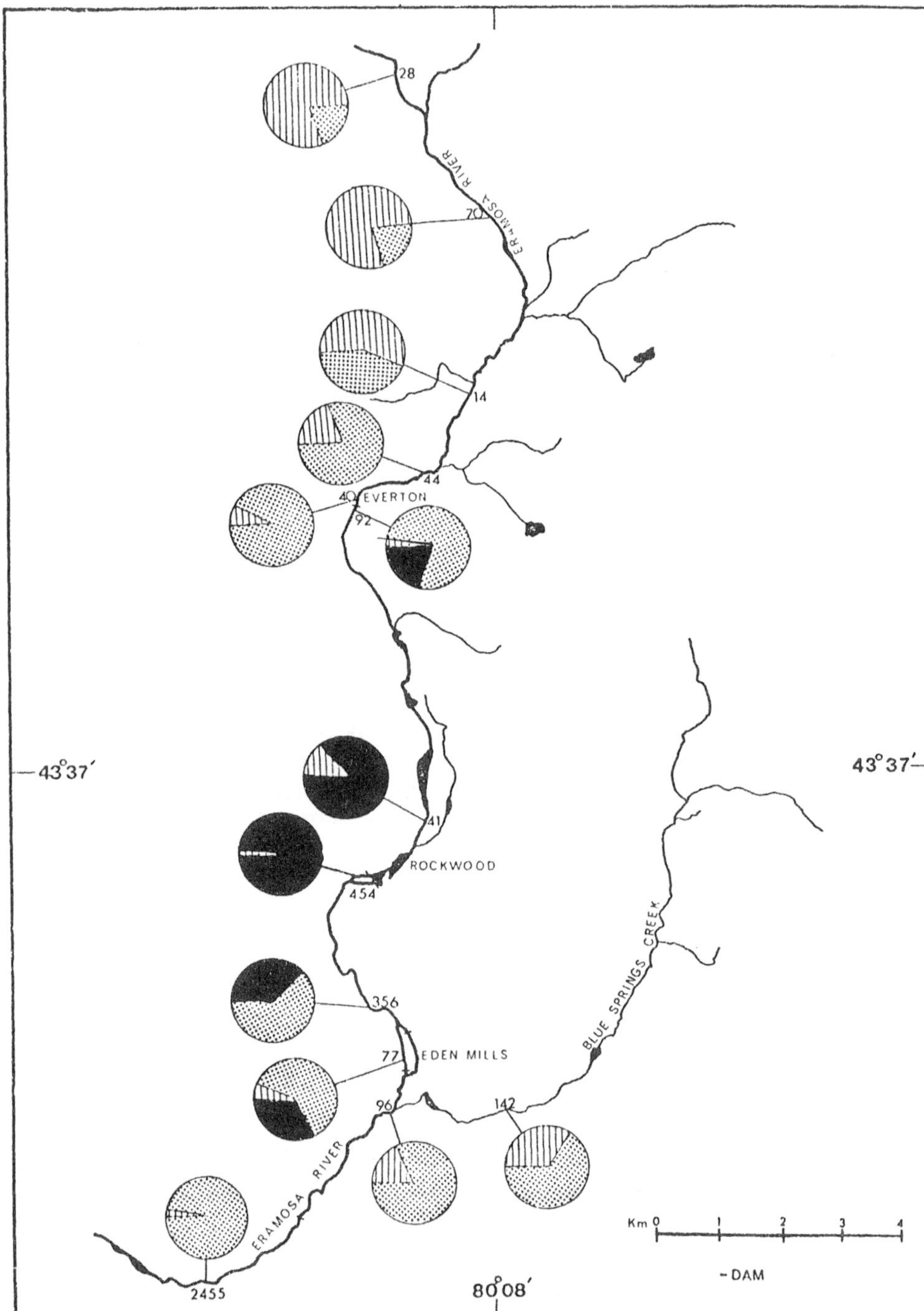

Figure 1. Relative distribution of *Cambarus robustus* (vertical lines), *Orconectes propinquus* (dots) and *O. rusticus* (solid) in the Eramosa River, near Guelph, Ontario.

Momot & Gowing 1977). Differences in realized reproduction between populations of the same species have been noted for *O. virilis* and *O. propinquus* (Momot & Gowing 1977, Corey 1987b). Reproductive efficiency, percentage of brood produced as stage II young, was determined in the laboratory by Mason (1977) for *Pacifastacus leniusculus*.

Young crayfish become truly independent at stage III (Mason 1977). Thus a more accurate measurement of recruitment of young-of-the-year into the population would be production, the number of stage III larvae carried per female. Production has been estimated for different populations of *O. propinquus* (Corey 1988a).

Cambarus robustus occurs in a broad geographic band from the Georgian Bay to the Niagara peninsula in Ontario as well as in New York, Pennsylvania, Ohio and West Virginia (Crocker & Barr 1968). This species is typically found in lakes, large rivers and swiftly flowing streams with stony bottoms; it frequently occurs along with *O. propinquus, O. virilis* or *O. rusticus*, being relatively more abundant in the headwaters and in the central part of rivers and streams (Fig. 1).

During the past few years, *C. robustus, O. propinquus* and *O. rusticus* were extensively collected throughout the Eramosa-Speed river system near Guelph and in the Nottawasaga river northwest of Toronto, Ontario. Interpopulation differences in reproduction and production are compared here with the co-inhabitors *O. propinquus* or *O. rusticus*; they are correlated with certain parameters such as water depth and velocity, presence or absence of internal parasites and duration of the incubation period.

2 MATERIALS AND METHODS

Specimens of *Cambarus robustus* and *Orconectes rusticus* were collected from the Eramosa River; *C. robustus* and *O. propinquus* were obtained from the Speed and Nottawasaga rivers. Non-ovigerous and/or ovigerous females were collected monthly from April through November. Carapace lengths were measured to the nearest 0.1 mm from the base of the orbit to the posterior mid-dorsal margin of the carapace. The reproductive potential was determined by counting the number of ova in the ovaries during the latter stages of development. In the two *Orconectes* species, ova were 1.5 mm in diameter (see Corey 1987a) and in *Cambarus robustus*, when ova were 1.8 mm. The presence or absence of internal parasites was recorded. Ovigerous females were classified according to stage of egg or larval development: 1) realized reproduction or early egg (within 1 week of oviposition) and 2) potential production or late egg stage (within one week of hatching). Larval development was categorized into stage I, stage II and stage III. Stage III larvae were counted only if they were still closely associated with the female.

Since female sizes between samples and between populations of the same species varied, all analyses of numbers of ova, eggs and larvae included the covariate, carapace length.

3 RESULTS AND DISCUSSION

3.1 *Reproductive potential*

There is a positive relationship between the number of mature ova and carapace length of the female in all species of crayfish studied to date (Table 1). The rate of increase of ova (slope) is

Table 1. Comparison of the regression lines of the number of ovarian eggs (y) as related to carapace length (x) in North American crayfish. [1] Eramosa-Speed, [2] Nottawasaga.

Species	Equation	r	(n)	Source
Procambarus hayi	$y = 7.87x + 8.33$	.630	(?)	Payne 1971
Cambarus longulus longulus	$y = 8.63x - 124.07$	.821	(91)	Smart 1962
C. robustus	$y = 5.43x - 30.16$[1]	.391	(14)	Present study
Orconectes rusticus	$y = 8.31x - 91.49$	.860	(49)	Prins 1968
O. propinquus	$y = 12.11x - 125.06$[1]	.829	(27)	Present study
	$y = 10.26x - 17.29$[2]	.607	(22)	Present study

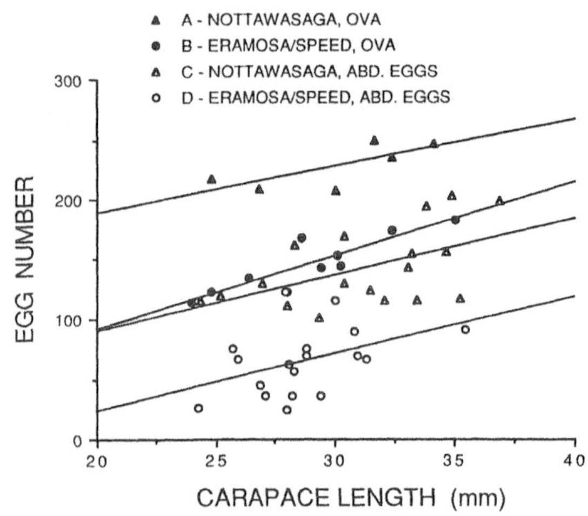

Figure 2. The ova and abdominal egg counts in relation to carapace length for *Cambarus robustus* in the Nottawasaga and Eramosa/Speed Rivers.

highest in *O. propinquus* and lowest in *C. robustus*. This is partly due to the larger ova (2.6 mm diam.) in *C. robustus* compared to *O. propinquus* (2 mm diam.).

The number of ova is approximately double for *C. robustus* and from the Nottawasage river compared to that of Eramosa-Speed river system (Fig. 2). This occurs especially in the younger females. A parasite, *Crepidosomum cornutum*, is very prevalent (80-100%) in the Eramosa-Speed crayfish *O. propinquus*, but absent in this species from the Nottawasaga River. this nematode inhabits the area around the heart and the surface of the gonads. A study on *O. propinquus* from the Sauble river, which has crayfish (of similar sizes) parasitized and non-parasitized in a ratio of 50:50, reveals that female crayfish infected with *C. cornutum* have a significantly ($P = .0001$) lower number of ova (mean 145.9 ± 40.5, $n = 35$) as compared to crayfish with no *C. cornutum* (mean 207.1 ± 61.3, $n = 18$) (Corey 1988a). Corey (1987b) found that separate populations of *O. propinquus* that are not infected have significant differences in reproductive potential. Another cause of differences in reproductive potential could be density of animals, which apparently affect reproductive potential in an *O. virilis* lake population (Momot & Gowing 1977). However, no such affect was found by Corey (1988a) for river populations of *O. propinquus*. Water quality (oxygen and temperature) and food (quality and quantity) may affect the bioenergetic capabilities of the species; energy input may be diverted from growth to reproduction (Momot 1984).

Table 2. Comparison of regression equations of eggs carried on pleopods (y) versus carapace length (x) for several species of North American crayfish. [1] Eramosa-Speed, [2] Nottawasaga.

Species	Equation	r	(n)	Source
Procambarus clarkii	$y = 29.00x - 916.42$	.730	(10)	Penn 1943
P. hayi	$y = 5.91x - 34.31$	.663	(?)	Payne 1971
Cambarus longulus longulus	$y = 8.63x - 124.07$	.821	(?)	Smart 1962
C. shufeldtii	$y = 9.23x - 52.89$	.690	(98)	Penn 1942
C. robustus	$y = 4.98x - 78.86$[1]	.512	(19)	Corey 1987a
	$y = 5.64x - 15.34$[2]	.729	(8)	Present study
Orconectes kentuckiensis	$y = 9.56x - 93.60$	.037	(15)	Boyd & Page 1978
O. propinquus	$y = 6.95x - 59.01$[1]	.639	(89)	Present study
	$y = 10.26x - 17.29$[2]	.607	(22)	Present study
O. rusticus	$y = 8.31x - 91.49$	.860	(49)	Prins 1968
	$y = 12.51x - 135.36$	.736	(52)	Corey 1987a
O. virilis	$y = 12.51x - 239.00$	.894	(16)	Weagle & Ozburn 1972
	$y = 13.53x - 154.02$	.749	(38)	Corey 1987a

3.2 *Actual reproduction and egg loss during oviposition*

Actual reproduction, the number of eggs carried per female, has been most frequently reported in fecundity studies (Table 2). The greatest slope was found in *Procambarus clarkii* and the least slope in *Cambarus robustus*. As was true for the ovarian counts, the abdominal egg counts are almost double in the Nottawasage populations of *C. robustus* and *O. propinquus* compared to the Eramosa-Speed populations of the same species (Corey 1987b). See proposed explanation for the difference for the ovarian egg counts.

All species of crayfish for which data are available show a loss of eggs during oviposition and brooding, i.e. from reproductive potential to realized reproduction (Table 3) which range from 0 to 69.8% (e.g. Penn 1943, Momot 1967, Corey 1988a, b). These losses have been attributed to many factors: strength of river current (Corey 1988a, b), lack of fertilization, lack of proper attachment, certain percentage of ova reabsorbed and size of female (such that, with excess release of eggs, some eggs protrude outside the protective cover of the abdomen and are lost).

Corey (1987b) found an inverse relationship between the number of newly released eggs and the velocity of the river current in *O. propinquus*. Strength of river current, especially during the spring runoff can have two effects. First, the substrate is shifting and causes increased movement of ovigerous females, resulting in a loss of eggs (Prins 1968, Corey 1987b). Second, the current may be so strong as to prevent spring mating which is crucial to the successful fertilization and subsequent attachment of eggs. The lack of spring mating associated with a high number of unfertilized eggs has been noted in *O. rusticus* (Berrill & Arsenault 1982). The unfertilized eggs become covered with fungus and are eaten by the female.

In some species of crayfish, incomplete extrusion followed by resorption of eggs occurs. In *O. propinquus*, the percentage of females actively resorbing ova ranged from 5.0% to 75.0% depending on the population studies. The mean number of eggs reabsorbed per female was 2.5 (with a range of 1 to 55) (Corey 1987a). Corey (1987a) also noted that for rivers with two or more sites the site furthest down river had a higher percentage of females reabsorbing eggs and lower mean number of eggs per female. This loss possibly could be due to river velocity.

Incomplete extrusion did not occur in *O. rusticus* nor in *C. robustus*. *O. rusticus* extrudes

Table 3. Interspecific comparisons of the percentage loss between the number of ova in the ovary and the number of abdominal eggs or stage III young. [1] Eremosa-Speed, [2] Nottawasaga, [3] range of losses due to different years and/or localities, [4] range of losses due to different sizes of females, [5] loss abdominal egg to stage II young, animals held in aquaria.

Species	Percent loss of eggs	Source
Cambarus longulus longulus	Approx. 50[1]	Smart 1962
C. robustus	58.9[1]	Current study
	69.7[2]	Current study
Procambarus clarkii	38.3[2]	Penn 1943
P. hayi	41.3[1]	Payne 1971
Orconectes virilis	42[1]	Momot 1967
O. virilis	0 – 48.5[1,3]	Momot & Gowing 1977
O. rusticus	0 – 30[3]	Prins 1968
O. rusticus	0 – 23.1[1,3]	Reid & Corey (pers. comm.)
	39 to 41[2,3]	
O. propinquus	13.8 – 47.0[1,3]	Corey 1987b
	44.0 – 69.8[2,3]	
O. kentuckiensis	6 – 15[1,4]	Boyd & Page 1978
Pacifastacus leniusculus	Approx. 27[5]	Mason 1977

eggs about one week earlier in the spring than *O. propinquus* (Corey 1988b). However, Maude & Williams (1983) found that *O. rusticus* is more streamlined and resists a current better than *O. propinquus*. On the other hand, *C. robustus* does not extrude eggs until July, when river currents are much slower than in spring.

Older and larger females lose more eggs than younger and smaller females in *O. rusticus* (Prins 1968), *O. kentuckiensis* (Boyd & Page 1978) and *Procambarus hayi* (Payne 1971).

In some females of *C. robustus, O. rusticus* and *O. propinquus*, too many eggs may be extruded; the eggs are inadequately protected by the abdomen and are lost (personal observations).

3.3 *Production and reproductive efficiency*

Production can be the number of stage II (Mason 1977) or the number of early stage III young (Corey 1987b & present study). Prematurely separated young grow more slowly (Mason 1977). Presumably such young would have a higher mortality than young which remain with the female until independence at later stage III.

Stage III larvae were obtained while still dependent on the female (within 3 days of ecdysis-pan stage II) for *O. propinquus* (Corey 1988a) and *O. rusticus* (Corey 1988b). *Cambarus robustus* tends to carry the young for a much longer period. In fact, some females did so from August through the fall and winter. These young became independent in April (Corey in press).

There has been a loss between potential reproduction and actual production in every species studied to date (Table 3). This loss ranged from 38.3% in *P. clarkii* to 69.8% in *O. propinquus*. This mean reproductive efficiency is 30.3%, 60.6%, 61.7%, 65.8% and 73% in *C. robustus, O. rusticus, P. clarkii, O. propinquus* and *P. leniusculus*, respectively. The data for *P. leniusculus* were based on females held in aquaria, whereas, for all other species, the data were based on field collections.

Although every species loses eggs and or young during development, the timing of these losses differ. A complete set of data, i.e. for all stages of development, are available only for *O. propinquus* and *C. robustus* (Corey 1987a, present study).

The greatest percentage loss occurred at oviposition for all populations in rivers with high current velocity. In rivers of low velocity, the greatest losses occurrred between early abdominal egg and stage I or between stages I and II, or between stages II and III. In these latter cases the actual increase in size of the larvae at molt caused many larvae to be dislodged prematurely. When the nine populations of *O. propinquus* are compared, the actual production falls into two significantly different categories. The population parasitized by *C. cornutum* had a mean actual production of 40 young, whereas the non-parasitized populations' mean production varied insignificantly from 60 to 69 young (see Corey 1987b).

The lake population of *O. rusticus* had the greatest loss in numbers between early abdominal eggs and stage III larvae, whereas the river population sustained its greatest loss during oviposition. The increase in size of developmental stages of young and the river current velocity, may respectively account for the differences.

During oviposition *C. robustus* loses about 60% of the eggs. No significant difference in loss occurs between any of the egg or young stages while attached to the pleopods. The mean actual production is 51 young per female. There is no evidence of ovarian resorption; the current is not a factor. Perhaps these apparent losses at oviposition are due to infertility or poor attachment of eggs.

3.4 *Conclusions*

In general, the greatest reproductive losses occur either at oviposition or during growth and molting of attached young.

I postulate that the five main causes for reproductive losses in crayfish are: 1) high river current velocities, 2) parasitism by *Crepidostomum cornutum*, 3) infertile eggs, 4) poor attachment to the pleopods and 5) growth of larval stages during development causing some young to become dislodged.

ACKNOWLEDGEMENTS

The assistance of D. Reid, B. Buckland and M. Vair in collecting is greatly appreciated. The computer analysis was done by D. Reid. This research was made possible by a grant in aid of research from the National Science and Engineering Council of Canada to the author.

REFERENCES

Abrahamsson, S.A.A. 1971. Density, growth and reproduction in populations of *Astacus astacus* and *Pacifastacus leniusculus* in an isolated pond. *Oikos* 22: 373-380.

Abrahamsson, S.A.A. & C.R. Goldman 1970. Distribution, density and reproduction of the crayfish *Pacifastacus leniusculus* Dana in Lake Tahoe, California – Nevada. *Oikos* 21: 83-91.

Berrill, M. & M. Arsenault 1982. Spring breeding of a northern temperate crayfish, *Orconectes rusticus*. *Can. J. Zool.* 60: 2641-2645.

Boyd, J.A. & L.M. Page 1978. The life history of the crayfish *Orconectes kentuckiensis* in Big Creek, Illinois. *Amer. Midl. Nat.* 99: 398-414.

Corey, S. 1987a. Comparative fecundity of four species of crayfish in southwestern Ontario, Canada (Decapoda, Astacidea). *Crustaceana* 52: 276-282.

Corey, S. 1987b. Intraspecific differences in reproductive potential, realized reproduction and actual production in the crayfish *Orconectes propinquus* (Girard 1852) in Ontario. *Amer. Midl. Nat.* 118: 424-432.

Corey, S. 1988a. Comparative life histories of four populations of *Orconectes propinquus* (Girard 1852) in southwestern Ontario. *Crustaceana* 54: 129-138.

Corey, S. 1988b. Comparative life histories of two populations of the introduced crayfish *Orconectes rusticus* (Girard 1852) in Ontario. *Crustaceana* 55: 29-38.

Corey, S. in press. Life history of *Cambarus robustus* in the Eramosa-Speed River system of southwestern, Ontario. *Crustaceana.*

Crocker, D.W. & D.W. Barr 1968. *Handbook of the Crayfishes of Ontario.* University of Toronto Press, Toronto.

Dean, J.L. 1969. Biology of the crayfish *Orconectes causeyi* and its use for control of aquatic weeds in trout lakes. *Bureau Sport Fish. Wildl. Tech. Paper* 24: 3-15.

Mason, J.C. 1970. Maternal-offspring behavior of the crayfish *Pacifastacus trowbridgii* (Stimpson). *Amer. Midl. Nat.* 84: 463-473.

Mason, J.C. 1977. Reproductive efficiency of *Pacifastacus leniusculus* (Dana) in culture. *Freshwater Crayfish* 3: 101-107.

Maude, S.H. & D.D. Williams 1983. Behaviour of crayfish in water currents: hydrodynamics of eight species with reference to their distribution patterns in Southern Ontario. *Can. J. Fish. Aquat. Sci.* 40: 68-77.

Momot, W.T. 1967. Population dynamics and productivity of the crayfish, *Orconectes virilis* in a marl lake. *Amer. Midl. Nat.* 78: 55-81.

Momot, W.T. 1984. Crayfish production: a reflection of community energetics. *J. Crust. Biol.* 4: 35-54.

Momot, W.T. & H. Gowing 1977. Production and population dynamics of the crayfish *Orconectes virilis* in three Michigan lakes. *J. Fish. Res. Board Can.* 34: 2041-2055.

Payne, J.F. 1971. Fecundity studies on the crawfish Procambarus hayi. *Tulane Studies in Zool. Bot.* 17: 35-37.

Penn, G.H. 1942. Observations on the biology of the dwarf crawfish, *Cambarellus shufeldtii* (Faxon). *Amer. Midl. Nat.* 28: 644-647.

Penn, G.H. 1943. A study of the life history of the Louisiana red-crawfish, *Cambarus clarkii* Girard. *Ecol.* 24: 1-18.

Prins, R. 1968. Comparative ecology of the crayfishes *Orconectes rusticus rusticus* and *Cambarus tenebrosus* in Doe Run, Meade County, Kentucky. *Int. Revue ges. Hydrobiol.* 53: 667-714.

Smart, G.C. 1962. The life history of the crayfish *Cambarus longulus longulus*. *Amer. Midl. Nat.* 68: 83-94.

Smith, D.G. 1981. Life history parameters of the crayfish *Orconectes limosus* (Ref) in southern New England. *Ohio J. Sci.* 81: 169-172.

Weagle, K.V. & G.W. Ozburn 1972. Observations on aspects of the life history of the crayfish *Orconectes virilis* (Hagen), in northwestern Ontario. *Can. J. Zool.* 30: 366-370.

Scheduling of reproduction in relation to molting and growth in malacostracan crustaceans

Keith Nelson
Bodega Marine Laboratory, University of California, Bodega Bay, USA

ABSTRACT

Molting (or growth) and reproduction in female crustaceans are often considered in some way to be antagonistic processes. A review of physiology, especially endocrinology, reveals that the interaction between molting and growth cycles in iteroparous malacostracans can range from relative independence in non-brooding types to close synchrony in peracarids. In the latter, secondary vitellogenesis proceeds in step with the molt cycle; both beginning and end of the former, as well as ovulation, are bound tightly to the endocrine events of the molt. The molt in turn is usually delayed by reproduction. In reptants the coupling of oscillations is different, with reproduction and molting alternating; again, molt delay is often consequent upon egg extrusion or brooding.

Scheduling of growth, molting, and reproduction does not appear to be governed by potential competition for resources or by other physiological antagonisms within the female. The patterning instead seems to enhance brood survival – that is, that the brood not be lost with the molt and that the brood hatch with a good prospect for successful development. Resources must be allocated by the long-lived iteroparous female to ensure future survival and enhance future reproduction. The appropriate norm of reaction pattern for a young female in a species with low adult mortality may be a shift in allocation – toward reproduction in times of scarcity and toward growth in times of abundance. If this reaction norm characterizes exploited species, the selection regime imposed by the fishery may be exacerbated.

1 INTRODUCTION

Most malacostracan Crustacea are similar to the apterygote insects and myriapods in that they begin reproduction before somatic growth is completed (Hartnoll 1985, Sareen & Adiyodi 1983). Since continued somatic growth can be accomplished only by means of a periodic shedding of the exoskeleton, special problems of temporal interaction or scheduling may arise beyond those faced by other groups with post-reproductive somatic growth, such as fishes. In this chapter, I survey the variety of molting and reproductive schedules found within the iteroparous Malacostraca of indeterminate growth, discuss the physiological and particularly neuro-endocrinological bases of scheduling, compare the temporal and energetic relationships of molting and reproduction to somatic growth, and finally examine the role of these schedules in the life history of their possessors.

Aspects of scheduling have been addressed in recent reviews (Adiyodi & Adiyodi 1970, Adiyodi & Subramoniam 1983, Adiyodi 1985, Blanchet-Tournier 1982, Botsford 1985, Chang 1985, Chang & O'Connor 1988, Charniaux-Cotton 1985, Charniaux-Cotton & Payen 1988, Conan 1985, Fingerman 1987, Hartnoll 1982, 1983, 1985, Legrand et al. 1982, Lipcius 1985, Meusy & Charniaux-Cotton 1984, Quackenbush 1986, Sastry 1983, Schram 1986, Skinner 1985a, b). These reviews, rather than primary sources, are cited wherever they cover a subject adequately. The field was profoundly influenced by the earlier review of endocrine control of reproduction by the Adiyodis (Adiyodi & Adiyodi 1970). Hartnoll's (1985) recent and excellent review has covered much other material. This chapter will be concerned with seasonal scheduling and environmental cuing only insofar as they modify the phasing of the various processes with respect to one another.

Somatic and reproductive processes are often conceived of as somehow antagonistic or competitive. When the mutual scheduling of these processes is made the focus, inadequacies of that conceptual framework become apparent. A more inclusive interpretation would regard growth, molting and reproduction along with differentiation as coupled processes, mutually interdependent aspects of individual and population development.

2 PROCESSES

The discussion of life-history scheduling is simplified if points can be found to delineate each component process in time, even though they are often essentially continuous processes. There is always a danger that these markers once identified may be reified, but we can hardly speak of the scheduling of these processes without defining at least their beginnings.

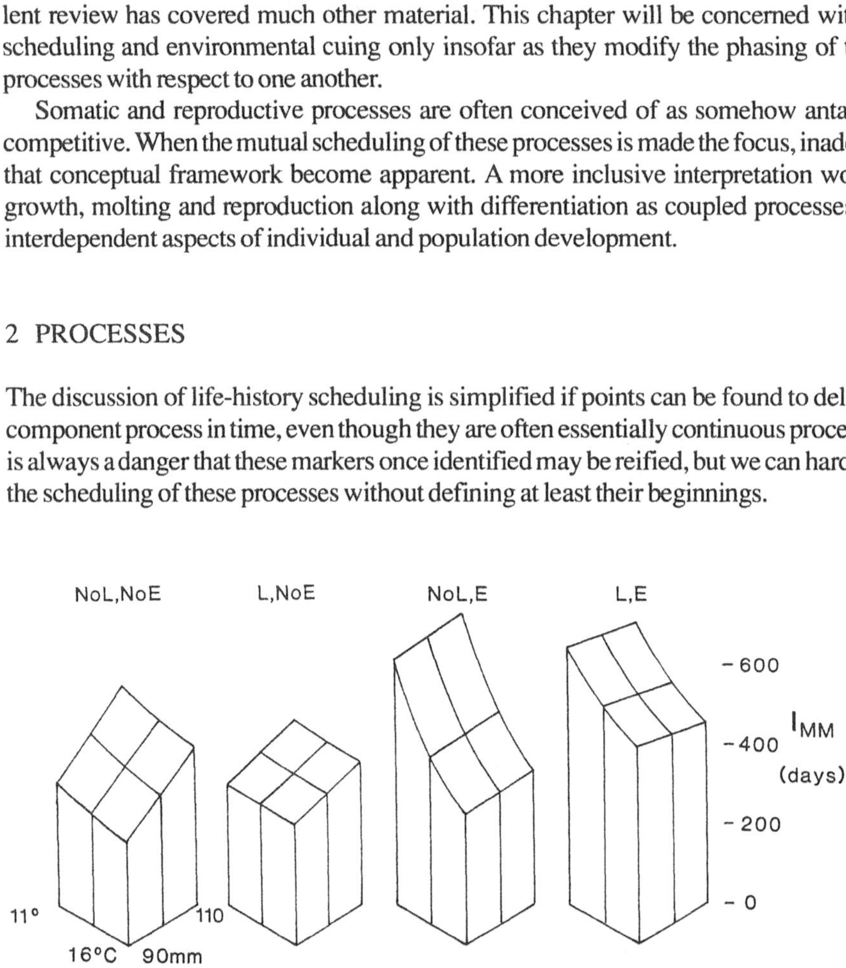

Figure 1. *Homarus americanus* females. Dependence of intermolt interval duration (I_{mm}, ordinate) upon temperature and carapace length, for four molt cycle categories of reproduction. Dependence of I_{mm} on carapace length in the absence of extrusion shifts to dependence upon temperature when extrusion occurs. NoL, NoE: molt cycles with neither long day onset (L) nor extrusion (E). Vitellogenesis not activated. $n = 27$. L, NoE: an L and probably some vitellogenesis, but no E during cycle. $n = 54$. In some, vitellogenesis and E occur immediately in following cycle, of category NoL, E (n = 16). L,E: A delayed L is followed by a delayed E and a delayed molt (See Fig. 5). $n = 63$. Regressions from Nelson et al. (1988a).

2.1 *The molt cycle*

Skinner (1985a, b) recently discussed the organization of the crustacean molt cycle, a sequence of processes usually enumerated in terms of the integumental changes described by Drach (1939). One can distinguish postmolt (metacdysis; Drach stages A-C_3), premolt (proecdysis; D_1-D_4), and a hiatus between the two called intermolt if prolonged (anecdysis; C_4) and diecdysis if transitory (Carlisle & Dohrn 1953; Stage C in amphipods and natant decapods, Passano 1960). The beginning of premolt, Drach stage D_0, is often not easily detected or defined (Passano 1960). Assigning a beginning to the molt cycle is problematic and perhaps unnecessary. Investigators from Carlisle to the present have acknowledged that the beginning of premolt (D_0-D_1) terminates the plateau of intermolt and marks the beginning of the molt cycle. With respect to the interaction of reproduction and growth, however, ecdysis is often a logical time at which to punctuate the continuous flow of events.

The prepuberal and puberal molts of females are usually characterized by morphometric changes signalling reproductive maturity; some of these changes are permanent but others may be reversed at the molt beginning the next anovulatory molt cycle (Hartnoll 1985). In certain terrestrial isopods the changes of the first but not the subsequent parturial molts are dependent upon prior copulation (Mocquard et al. 1976). Coordination of these processes is clearly important.

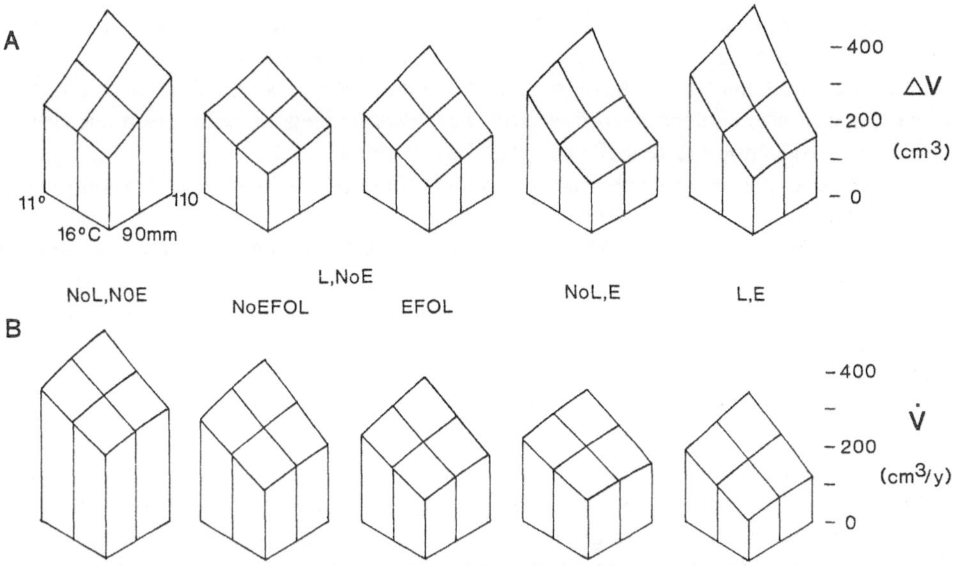

Figure 2. *Homarus americanus*. Dependence of A, volume increment at molt (ΔV) and B, the corresponding instantaneous rate of volume increase ($\dot{V}$) upon temperature and carapace length for five molt cycle reproductive categories. The categories from left to right represent increasing commitment to reproduction. In A, volume increment at ΔV shifts from length dependence to temperature dependence with increasing commitment to reproduction. In B, all categories of commitment share the same relationships of rate of volume increase $\dot{V}$ to length and temperature. NoL, NoE: no vitellogenesis. L, NoE: probably some vitellogenesis but no extrusion (E) in the molt cycle. There are two subcategories, depending upon whether (EFOL) or not (NoEFOL) E occurs early in the next molt cycle, which is of category NoL, E. L,E: delayed extrusion later in the cycle. See Figure 1 for further details. Recalculation from linear regressions in Nelson et al. (1988a).

2.2 *Somatic growth*

Molting and growth are entirely different phenomena, yet molting rate (e.g. the reciprocal of the intermolt interval) and growth rate (e.g. increase in length per unit time) are often used interchangeably. Although in larvae and in juvenile Crustacea, molting rate and growth rate often covary under different thermal, nutritional, and social conditions, in adult female lobsters (*Homarus americanus*) molting rate and growth rate have become uncoupled. Whereas the former still increases with temperature, the latter now decreases (Figs. 1, 2B; Nelson et al. 1988a; Kurata 1962 and Vidal 1980 noted similar phenomena in other crustaceans).

Overall growth rate estimates may yield contradictory results. In animals with indeterminant growth, rate of increase in a linear dimension usually declines monotonically with increasing size (e.g. Hartnoll 1983). Yet this must mean that rate of increase in volume is a convex (dome-shaped) function of size. For example, young adult *Homarus americanus* females display the expected decreasing rate of carapace lengthening with increasing size (Nelson et al. 1988a), while their rate of increase in volume is still increasing, at least below more extreme temperatures (Fig. 2B). Conversely, animals of different sizes with quite different linear growth rates might have identical growth rates in volume or mass; the function relating length growth rate to volume growth rate is not single-valued.

Perhaps in no organism, certainly not in crustaceans, is there a universally satisfactory definition of growth, nor is there a single adequate measure. The periodic shedding of an exoskeleton complicates matters excessively. At the molt there is usually an increment in linear dimension, often an increment in volume, but universally a transitory decrement in dry weight or energy content with the loss of the exuvia (e.g. Heath & Barnes 1970). Growth in certain tissues is often accompanied by degrowth of others. In particular, depot organs such as the midgut gland, the hemolymph, the muscles, and the often diffuse and usually neglected adipose tissue change in size, content, and activity with the metabolic needs of molting, somatic growth, inanition, reproduction, etc. (e.g. Stewart et al. 1967, Durliat 1983).

Even when expressed in absolute terms (grams, kilojoules), each of the various components of somatic growth changes at its own rate, specific to the phase of the molting or vitellogenic cycle (Fig. 3; Read & Caulton 1980). Thus, during a particular phase there is not simply a redistribution of the relative amounts of each component within the various compartments of the body, but rather component-specific change in the rate of appearance or loss of that component in the body as a whole. Similar shifts in component-specific rates of accrual may be expected to occur with temperature and availability of food and over ontogeny as a whole; the patterns, of course, change with the taxon (Heath & Barnes 1970, Pillay & Nair 1973).

Nonetheless, there is general agreement that there is a temporal association of somatic growth with particular phases within the molt cycle. There is no feeding and hence no increment of organic material near the time of the molt, for a period which is brief (D_4-A) in peracarids and natant decapods, and quite prolonged in some of the Brachyura (D_2-B; Scheer 1960). Volume change is usually thought to be limited, by the rigid exoskeleton, to the period at and just after the molt. Copious amounts of water are taken up at the molt; when feeding begins again and tissue growth resumes this water is usually considered to be gradually replaced by organic matter. As the molt cycle progresses, rapid tissue growth in stage C is thought to be superceded by storage of organic materials in the midgut gland and elsewhere. These reserves would then be mobilized for energy and tissue formation during the aphagic period near the following molt (Passano 1960, Gibson & Barker 1979). But in natant decapods and peracarids, intermolt growth in fresh mass or volume (Read & Caulton 1980) and even in length (Mauchline 1973)

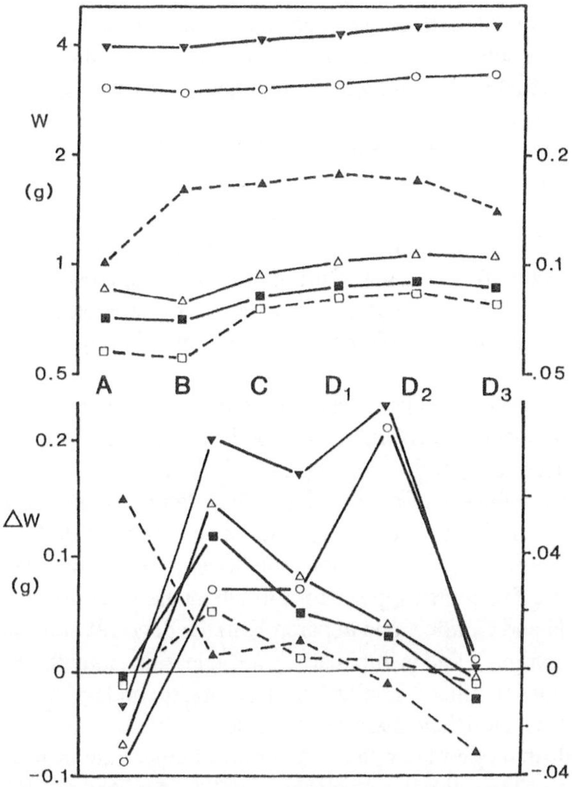

Figure 3. *Penaeus indicus*. (Above) absolute mass in g of constituents in molt cycle stages. Solid lines, left ordinate: inverted triangles, fresh mass; circles, water; open triangles, organic content; solid squares, protein. Dashed lines, right ordinate: solid triangles, ash; open squares, lipid. (Below) changes in absolute mass in grams from stage to stage. Symbols as above. Redrawn from Read & Caulton (1980).

may be considerable. Following actual loss from stage A to B, water content may increase over the rest of the molt cycle, but especially during the latter half (open circles, Fig. 3; Read & Caulton 1980). Even in some brachyurans, there may be little depletion of stored midgut gland material over the molt (stage D_2 to A_2), and the major accretion in the dry weight of the midgut gland may occur with general tissue growth in the earlier substages of stage C (Heath & Barnes 1970). Furthermore, the distinction between 'tissue' and 'organic reserves' is not absolute, and muscle protein for example is mobilized readily if need be (Stewart et al. 1967).

It is possible that no tissue is exempt from a role as a storage site. Overall somatic growth may be much more continuous than its components. Even if we restrict our definition to exclude accretion of reserves, it may be difficult to find a period in the molt cycle in which no tissue is growing. The focus of growth may be now on the muscle, hemolymph or connective tissue; but at another time, during premolt, it shifts to the epidermis. With the techniques of biochemical analysis 'somatic growth' may be found to be confined to the substages of stage C (Renaud 1949, Passano 1960); if attention shifts to the effects of limb or eyestalk ablation, 'somatic growth' only begins with premolt (Démeusy 1965, Kurup & Adiyodi 1981, Skinner 1985a, b).

2.3 *The female reproductive cycle*

2.3.1 *Vitellogenesis*

In this paper first reproductive maturity is regarded as a primarily ovarian event and for most purposes is not differentiated from subsequent maturations. Size and age at first reproduction are regarded as an important but separate subject and are not discussed here. I shall take the interaction of molting, growth and reproduction as beginning with first reproductive maturity, although I know better.

The processes of oöcyte maturation have been reviewed by Adiyodi & Subramoniam (1983), Adiyodi (1985), Charniaux-Cotton (1985) and Meusy & Charniaux-Cotton (1984). Following proliferation in the germinative zone, oögonia leave this zone and may enter a resting phase as primary oöcytes. Cohorts of primary oöcytes are then somehow selected to undergo further development. Vitellogenesis occurs in two stages. During primary vitellogenesis the lipoproteins and glycoproteins which appear in the oöcytes seem to be autochthonous, that is, synthesized within the oöcyte itself. At the end of this phase the cohorts of oöcytes may accumulate at a second plateau.

In peracarids and carideans at least, secondary vitellogenesis is marked by the appearance of secondary follicle cells completely surrounding each oöcyte of a secondary cohort. The oöcytes in secondary vitellogenesis are characterized by endocytotic uptake of allochthonous vitellogenin, transported in the hemolymph from the site of manufacture. The relative contributions of autochthonous and allochthonous material to the oöcyte lipovitellin has been debated and may vary between groups; however, sources of allochthonous vitellogenin have now been identified in all major malacostracan taxa. The source appears to be in subcutaneous connective tissue ('fat body') at least in Peracarida and Caridea, and apparently in the dendrobranchiate shrimp as well (Tom et al. 1987). The midgut gland is implicated also, at least in stomatopods (Marzari & Ferrero 1986) and brachyurans (Paulus & Laufer 1987). During secondary vitellogenesis the oöcytes rapidly enlarge and develop their characteristic color.

After secondary vitellogenesis and final egg maturation the cohort of eggs may in some cases again enter a plateau, awaiting some final stimulus (such as copulation; McMullen 1969, Blanchet-Tournier 1982) for ovulation and extrusion. In the dendrobranchiates the follicle cells are shed with the eggs (W.H. Clark Jr. pers. comm.); in other taxa studied they appear to be retained within the ovary at least for a time.

Thus for each cohort of oöcytes there appear to be three salient points, corresponding to the termination of three more or less brief plateaus: the beginning of primary vitellogenesis, the beginning of secondary vitellogenesis and ovulation and extrusion. As with the molt cycle, the reproductive cycle of the ovary has the general appearance of a relaxation oscillation: following a slow buildup there is a sudden release, followed by a recovery period.

2.3.2 *Brooding*

Care of the embryos appears to be absent in Syncarida, Euphausiacea, and most dendrobranchiate decapods (Schram 1986). In all other malacostracans, some form of brooding arrangement is developed typically at a parturial molt. In stomatopods such as *Squilla* the egg mass is held by thoracopods 3-5 (Kaestner 1970), by more posterior thoracopods in the brooding dendrobranchiates such as *Lucifer* (Schram 1986). In the remaining decapods the embryos are attached to long pleopodal setae. Other malacostracans including the phyllocaridan *Nebalia* develop one form or another of brood pouch.

Even in the non-brooding types the eggs are rather large. In the euphausiids and dendrobranchiates, the eggs hatch as non-feeding nauplii and only begin to feed after several molts; in other malacostracans naupliar and often later stages are passed in the egg (references in Newman 1983). In peracarids, particularly in isopods, tanaids, and cumaceans, the young are retained in the brood pouch for a time after hatching (Kaestner 1970). We shall see that release of the young, and not necessarily hatching, is usually the salient event in scheduling interactions.

3 SCHEDULES

3.1 *Life histories*

Hartnoll (1985) divided the great variety of lifetime schedules of molting, growth and reproduction in female crustaceans initially according to whether growth is determinate or indeterminate, that is whether molting continues indefinitely past maturity or 'ceases definitively' at some point, with a clear terminal anecdysis. In this chapter, our interest lies with the relation between multiple brood production or iteroparity and continued post-maturational molting and growth. Neither the timing of first reproductive maturity per se nor reproduction past a terminal molt will concern us except by way of contrast. Hartnoll (1985) demonstrated that there were striking correlations of variety of schedule types with taxonomic grouping, and I take this as a point of departure. Restricted as it is to the Malacostraca that continue to molt and reproduce our classification of schedule types must depart from that of Hartnoll, to whom it nevertheless owes much.

3.2 *The non-brooding Eucarida*

Brooding of embryos is found throughout the Malacostraca except in the poorly-known Syncarida and two groups of Eucarida, the Euphausiacea and the dendrobranchiate shrimps (penaeoids, some sergestoids; Schram 1986).

In the eucarid groups rapid increase in size of oöcytes (secondary vitellogenesis) may begin very soon after the molt (Fig. 4A; Aquacop 1977, Emmerson 1980). Eggs are usually shed freely and soon hatch and pass through several non-feeding naupliar instars before arriving at the feeding protozoeal stage. There may be more than one oöcyte maturation and spawning within a molt cycle (Aquacop 1977, Emmerson 1980; in euphausids it is not clear whether several complete vitellogenic cycles may occur within an instar, or whether instead groups of eggs of the same cohort successively undergo final development and release (Harrington & Ikeda 1986). In *Penaeus indicus* spawning occurs about equally frequently in intermolt (stage C) and early premolt (D_0-D_1), with a few in stage D_2; in later premolt vitellogenesis may not begin (Fig. 4A), or if begun the oöcytes may be resorbed (Fig. 4B; Emmerson 1980).

Euphausiid and peneoid females are similar in possessing a thelycum or midventral seminal receptacle. Within the genus *Penaeus* the species can be divided into two groups according to the timing of copulation (Aquacop 1977). In the closed-thelycum group including *P. indicus, P. merguiensis, P. aztecus, P. monodon* and *P. japonicus*, copulation occurs immediately after molting, when no development of the gonads is visible (Fig. 4A). Spermatophores are deposited deep within the soft thelycum. In species with an open thelycum such as *P. vannamei* and *P. stylirostris*, spermatophores are applied superficially and only a few hours before spawning (Fig. 4B); the parts containing the sperm overlap the opening of the oviducts.

3.3 *Vitellogenesis and the molt in brooding malacostracans*

In the stomatopods, the clutch is a ball of eggs continuously manipulated by the last three

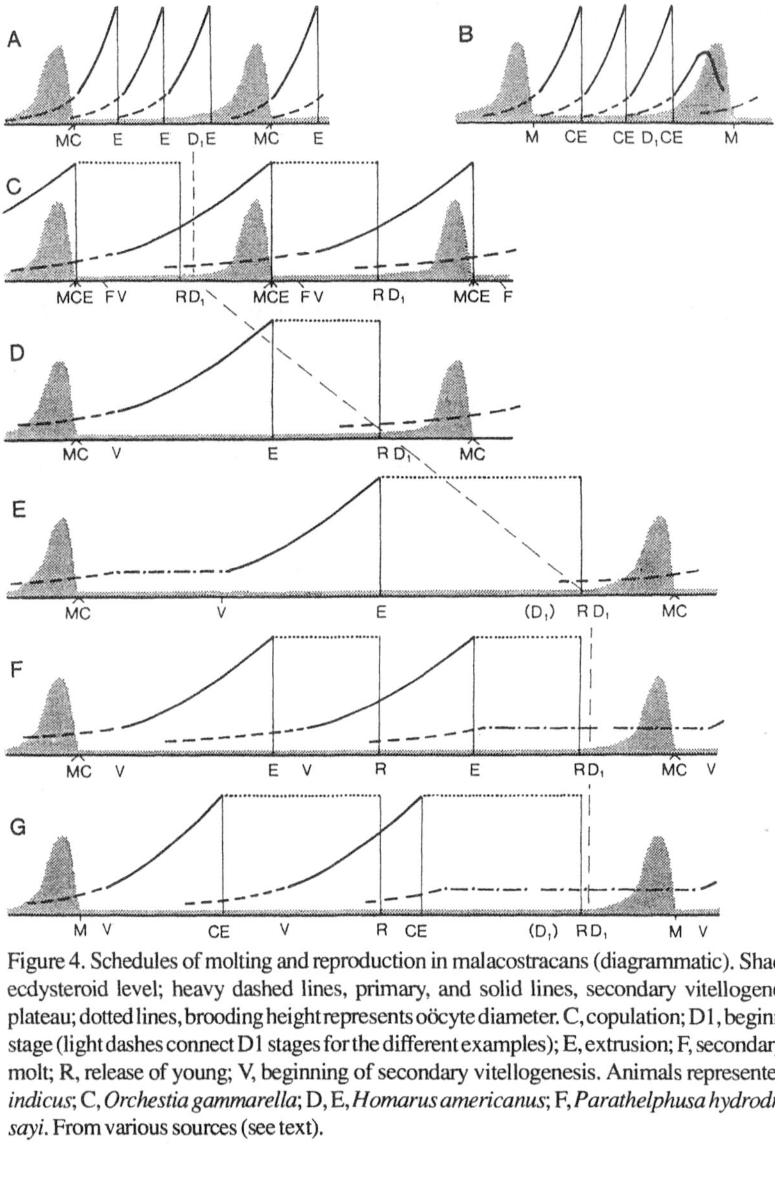

Figure 4. Schedules of molting and reproduction in malacostracans (diagrammatic). Shaded area, hemolymph ecdysteroid level; heavy dashed lines, primary, and solid lines, secondary vitellogenesis; dot-dashed line, plateau; dotted lines, brooding height represents oöcyte diameter. C, copulation; D1, beginning of that molt cycle stage (light dashes connect D1 stages for the different examples); E, extrusion; F, secondary folliculogenesis; M, molt; R, release of young; V, beginning of secondary vitellogenesis. Animals represented are Ab, B, *Penaeus indicus*; C, *Orchestia gammarella*; D, E, *Homarus americanus*; F, *Parathelphusa hydrodromus*; G, *Neopanope sayi*. From various sources (see text).

maxillipedes of the female; she may set it aside to engage in shelter defense or, presumably, to molt. Neither copulation nor ovulation have a noticeable relationship to the molt cycle; the female becomes receptive as the ovaries ripen, and a recently-molted *Gonodactylus* has been found with a clutch (Dingle & Caldwell 1972, Reaka 1976). In the other brooding malacostracans there is a danger that a premature molt will result in loss of the clutch with the exuvia. Coordination between the molting and reproductive cycles would then appear to be essential.

3.3.1 *Synchronization in diecdysic groups*

In most peracaridans, carideans and anomurans, the solution (in part) to the coordination problem was to position extrusion of the clutch as early in the molt cycle as possible, to molt stage A or B (Fig. 4C; Wallace et al. 1949, Pike & Williamson 1959, Cox & Dudley 1968, Kaestner 1970, Bückle-Ramirez 1965). In this pattern the molt lies between secondary vitellogenesis and extrusion. In the amphipod *Orchestia gammarella* (Charniaux-Cotton 1974) and the carideans *Macrobrachium rosenbergii* (Fauvel 1981) and *Palaemon serratus* (Faure et al. 1981), secondary folliculogenesis also occurs very early, in molt stage B. Hence the secondary vitellogenesis occupies virtually the entire molt cycle in *Orchestia* (Fig. 4C), or continues through two or more successive molt cycles in some carideans.

In mysids, amphipods, terrestrial isopods and carideans, primary vitellogenesis of the next cohort of oöcytes can occur simultaneously with secondary vitellogenesis of the current cohort (Fig. 4C), and thus successive cohorts can be brooded in successive molt cycles. In cumaceans, tanaids and aquatic isopods parturial molt cycles alternate obligatorily with non-parturial cycles and primary and secondary vitellogeneses do not co-occur (Hartnoll 1985).

In all these groups, other reproductive events are often closely tied to molt cycle events. In mysids, tanaids, many amphipods, and carideans, copulation occurs only between molting and ovulation (Kaestner 1970, Bückle-Ramirez 1965, Jefferies 1964, etc. Amphipods, like isopods, have internal fertilization and hence some at least may have the potential for sperm carryover beyond the molt). In most aquatic isopods and in the Tylidae among terrestrial isopods, copulation similarly occurs only between molting and ovulation; in *Asellus aquaticus* and others copulation must occur during the brief interval after the posterior pre-parturial exuvia is cast but prior to the anterior molt (Kaestner 1970). In other terrestrial isopods initial copulation may occur variously during preparturial or even earlier instars (see Jassem et al. 1982). Similar variation exists among anomurans (Kaestner 1970).

In some groups eclosion, or alternatively release of the young, is keyed to events of the maternal molt cycle. In *Orchestia gammarella*, eclosion of the brooded young occurs at molt stage D_0, but they remain in the marsupium for a few days before leaving (Blanchet-Tournier 1982). In the terrestrial isopods, the parturial molt cycle is prolonged; the brooded hatchlings leave at the end of molt stage C (Mocquard et al. 1976, Steel 1980). In carideans, the parturial molt cycle is often prolonged but (as discussed below) the relationship to brooding and hatching is often more subtle and variable.

In certain mysids (Clutter & Theilacker 1971, Wittmann 1981, Cuzin-Roudy & Tchernigovtzeff 1985) synchronization of molting and reproduction is perhaps complete. Release of the young and their juvenile molt occur at the time of the maternal molt. Thus in such species primary vitellogenesis, secondary folliculogenesis, secondary vitellogenesis, copulation, extrusion, brooding and a critical developmental stage of the young are all keyed to the timing of ecdysis (Cuzin-Roudy & Saleuddin 1985).

3.3.2 *Asynchronous schedules in reptants*

A sufficiently long parturial molt cycle is another means of ensuring that the clutch is not lost with the exuvia. Such a 'strategy' is already available to all crustaceans whose intermolt intervals increase in duration with age or size (Kurata 1962, Mauchline 1977, Hartnoll 1982), to the extent that puberty can be postponed until a later molt. In such cases parturial molt cycles are not necessarily longer than those of similar-sized non-ovigerous females. Such is the case in the sand crab *Emerita analoga* (Siegel 1984). Some of the confusion surrounding the duration of parturial molt cycles and its control (e.g. Kamiguchi 1971a) may result from inattention to size-related variation in duration.

'True' molt postponement occurs in addition to preecdysial vitellogenesis in some species of the peracarid and caridean groups of the previous section (e.g. amphipods, see Blanchet-Tournier 1982, isopods, Mocquard et al. 1976, Steel 1980, mysids, Wittmann 1981, carideans, Hess 1941, Jefferies 1964, Ranga Rao & Inkel 1975, Berglund & Rosenqvist 1986). In the antarctic isopod *Serolis polita* incubation endures for 20 months and the parturial molt cycle for 24 (Luxmoor 1985). In these groups no one has described more than a single incubation within an instar. Note that demonstration of post-parturial molt postponement alone is often insufficient to assign to it a likely function, especially in forms with successive synchronous molting and secondary vitellogenesis cycles. Thus in *Oniscus asellus* exposed to long days, the preparturial premolt is lengthened as well, evidently to accommodate secondary vitellogenesis (Steel 1980). In the following parturial molt cycle the increase in duration is not demonstrable in premolt but rather in intermolt (stage C). Steel could only establish the relationship of the increase to brooding by further experimentation (see next section).

It is in the reptant decapods that molt postponement becomes the major insurance of successful brooding. The variety of patterning is bewildering. In contrast to peracarids and carideans, preecdysial vitellogenesis is rare. A certain amount of egg development prior to the parturial molt has been observed in a number of crabs (Hartnoll 1963, 1965, Kuris 1971, Swartz 1978), and even fully-developed ovaries in pre- or postmolt have occasionally been noted (Vernet-Cornubert 1958, Cheung 1966, Adiyodi 1968). But complete preparturial molt cycle vitellogenesis with extrusion soon after the molt seems to be the rule only in certain anomurans (Wallace et al. 1949, Pike & Williamson 1959, Cox & Dudley 1968) and at the puberal (and terminal) molt of the majid *Hyas coarcticus* (Hartnoll 1963).

Hartnoll (1985: Table 1) noted that within the category of 'determinate growth, only one post-puberty instar' the malacostracan taxa are clearly split between semelparous (only one egg-laying in final instar) and iteroparous species. Whereas those decapods (all Brachyura) that fall in the category are all iteroparous, the peracarids are all semelparous: mysids, amphipods, isopods, and one might add the 'effectively semelparous' cumaceans (Corey 1981). Iteroparity confined to a terminal anecdysis (C_4T, Hiatt 1948) represents the ultimate in postparturial molt postponement, and is achieved in no peracarid.

Carlisle (1957) made a distinction between C_4T's maintained by secretion of the eyestalk molt-inhibiting hormone (MIH; see next section), as in *Carcinus maenas*, and those characterized (as in the majid *Maja squinado*) by reduction in the size or secretory activity of both the MIH source and the Y-organ, which produces molting hormone (ECD). Kuris (1971) concluded that evidence for true C_4T–in particular, absence of limb bud regeneration–is confined to a few brachyuran families, notably the Majidae and a few others. He discounted reports of C_4T in Grapsidae, Xanthidae and in the portunid *Carcinus maenas*. Skinner (1985b) cautions that age of epibionts and other evidence of extended absence of molting, even upon multiple limb autotomy, and even basal limb bud calcification itself, may not be indicative of C_4T.

A more finite molt postponement characterizes most reptants. Secondary vitellogenesis is arrested during premolt and often begins in stage B or early substages of C (Fig. 4D; e.g. Cheung 1966, Adiyodi 1968, Kuris 1971, Nelson et al. 1988c). Environmental conditions (temperature or photoperiod) may postpone secondary vitellogenesis and extrusion for several months following the molt; in *Homarus americanus* the following molt is postponed accordingly (Figs. 4E, 5A; Nelson et al. 1988a). With or without post-molt delay, extrusion generally occurs during anecdysis, molt stage C_4 (Cheung 1966, Kuris 1971, Aiken & Waddy 1980), although in *Parathelphusa hydrodromus* (On the identity of *Parathelphusa hydrodromus*: cf. Bliss & Mantel 1985, p. 531, footnote) the first extrusion in a molt cycle happens in C_3 (Adiyodi 1968). Whereas in the peracaridans and natant decapods premolt occupies close to half the molt cycle

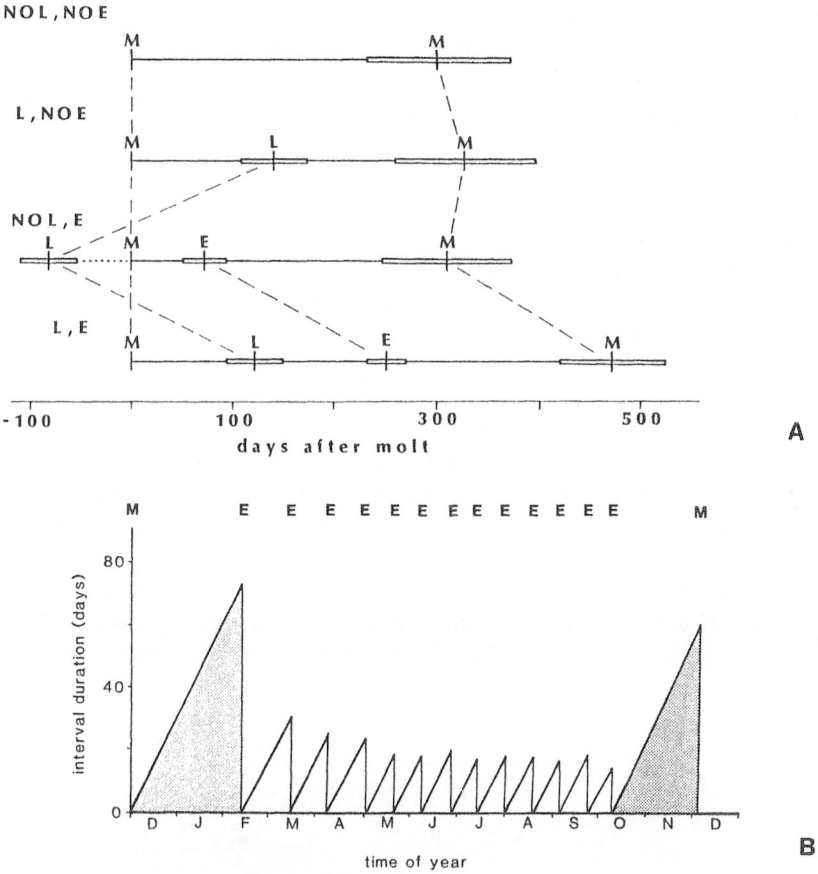

Figure 5. A. *Homarus americanus* females. Long day onset (L) results in delay of molt (M) only when followed in same molt cycle by extrusion (E). The mean time between the initial M and the event indicated is shown by a short vertical line. Bars are two standard deviations wide. Events belonging to the same class are connected by dashed lines. Four categories of molt cycle represent different conditions of reproduction. NoL, NoE: long days throughout previous and current molt cycles, no vitellogenesis or E. L, NoE: L, possibly followed by some vitellogenesis, but no extrusion in molt cycle. NoL, E: L late in the previous cycle elicits vitellogenesis and E early in the current one. L, E: A delayed L elicits a delayed E followed in turn by a delayed M. From Nelson et al. (1988a). B. *Menippe mercenaria*. 'Cheung's champion': an exceptionally long molt cycle (slightly greater than one year in duration) containing a total of 13 vitellogeneses. Molt-extrusion and extrusion-molt intervals are differentially shaded. Data from Cheung (1969).

(Fig. 4A-C), in the reptant decapods it may occupy only a small fraction of that; most of the molt cycle is now taken up by the intercalated anecdysis (Fig. 4D-G; Carlisle 1960, Adiyodi 1985).

Typically only one cohort of eggs may be incubated during a molt cycle but there may be as many as 13 broods (Fig. 5B; Cheung 1969, Hartnoll 1985). Multiple broods within an instar are more common in certain groups than in others (Kuris 1971, Hartnoll 1985); the number may vary with season (Kuris 1971) or with age, as when molt cycle duration is size-dependent (Démeusy 1963, Lipcius 1985, Waddy & Aiken 1986). A subsequent secondary vitellogenesis must await extrusion (Fig. 4F); the next cohort of oöcytes then can move into the space vacated by the earlier brood (Ryan 1967). Ryan's description of this in *Portunus sanguinolentis* suggests that the septa left by the evacuated eggs become the follicles of the next oöcyte generation, as in the caridean *Macrobrachium rosenbergii* (Fauvel 1981). In some species the next secondary vitellogenesis may be postponed even longer, in the case of *P. hydrodromus* until after hatching of the first cohort (Adiyodi 1968). Alternatively, another extrusion may follow shortly after hatching, as in xanthids (Fig. 4G; Swartz 1978).

In reptants, premolt may begin either sometime during or after the final incubation in a molt cycle as in *Homarus americanus* (Fig. 4e; Aiken & Waddy 1980, Nelson et al. 1988a, unpubl. observ.) and *Neopanope sayi* (Fig. 4G; Swartz 1978), or only near release of the young, as appears to be the case in the crayfish *Cambarus propinquus* (Fig. 4D; Scudamore 1948) and possibly *P. hydrodromus* (Fig. 4F; Adiyodi 1968).

The schedules of mating in the reptants (reviewed by Hartnoll 1969, 1985) are equally various. The 'typical' pattern is one of post-ecdysial mating, found in at least some species of each infraorder and often associated with a consort relationship in which the male protects the soft female. Strict confinement of mating to the post-ecdysial period may not be as common as once thought; in a frequent variant most mating occurs during early postmolt, but hard-shelled females may mate as well. Thus the female puberal molt and mating in *Maja squinado* occur in the center of large mounds of crabs, containing hard mature males as well as immature males and females, but hard-shelled females have also been observed to mate. Mating of hard-shelled females may occur practically anywhere in the molt cycle, as in *Neopanope sayi* (Swartz 1978), or it may be restricted to a short period, as in those brachyurans in which decalcification of the

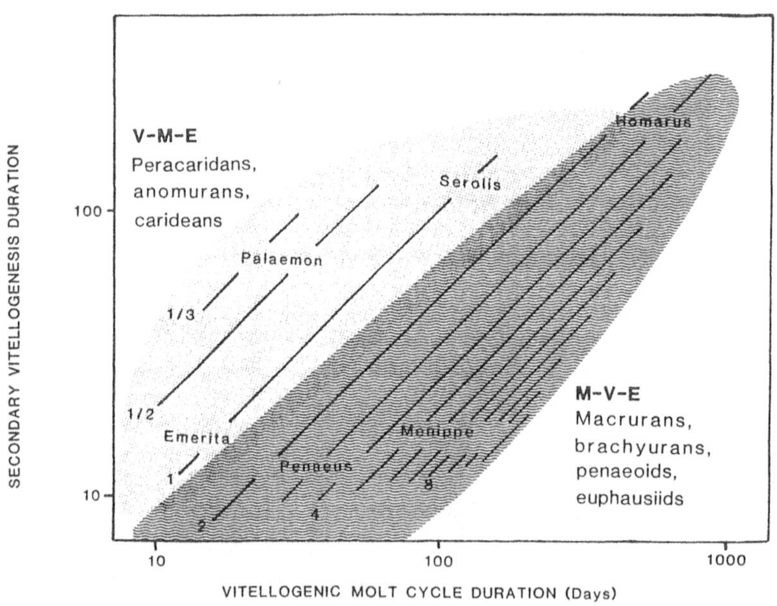

Figure 6. Duration of secondary vitellogenesis and the vitellogenic molt cycle in synchronous (Vitellogenesis-Molt-Extrusion, light shading) and asynchronous reproducers (Molt-Vitellogenesis-Extrusion, heavy shading). Lines indicate ratios of intermolt interval duration to the contained secondary vitellogenesis duration, as indicated. Shown are a few more extreme examples discussed in the text, from multiple-molt-cycle vitellogeneses (*Palaemon*) to multiple-vitellogenesis molt cycles (*Menippe*, see Fig. 5B).

hinge on the genital pore operculum is confined to the few days prior to egg extrusion. In the majid *Libinia emarginata*, mating often occurs at the time of release of the young, when it is associated with 'obstetrical' behavior consisting of carrying of the female by the male during eclosion (Hinsch 1968).

Figure 6 contrasts the synchronous (V-M-E) and asynchronous (M-V-E) schedule categories, in terms of the relationship between molt cycle duration and duration of secondary vitellogenesis. The categories appear to overlap very little.

4 CONTROLS

The malacostracan endocrine system is characterized by many neuro-secretory principles, elaborated by various groups of central nervous system cells, some of which are known to affect aspects of molting, growth, and reproduction in part via more peripheral endocrine organs (Fig. 7). The subject as a whole has been reviewed recently by Fingerman (1987).

4.1 *Hormones and molting*

Preparation for ecdysis in most groups of malacostracans is known to be associated with the

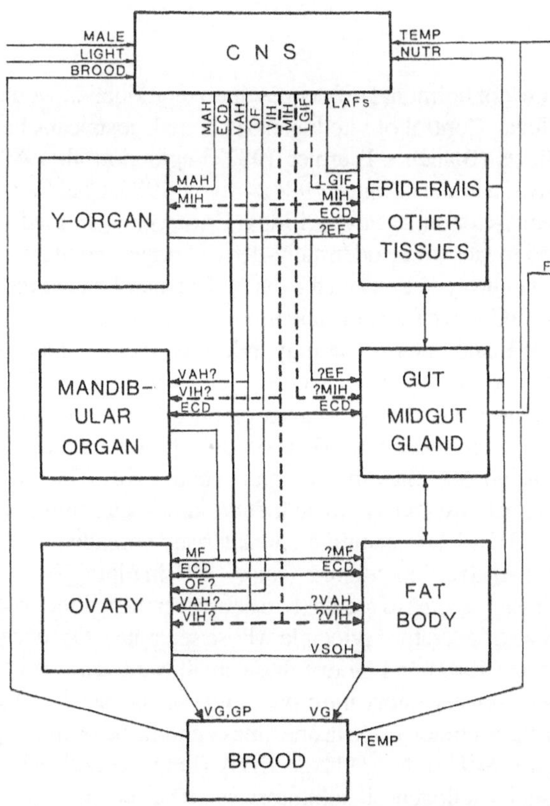

Figure 7. Possible pathways of control of molting and reproduction in malacostracans. Dashed lines indicate those in which the presumed major effect is inhibitory. Most factors have been named as hormones though hardly any have been characterized. ECD, ecdysteroid; EF, exuviation factor; LAFs, limb autotomy factors; LGIF, limbbud growth inhibiting factor; MAH, MIH, molt accelerating and inhibiting hormones; MF, mandibular organ factor; OF, ovulation factor; VAH, VIH, vitellogenesis accelerating and inhibiting hormones; VSOH, vitellogenesis stimulating ovarian hormone. Other abbreviations are GP, glycoprotein, NUTR, nutritional state; TEMP, temperature; VG, vitellogenin. From various sources (see text).

secretion of molting hormone (ecdysteroid, ECD) by the ectodermally-derived Y-organ (Fig. 7). A slow rise in hemolymph titer of ECD from the base level of intermolt is followed by a characteristic peak and sharp dropoff just before the molt; there may also be a small preliminary peak at the beginning of premolt (not shown in Fig. 4; reviews in Skinner 1985a, b, Chang 1985, Chang & O'Connor 1988).

During intermolt, both Y-organ ECD secretion and possibly the action of ECD on epidermal and other target tissues are countered by a molt-inhibiting neurosecretory peptide principle (MIH), released from the X-organ-sinus gland complex in stalk-eyed forms and from cells in the protocerebrum in isopods and amphipods. A molt-accelerating principle (MAH), inferred in various species, may be produced in eyestalk, brain or thoracic ganglia (Legrand et al. 1982). Several additional principles have been proposed from time to time (Skinner 1985a, b). There is some evidence for an exuviation factor also produced by the Y-organ without which ecdysis will not occur.

Premolt-excitatory and molt-inhibitory limb autotomy factors have been suggested, not necessarily hormonal in nature, possibly but not necessarily acting via the central nervous system as depicted in Figure 7. Inhibition of limb bud growth by eyestalk extracts implies the existence of a neurosecretory limb growth inhibiting factor, LGIF, not necessarily distinct from MIH. Finally, the ectodermally-derived mandibular organ (see below) appears to shorten the molt cycle in a shrimp (Yudin et al. 1980) and the vitellogenic ovary has been postulated to lengthen it in an amphipod (Blanchet-Tournier 1982), these possibilities are not shown in Figure 7 but should not be forgotten.

4.2 *Hormones and reproduction*

Until recently the representation in Figure 7 of hormonal influences upon reproduction would have included fewer and more certain links. Control of vitellogenesis in malacostracans has been the subject of several recent reviews (Blanchet-Tournier 1982, Legrand et al. 1982, Adiyodi & Subramoniam 1983, Meusy & Charniaux-Cotton 1984, Adiyodi 1985, Charniaux-Cotton 1985, Fingerman 1987, Charniaux-Cotton & Payen 1988). Crustaceans seemed to differ from insects, in that they appeared to lack an ectodermally-derived organ producing a gonadotropic link between the product of neurosecretion and the ovary. The mandibular organ had been suggested to have a possible reproductive function, and more recent evidence (references in Laufer et al. 1987) indicates the mandibular organ may indeed be the hypothetical analog of the insect corpora allata sought years ago by Passano (1960, see Byard et al. 1975). If so, then in crustaceans as in insects ectodermally-derived organs produce tropic hormones involved in molting and male and female reproduction. In Malacostraca, these would be the Y-organ, the androgenic gland in the male and the mandibular organ in both sexes. The mandibular organ product with gonadotropic activity appears to be methyl farnesoate, similar in structure to the insect juvenile hormone JH III but lacking a characteristic epoxide moiety (Laufer et al. 1987). It is conservatively labelled as 'mandibular factor' (MF) in Figure 7.

Control of vitellogenesis in malacostracans seems to require two neurosecretory principles, a well-established inhibitory factor and an acceleratory principle whose separate existence is less secure (Legrand et al. 1982, Fingerman 1987). In Figure 7 these are illustrated as vitellogenesis-inhibitory (VIH) and vitellogenesis-acceleratory hormones (VAH), but each has had other names (e.g. GIH, GAH), and each has been suggested at one time or another to be nothing other than its molting-control counterpart, MIH or MAH respectively (Legrand et al. 1982, Adiyodi & Subramoniam 1983). A separate vitellogenesis-inhibitory factor had been sought in amphipods without success, but recently Blanchet-Tournier (1987) obtained such evidence. In the caridean *Palaemon serratus*, eyestalk removal shortens the duration of vitellogenesis, which normally takes three molt cycles, to two or even one molt cycle, strongly suggesting that MIH and VIH are different (Faure et al. 1981). In the same study extirpation of the eyestalk medulla externa X-organ resulted in cessation of vitellogenesis and drawing-back of follicular tissue from the oöcytes. These effects appeared to be mediated indirectly by removal of inhibition of VIH production, rather than by removal of a source of a more direct vitellogenesis-accelerating principle (see also Cuzin-Roudy & Saleuddin 1985). Such problems abound in the literature on control of malacostracan vitellogenesis and are now further confounded by the question of whether excitatory and inhibitory neurosecretory effects on vitellogenesis are mediated directly or via the mandibular organ.

Secondary vitellogenesis involves both secretion and release of vitellogenin by a source organ and uptake of vitellogenin by the ovary (see Section 2.3, above). In the amphipod *Orchestia gammarella*, but not in the isopod *Porcellio dilatatus*, the former process will not occur without the presence of the ovary; this and other results imply follicular production, in amphipods at least, of a vitellogenesis-stimulating ovarian hormone (VSOH, Fig. 7; Charniaux-Cotton 1985).

The site of production and role of the molting hormone ECD during reproduction again seem to vary among taxa (references in Blanchet-Tournier 1982). Vitellogenin synthesis in isopods and amphipods is dependent upon presence of the Y-organ and hence low titers of ECD, but in *Orchestia* injections of exogenous ECD may either shorten or prolong the molt cycle in which vitellogenesis is occurring. In the former case, vitellogenin synthesis is suppressed, and the expected ovulation does not follow the molt. During the normal synchronous molt and secondary vitellogenesis cycles the total amount of ovarian ECD peaks with the circulating hemolymph titer; this ECD then leaves the ovary in the eggs. In crabs, the relationships appear to be somewhat different. A similar increase in ovarian ECD during vitellogenesis occurs in crabs without a concomitant increase in hemolymph titer, and even in the absence of a functioning Y-organ (Lachaise & Hoffmann 1977, Chaix & De Reggi 1982), this ecdysteroid again leaves the ovary in the eggs. The question of the generality of ovarian ECD synthesis in the Malacostraca is an open one.

The normal termination of vitellogenin synthesis and subsequent ovulation may be hormonally-controlled, but little is known. *Orchestia* females may be blocked in stage D_0 of the molt cycle by ablation of the median protocerebral region; oöcyte maturation, follicular dissociation and termination of vitellogenin synthesis all appear to proceed normally. Blanchet-Tournier (1982) suggested that oöcyte uptake of vitellogenin and follicular VSOH production both cease with separation of the follicular envelope from the oöcytes. In the crab *Rhithropanopeus harrisi* during the breeding season, the main effect of eyestalk ablation seems to be inhibition of oviposition, suggesting but not demanding an eyestalk oviposition factor (Fig. 7, OF; Bomirski & Klek 1974).

The ovary has been associated with other endocrine functions not shown in Figure 7. Permanent secondary sexual characters in female amphipods, and temporary characters developed at parturial molts in both amphipods and isopods, have been shown to be dependent upon 'permanent' and 'temporary' ovarian hormones respectively (Charniaux-Cotton & Payen 1985). The ovary has the capacity to synthesize steroids other than ECD. Vertebrate reproductive steroids (and also pituitary gonadotropins) have been shown to have effects upon oögenesis and maturation (Fingerman 1987). Whether these are physiological or pharmacological effects is often not clear.

4.3 *Stimulation and restraint*

Our survey of malacostracans emphasizes that there are marked similarities between the cycles of molting and vitellogenesis. Each process has at least the appearance of a relaxation oscillation (Fig. 4). There is a slow rise to a peak, e.g. in oöcyte size or ECD titer, an abrupt fall signalled by an output (molt or ovulation), and finally a recovery period.

Each process is modulated by a pair of neurosecretory principles, with one member of each pair stimulating or accelerating its cycle and the other inhibiting, decelerating, or 'restraining' it (Fig. 7; Adiyodi & Subramoniam 1983). While each process may appear to reside chiefly in a peripherally-located endocrine organ – the Y organ on the one hand, the ovarian follicles or stroma or perhaps the mandibular organ on the other – probably in each case the peripheral secretory product feeds back on the CNS, and thereby influences the periodicity of the process as a whole (Blanchet-Tournier 1982). It is often unclear whether 'excitation' is just that, or rather 'inhibition of inhibition' as suggested by the studies of Faure et al. (1981) and Cuzin-Roudy & Saleuddin (1985). Again, either may act in an additive, multiplicative, or in a more complex fashion, with quite different results.

Each peripheral endocrine organ in turn appears, through its secretion products, to influence in different ways more than one target organ, both 'sources' such as midgut gland or fat body and 'sinks' such as epidermis or oöcyte. In this way, in each case is effected the complex shifting pattern of material transfers that constitutes the aim of the process as a whole. In turn, we must not overlook the possibility that each target organ may influence the periodicity, perhaps by the effect of nutrient levels upon the CNS (Fig. 7; Hubbell 1971).

The phase and frequency of each cycle must be attuned to, but not necessarily governed by, the local cycles of resource availability, temperature and photoperiod (Sastry 1983, Conan 1985), and thus the CNS integrates appropriate environmental cues. Brooding adds a further complication to the vitellogenic cycle: its output becomes as it were a potential tonic stimulus, or hatching or release of the young provides a delayed phasic stimulus which may be integrated into the pattern where appropriate (Figs. 4, 7). The stimulus pattern provided by the brood may be complex. The mysid *Leptomysis lingvura* can be induced to 'adopt' into her brood pouch larvae or even pieces of larvae from another brood, but usually only if they are the same age or older than her own (Wittmann 1978). Finally, each cycle must be attuned to the other.

4.4 *Reciprocal interactions*

In this section I examine, first, the influence of the events and processes of the molt cycle upon reproduction, and second, the effect of reproduction on the timing of the molt. We must use such a piecemeal method of analysis, but we should keep in mind that if molt and reproductive cycles are indeed the output of a system comprised of a pair of coupled nonlinear oscillators, such analysis may provide only part of the story (Gollub et al. 1978).

4.4.1 *Molt cycle timing of vitellogenesis*

The striking synchrony of molt and vitellogenic cycles in peracarids has been most thoroughly studied in the amphipod *Orchestia gammarella* (Blanchet-Tournier 1982). Here and in isopods the dependence of folliculogenesis and secondary vitellogenesis upon low titers of ECD has been noted above. The high titers of ECD during later premolt (stages D_2-D_3) or during exogenous ECD administration, and equally the very low levels in the absence of the Y-organ, prevent both secondary folliculogenesis and fat body synthesis of vitellogenin.

Thus the timing of both the beginning and the termination of the vitellogenic cycle may depend upon ECD titer. Nevertheless, in *Orchestia* folliculogenesis and vitellogenesis will slowly proceed even when the molt is blocked by decerebration. In the mysid *Siriella armata* destruction of the medulla interna-medulla externa X-organ results in inhibition of preparation for the molt; reproducing females show inhibition of secondary vitellogenesis, of liberation of the young, and of the juvenile molt of the latter even when they are artificially freed from the marsupium (Cuzin-Roudy & Saleuddin 1985). In the isopod *Oniscus asellus*, the first vitellogenesis begins during a prolonged premolt and the next during early intermolt (Steel 1980), it is unlikely that ECD titer is the cue in both cases.

In carideans, secondary folliculogenesis and vitellogenesis begin in postmolt, at the low point in the ECD cycle, and are prevented by ECD injection (Charniaux-Cotton 1985). In *Macrobrachium* as in the peracarids, vitellogenesis occupies one molt cycle, with spawning just after the molt (Fauvel 1981, Rao et al. 1981). But in *Palaemon serratus* the vitellogenic period which ordinarily lasts three full molt cycles can be accelerated into one by eyestalk ablation (Faure et al. 1981). Whether the intervening premolt peaks of ECD affect vitellogenesis in the normal pattern does not seem to have been investigated, but ECD titer appears to regulate the timing of only the beginning of vitellogenesis in this species.

In the lobster *Homarus americanus* short-day photoperiod appears to promote primary maturation but prevents secondary vitellogenesis (Nelson 1986). The long-day cue can result in vitellogenesis initiation at any time in the molt cycle, but allochthonous vitellogenin release into the hemolymph appears to be suppressed during the ECD peak of premolt (Fig. 8: Nelson et al. 1988c). If the photoperiodic stimulus arrives at this time, vitellogenesis initiation is suppressed until after the molt, and extrusion is thereby delayed. Later egg maturation and extrusion, however, will only occur if long day onset falls within ± 1400 day-degrees of the molt (in this example, day-degrees are the cumulated daily temperatures above a 6° threshold; Nelson et al. 1988b). Oöcyte maturation might in such cases be a function of ovarian sequestration of the fat-soluble and very mobile ECD during the preceding premolt and its gradual loss following the molt.

The point of such sequestration of ECD (if it occurs) might be to provide the embryos with sufficient ECD to carry through their early development (Blanchet-Tournier 1982). In the portunid *Carcinus maenas* and the majid *Acanthonyx lunulatus* ECDs appear to be manufactured by the ovary for sequestration in the eggs; in the latter species terminal anecdysis has set in and the Y-organ has degenerated (Lachaise & Hoffmann 1977, Chaix & De Reggi 1982). Thus in these reptants the innovation of ovarian ECD synthesis may have contributed to freeing oöcyte maturation from temporal dependence upon the molt. Vitellogenesis might still be inhibited by high titers of circulating ECD during premolt, as in other malacostracans; in reptants vitellogenesis is generally absent at this time, and in *Carcinus maenas* the Y-organ has been suggested to inhibit ovarian development (Démeusy 1962).

It would be interesting to determine if a similar situation prevailed among other groups which have achieved multiple vitellogeneses within a molt cycle, such as penaeid shrimp, certain palinurid lobsters, and even very large American lobsters (Waddy & Aiken 1986) or whether in fact ECD is even essential for early development in such groups.

4.4.2 *Reproduction and timing the molt*

In the non-brooding eucarids, and in the stomatopods that may be able to recover the brood following the molt (Section 3, above), there is no experimental evidence of molt delay by reproduction (but see Anderson et al. 1985). In all other brooding groups, such a delay has been

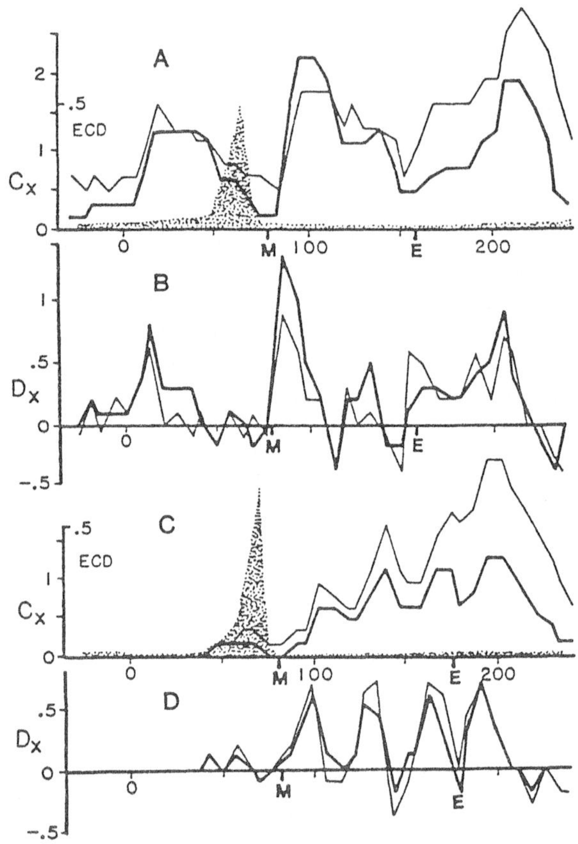

Figure 8. *Homarus americanus* females. Input of vitellogenin into hemolymph is suppressed during premolt ecdysteroid (ECD) peak. A, C: course of two hemolymph vitellogenins (heavy and light lines; C_X, concentration of vitellogenin X in mg/ml obtained from rocket immunoelectrophoresis, outer ordinate), in days following long day onset (O), for two extruding females, and of hemolymph ecdysteroids (ECD, shaded in μg/ml obtained from radioimmunoassay, inner ordinate). E, extrusion; M, molt. B, D: course of changes of hemolymph concentration of vitellogenins (D_X, in mg/ml/week, adjusted for constant clearance estimated from terminal decline in C_X). From Nelson et al. (1988c).

demonstrated in at least some species, but the mechanism is not always the same.

Both the pre-parturial (i.e. vitellogenic) molt cycle and the parturial cycle are affected, but differently. Thus in the terrestrial isopod *Oniscus asellus* the preparturial premolt is prolonged by the photoperiodic cue initiating vitellogenesis, whereas in the following cycle it is intermolt that is prolonged; experimental removal of the brood is followed by the initiation of premolt, which can also be delayed by stuffing the empty marsupium with a sponge (Steel 1980). Such a direct effect may also be found in some carideans (Hess 1941), but in other species studied the effect of the brood upon molt delay is more subtle or even absent as in palaemonids. In these species eye-stalk ablation does not shorten intermolt interval but ECD injection may (Kamiguchi 1971a, b: Ranga Rao & Inkel 1975, Pandian & Balasundaram 1982, Rao et al. 1981).

Blanchet-Tournier (1982) presents evidence that in *Orchestia gammarella* the molt cycle is prolonged by secondary vitellogenesis, and deduces that the ovary stimulates production of an

MIH by a hypothetical CNS center. In the lobster *Homarus americanus* the occurrence of vitellogenesis does not result in molt delay unless 1) its onset has been delayed well past the preceding molt by inappropriate photoperiod and 2) then only if it leads to ovulation and extrusion (Fig. 5A; compare Fig. 1; Nelson et al 1988a). A small additional delay is then contributed by the presence of the clutch on the pleopods, but this effect is indirect, adding in some fashion to the total molt delay. Premolt may begin before or after hatching, and clutches are even lost with the molt.

Similar combinations of molt delay contingent upon mating, vitellogenesis, or extrusion, and upon continued presence of the clutch appear in another terrestrial isopod, *Porcellio dilatatus* (Mocquard et al. 1976), in *Leptomysis lingvura* (Wittmann 1981), and in *Parathelphusa hydrodromus* (Adiyodi & Subramoniam 1983). In *Porcellio*, the period between release of young and the molt is highly variable, as in *Homarus*. *L. lingvura* females occasionally accepted larvae younger than their own; in this case the molt would be delayed beyond its normal time by two to five days, but the molt might occur while the added larvae were still premature. Wittmann (1978, 1981) could not control for the possibility that acceptance of prematures and molt delay were both functions of a retarded molt cycle in those particular females that adopted younger larvae. In *P. hydrodromus*, the 'brood effect' is abolished by eyestalk ablation, suggesting MIH as the agent of maintenance.

In *H. americanus*, a shift upon extrusion of eggs from dependence of molt-molt interval duration on female size to dependence on temperature (Fig. 1; Nelson et al. 1988a) suggests a shift in control of intermolt interval from a 'somatic' program to a 'reproductive' program. In ovigerous molt cycles, the Q_{10} of I_{mm} duration increases from less than 1.5 to a range of approximately 2-3.5, which is similar to the Q_{10} range of *H. americanus* embryonic development duration at the same temperatures (calculations from data in Perkins 1972). Thus under the reproductive program, intermolt intervals are attuned to the temperature dependence of development of the young.

5 THE CONFLICT BETWEEN GROWTH AND REPRODUCTION

The common premise in studies of reproduction and life history patterns is that reproduction is associated with a cost. In life history theory, the cost is in terms of future survival or fecundity (Bell 1980). A related premise is that there exists some sort of fundamental antagonism between reproduction and growth which must be resolved; this is often used as an explanatory principle to demonstrate the adaptiveness of a particular relationship between the two – never to demonstrate its inadaptiveness. In crustacean studies, the premise takes one of several forms: competition of the processes for resources or reserves, physiological (e.g. endocrinological) incompatibility, the negative effect of brooding upon growth rate. The present chapter has emphasized the possibility of loss of the brooded clutch through molting. The following quotations from recent reviews will convey the flavor:

> All three strategies [diecdysic and anecdysic Malacostraca, cirripedes] involving differential programming of growth and reproduction have evolved in the Crustacea in response to the need that the combined demands thereof should not overstrain the energy resources of the individual (Adiyodi & Subramoniam 1983, p. 476).
>
> The antagonism that exists between ovarian growth and somatic growth and that between MIH and GIH [=VIH] are too obvious. (ibid., p. 478).
>
> Coordination of breeding and molt cycles of the females seems necessary in order to minimize metabolic competition between the two processes...[and also] molt should not normally occur during incubation...to the expense of an unnecessary loss of eggs (Conan 1985, pp. 92-93).

...Growth and reproduction are antagonistic. Growth requires energy resources before ecdysis and to complete tissue growth after ecdysis, resources which will not be available for egg production...reproduction can never enhance growth but must always restrict it. Resources are required for ovary maturation, the intermolt may be prolonged by...brooding...and the incubatory behavior may restrict...resource acquisition (Hartnoll 1985, pp. 114-115).

In the other brooding malacostracans there is a danger that a premature molt will result in loss of the clutch with the exuvia. Coordination between the molting and reproductive cycles would then appear to be essential (Nelson, this chapter, Section 5.2).

The various forms of the premise often seem self-evident: the energy taken in by an organism can be partitioned into its various uses; if future fecundity is size-dependent, reproduction instead of growth represents a cost; a molt during incubation will surely result in loss of the clutch. But tests of the extent to which such principles have guided scheduling or its evolution are rare and must always be indirect, and their relevance and range of applicability may be limited by other factors.

5.1 *The threat of clutch abortion by premature molt*

In the terrestrial isopod *Oniscus asellus*, deflation of the brood pouch is directly followed by premolt. In other cases a 'brood effect' may be present, but does not account entirely for the increased duration of the intermolt interval. In *Homarus americanus*, the major delay is attributable to the occurrence of ovulation or extrusion; in other cases to copulation or vitellogenesis (Section 4.4.2). At first such findings may suggest that the function of the delay is in fact to protect the clutch from premature abortion, as we have assumed in our classification above (Section 3). But even in the direct inhibition of premolt by the clutch there exists also the strong possibility that the delay in molting is imposed by a reduced growth rate caused by some aspect of the metabolism of brooding. For example, ovigerous lobsters (*H. gammarus*) reportedly exhibit similar activity but reduced ingestion when compared to females not bearing eggs (Branford 1979). This may be an artifact of size or captivity, for example if the females were provided with inadequate shelter. The fact that in lobsters the delay consequent upon brooding is generally but a fraction of the total delay lends a certain credence to such an economic interpretation.

5.2 *Limitation of growth rate by brooding*

Hartnoll (1985) detailed the ways in which growth rate might be so constrained. I have found no clear-cut case of this. *Macrobrachium nobilii* females apparently display significantly higher percentage extrusion in the following molt cycle if the prior clutch has been stripped, even though no molt delay reduction results from the stripping (Pandian & Balasundaram 1982). In *Homarus americanus* females, ovulation and extrusion are followed by a decrease in growth rate for the molt cycle relative to that shown by females undergoing a certain amount of vitellogenesis but not extruding (Fig. 2). But in the absence of extrusion the potential exists that at least part of the investment in eggs may be recouped. Neither molt increment nor growth rate was lower in lobsters allowed to retain their eggs than in those that were stripped (Nelson et al. 1988a).

5.3 *Physiological incompatibility of molting and vitellogenesis*

In most peracarids and reptants, allochthonous vitellogenin secretion appears to be suppressed or at least absent during the peak of circulating ECD in the latter part of premolt (Section 4.4.1). The endocrinological environment of this period may be one of general suppression of competing activities, or the suppression may be specific to vitellogenin secretion. The apparent exceptions are seen in certain palaemonid shrimp in which vitellogenesis is said to continue over several molt cycles (Section 3.3.1) and in a few crabs and anomurans in which maturing ovaries have been found in late premolt (Section 3.3.2). But there appear to have been no studies detailed enough to demonstrate that in these cases vitellogenin secretion actually continues during this period. Even in the penaeids with no constraints imposed by brooding, egg production is reported to be either absent or oöcytes are resorbed during late premolt (Emmerson 1980; for groups such as the euphausids and stomatopods we simply have no information). In this respect our two major groupings, the peracarids and the reptant decapods seem to be in agreement: high titers of circulating ecdysteroids inhibit vitellogenesis.

In aquatic isopods, tanaids, and cumaceans primary and secondary vitellogenesis apparently are asynchronous, resulting in an alternation of parturial and anovulatory molt cycles (Section 3.3.1) and this may be an example of physiological incompatibility determining patterning in these groups.

Otherwise, direct interference between processes has not been demonstrated. Through the years, eyestalk ablation and limb autotomy have been used in a number of studies, particularly of crabs, as probes to investigate the underlying physiological state (reviews in Adiyodi & Adiyodi 1970 and Adiyodi & Subramoniam 1983). In general, the results have been interpreted supporting an alternation between reproductive and somatic growth potentials. Eyestalk hormone (MIH, GIH=VIH) appears to suppress one or the other tendency, depending upon which is dominant. This is the context of the second quotation from Adiyodi & Subramoniam, and might be interpreted (as they do) as a physiological incompatability between the production or the effects of MIH and GIH. But as Adiyoda & Subramoniam themselves suggested (ibid. p. 282), 'the similarity in the actions of MIH and GIH thus becomes more and more apparent, and one is naturally tempted to suspect that MIH and GIH represent a single hormone, say a 'growth-restraining hormone' (Adiyodi & Adiyodi 1981) that exercises its influence upon two 'target processes,' namely growth and reproduction, as in polychaetes.' Then, wherein lies the incompatibility of the processes?

5.4 *Resource allocation*

The question of resource partitioning is multifaceted and ultimately must include most of the life history of an organism. Here we can hope to touch only a few salient points. First, there is the question of whether the scheduling within a molting or reproductive cycle is such that competition between the processes for resources or reserves is ameliorated or avoided. Next, we must ask to what extent the distribution of molting, growth, vitellogenesis, and brooding over the year avoids such competition. Then there is the problem of assessing the cost of reproduction to an iteroparous organism. Finally, there is the 'life history problem,' the question of the nature of appropriate lifetime allocation of resources.

5.4.1 *Allocation within the molt cycle*

We have contrasted the scheduling of molting and vitellogenesis in peracarids and carideans with that of the majority of reptants, and found a series of correlations (Section 3). The former have a diecdysic type molt cycle organization, the latter intercalate an anecdysic period after stage C_3; in some brachyuran families this anecdysis becomes much prolonged (C_4T). In the former, secondary vitellogenesis terminates before the molt and extrusion immediately follows it; in reptants vitellogenesis rarely precedes the molt. In peracarids and carideans, multiple vitellogeneses within a molt cycle do not occur but such are not infrequent among the reptants (Fig. 5B, 6).

Possibly all of these features are correlated with a fundamental difference between reptants and the other malacostraca. The usually well-sclerotized and often mineralized cuticle of the reptants, and their elaborate endophragmal skeletal system (Kaestner 1970, Schram 1986), may make molting a more disruptive and energetically costly event. Whereas peracarids and carideans tend to synchronize the major reproductive events with those of the molt cycle, the usual reptant does the opposite (Adiyodi & Subramoniam 1983). However, the heavy-shelled lithodid anomurans appear to have synchronous molting-reproduction schedules, so the correlation is not complete (Wallace et al. 1949, Pike & Williamson 1959).

In the peracarids and carideans, the major energy and material demands of molting and reproduction appear to be synchronously met. There is little indication that the timing of molting and reproductive events is dictated by avoidance of competition for resources and reserves. Even in the case of the isopod *Oniscus* in which the vitellogenesis-accommodating preparturial premolt is prolonged (Section 4.4.3), it seems as likely that the prolonged premolt provides the time as it is that it provides the material for oöcyte maturation.

In reptants, there is more reason to suppose that considerations of resource allocation play an important role in scheduling the events of the molt and reproductive cycle, as discussed above. But vitellogenesis is often scheduled during C_1-C_3, the major period of growth of new somatic tissue (Section 3.3.2).

The issue turns upon the question of storage of organic reserves. The initial impression of the brachyuran molt cycle derived from Drach's and particularly Renaud's (1949) studies of *Cancer pagurus* was that of an organism living 'close to the edge,' like an undercapitalized business (Passano 1960). The major storage depot was the midgut gland or hepatopancreas, and the major organic reserve was lipid. A quotation from Passano (1960) is relevant:

> During the period between D_2 and C_1 the decapod is usually unable to feed. The organic reserves are almost entirely used up during this fasting period, being utilized both for maintenance of the animal and for extraordinary demands associated with construction of a new skeleton (p. 487).

In recent years the impression has been changing. For example, Heath & Barnes (1970) found that in *Carcinus maenas* the midgut gland both stored during intermolt and released over the molt much less of each category of nutrient than Renaud (1949) found with *Cancer*. Glynn (1968) obtained similar results in *Homarus gammarus*. Furthermore, muscle and subepidermal adipose tissue appear to play important roles as deposits, in both reptants and natants, and protein may be the largest organic reserve (Stewart et al. 1967, Barclay et al. 1983, Durliat 1983). Storage of organic reserves no longer seems to be specifically 'for' the molt, but rather for utilization as the occasion arises (Heath & Barnes 1970). In the crab *Parathelphusa hydrodromus* females starved for three months carried on normal vitellogeneses – although at the end of the period their midgut glands had lost most of their lipid content (Adiyodi 1968).

Thus, avoiding competition for nutrients between organs no longer seems as credible a reason for scheduling interactions, even in reptants. Reserves are stored in times of abundance and utilized in times of dearth. The evidence for their depletion forcing scheduling changes in the natural environment is meager. Siegel (1984) was able to trigger synchronized molting in a group of starved sand crabs (*Emerita analoga*) by supplying them with food; upon cessation of feeding, molting became infrequent and irregularly-spaced. Siegel postulates that this mechanism may initiate such synchronized molting in the spring. But molting and vitellogenesis appear to be synchronous processes in *Emerita* (Cox & Dudley 1968) and must make simultaneous demands upon the resources.

5.4.2 *Scheduling and the annual cycle*

For species which molt but once a year or even less as adults, the question of whether scheduling avoids resource competition between the processes of molting and reproduction is similar to the question addressed in the previous paragraphs. Most malacostracans probably live hardly a year, and for them the question of annual scheduling may be moot. What most concerns us in this section are the species that molt several times a year as adults.

A common temperate to subtropical pattern is one in which one or several long molt cycles with reproduction usually during the season of high productivity, alternate with several shorter avitellogenic cycles in the low-productivity seasons (e.g. Cheung 1969, Anderson et al. 1985). Are such patterns examples of scheduling allocation of resources alternately to growth and reproduction, or do they reflect the need to avoid premolt (ECD) interference with vitellogenesis or brooding? If the resources for growth are gathered at or near the time of reproduction as in these patterns, how is allocation to be scheduled? Species of the same group with generally similar habits, from environments that are climatically and edaphically similar, often differ enigmatically in their annual molting, migration and reproductive patterns (Hartnoll 1963, Allen 1966, Kuris 1971). Sastry (1983) summarized the literature on seasonal distribution of reproductive patterns in different habitats. The impression is one of confusion, but one may gather that reproduction is timed for the benefit of the young and not the parent. The distribution of molting events is equally difficult to understand. Molts may be rare in the season of reproduction, as discussed above, or most frequent at that time, as in *E. analoga* (Siegel 1984). This latter example suggest that the more likely growth is to be food-limited, the more likely somatic and reproductive growth are to co-occur. In either event, the avoidance of allocation conflict appears unimportant in annual scheduling.

5.4.3 *The cost of reproduction*

Much of the evidence which has been adduced concerning the 'conflict between growth and reproduction' has to do with the decrement in growth rate associated with female reproduction. Much of this evidence is circumstantial and has other interpretations. For example, Hartnoll (1985) points to the decreased percentage growth increment in reproductive female *Cancer magister* and other decapods, when compared with the males. However, in *Cancer* and other brachyurans the male carries the female beneath him during the consort relationship. Majid males are larger than females, notwithstanding that the latter reproduce only after a terminal molt. It is perhaps more likely that in *Cancer* females the decreased growth increment has a social function. In *Neopanope sayi* the sexes overlap in size, but females are rarely found in final copulatory position with smaller males (Swartz 1976). Hartnoll (1985, Fig. 6) cited the energy budget of the female *Metamysidopsis elongata* (Clutter & Theilacker 1971) as an example of energy partitioning between growth and reproduction; but, in reality, the male *Metamysidopsis* devotes less energy to growth than does the female (Fig. 9). In this case and in mysids generally the female is larger than the male; clearly one should not then deduce that therefore the male mysid expends more upon reproduction.

Even in cases where the female is not appreciably larger than the male, as in the antarctic isopod *Serolis polita* (Luxmoore 1982), she may expend more resources upon *both* growth *and* reproduction than does the male; not only that, she usually lives several years longer than him

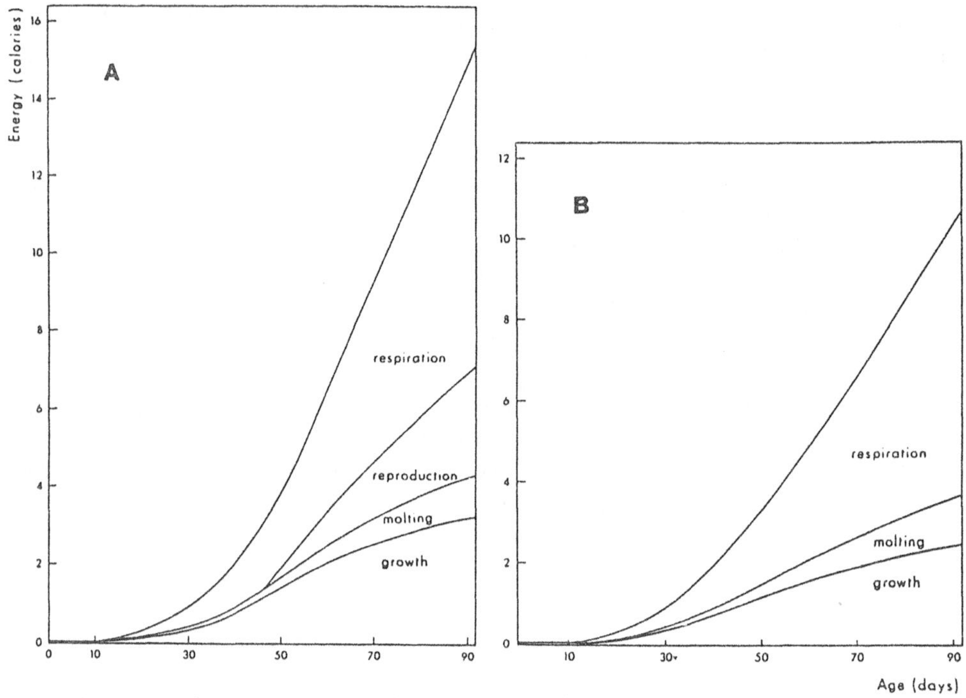

Figure 9. *Metamysidopsis elongata.* Cumulative energy used by individual females (A) and males (B). The space between two curves represents the energy devoted to the function. From Clutter & Theilacker (1971).

(Fig. 10; Luxmore 1985). The concept of reproductive cost is closely associated in the life-history theory literature with the concept of reproductive effort (E). Tuomi et al. (1983) point out that the meaning of reproductive effort must vary with the allocation system, that is, with the way in which somatic and reproductive investments by the reproductive individual vary in relation to the somatic investment of the non-reproductive (Fig. 11). The somatic investment by the reproductive may be smaller, exactly the same, or even larger than that of the non-reproductive. The situations with *Metamysidopsis* and *Serolis* (Figs. 9 and 10) indicate that the latter alternative is not necessarily unrealistic.

Comparisons of growth or energy budgets of males and females may conceal other sex-specific differences, and it is more germane to compare reproductive and non-reproductive females of the same size. Hartnoll (1985) notes that Lawlor (1976) did this for the isopod *Armadillidium vulgare*; he found that the reproductive females, with nearly double the intermolt interval of the non-reproductives and smaller growth increments as well, grew considerably more slowly. However, he also found that within the class of reproductive individuals, reproductive effort was significantly positively correlated with weight gain, however measured! This suggests that the lower growth of reproductive females compared with non-reproductives may not be dictated by competition for *currently* limited resources, but may for example represent a hedge against the potential of future resource limitation, that is an allocation not strictly dictated by short-term economics.

With female *H. americanus*, one can manipulate photoperiod cues to obtain different levels of 'commitment to reproduction' (Fig. 5; Nelson et al. 1988a). In each reproductive category,

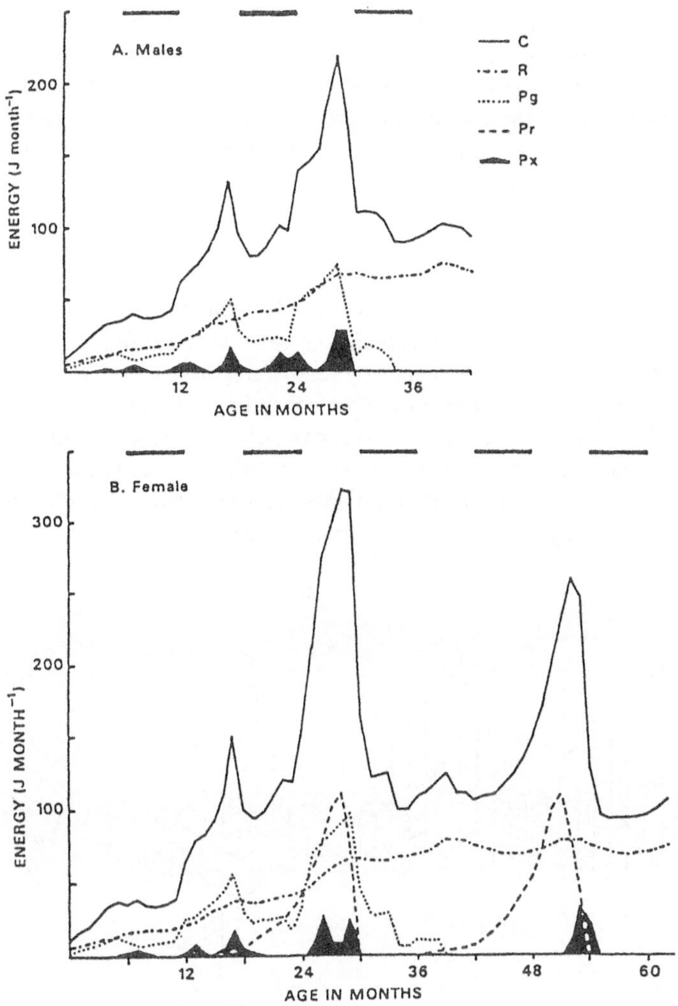

Figure 10. *Serolis polita.* Monthly energy budgets for an average male (A) and female (B) throughout their lives. Black bars, winter periods; C, total energy consumption; R, respiration; Pg, somatic production; Pr, gonad and gamete production; Px, exuvia. From Luxmoore (1985).

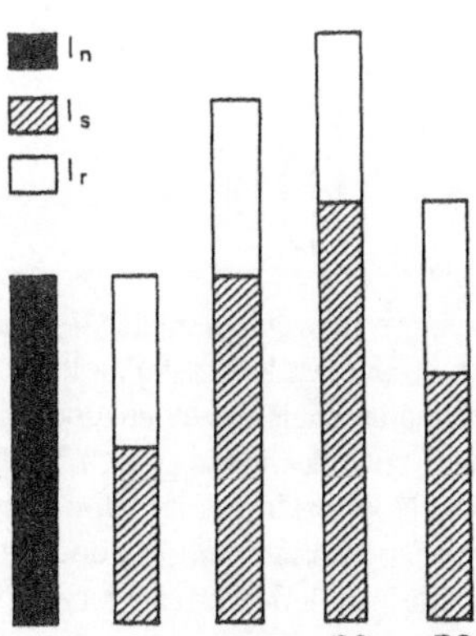

Figure 11. Allocation systems. I_n, somatic investment by nonreproductive individual. I_s and I_r, somatic and reproductive investments by reproductive individual. The somatic investment by the reproductive can be less than, the same as or greater than that by the non-reproductive individual, with different implications for reproductive effort. From Tuomi et al. (1983).

volume increment at the molt, ΔV, increases with intermolt interval duration; as a result, the dependence of volume increment upon carapace length and inversely upon temperature (Fig. 2A) is similar to the corresponding dependence of intermolt interval duration upon those two variables (Fig. 1). Thus, with increasing commitment to reproduction, dependence of growth increment upon size decreases and dependence upon temperature increases. When molt interval and molt increment are combined and expressed as instantaneous growth rate in volume, $\dot{V}$, the different categories of molt cycle display a decrease in rate with increasing commitment to reproduction, but no differences in dependence upon temperature and carapace length (Fig. 2B).

The meaning of this may be seen in Figure 12. There the 'costs' in volume increment at the molt (V_G, Fig. 12A) with different commitments to reproduction are shown for different values of temperature and carapace length. These were obtained by subtracting the value of ΔV under a condition of commitment, temperature and size from the corresponding value for the non-reproductive condition (No L, No E; Fig. 2A). 'Cost' is seen to be positively correlated with both temperature and size, both of which perhaps seem reasonable (fecundity for example doubles between 90 and 110 mm carapace length. For clarity, the polarity of the temperature coordinate is reversed from Figs. 1 and 2). The negative values of V_G for lower temperatures and sizes in

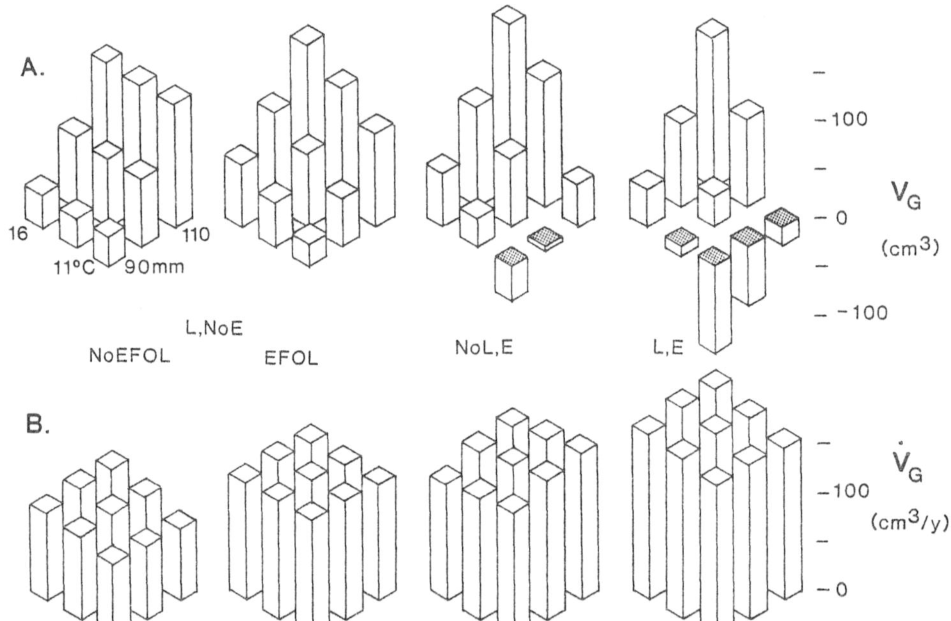

Figure 12. *Homarus americanus* females. Dependence of A, reduction in volume growth increment (V_G), and B, reduction in instantaneous growth rate in volume ($\dot{V}_G$), upon temperature and carapace length, for molt cycles representing increasing commitments to reproduction. Reduction in growth increment is dependent upon size (in the absence of extrusion) or upon both size and temperature (in its presence), but reduction in growth rate the 'cost of reproduction' is dependent only upon the commitment to reproduction. Note that for clarity, temperature axis is reversed from Figures 1 and 2. Values were obtained by subtracting the appropriate value for one of the four right-hand categories of molt cycle in Figure 2 from the corresponding value in the left-hand category (NoL, NoE, representing the condition of no reproduction). Other symbols as in Figure 2.

the categories of molt cycle containing extrusion are attributable to the relatively longer intermolt intervals in those categories (Fig. 1).

But in fact when 'cost' is expressed in terms of the fraction of instantaneous growth in volume which is devoted to reproduction ($\dot{V}_G$, Fig. 12B), it is seen to be essentially independent of both size and temperature and dependent only upon degree of commitment to reproduction. Thus once again, the reduction in instantaneous growth rate with reproduction seems not to be dictated by the realities of size-dependent fecundity and competitive resource utilization but seems rather to reflect an allocation based upon commitment to reproduction alone. The 'cost of reproduction' is here an allocation which appears to serve as insurance against unforeseen circumstances. A long-lived iteroparous organism must be buffered against exigencies if it is to survive to reap the advantages of iteroparity.

Iles (1984) came to similar conclusions in his study of herring growth and reproduction: 'The key point to emphasize is that the animal is to a remarkable degree independent of the environment. At no time is the metabolic activity, or even levels of the metabolic pool, entirely dependent upon incoming material; metabolism has been effectively decoupled from assimilation and it is this decoupling that defines the true function of the hormonal system' (Iles 1984: 346).

Tuomi et al. (1983) list several ways by which allocation conflict may be avoided: by investing more resources in somatic growth (i.e. storage prior to reproduction), or by increasing resource intake and ingestion during reproduction. We can add with Iles (1984) that they might alter other components of their energy budgets as well. Reznick (1983) found that reproducing female guppies (*Poecilia reticulata*) grew as much as ones prevented from reproducing. There was little 'tradeoff' evident between growth and reproduction within a population.

The requirement of a balanced energy budget may have led some investigators to the impression that growth and reproduction must always be in competition for resources. Such might be the case if in the absence of reproduction utilization of available resources were already maximal. That this is not necessarily so can be easily shown by eyestalk ablation. We found that 'naturally' ablated juvenile lobsters found among lab-reared families were characterized not only by shorter intermolt intervals but by larger molt increments; their growth rate might be twice that of their sibs (Hedgecock et al. 1976, and unpubl. observ.). Similarly, eyestalk ablation in adult *Palaemon* allows the females to complete vitellogenesis in a single molt cycle instead of the normal three. In each of these examples the inhibitory hormone appears to ensure that unrestrained growth or reproduction does not erode the reserves of 'capital' sequestered for future exigencies. The habit of iteroparity might not otherwise survive.

5.4.4 *Changing allocation schedules*

Several authors have noted the presence of 'alternative reproductive strategies' in young iteroparous malacostracans (Lawlor 1976, Aiken & Waddy 1980, Lipcius 1985, Berglund & Rosenqvist 1986). Specifically, among females of the same size, some are observed to mature and reproduce during a molt cycle and some are not. Partly this may be fortuitous natural variation: thus in *H. americanus* females, extrusion in response to photoperiodic cue is dependent upon the phase of the cue in the current molt cycle. If the puberal molt occurs too late in the summer, extrusion may be foregone until the following spring (Nelson et al. 1983, 1988a). Nevertheless, such cases suggest the 'life history problem': what is the effect of current reproduction upon the potential for future reproduction?

Lawlor (1976) noted that greater weight of sowbugs (*Armadillidium*) that had forgone early summer reproduction should result in greater later fecundity, but he suggested that this must be understood in the context of survivorship in the face of unpredictable summertime humidity. While cryptic behavior and less frequent molting by gravid females might through lower mortality offset later reproductive disadvantage, Brody et al. (1983) added that the early reproducers might be more vulnerable to extremes of temperature and desiccation because of their smaller size. Other experimental studies are less detailed. Berglund & Rosenqvist (1986) found reproduction in *Palaemon adspersus* to be associated with lower growth and higher fish predation. Steele & Steele (1986) found photoperiod-induced termination of reproduction to be associated with greater survival in the amphipod *Gammarus lawrencianus*.

Variation in allocation schedules may be genetically maintained or only phenotypic in nature; if the latter, it may be fortuitous or undirected as discussed above, or else a directed response to environmental variation. Such directed responses have been much discussed; plasticity of a trait 'is itself a trait and may be under genetic control' (Caswell 1983). These directed responses have been called 'norms of reaction' (e.g. Caswell 1983, Stearns & Koella 1986). If the response is in the same direction as the expected evolutionary response to the environmental change it is called 'cogradient variation'; otherwise, it is 'countergradient.' If the response is cogradient with respect to the current selective regime, the selective pressures are ameliorated; if the reaction norm is countergradient, those selective pressures become more severe.

Variation in allocation schedule in response to changing conditions in malacostracans may represent such a cogradient reaction norm. Sowbug females reproduced at near-normal levels but did not grow when food was scarce (Brody et al. 1983); female hermit crabs (*Clibanarius* and *Pagurus*) shifted from somatic growth to reproduction when larger snail shells were

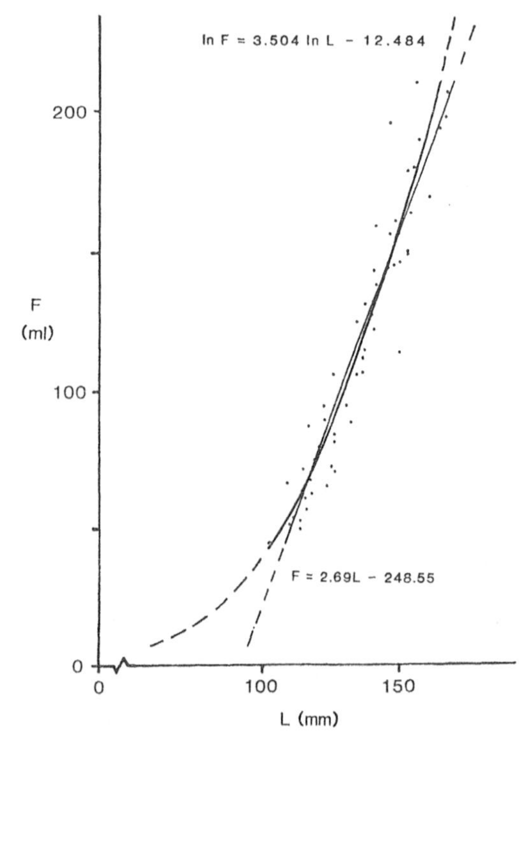

Figure 13. *Paralithodes camtschatica*. Fecundity F (as clutch volume in ml), as a function of carapace length L, with fitted linear and logarithmic (power function) regressions. Reproductive investment increases as a more-than-cubic function of length. From McMullen (1969).

removed (Bertness 1981). These examples appear to be cogradient reactions; continued exposure to such conditions might lead eventually to evolution of earlier reproduction.

Although the allocation of resources to reproduction ultimately may be limited by the rigid crustacean integument (Hines 1982), there evidently exists wide latitude for variation in response to genetic and environmental factors. The relationship between fecundity and a measure of size may be used to estimate change in the allocation to reproduction by an iteroparous organism during its lifetime (Fig. 13). The example displays both linear and logarithmic (power function) regressions of fecundity (F) upon carapace length (L). Comparative study may reveal variation in the slopes of such regressions within a group. Thus in cumaceans Corey (1981) found that the linear coefficient b relating brood volume to carapace length within a species ranged from 0.3 to 3.4 among species. In Figure 14, with the exception of one species, the intercept a is highly negatively correlated with the coefficient b (r =-0.899, p 0.001). Such a relationship was first commented upon by Roff (1982) in a study of flatfish reproductive 'strategies'. Roff noted that part of such a correlation was a result of non-independence of the parameters; in fact, the principal axis of the joint confidence interval of b and a for any one species can be shown to have a similar slope to the regression in Figure 14 (by the methods in Draper & Smith 1981).

But part of the joint variation seems to be real, and may represent variation in life-history allocation schedule among these species of cumaceans. The within-species coefficient b as calculated by Corey (1981) is significantly correlated with average carapace length in the sample ($b = 0.316L + 0.548$, $r = +0.755$, $p < 0.001$). If each estimate of b was of a region of an underlying logarithmic relationship such as in Figure 13, a positive correlation with average length might be expected. But when the data are transformed to their logarithms, the exponent b remains correlated with average length ($b = 1.014L + 0.804$, $r = 0.492$; $p < 0.05$). The exponent b represents the shift toward greater reproductive investment with increasing size; thus, the larger the species, the greater the allocation at larger sizes, or the less at smaller sizes. Corey (1981 and elsewhere) considers these small interstitial cumaceans with alternating parturial and

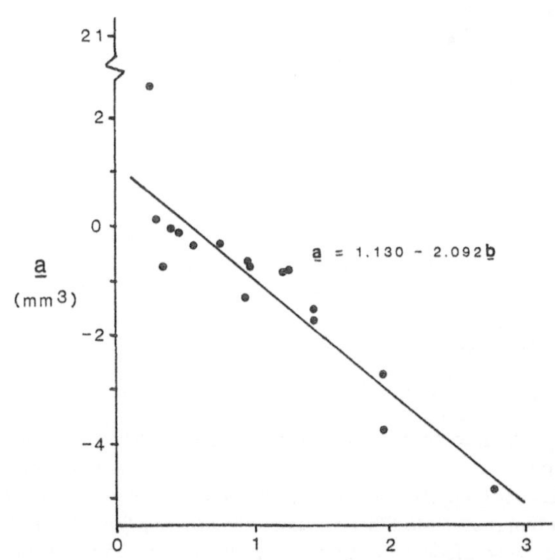

Figure 14. Various cumaceans. Intercept (a, brood volume in ml) in fecundity-length linear regression as a function of carapace length coefficient b, for 17 species. Data and regressions from Corey (1981). The derived regression excludes the aberrant point at the upper right (representing *Diastylis goodsiri*, a much larger species).

anovulatory molt cycles to be effectively semelparous, but it is reasonable to suggest that the larger species perhaps live longer and have more chance to reproduce again. The regression of *b* upon species size indicates the evolutionary relationship between size and allocation schedule we might expect to find with groups of iteroparous crustaceans.

As Hines (1982) pointed out, it is usually more appropriate for comparative purposes to fit the size-fecundity relationship with such a power function. A common assumption is that fecundity ought to be a cubic function of length ($b = 3$, e.g. Jensen 1958, see also Somers, this volume). When the exponent *b* is allowed to 'float' it may become apparent (Fig. 15) that considerable variation exists between populations within a species, at least in the larger decapods. Again, some of this variation may be explained as sampling error. As with the linear regressions, the straight line relating *ln a* to *b* for different populations of the same species is

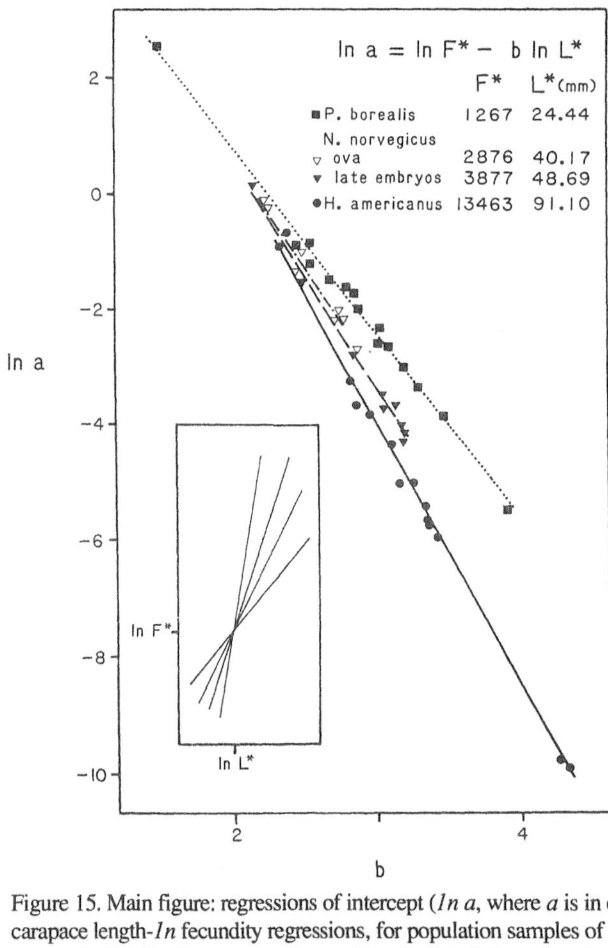

Figure 15. Main figure: regressions of intercept (*ln a*, where *a* is in egg numbers) upon coefficient (b) of *ln* carapace length-*ln* fecundity regressions, for population samples of three decapod species: *Homarus americanus*, circles; *Nephrops norvegicus*, open triangles (oöcytes) and solid triangles (eggs); *Pandalus borealis*, squares. Each derived regression represents a family of *ln* length-*ln* fecundity regressions (*ln L**, *Ln F**). *Pandalus* regressions from Parsons & Tucker (1986); *Homarus* and *Nephrops* regressions compiled from various sources (available from author upon request).

close to the major axis of the joint confidence ellipse of *ln a* and *b* for any population; its slope represents a family of regressions of log fecundity (*ln F*) upon log length (*ln L*) that seesaw about a common point (*ln L**, *ln F**; inset). But the variation is too great to be explained entirely as sampling error, and some of it may represent reaction-norm variation.

If such reaction-norm variation exists, how should the allocation schedule shift with fluctuation in resources? What kind of reaction norm, if any, obtains for a population may be tested by comparing size-specific growth increments for different years with the corresponding size-specific fecundities. If simple competitive economics govern the allocations to growth and reproduction, and they remain proportional, we might expect a change in *ln a* without a corresponding change in *b*. This does not appear to be the case (Fig. 15). Alternatively, if somatic growth takes precedence, and the reproductive allocation is of the surplus after growth and metabolic needs are satisfied, *b* might be expected to decrease in times of abundance and to increase in times of hardship. However, if iteroparity evolved as a 'bet-hedging' response to highly variable juvenile mortality but relatively low adult mortality (Murphy 1968, Schaffer 1974), resources for the adults may have fluctuated inversely with longer-term fluctuations in cumulated recruitment and therefore in adult density. The 'appropriate' (i.e. cogradient) reaction norm for a smaller adult in such a case may be a greater allocation to growth in times of abundance to gain the benefits in fecundity of growing to a larger size, and in times of dearth or other stress a greater relative allocation to reproduction, as in the sowbugs and hermit crabs discussed above. We would expect *b* then to increase with decreasing hardship.

What will happen to such a species if it is subjected to severe and prolonged adult mortality, as with heavy fishing pressure? Even when not size-selective, the effect is to change the age and size structure by progressive elimination of the larger animals, to the point that with size-selection the population may become effectively semelparous (Nelson & Soulé 1987). With decrease in adult population density, the remaining small adults may experience a greater supply of resource and allocate less to reproduction and more to growth, in the expectation of enhanced later reproduction. But the phenotypic response is now no longer cogradient but countergradient, and exacerbates the heavy selective pressure toward egg production at smaller sizes imposed by exploitation. Whether this scenario accurately describes our crustacean fisheries remains to be seen.

6 CONCLUSIONS

Upon closer inspection, the malacostracan molting and reproductive cycles take on the aspect of more-or-less tightly-coupled oscillators, with the interaction ranging from the relative independence we see in non-brooding types to synchronous cycles in the peracarids. In the latter, cohorts of oöcytes acquire follicular envelopes after ecdysis and undergo secondary vitellogenesis in step with the molt cycle. Extrusion of eggs follows the molt. Not only both the beginning and the end of vitellogenesis but the following ovulation as well appear ordinarily to be signalled by the peak and sharp drop in circulating ecdysteroids of later premolt. Extrusion is typically accompanied by molt delay, which most often takes the form of a prolonged premolt.

In most reptant decapods the coupling is looser but still apparent; reproduction and molting tend to alternate. Instead of closely following the molt, extrusion must usually await molt stage C_3 or C_4, accompanied by a more-or-less prolonged period of delay of premolt which may be enhanced by presence of the brood. The increase in intermolt duration appears designed to benefit the brood, not the parent. In some brachyurans there seems to be a cycle of predisposition alternating between molting and reproduction; the inhibitory hormone(s) of the eyestalk modulate this cycle into an annual cycle appropriate to the environment (Kurup & Adiyodi 1981).

Scheduling of molting and reproduction does not appear to be governed by competition for resources nor particularly by the need to avoid it, nor usually by other antagonisms between the processes. Instead, while molting or reproductive processes may each be postponed and thereby accommodate the other, the impression is rather one of mutual interdependence and contingency. Vitellogenesis becomes contingent upon successful ecdysis, molting upon successful release of the brood, or successful release of the brood upon the processes of premolt. The peculiarity that in malacostracans the major endocrine modulators are often inhibitory in action may have fostered the impression of antagonism (one may as well say that growth of the epidermis is antagonistic to growth of the muscles!).

The iteroparous malacostracan must allocate resources to maintenance, growth and reproduction according to its current and projected needs and goals. The size of the cohort of eggs is predicted well in advance by the initial allocation of oöcytes to that cohort and need not reflect current resource availability. The allocations to growth and reproduction may be made according to a kind of projection of future growth and reproduction gleaned from the history of the population and current conditions, which perhaps we may understand by study of variation in length-growth and length-fecundity schedules.

If variation in the length-fecundity relationship indeed represents a norm of reaction to the availability of resources by such iteroparous species, this norm of reaction may have evolved in conditions of low adult mortality and high and variable juvenile mortality. The appropriate reaction for smaller adults of such a species may be a shift in the allocation toward reproduction in times of scarcity and toward growth in times of abundance. If this is the case in commercially-important species, this reaction norm may be counter-gradient to the new selection regime imposed by intensive fishing, and exacerbate it.

ACKNOWLEDGEMENTS

This work is a result of research sponsored in part by NOAA, National Sea Grant College Program, Department of Commerce, under Grant No. 04-8-M01-189 (to D. Hedgecock) and Grant NA80AA-D-00120, Project R/A-61 (to E.S. Chang) through the California Sea Grant College Program. The US Government is authorized to reproduce and distribute for government purposes. I thank Diane Cosgrove and Carol Culp for manuscript preparation and E.S. Chang, D. Hedgecock, M.J. Snyder and the editors for comments.

REFERENCES

Adiyodi, K.G. & R.G. Adiyodi 1970. Endocrine control of reproduction in decapod Crustacea. *Biol. Rev.* 45: 121-165.

Adiyodi, K.G. & R.G. Adiyodi 1981. Reproduction vs. growth: endocrine programming in the Brachyura (Crustacea: Decapoda). *Proc. 9th Int. Symp. Comp. Endocrinol.*, Hong Kong, p. 238.

Adiyodi, R.G. 1968. On reproduction and molting in the crab *Parathelphusa hydrodromus*. *Physiol. Zool.* 41: 204-209.

Adiyodi, R.G. 1985. Reproduction and its control. In D.E. Bliss (ed.) *The Biology of Crustacea* vol. 9, pp. 147-215. New York: Academic Press.

Adiyodi, R.G. & T. Subramoniam 1983. Arthropoda-Crustacea. In K.G. Adiyodi & R.G. Adiyodi (eds), *Reproductive Biology of Invertebrates* vol. 1, pp. 443-495. Chichester: Wiley.

Aiken, D.E. 1980. Molting and growth. In J.S. Cobb & B.F. Phillips (eds), *The Biology and Management of Lobsters* vol. 1, pp. 91-163. New York: Academic Press.

Aiken, D.E. & S.L. Waddy 1980. Reproductive biology. In J.S. Cobb & B.F. Phillips (eds), *The Biology and Management of Lobsters* vol. 1, pp. 215-276. New York: Academic Press.

Allen, J.A. 1966. The dynamics and interrelationships of mixed populations of Caridea found off the northeast coast of England. In H. Barnes (ed.), *Some Contemporary Studies in Marine Science*. pp. 45-66. London: Allen & Unwin.

Anderson, S.L., W.H. Clark, Jr & E.S. Chang 1985. Multiple spawning and molt synchrony in a free spawning shrimp (*Sicyonia ingentis*: Penaeoidea). *Biol. Bull.* 168: 3377-394.

Aquacop 1977. Observations sur la maturation et la reproduction in captivité des crevettes Pénéides in milieu tropical. *Third meeting of ICES Working Group on Mariculture*, Brest, France, 10-13 May, 1977. *Actes de Colloques du CNEXO* 4: 157-178.

Barclay, M.C., W. Dall & D.M. Smith 1983. Changes in lipid and protein during starvation and the molting cycle in the tiger prawn, *Penaeus esculentus* Haswell. *J. Exp. Mar. Biol. Ecol.* 68: 229-244.

Bell, G. 1980. The costs of reproduction and their consequences. *Amer. Nat.* 116: 45-76.

Berglund, A. & G. Rosenqvist 1986. Reproductive costs in the prawn *Palaemon adspersus*: effects upon growth and predator vulnerability. *Oikos* 46: 349-354.

Bertness, M.D. 1981. Pattern and plasticity in tropical hermit crab growth and reproduction. *Amer. Nat.* 117: 754-773.

Blanchet-Tournier, M.-F. 1982. Quelques aspects des interactions hormonales entre la mue et la vitellogenèse chez le Crustacé Amphipode *Orchestia gammarellus* (sic) (Pallas). *Reprod. Nutr. Dévelop.* 22: 325-344.

Blanchet-Tournier, M.-F. 1987. Mise en évidence d'une activité neuro-hormonal inhibitrice de la vitellogenèse chez l'amphipode *Orchestia gammarella*. *Can. J. Zool.* 65: 1917-1922.

Bliss, D.E. & L.H.Mantel (eds) 1985. *The Biology of Crustacea* vol. 9. New York: Academic Press.

Bomirski, A. & F. Klek 1974. Action of eyestalks on the ovary in *Rhithropanopeus harrisii* and *Crangon crangon*. *Mar. Biol.* 24: 329-337.

Botsford, L.W. 1985. Models of growth. In A.M. Wenner (ed.), *Factors in adult growth, Crustacean Issues* 3: pp. 171-188. Rotterdam: Balkema.

Branford, J.R. 1979. Locomotor activity and food consumption by the lobster *Homarus gammarus*. *Mar. Behav. Physiol.* 6: 13-24.

Brody, M.S., M.H. Edgar & L.R. Lawlor 1983. A cost of reproduction in a terrestrial isopod. *Evol.* 37: 653-655.

Bückle-Ramírez, L.F. 1965. Untersuchungen über die Biologie von *Heterotanais oerstedi* Kröyer (Crustacea, Tanaidacea). *Z. Morph. Ökol. Tiere* 55: 714-782.

Byard, E.H., R.R. Shivers & D.E. Aiken 1975. The mandibular organ of the lobster, *Homarus americanus*. *Cell Tissue Res.* 162: 13-22.

Carlisle, D.B. 1957. On the hormonal inhibition of molting in decapod Crustacea. II. The terminal anecdysis in crabs. *J. Mar. Biol. Ass. UK* 36: 291-307.

Carlisle, D.B. 1960. Molting cycles in Crustacea. *Symp. Zool. Soc. Lond.* 2: 109-120.

Carlisle, D.B. & P.F.R. Dohrn 1953. Studies on *Lysmata seticaudata* Risso (Crustacea Decapoda). II. Experimental evidence for a growth- and moult-accelerating factor obtainable from eyestalks. *Pubbl. Staz. Zool. Napoli* 24: 69-83.

Caswell, H. 1983. Phenotypic plasticity in life-history traits: demographic effects and evolutionary consequences. *Amer. Zool.* 23:35-46.

Chaix, J.-C. & M. De Reggi 1982. Ecdysteroid levels during ovarian development and embryogenesis in the spider crab *Acanthonyx lunulatus*. *Gen. Comp. Endocrinol.* 47: 7-14.

Chang, E.S. 1985. Hormonal control of molting in decapod Crustacea. *Amer. Zool.* 25: 179-185.

Chang, E.S. & J.D. O'Connor 1988. Crustacea: Molting. In H. Laufer & R.G.H. Downer (eds), *Endocrinology of selected invertebrate types*. pp. 259-278. New York: Alan R. Liss.

Charniaux-Cotton, H. 1974. Données nouvelles concernant la vitellogenèse des Crustacés Malacostracés obtenues chez l'Amphipode *Orchestia gammarellus* (sic) (Pallas): folliculogenèse à partir d'un tissu permanent; action du busulfan; action inhibitrice de l'hormone juvénile. *C.R. Acad. Sci. Paris* 279: 563-566.

Charniaux-Cotton, H. 1985. Vitellogenesis and its control in malacostracan Crustacea. *Amer. Zool.* 25: 197-206.

Charniaux-Cotton, H. & G. Payen 1985. Sexual differentiation. In D.E. Bliss & L.H. Mantel (eds), *The Biology of Crustacea* vol. 9, pp. 217-300. New York: Academic Press.

Charniaux-Cotton, H. & G. Payen 1988. Crustacean reproduction. In H. Laufer & R.G.H. Downer (eds), *Endocrinology of Selected Invertebrate Types*. pp. 279-303. New York: Alan R. Liss.

Cheung, T.S. 1966. The interrelations among three hormonal-controlled characters in the adult female shore crab, *Carcinus maenas* (L.) *Biol. Bull.* 130: 59-66.

Cheung, T.S. 1969. The environment and hormonal control of reproduction in the adult female stone crab, *Menippe mercenaria* (Say). *Biol. Bull.* 136: 327-346.

Clutter, R.I. & G.H. Theilacker 1971. Ecological efficiency of a pelagic mysid shrimp, estimates from growth, energy budget, and mortality studies. *Fish. Bull. (US)* 69: 93-115.

Conan, G.Y. 1985. Periodicity and phasing of molting. In A.M. Wenner (ed.), *Factors in adult growth, Crustacean Issues* 3: 73-99. Rotterdam: Balkema.

Corey, S. 1981. Comparative fecundity and reproductive strategies in seventeen species of the Cumacea (Crustacea: Peracarida). *Mar. Biol.* 62: 65-72.

Cox, G.W. & G.H. Dudley 1968. Seasonal patterns of reproduction of the sand crab, *Emerita analoga*, in Southern California. *Ecol.* 49: 746-751.

Cuzin-Roudy, J. & A.S.M. Saleuddin 1985. A study of the neurosecretory centers of the eyestalks in *Siriella armata* M. Edw. (Crustacea: Mysidacea): their involvement in molting and reproduction. *Can. J. Zool.* 63: 2783-2788.

Cuzin-Roudy, J. & C. Tchernigovtzeff 1985. Chronology of the female molt cycle in *Siriella armata* M. Edw. (Crustacea: Mysidacea) based on marsupial development. *J. Crust. Biol.* 5: 1-14.

Démeusy, N. 1962. Rôle de la glande de mue dans l'evolution ovarienne du crabe *Carcinus maenas* Linné. *Cah. Biol. Mar.* 3: 37-56.

Démeusy, N. 1963. Étude d'une population de *Carcinus maenas* L. des côtes de la Manche: cycle génital ovarien. *C.R. Acad. Sci. Paris* 256: 4095-4097.

Démeusy, N. 1965. Croissance somatique et fonction de reproduction chez la femelle du décapode brachyoure *Carcinus maenas* Linné. *Arch. Zool. Exp. Gen.* 106: 625-663.

Dingle, H. & R.L. Cladwell 1972. Reproductive and maternal behavior of the mantis shrimp *Gonodactylus bredini* Manning (Crustacea: Stomatopoda). *Biol. Bull.* 142: 417-426.

Drach, P. 1939. Mue et cycle d'intermue chez les Crustacés Décapodes. *Ann. Inst. Océanogr. (Paris)* [N.S] 19: 103-391.

Draper, N. & H. Smith 1981. *Applied Regression Analysis*, 2nd ed. Wiley, New York.

Durliat, M. 1983. Etude immunochimique des protéins de divers tissus chez les Crustacés Décapodes: nature, rôle, origine. *Arch. Int. Physiol. Biochim.* 91: 297-332.

Emmerson, W.D. 1980. Induced maturation of prawn *Penaeus indicus*. *Mar. Ecol. Prog. Ser.* 2: 121-131.

Faure, Y., C. Bellon-Humbert & H. Charniaux-Cotton 1981. Folliculogenèse et vitellogenèse secondaires chez la Crevette *Palaemon serratus* (Pennant); contrôle par les pédoncules oculaires et l'organe X de la medulla externa (MEX). *C.R. Acad. Sci. Paris* 293: 461-466.

Fauvel, C. 1981. Etude de l'ovaire de la crevette d'eau douce *Macrobrachium rosenbergii* (de Man) au cours de cycle de reproduction. Première description de la folliculogenèse secondaire chez un Crustacé Décapode. *C.R. Acad. Sci. Paris* 292: 547-552.

Fingerman, M. 1987. The endocrine mechanisms of crustaceans. *J. Crust. Biol.* 7: 1-24.

Gibson, R. & P.L. Barker 1979. The decapod hepatopancreas. *Oceanogr. Mar. Biol. Ann. Rev.* 17: 285-346.

Glynn, J.P. 1968. Studies on the ionic, protein and phosphate changes associated with the molt cycle of *Homarus vulgaris*. *Comp. Biochem. Physiol.* 26: 937-946.

Gollub, J.P., T.O. Brunner & B.G. Danly 1978. Periodicity and chaos in coupled nonlinear oscillators. *Science* 200: 48-50.

Harrington, S.A. & T. Ikeda 1986. Laboratory observations on spawning brood size and egg hatchability of the Antarctic krill *Euphausia superba* from Prydz Bay, Antarctica. *Mar. Biol.* 92: 231-235.

Hartnoll, R.G. 1963. The biology of Manx spider crabs. *Proc. Zool. Soc. London* 141: 423-496.

Hartnoll, R.G. 1965. The biology of spider crabs: a comparison of British and Jamaican species. *Crustaceana* 9: 1-16.

Hartnoll, R.G. 1969. Mating in the Brachyura. *Crustaceana* 16: 161-181.

Hartnoll, R.G. 1982. Growth. In L.G. Abele (ed.), *The Biology of Crustacea*, vol. 2, pp. 111-196. New York: Academic Press.

Hartnoll, R.G. 1983. Strategies of crustacean growth. *Mem. Austr. Mus.* 18: 121-131.

Hartnoll, R.G. 1985. Growth, sexual maturity and reproductive output. In A.M. Wenner (ed.), *Factors in adult growth, Crustacean Issues* 3: pp. 101-128. Rotterdam: Balkema.

Heath, J.R. & H. Barnes 1970. Some changes in biochemical composition with season and during the molt cycle of the common shore crab, *Carcinus maenas* (L.) *J. Exp. Mar. biol. Ecol.* 5: 199-233.

Hedgecock, D., K. Nelson & R.A. Shleser 1976. Growth differences among families of the lobster, *Homarus americanus*. *Proc. World Maricult. Soc.* 7: 347-361.

Hess, W.N. 1941. Factors influencing moulting in the crustacean, *Crangon armillatus*. *Biol. Bull.* 81: 215-220.

Hiatt, R.W. 1948. The biology of the lined shore crab, *Pachygrapsus crassipes* Randall. *Pacif. Sci.*: 2: 135-213.

Hines, A.H. 1982. Allometric constraints and variables of reproductive effort in brachyuran crabs. *Mar. Biol.* 69: 309-320.

Hinsch, G.W. 1968. Reproductive behavior in the spider crab *Libinia emarginata*. *Biol. Bull.* 135: 273-278.

Hubbell, S.P. 1971. Of sowbugs and systems: the ecological bioenergetics of a terrestrial isopod. In B.C. Patten (ed.), *Systems Analysis and Simulation in Ecology* vol. 1. 269-324. New York: Academic Press.

Iles, T.D. 1984. Allocation of resources to gonad and soma in Atlantic herring *Clupea harengus* L. In G.W. Potts & R.J. Wootton (eds), *Fish Reproduction: Strategies and Tactics*, pp. 331-347. London: Academic Press.

Jassem, W., P. Juchault & J.-P. Mocquard 1982. Déterminisme de la reproduction saisonnière des femelles d'*Armidillidium vulgare* Latr. (Crustacé, Isopode, Oniscoïde). *Ann. Sci. Nat. Zool. Biol. Anim.* 4: 195-201.

Jefferies, D.J. 1964. The moulting behaviour of *Palaemonetes varians* (Leach) (Decapoda; Palaemonidae). *Hydrobiol.* 24: 457-488.

Jensen, J.P. 1958. The relation between body size and number of eggs in marine malacostrakes. *Medd. Dan. Fisk- og Havunders* (N.S.) 2, 11: 1-25.

Kaestner, A. 1970. *Invertebrate Zoology* Vol. III. New York: Interscience Publishers.

Kamiguchi, Y. 1971a Studies on the molting in the freshwater prawn, *Palaemon paucidens*. I. Some endogeneous and exogenous factors influencing the intermolt cycle. *J. Fac. Sci. Hokkaido Univ. Ser. IV, Zool.* 18: 15-23.

Kamiguchi, Y. 1971b. Studies on the molting in the freshwater prawn *Palaemon paucidens*. II. Effects of eyestalk removal in relation to the state of ovarian growth. *J. Fac. Sci. Hokkaido Univ. Ser. VI, Zool.* 18: 24-31.

Kurata, H. 1962. Studies on the age and growth of Crustacea. *Bull. Hokkaido Reg. Fish. Res. Lab.* 24: 1-115.

Kuris, A.M. 1971. Population interactions between a shore crab and two symbionts. Ph.D. thesis, University of California, Berkeley.

Kurup, K.N.P. & R.G. Adiyodi 1981. The programming of somatic growth and reproduction in the crab, *Parathelphusa hydrodromus* (Herbst). *Int. J. Invert. Repro.* 3: 27-39.

Lachaise, F. & J.A. Hoffmann 1977. Ecdysone et développement ovarien chez un Décapode, *Carcinus maenas*. *C.R. Acad. Sci. Paris* (D.) 285: 701-704.

Laufer, H., D. Borst, F.C. Baker, C. Carrasco, M. Sinkus, C.C. Reuter, L.W. Tsai & D.A. Schooley 1987. Identification of a juvenile-hormone-like compound in a crustacean. *Science* 235: 202-205.

Lawlor, L.R. 1976. Molting, growth and reproductive strategies in the terrestrial isopod, *Armadillidium vulgare*. *Ecology* 57: 1179-1194.

Legrand, J.-J., G. Martin, P. Juchault & G. Besse 1982. Contrôle neuro-endocrine de la reproduction chez les Crustacés. *J. Physiol., Paris* 78: 543-552.

Lipcius, R.N. 1985. Size-dependent reproduction and molting in spiny lobsters and other long-lived decapods. In A.M. Wenner (ed.), *Factors in adult growth, Crustacean Issues* 3: pp. 129-148. Rotterdam: Balkema.

Luxmoore, R.A. 1982. Moulting and growth in serolid isopods. *J. Exp. Mar. Biol. Ecol.* 56: 63-85.

Luxmoore, R.A. 1985. The energy budget of a population of the antarctic isopod *Serolis polita*. In W.R. Siegfried, P.R. Condy & R.M. Laws (eds), *Antarctic nutrient cycles and food webs:* pp. 389-396. Berlin: Springer.

Marzari, R. & E.A. Ferrero 1986. Different tissue compartments cooperate in producing yolk proteins in a crustacean. *J. Cell Biol. Suppl.* 16: A-113.

Mauchline, J. 1973. Inter-moult growth of species of Mysidacea (Crustacea). *J. Mar. Biol. Assoc. UK.* 53: 569-572.

Mauchline, J. 1977. Growth of shrimps, crabs and lobsters – an assessment. *J. Cons. Int. Explor. Mer.* 37: 162-169.

McMullen, J.C. 1969. Effects of delayed mating on the reproduction of king crab, *Paralithodes camtschatica. J. Fish. Res. Bd. Canada* 26: 2737-2740.

Meusy, J.-J. & H. Charniaux-Cotton 1984. Endocrine control of vitellogenesis in Malacostraca crustaceans. In W. Engels et al. (eds), *Advances in Invertebrate Reproduction* vol. 3. pp. 231-241. Amsterdam: Elsevier.

Mocquard, J.-P., G. Besse, P. Juchault, J.-J. Legrand, J. Maissait, G. Martin & J.-I. Picaud 1976. Durées des cycles de mue chez les femelles de l'oniscoide *Porcellio dilatatus* Brandt, suivant leur état sexuel et les conditions d'élevage: température, photopériode et groupement. *Bull. Ecol.* 7: 297-314.

Murphy, G.I. 1968. Pattern in life history and the environment. *Amer. Nat.* 102: 391-403.

Nelson, K. 1986. Photoperiod and reproduction in lobsters (*Homarus*). *Amer. Zool.* 26: 447-457.

Nelson, K., D. Hedgecock & W. Borgeson 1983. Photoperiodic and ecdysial control of vitellogenesis in lobsters (*Homarus*) (Decapoda, Nephropidae). *Can. J. Fish. Aquat. Sci.* 40: 940-947.

Nelson, K., D. Hedgecock & W. Borgeson 1988a. Effects of reproduction upon molting and growth in female lobsters (*Homarus americanus*). *Can. J. Fish. Aquat. Sci.* 45: 805-821.

Nelson, K., D. Hedgecock & W. Borgeson 1988b. Factors influencing egg extrusion in the American lobster (*Homarus americanus*). *Can. J. Fish. Aquat. Sci.* 45: 797-804.

Nelson, K., B. Heyer, E. Johnson, D. Hedgecock & E.S. Chang 1988c. Photoperiod-induced changes in hemolymph vitellogenins in female lobsters (*Homarus americanus*). *Comp. Biochem. Physiol.* 90B: 809-821.

Nelson, K. & M. Soulé 1987. Genetical conservation of exploited fishes. In N. Ryman & F. Utter (eds), *Population Genetics and Fisheries Management.* pp. 345-368. Seattle: University of Washington Press.

Newman, W.A. 1983. Origin of the Maxillopoda; urmalocostracan ontogeny and progenesis. In F.R. Schram (ed.), *Crustacean phylogeny. Crustacean Issues* 1: pp. 105-119. Rotterdam: Balkema.

Pandian, T.J. & C. Balasundaram 1982. Moulting and spawning cycles in *Macrobrachium nobilii* (Henderson & Mathai). *Int. J. Invert. Repro.* 5: 21-30.

Parsons, D.G. & G.E. Tucker 1986. Fecundity of northern shrimp, *Pandalus borealis* (Crustacea, Decapoda) in areas of the Northwest Atlantic. *Fish. Bull. (US) 84: 549-558.*

Passano, L.M. 1960. Molting and its control. In T.H. Waterman (ed.), *The Physiology of Crustacea* vol. 1. pp. 473-536. New York: Academic Press.

Paulus, J.E. & H. Laufer 1987. Vitellogenocytes in the hepatopancreas of *Carcinus maenas* and *Libinia emarginata* (Decapoda brachyura). *Int. J. Invert. Repro. Devel.* 11: 29-44.

Perkins, H.C. 1972. Developmental rates at various temperatures of embryos of the northern lobster *Homarus americanus* Milne Edwards. *Fish. Bull.* 70: 95-99.

Pike, R.B. & D.I. Williamson 1959. Observations on the distribution and breeding of British hermit crabs and the stone crab (Crustacea: Diogenidae, Paguridae and Lithodidae). *Proc. Zool. Soc. London.* 132: 551-567.

Pillay, K.K. & N.B. Nair 1973. Observations on the biochemical changes in gonads and other organs of *Uca annulipes, Portunus pelagicus* and *Metapenaeus affinis* (Decapoda; Crustacea) during the reproductive cycle. *Mar. Biol.* 18: 167-198.

Quackenbush, L.S. 1986. Crustacean endocrinology: a review. *Can. J. Fish. Aquat. Sci.* 43: 2271-2282.

Ranga Rao, K. & M.J. Inkel Jr 1975. Hormonal control of molting in ovigerous *Palaemonetes pugio. Florida Scientist* 38 (suppl. 1):10.

Rao, Ch.N., K. Shakuntala & S.R. Reddy 1981. Molt-reproduction relationship in the freshwater prawn *Macrobrachium lanchesteri* (de Man). *Proc. Ind. Acad. Sci. (Anim. Sci.)* 90: 39-52.

Read, G.H.L. & M.S. Caulton 1980. Changes in mass and chemical composition during the molt cycle and ovarian development in immature and mature *Penaeus indicus* Milne Edwards. *Comp. Biochem. Physiol.* 66A: 431-437.

Reaka, M.L. 1976. Lunar and tidal periodicity of molting and reproduction in stomatopod Crustacea: a selfish herd hypothesis. *Biol. Bull.* 150: 468-490.

Renaud, L. 1949. Le cycle des réserves organiques chez les Crustacés Décapodes. *Ann. Inst. Océanogr. (Paris) [N.S.]* 24: 259-357.

Reznick, D. 1983. The structure of guppy life histories: the trade-off between growth and reproduction. *Ecology* 64: 862-873.

Roff, D.A. 1982. Reproductive strategies in flatfishes. A first synthesis. *Can. J. Fish. Aquat. Sci.* 39: 1686-1698.

Ryan, E.P. 1967. Structure and function of the reproductive system of the crab *Portunus sanguinolentis* (Herbst) (Brachyura: Portunidae). II. The female system. *Proc. Ernakulam Symp. Crustacea* 1965, part II. Mandapam Camp, India.

Sareen, M.L. & K.G. Adiyodi 1983. Arthropoda-Myriapoda. In K.G. Adiyodi & R.G. Adiyodi (eds), *Reproductive Biology of Invertebrates* vol. 1. pp. 497-520. Chichester: Wiley & Sons.

Sastry, A.N. 1983. Ecological aspects of reproduction. In D.E. Bliss & W.B. Vernberg (eds), *The Biology of Crustacea* vol. 8. (F)179-270. New York: Academic Press.

Schaffer, W.M. 1974. Optimal reproductive effort in fluctuating environments. *Amer. Nat.* 108: 783-790.

Scheer, B.T. 1960. Aspects of the inter-molt cycle in natantians. *Comp. Biochem. Physiol.* 1: 3-18.

Schram, F.R. 1986. *Crustacea.* Oxford University Press, New York.

Scudamore, H.H. 1948. Factors influencing molting and the sexual cycles in the crayfish. *Biol. Bull.* 95: 229-237.

Siegel, P.R. 1984. Food-induced size-specific molt synchrony of the sand crab, *Emerita analoga* (Stimson). *Biol. Bull.* 167: 579-589.

Skinner, D.M. 1985a. Interacting factors in the control of the crustacean molt cycle. *Amer. Zool.* 25: 275-284.

Skinner, D.M. 1985b. Molting and regeneration. In D.E. Bliss & L.H. Mantel (eds), *The Biology of Crustacea* vol. 9. pp. 44-146. New York: Academic Press.

Stearns, S.C. & J.C. Koella 1986. The evolution of phenotypic plasticity in life-history traits: predictions of reaction norms for age and size at maturity. *Evol.* 40: 893-913.

Steel, C.G.H. 1980. Mechanisms of coordination between moulting and reproduction in terrestrial isopod Crustacea. *Biol. Bull.* 159: 206-218.

Steele, D.H. & V.J. Steele 1986. The cost of reproduction in the amphipod *Gammarus lawrencianus* Bousfield, 1956. *Crustaceana* 51: 176-182.

Stewart, J.E., J.W. Cornick, D.M. Foley, M.F. Li & C.M. Bishop 1967. Muscle weight relationship to serum proteins, hemocytes and hepatopancreas in the lobster, *Homarus americanus*. *J. Fish. Res. Bd. Canada* 24: 2339-2354.

Swartz, R.C. 1976. Agonistic and sexual behavior of the xanthid crab, *Neopanope sayi*. *Chesapeake Sci.* 17: 24-34.

Swartz, R.C. 1978. Reproductive and molt cycles in the xanthid crab *Neopanope sayi* (Smith, 1969). *Crustaceana* 34: 15-32.

Tom, M., M. Goren & M. Ovadia 1987. Localization of the vitellin and its possible precursors in various organs of *Parapenaeus longirostris* (Crustacea, Decapoda, Penaeidae). *Int. J. Invert. Repro. Devel.* 12: 1-12.

Tuomi, J., T. Hakala & E. Haukioja 1983. Alternative concepts of reproductive effort, costs of reproduction, and selection in life-history evolution. *Amer. Zool.* 23: 25-34.

Vernet-Cornubert, G. 1958. Biologie générale de *Pisa tetraodon*. *Bull. Inst. Océanogr. Monaco* 1113: 1-52.

Vidal, J. 1980. Physioecology of zooplankton. II. Effects of phytoplankton concentration, temperature, and body size on the development and molting rates of *Calanus pacificus* and *Pseudocalanus* sp. *Mar. Biol.* 56: 135-146.

Waddy, S.L. & D.E. Aiken 1986. Multiple fertilization and consecutive spawning in large American lobsters *Homarus americanus*. *Can. J. Fish. Aquat. Sci.* 43: 2291-2294.

Wallace, M.M., C.J. Pertuit & A.R. Hvatum 1949. Contribution to the biology of the king crab (*Paralithodes camtschatica* Tilesius). *US Fish. Wildlife Serv. Fishery Leaflet* 340: 1-50.

Wittmann, K.J. 1978. Adoption, replacement, and identification of young in marine Mysidacea Crustacea). *J. Exp. Mar. Biol. Ecol.* 32: 259-274.

Wittmann, K.J. 1981. On the breeding biology and physiology of marsupial development in mediterranean *Leptomysis* (Mysidacea: Crustacea) with special reference to the effects of temperature and egg size. *J. Exp. Mar. Biol. Ecol.* 53: 261-279.

Yudin, A.I., R.A. Diener, W.H. Clark Jr & E.S. Chang 1980. Mandibular gland of the blue crab, *Callinectes sapidus*. *Biol. Bull.* 159: 760-772.

2. Variations in life history patterns

A review of patterns and causes of crustacean brood mortality

Armand M. Kuris
Department of Biological Sciences & Marine Science Institute, University of California, Santa Barbara, USA

ABSTRACT

Brood mortality of crustaceans is reviewed, with emphasis on Malacostraca. There is considerable variation in size-specific probability of egg loss. About half the data sets exhibit a constant proportional loss of the clutch over the size range of brooding females; 30% show a decrease in brood loss with increasing size and 20% show a relative increase in egg mortality with increasing maternal size. These patterns are discussed in terms of stock-recruitment relationships and with respect to life history theory. No significant patterns were found between taxa, habitats, climates or intensities of brood loss.

The causes of brood loss are reviewed in sequence: from oviposition, fertilization and egg adhesion, through embryogenesis, to eclosion. Losses are most intense at oviposition or late in embryogenesis. The relative roles of mechanical losses, various diseases and egg predators is reviewed. Nemertean egg predators are often an important cause of mortality in decapod crustaceans. Fungal diseases of eggs are often a threat to cultured animals, but there is little evidence that they contribute substantially to egg mortality in nature.

1 INTRODUCTION

Brood mortality is of interest to ecologists, fisheries biologists and evolutionary biologists. For ecologists and fisheries biologists, this interest stems from efforts to quantitatively assess long-term population dynamics. This evaluation has occurred in the context of an examination of life cycles for predictive linkages between successive stages: adult stock size, reproductive output, larval abundance, post-larval recruitment and survivorship to the subsequent adult stock. Some generalities have emerged. Firstly, population studies, many of which are based on analyses of marine fisheries (including crustacean fisheries), have often demonstrated a clear association between the density of juvenile recruits and subsequent abundance of the stock of adults available to the fishery. Hence, strong year classes can often be followed through a fishery (Gulland 1974, Methot & Botsford 1982, Hayes 1983). The importance of recruitment to marine population and community ecology is presently a major research emphasis (Roughgarden et al. 1988, Underwood & Fairweather 1989). Secondly, abundance, size-structure and nutritional status of the adult population impact population fecundity (e.g. Saila & Flowers 1965, Wenner 1977, Wenner et al. 1987). Thirdly, the relationship between the reproductive output of the stock and the subsequent recruitment of postlarval forms to the juvenile population of that stock, termed the stock-recruitment relationship, typically is obscure (Ricker 1954, Gulland 1973, Tang 1985, Ennis 1986, Botsford this volume). Indeed, fisheries models typically assume recruitment to be constant (Botsford this volume). These temporally linked life cycle components are often treated as decoupled, with reproductive output providing little or no predictive value for the determination of juvenile abundance (see Jamieson 1986). However, some recent studies suggest that the relationship between reproduction and recruitment may be more important than previously suspected, particularly for intensive fisheries where the number of reproductive adults is driven to very low levels (Ennis 1986, Botsford this volume).

Reassessment of the importance of reproductive output to crustacean population dynamics should begin with a quantitative evaluation of mortality factors that follow the oviposition of a clutch of eggs. For a crustacean there are losses at oviposition, during embryogenesis, at eclosion, during larval development and at settlement. Larval mortality is always high and larval transport further reduces numbers available for recruitment (e.g. Botsford et al. 1982, Rothlisberg 1982, Johnson et al. 1986, Ennis 1986, Incze et al. 1986, 1987). Losses earlier in the life cycle occur while females brood the eggs. Mortality in these stages was generally ignored until studies showed that brood mortality could be substantial and might have a significant impact on crustacean fisheries (Wickham 1979, 1986). Recently, detailed studies of *Cancer magister* (Dungeness crab), *Paralithodes camtschatica* (red king crab) and *Hemigrapsus oregonensis* (green shore crab) demonstrated that brood mortality due to egg predation was variable, often high, and temporally associated with either the collapse or the nonrecovery of fished stocks (Kuris & Wickham 1987, Shields & Kuris 1988a, Kuris et al. 1990 in press.).

Age-structured population dynamics models have shown that brood mortality caused by egg predators, though density-dependent, could not cause the cyclical dynamics of fished Dungeness crab populations. However, brood mortality may significantly affect the amplitude or frequency of these cycles (Botsford 1986, Kuris & Wickham 1987, Hobbs & Botsford 1989). Models of stock management strategies also show that conservation of the reproductive output may be an effective management strategy (Botsford this volume).

From an evolutionary perspective, brood mortality is of considerable interest, because it may differentially impact reproductive success of individuals. Selection is presumed to maximize reproductive success over the lifetime of the individual. Age-specific fecundity and brood quality generally rise to a maximum and then decline with age (Medawar 1957, Emlen 1970), as reflected in data on human birth defects (Table 1). At older ages, the increase in frequency of defective or unviable young with increasing maternal age parallels the rising age-specific mortality rate (Emlen 1970). The shape of an age-fecundity/brood quality curve is influenced by survivorship, size-specific reproductive output and growth rate.

Organisms with indeterminate growth are particularly interesting. Selection pressure may still fall with advanced age because earlier broods, produced by younger, smaller females generally represent the bulk of the reproductive output of a population. Stronger selection pressure should be exerted on the quality of these broods. If selection weakens, older individu-

Table 1. Human birth defects per 10 000 births (data from Milham & Gittleson 1965).

	Maternal age (yr) 15-19	20-24	25-29	30-34	35-39	40-44	45+
Total defects	84.8	72.0	65.5	67.8	90.9	141.2	203.8

als might be expected to produce relatively low quality young or be unable to protect their young from extrinsic mortality causes. Alternatively, parental care could improve with age. On a per individual basis, larger, older individuals produce considerably more young (see Somers this volume). Late broods may carry more selective weight than early broods because they represent a major part of the reproductive success of these large females. Consequently, older, larger females might better protect their young and the rate of brood mortality should decline with age (size). One might expect variation in brood mortality to be most often associated with female size when mean egg mortality rates are high. Selection pressure should increase with increasing mortality rates.

Decapod and peracarid crustaceans are ideally suited for a comparative analysis of clutch mortality because reproductive females brood their eggs and embryogenesis is synchronous. Thus, egg development stage identifies the relative age of the clutch and permits mortality to be estimated by comparing the number of eggs in early and late clutches from females of similar size.

Here I have compiled literature on crustacean brood mortality where size data were also available and have analyzed these data by taxon, life history parameters, and environmental variables. I also review egg mortality in the sequence from oviposition to eclosion, emphasizing losses due to egg predators. Brood mortality is well recognized as an important factor in cladoceran population dynamics (Green 1974, Threlkeld 1979, Platt & Yamamura 1986) and will not be treated further here. This review focusses on malacostracans as there is little information on brood mortality of other crustaceans.

2 SIZE-SPECIFIC BROOD MORTALITY

Tables 2 and 3 summarize the available data on variation in brood mortality with maternal sizes for decapod and peracarid crustaceans, respectively. The details of these studies vary considerably. Size was based on either total length, carapace length, carapace width, other miscellaneous dimensions (for some peracarids), weight or age (rarely). Size-specific fecundity, an estimate of the number of eggs produced in the ovary for deposition in a single clutch, was based on counts of either ovarian oöcytes, embryos at oviposition, embryos in early development stages, or embryos early in a well-defined brooding season. For smaller crustaceans counts usually direct; for larger forms estimates employing volumetric or proportional weight of subsamples were often used. For some large crabs egg number estimates were based on eggs attached to a single pleopod rather than the entire egg mass. Size-specific natality, an estimate of the number of larvae that hatched from a clutch, was based on counts (or estimates) of hatchlings, embryos in late developmental stages, embryos late in the brooding season, or embryos in any developmental stage (if the fecundity estimate was based on ovarian egg counts). In some cases, brood mortality was directly estimated using counts of empty (dead) egg shells or obviously dead eggs. A few taxonomic designations have been updated. The table does not include data for species in which no mortality was reported nor species for which mortality data was available but not stratified by female size. Such additional mortality records are reviewed by Balasundaram & Pandian (1982) and Moore (1981).

Mortality was recorded as associated with maternal size (increasing or decreasing) if statistical significance was reported or could be estimated. In some cases, mortality is listed if clear trends were evident but insufficient information was available to calculate significance levels. In many cases the smallest and largest size classes were often represented by very small sample

Table 2. Mean brood mortality and association of brood mortality with maternal size in decapod crustaceans, arranged by taxa. *Proportion of infertile eggs.

Species	Mean brood loss (%)	Association with maternal size	Source
Brachyura			
Cancer anthonyi	22	Decreases	Shields & Kuris (unpubl. inform.)
Chionoecetes bairdi	22	Increases	Hilsinger 1976
C. opilio	68	Increases	Brunel 1963
C. opilio	44	Increases	Kon 1974
Hemigrapsus oregonensis	23	None	Kuris 1971
Heterozius rotundifrons	10	None	Jones 1978
Anomura (Lithodidae)			
Lithodes aequispina	11	Decreases	Jewett et al. 1985
Paralithodes camtschatica	58	None	Shields (in prep.)
Anomura (Porcellanidae)			
Petrolisthes elongatus	15	None	Jones 1977
Astacidea (marine)			
Homarus gammarus	38	Decreases	Hepper & Gough 1978
H. americanus	36	None	Perkins 1971
H. americanus	39	None	Campbell & Brattey 1986
Nephrops norvegicus	45	Decreases	Morizur et al. 1981
Astacidea (freshwater)			
Astacus kessleri	71	None	Cherkashina 1970
A. leptodactylus	58	Increases	Cherkashina 1970
Austropotamobius pallipes	33	Increases	Rhodes & Holdich 1982
A. pallipes	19	None	Woodlock & Reynolds 1988b
Orconectes rusticus	28	Increases	Prins 1968
Procambarus clarki	28	None	Penn 1943
P. hayi	37	None	Payne 1971
Palinura			
Panulirus cygnus	29	Decreases	Morgan 1972
P. cygnus	4*	None*	Morgan 1972
Thalassinidea			
Calocaris macandrae	66	None	Buchanan 1963
Caridea			
Macrobrachium nobili	44	Decreases	Balasundaram & Pandian 1982
Palaemon serratus	38	Decreases	Reeve 1969

sizes (less than 5). If data for such size classes conflicted with trends for adjacent size classes with adequate sample sizes such insufficiently sampled size classes were ignored. In a few cases, my interpretation of the data differs from that of the authors.

The principal conclusions drawn from Tables 2 and 3 are that considerable variation occurs in the percentage of brood mortality and that the relationship of this mortality to maternal size is highly variable. In a search for pattern, these data were summarized by taxon (Brachyura, Anomura, Astacidea, Caridea, Isopoda, Amphipoda), mean mortality (3-19%, 20-39%, 40-71%), habitat (marine subtidal, marine intertidal, marine pelagic, estuarine, freshwater, terrestrial) and climatic zones (polar, subarctic, cold temperature, warm temperate, tropical). In all

Table 3. Mean brood mortality and association of brood mortality with maternal size in peracarid crustaceans. *Proportion of infertile eggs (laboratory study).

Species	Mean brood loss (%)	Association with maternal size	Source
Mysidacea			
Metamysidopsis elongata	30	None	Clutter & Thielacker 1971
Mysidopsis bigelowi	16	None	Allen 1984
Neomysis mercedes	37	Increases	Daly & Damkaer 1986
Isopoda (terrestrial)			
Armadillidium album	19	Decreases	Vader & deWolf 1988
A. vulgare	3*	Increases*	Lawlor 1976
A. vulgare	18	Increases	Miller & Cameron 1983
Porcellio scaber	18	None	Brereton 1956
Isopoda (marine)			
Excirolana chiltoni	19	Decreases	Klapow 1970
Iais pubescens	8	Decreases	Marsden 1982
Idotea balthica	27	Decreases	Salemaa 1979
I. chelipes	12	Decreases	Betz 1974
Jaera nordmanni	26	Increases	Jones 1974
Sphaeroma hookeri	16	None	Kinne 1954
Amphipoda			
Amphiporeia lawrenciana	44	Decreases	Downer & Steele 1979
Bathyporeia pilosa	27	Increases	Fish 1975
Calliopus laeviusculus	5	None	Steele & Steele 1973
C. laeviusculus	15	None	Steele & Steele 1973
Corophium bonnellii	14	Increases	Moore 1981
Diogodas littoralis	30	None	Fenwick 1985
Gammarellus angulosus	5	None	Steele & Steele 1972a
G. angulosus	15	Decreases	Steele & Steele 1972a
Gammarus duebeni	23	Decreases	Steele & Steele 1969
G. duebeni	67	Decreases	Steele & Steele 1969
G. finmarchicus	21	None	Steele & Steele 1975b
G. finmarchicus	33	none	Steele & Steele 1975b
G. obtusatus	40	None	Steele & Steele 1970
G. obtusatus	57	Increases	Steele & Steele 1970
G. obtusatus	36	None	Steele & Steele 1970
G. obtusatus	32	None	Steele & Steele 1970
G. oceanicus	8	Increases	Steele & Steele 1972a, b
G. oceanicus	15	None	Steele & Steele 1972a, b
G. oceanicus	5	Decreases	Steele & Steele 1972a, b
G. tigrinus	5	Decreases	Steele & Steele 1972b
G. tigrinus	9	Decreases	Steele & Steele 1972b
G. wilkitzkii	48	None	Steele & Steele 1975a
Hippomedopon kergueleni	10	Increases	Bregazzi 1972
H. whero	14	None	Fenwick 1985
Lembos websteri	25	None	Moore 1981
Rhepoxynius abronius	45	Increases	Kemp et al. 1985
Talitrus saltator	28	None	Williams 1978

of these analyses, no significant differences were evident in the distribution of species exhibiting changes in mortality with maternal size.

A few remarks on the distribution of the available size-specific mortality data seem appropriate. By taxon the data are dominated by amphipods (thanks to the prodigious efforts of the Steeles). Particularly few data are available for the otherwise well-studied carideans and hermit crabs. Changes in egg mortality rates with female size are more frequent as mean mortality levels increase, but this trend is not significant. Few data are available for marine pelagic, freshwater and terrestrial habitats. Tropical species have been scarcely investigated with respect to brood mortality.

Several commercially important species exhibited changes in per capita brood mortality with increasing female size (Table 2). Botsford (this volume), modeling the impact of crustacean fisheries management strategies on future yields, shows that raising the legal size limits to exclude more and larger reproductive females from the fishery may increase future reproductive output of a stock and benefit subsequent yields. The changes in size-specific brood mortality documented herein for commercially fished species suggest that egg mortality may have to be included in such models. For species such as *Nephrops norvegicus* (Table 2), the lessened brood mortality seen in large females reinforces the conclusion that large sizes of mature females merit protection. On the other hand, the high per capita egg mortality seen in large female *Chionoecetes* spp. diminishes the need to protect large reproductive females.

3 TIMING OF BROOD MORTALITY

Some studies permitted an analysis of the timing of egg mortality during embryogenesis. Here too, considerable variation exists. For certain peracaridans, most egg mortality occurred late in embryogenesis (Holdich 1968, Bregazzi 1972, Jones 1974, Bishop 1982). However, for some species, brood losses were relatively evenly distributed over the embryogenic period (Fish 1975, Salemaa 1979, Wittmann 1981) or occurred early in development (Sheader & Chia 1970). Data for *Gammarus duebeni* indicate that there is a substantial loss of eggs very early in development (based on a comparison of counts of oögonia and eggs in early embryogenic stages), with negligible egg mortality thereafter until hatching, when egg mortality increased. Steele & Steele (1969) suggested that the poor survival of embryos early in development is due to the particular sensitivity of the early embryogenic stages to temperature stress (Kinne 1959).

For decapods, Balasundaram & Pandian (1982) provided careful quantification of brood losses during embryogenesis. They showed for the freshwater prawn, *Macrobrachium nobili*, that mortality was heaviest at the beginning and end of the 12-14 day brooding period. Mortality in the first three days was strongly size dependent, averaging 55% in the smallest prawn but only 21% in the largest size class. Mortality late in the brooding period was not dependent on maternal size. In contrast, Kon (1974) noted that egg mortalilty for *Chionoecetes opilio* was high early in the brooding season for small crabs, while for large crabs it was high late in the brood period. Brood mortality is greatest late in the egg development period of *Cancer anthonyi* (Shields et al. 1990b) and the American lobster (Aiken & Waddy 1986), but early losses are reported for *Austropotamobius pallipes* (Rhodes & Holdich 1982).

For decapod crustaceans the bulk of the detailed observations on brood mortality have been associated with the study of egg predators, in particular, symbiotic nemerteans. For most of these symbiotic predator-prey systems brood mortality followed the same temporal pattern through host embryogenesis. There was little detectable egg mortality for the first third to half of the brooding period; then the proportion of eggs killed rose steeply and sometimes reached a plateau in advanced embryogenesis. This pattern has been documented for *Hemigrapsus oregonensis* (Kuris 1971, Kuris & Wickham 1987, Shields & Kuris, 1988a) *Cancer magister* (Wickham 1979), *C. anthonyi* (Shields et al. 1990a) and *Paralithodes camtschatica* (Kuris et al. in press). This consistent pattern was apparently caused by three features of the biology of these worms. Firstly, they feed at low rates early in embryogenesis. Some of this early feeding may have been hard to detect because some eggs were partially consumed and sometimes could not be readily distinguished from intact embryos until later in the brooding period (Roe 1984). Secondly, in general, nemertean intensity increased during the brooding period, increasing the numerical response (Kuris 1978, Wickham 1979, Shields & Kuris 1988a, Shields et al. 1990a, Kuris et al. in press). Thirdly, nemerteans primarily eat yolk and kill more eggs to satisfy their yolk requirements as the amount of yolk in an egg decreases during development (Shields & Kuris 1988b). Interestingly, this temporal pattern of egg mortality was not seen in *Homarus americanus* infested with *Pseudocarcinonemertes homari* (Campbell & Brattey 1986). This may be due to the egg stripping behavior that some lobsters exhibit if infected with nemerteans (see 4.5). This host defense mechanism may have prevented a rise in infestation intensity, with a concomitant increase in mortality, later in the brooding season.

For about three-quarters of the species for which data on size-specific brood mortality are available, older, larger females present no evidence of a senescent reduction in egg quality (Tables 2 and 3). For one-third of the species the data suggest that older females either produce higher quality eggs, or better protect their eggs, as size-specific mortality declines with age. Thus, selection pressure generally does not decline with age and thereby decrease the quality of young in the older, larger females, supporting the hypothesis that parental care improves with age (see Shields this volume). A structural basis for improved protection may be the decreased surface to volume ratio of the egg mass of larger females. For example, in the rock lobster, the infertility rate per egg produced is constant over size. Thus, increased survival rate of eggs on larger females is due to improved brood protection of the larger females.

For one-quarter of the species in Tables 2 and 3 proportional egg survivorship declines with age. In these systems senescent processes may play a role. Poorer quality eggs are produced by larger females in at least *Armadillidium vulgare*. In other species, larger females may be less able to protect the eggs.

A number of size-fecundity relationships indicate a reduced rate of increase in egg output with body size for large females (e.g. Somerton & MacIntosh 1985, see Somers this volume for review). This has sometimes been considered evidence for senescence. However, these large females still produce many eggs. A determination of senescence is inappropriate unless environmental causes of reduced productivity are separated from the developmentally senescent, physiological causes of reduced fertility.

4 CAUSES OF BROOD LOSS

Here I summarize egg mortality observations over the reproductive sequence from oviposition, through embryogenesis, to eclosion. With the exception of the several quantified analyses of the impact of nemertean egg predation, relatively few field studies have attemped to determine or separate causes of egg loss. This section will focus on the mechanisms of egg mortality that operate at different phases of the reproductive process.

4.1 *Retention of oöcytes in the ovaries*

A few atretic oöcytes are always retained in the ovary after extrusion of the egg mass onto the pleopods. Losses to atresia in shore crabs, *Hemigrapsus oregonensis*, are usually much less than 1% of the oöcytes (Kuris 1971). In *H. oregonensis* and the rock lobster, *Jasus novaehollandeae*, the distal parts of the anterior and posterior ovarian horns and the vicinity of the seminal receptacle near the oviduct are the most common sites of atretic oöcytes (Fielder 1964, Kuris 1971). Post-spawning atresia was common in natural populations of *Penaeus setiferus* (King 1948), but was not observed in crayfish (Penn 1943), *Nephrops norvegicus* (Thomas 1964), *Calacaris macandreae* (Buchanan 1963) or several species of caridean shrimp (J.W. Born in litt.).

Occasionally, a substantial portion of the clutch fails to exit the ovary at oviposition. In *H. oregonensis*, this was more likely to occur in large crabs. It was sometimes associated with an external injury to the carapace sufficiently severe to block the lumen of the ovary or oviduct, forcing ripe oöcytes to be retained in the distal horns (Kuris 1971). Ryan (1967) experimentally blocked one of the gonopores of the swimming crab, *Portunus sanguinolentus* and nearly halved the brood size. Ripe oöcytes were retained in the ovary on the blocked side posterior to the middle commissure.

Failure to extrude the entire clutch may occur in unmated red king crabs, *Paralithodes camtschatica* (McMullen & Yoshihara 1971). In apparent response to heavy fishing of male crabs in the vicinity of Kodiak Island, 5.8% of 2,475 females retained their entire brood. In other decapods, nonretention of the entire brood has been reported at low frequencies (0.4-2.0%) (Tack 1941, Kuris 1971). Laboratory studies of crayfish, however, found frequent failure to extrude the clutch. Such females sometimes died while resorbing the ripe oöcytes (Penn 1943, Mason 1970).

4.2 *Egg loss at oviposition*

Failure of eggs to adhere to the pleopodal setae, following extrusion of the egg mass through the gonopores, can reduce brood size. Formation of the egg membranes and attachment strands (funiculi) is a complex process (Goudeau & Lachaise 1983, Fisher & Clark 1983, Talbot et al. 1984). Improper formation of the funiculi is a major cause of brood loss in captive *Homarus americanus* (Talbot & Harper 1984). In some crabs a suitable environment appears to be a requisite for the ovipositing female. *Corystes cassivelaunus, Portunus sanguinolentus, Cancer* spp., *Carcinus maenas* and *Homarus americanus* all required appropriate substrate to successfully oviposit. In bare containers all but a few eggs were sloughed (Hartnoll 1968, Broekhuysen 1936, Ryan 1967, Edwards & Meaney 1967, Kuris 1971, Shields this volume). Early brood losses in *Austropotamobius pallipes* may be due to improper positioning of the female at oviposition (Ingle & Thomas 1974). Other species, however, are able to successfully oviposit under a wide variety of circumstances. *Hemigrapsus oregonensis*, isolated and permitted to oviposit in bare plastic containers, lost less than 0.1% of the clutch (Kuris 1971).

Rare errors in vitellogenesis may cause all or part of a clutch to be sloughed. *Hemigrapsus oregonensis* normally has chocolate brown yolk. However, orange eggs are sometimes seen. These are sloughed within a few days of oviposition (Kuris 1971). In one case such a crab proceeded to develop normal brown ova and successfully oviposited the next clutch. Fertilization is required for attachment in at least some species (Hoglund 1943, Cheung 1966). In others, attachment occurs but eggs are subsequently sloughed (Jefferies 1964) or stripped from the pleopods by the female (Burkenroad 1947, Woodlock & Reynolds 1988a). Interestingly, stripped females continued to exhibit brooding behavior such as grooming (Woodlock & Reynolds 1988a). Most of the unfertilized broods observed by McMullen & Yoshihara (1971) were sloughed. In the American lobster, infertility does not cause improper formation of the *funiculi* (Talbot & Harper 1984).

Physical injuries to the abdomen or pleopods are often associated with partial clutches. Partial clutches are also associated with the onset of parasitic castration by entoniscid isopods (Kuris et al. 1980) or by infestations of bopyrid isopods that may cause substantial reductions in brood size (Van Wyk 1982). Brood size is also small following recovery from parasitic castration (Pike 1960, Kuris et al. 1980).

Among peracaridans infertile eggs are detected at a low rate, generally less than 1% (Paris & Pitelka 1962, Sutton 1968, Sheader 1983). These eggs usually deteriorate (Sheader 1983).

4.3 *Aborted development and retention of infertile eggs*

Observations of egg masses with advanced broods usually reveal a few eggs that have arrested development. These eggs fail to hatch. Losses due to aborted development rarely exceed 1% (Haynes & Wigley 1969, Kuris 1971) but may reach 3-4% (Lochhead & Newcombe 1942, Rasmussen 1953, Morizur et al. 1981). The reasons for these rare developmental failures are unknown. Entire unfertilized broods of the red king crab may remain attached to the pleopods. McMullen & Yoshihara (1971) reported a high frequency of this phenomenon (8.3% of 2,332 ovigerous crabs) in association with heavy fishing pressure on male crabs. Retention of entire unfertilized broods of the snapping shrimp, *Alpheus dentipes*, occurs seasonally (J.W. Born pers. comm.).

Morgan (1972) noted that infertility accounted for 4% of the egg losses for rock lobsters at two different localities and showed that the infertility rate was not associated with maternal size. The rate of egg mortality attributed to other (unrecognized) causes declined with increasing lobster size. In contrast, in a careful laboratory study, Lawlor (1976) showed that the proportion of infertile eggs increased significantly with maternal size of the terrestrial isopod, *Armadillidium vulgare*.

4.4 *Mechanical losses*

Decapod eggs are attached to the setae of the pleopods by funiculi, twisted strands contiguous with the outer membranous coat of the eggs. These may break during the brooding period, causing detachment and loss of the embryo. In some species, mechanical losses increase in the latter half of embryogenesis and such losses may be substantial (Balasundaram & Pandian 1982, Kuris & Wickham 1987, Shields et al. 1990a). These losses may be a necessary consequence of normal egg development as, for most crustacean species, egg size increases substantially during development. For example, the diameter of *H. oregonensis* embryos increases from 0.3 mm at oviposition to 0.4 mm prior to hatching (Kuris 1971). Hence, if no eggs were lost, the volume of the entire egg mass would increase more than two-fold by the time of eclosion. However, the mean clutch of a 15.0 mm crab decreases from 8,800 to 6,700 during embryogenesis in the *absence* of nemertean egg predators (from Kuris & Wickham 1987). A predictable mechanical loss of eggs may compensate for the increased volume of the egg mass and prevent the egg mass from outgrowing available brood space. Of course, this begs the question: why not start with fewer eggs?

In a detailed topographic analysis of egg mortality within the clutch of the yellow rock crab, *Cancer anthonyi*, Shields et al. (1990a) noted that losses due to mechanical abrasion occurred among eggs at the tip of the pleopods. By the end of the brooding period these losses included 4-6% of the clutch. In contrast, brood mortality caused by nemertean egg predators was concentrated among eggs adhering to the base of the pleopod and ranged from 1-30% of the clutch, depending on worm intensity. Other known causes of mechanical loss include abrasion of the egg mass in burrows or other confined habitats, and handling of ovigerous females captured and released, or temporarily confined, in fisheries (Schlieder 1980, Balasundaram & Pandian 1982).

As Aiken & Waddy (1986) observed, in at least some species, brood mortality due to mechanical losses is not a necessary feature of the development process. They note that losses of eggs from *Homarus americanus* held under excellent laboratory conditions are much less than the 36% losses observed in nature (Perkins 1971). This has also been reported for *Austropotamobius pallipes* (Woodlock & Reynolds 1988b).

Studies of peracaridan reproduction often report loss of eggs from marsupia late in development (Paris & Pitelka 1962, Fish 1975). As with decapods, development of peracaridan eggs is accompanied by a large increase in egg volume. Peracaridan eggs also change from oblate spheroids to kidney-bean shapes. Again, brood reduction may be necessary to provide sufficient space for continued growth and development of surviving embryos. Fish (1975) observed that large, late stage embryos were held less securely within the marsupium. Structural features of the marsupia of certain taxa may predispose some species to mechanical brood loss (Fenwick 1985). However, Moore (1981) noted that the swelling of egg volume during embryogenesis is universal while high egg mortality is not. Thus, brood mortality is not a necessary consequence of the increase in size of eggs during development.

Spillage of embryos, particularly from amphipods with relatively open brood pouches, may occur in certain highly active species (Moore 1981). Shillaker (1977 cited in Moore 1981) even observed gravid female *Corophium bonnellii* replacing lost eggs into the marsupium.

4.5 *Egg stripping and maternal cannibalism*

Specialized grooming appendages help maintain the egg masses of carideans and anomurans (Bauer 1979). Lobsters and brachyurans use their chelipedes and walking legs for the same purpose. Schlieder (1980) showed that declawed stone crabs, *Menippe mercenaria*, decreased grooming and declawed females suffered increased egg mortality. Sometimes grooming activity results in detachment of viable embryos. Aiken et al. (1985) have observed that some female lobsters strip all or part of the egg mass from the pleopods if infested with nemertean egg predators. They often eat the detached eggs. This behavior seems to be idiosyncratic; only certain individuals are strippers. While stripping may be an effective anti-egg predator adaptation, it often results in the loss of far more eggs than would be removed by the egg predators. In some individuals stripping did not cease until the entire clutch was destroyed (Aiken et al. 1985). In the freshwater crayfish, *Austropotamobius pallipes*, grooming females always ate dislodged normal eggs but never consumed microbially infested eggs (Woodlock & Reynolds 1988a).

Certain amphipods have also been observed to remove and eat eggs from their own marsupium (Moore 1981). Sheader (1983) experimentally demonstrated that maternal cannibalism increased if gravid *Gammarus duebeni* were fed inadequate diets. Egg consumption rates were highest during the middle to late stages of embryogenesis. At that time the gravid females were also ripening oöcytes in preparation for the next brood; such females presumably have high energy demands. Analyses of field samples indicate that, in nature, rates of egg mortality also increase late in embryogenesis.

Social interactions appear to intensify egg losses and egg cannibalism in freshwater crayfish. Such losses may be considerable (15-60%) (Mason 1970). Brooding females held in isolation exhibited significantly less loss of eggs than did females either held in groups or collected from nature (Woodlock & Reynolds 1988a).

4.6 *Hatching failure*

Hatching is remarkably synchronous in crustaceans, occurring within a few hours if the embryogenic period is relatively short (Kuris 1971) to several days if the brood period is several months to a year (Pandian 1970). Thus judgments as to whether hatching has been completed are easily made. Less than 1% of a clutch in *H. oregonensis* completes development but fails to hatch at the appropriate time (Kuris 1971).

4.7 *Environmental factors*

Cold temperature, salinity changes and dessication can kill broods in the marsupium of amphipods in nature (Vlasblom 1969, Sheader 1983). Laboratory observations (Wild 1983, Shirley et al. 1987) indicate that both high and low temperatures can increase mortality of *Cancer magister* eggs. However, in this crab, the increased mortality at higher temperatures may be due to increased egg predation by nemertean worms at elevated temperatures. Thus, although changing oceanographic conditions may be the ultimate cause of the collapse of the Dungeness crab fishery, in central California (Johnson et al. 1986), the proximate mechanisms may include increased worm predation (see below, Shields & Kuris 1988b).

Siegel & Wenner (1984) report an unusual localized loss of eggs from most clutches of the mole crab, *Emerita analoga*. Environmental degradation due to pesticides, increased turbidity or release of metals from a nearby nuclear power plant were possible causes. Elevated cadmium concentration caused eggs of the amphipod, *Pontoporeia affinis*, to die or develop abnormally (Sundelin 1983). Shields (this volume) reviews effects of environmental pollutants on egg mortality of *Cancer* crabs.

4.8 *Egg predation and parasitism*

The consumption of part or all of a crustacean brood by specialized symbiotic egg predators or parasites is more common than has generally been recognized. The term 'parasite' should be restricted to those agents that infect a single host egg and for which damage to the egg depends on the intensity of parasites within that egg. If infected eggs usually die, the infectious agents are best termed parasitoids (Kuris 1974). 'Egg predator' is used for those symbionts that feed on more than one egg. Each egg is a genetically distinct individual, albeit young and small. Thus, it is clear that consumers that kill more than one egg are predators, paralleling Janzen's (1969) usage for seed predators. Forms such as the nemertean egg predators are termed symbiotic since, in the original sense of De Bary (1879), they live intimately with their hosts; the nature of the trophic interaction with the host is not specified by the term 'symbiont.'

Table 4 summarizes the organisms reported as egg predators or egg parasites of crustaceans. It includes the principal host records and some remarks on the nature of the association. Some further observations follow.

Table 4. Summary of the available information on suspected and documented predators and parasites of crustacean eggs.

Predator (parasite)	Prey (host)	Remarks	Reference
Bacteria			
Leucothrix	*Cancer magister*, *C. anthonyi*, *Homarus americanus*	Impact of bacterial fouling controversial, chiefly a problem under culture conditions; increases in presence of nemertean egg predators, increases through brooding period	Fisher & Wickham (1976), Fisher (1983a), Fisher et al. (1978), Shields & Kuris (1988b)
Protozoa			
Ciliata			
Ephelota gemmipara	*Homarus gammarus*	Suspect cause of tank mortality	Dannevig (1928 cited in Sindermann & Rosenfeld 1967)
Sarcodina			
'Amoeba' paedophthora	*Peltogaster curvatus*	A hyperparasite; not all eggs of the rhizocephalan are infected	Caullery (1906)
Mastigophora			
Chytriodinium	Penaeids, euphausiids, copepods	Attacks free-floating eggs, egg masses, ovisacs	Cachon & Cachon (1968)
Euglenoid flagellates (several spp.)	Freshwater cyclopoid copepods (several spp.)	Destroy eggs; many life cycles completed; parasites	Michajłow (1968)
Blastulidium paedophthorum	*Daphnia obtusa*	Only in eggs and embryos of parthenogenetic females	Pérez (1903), Green (1974)
Peridinian dinoflagellate	*Pandalus borealis*	Enlarges eggs, may kill up to 30% of clutch	Stickney (1978)
Fungi			
Lagenidium callinectes	*Callinectes sapidus*, *Palaemon macrodactylus*, barnacles	Rapidly destroys eggs, limited to eggs on the periphery of egg mass; unfertilized eggs more susceptible, percentage and intensity of infestation increase during brood period; a common parasite in Chesapeake Bay populations of blue crab; also parasitizes larval and adult crustaceans in culture; a substance released by an egg surface bacterium, *Alteromonas* sp., protects eggs from fungal infection	Couch (1942), Rogers-Talbert (1948), Johnson & Bonner (1960), Bland & Amerson (1973), Lightner (1981), Fisher (1983b), Shields & Kuris (1988b), Gil-Turnes et al. (1989)

Lagenidium-like species	*Cancer anthonyi*	Kills a small proportion of eggs on the periphery of the egg mass, prevalence low	Shields & Kuris (1988b)
Rhizophydium littoreum	*Cancer anthonyi*	Associated with egg mortality but experimentally shown to be a saprobe	Shields (in prep.)
Haliphthoros milfordensis	*Artemia, Callinectes sapidus*	Kills eggs	Tharp & Bland (1977)
Olpidiopsis sp.	Copepods	Destroys clutches	Michel (1971, cited in Manier 1977)
Schizochytrium sp.	*Homarus americanus*	Heavily infests crab eggs, consistent association with nemertean egg predators suggests it is a saprobe in damaged eggs	Miller & Fleming (1983)
Atkinsiella dubia	*Pinnotheres pisum, Dynamene bidentata*, other crustaceans	Parasites of eggs and adult crustaceans in laboratory, affects only a few eggs per clutch, transmission in nature uncertain	Atkins (1954), Holdich (1968), Sparrow (1973)
Pythium thalassium	*P. pisum*	Parasites	Atkins (1955)
Coelenterata			
Hydrozoa			
Podocoryne sp.	*Eupagurus prideauxi*	Observed to eat eggs	Mathews (1959)
Anthozoa			
Adamsia palliata	*Eupagurus prideauxi*	Observed to eat eggs	Mathews (1959)
Platyhelminthes			
Turbellaria			
Coleophora chionoecetes	*Chionoecetes opilio*	On eggs	Gordon (1966)
Peraclistus oofagus	*Hyas coarctatus*	Suspected egg predator	Gordon (1966)
cf. *Monocelis*	*Pugettia producta*	Egg contents observed in gut; may be a scavenger, does not feed on intact eggs in vitro; 100% of ovigerous females infested, often at high intensities	Kuris (unpubl. observ.)

Table 4 Continued.

Predator (parasite)	Prey (host)	Remarks	Reference
Nemertea			
Carcinonemertes spp.	Many Brachyura, Anomura, Palinura	Penetrates egg membranes with stylet; egg contents observed in gut; feeding observed *in vivo* and *in vitro*; prevalence and intensity variable, often high. Outbreaks may temporarily eliminate reproductive output of hosts over large geographic areas; life cycle cued by host molting and reproduction	Humes (1942), Hopkins (1947), Rogers-Talbert (1948), Kuris (1978), Roe (1984), Wickham et al. (1984), Wickham (1986), Kuris & Wickham (1986), Shields & Kuris (1988a), Kuris et al. (in press)
Pseudocarcinonemertes homari	*Homarus americanus*	Feeds on eggs; facultative parasite, may feed through host cuticle on non-ovigerous hosts	Aiken et al. (1985), Uhazy et al. (1985), Campbell & Brattey (1986)
Alaxinus oclairi	*Paralithodes camtschatica*	Feeding not confirmed, only one specimen collected	Gibson et al. (1990)
Undescribed heteronemertean	*Blepharipoda occidentalis*	Engulfs whole embryos	Lafferty & Kuris (unpubl. observ.)
Rotifera			
Seison spp.	*Nebalia* sp.	Pierces eggs, also on exoskeleton	Hyman (1951)
Mollusca			
Gastropoda			
Unidentified	*Argis lar*	Snail egg capsules attached to egg mass, newly hatched snails may be egg predators	McCauley (1964)
Tricolia?	*Cancer magister*	Eggs in gut; associated with heavy infestations of nemertean egg predators	Wickham (pers. comm.)
Annelida			
Polychaeta			
Nereis fucata	*Eupagurus prideauxi*	Observed to eat eggs	Mathews (1959), Goerke (1971)
Oligochaeta			
Cambarincola aliena	*Asellus* sp.	Egg contents observed in gut	Holt (1963)
Nais elinquis	*Gammarus duebeni*	Marsupium empty if worms present; suspected facultative egg predator	Ward (1985)

Copepoda (Nicothoidae)			
Choniosphaera spp.	*Cancer irroratus*, *C. borealis*, *Carcinus maenas*, *Portunus* sp.	Suck egg contents, copepodid II on gills of both host sexes; an egg mimic	Connolly (1929), Gallien & Bloch (1936), Gnanamuthu (1954), Johnson (1957)
Hansenulus trebax	Mysids	Makes a characteristic incision and feeds on eggs, also feeds on brooded larvae, may reduce host population density	Heron & Damkaer (1986)
Spaeronella spp.	Amphipods	Suggested predator on eggs at or soon after oviposition (but see text)	Della-Valle (1893)
Paranicothoe sp.	Bopyrids	Egg predator of a parasite	Carton (1970)
Isopoda			
Ancyroniscus bonnieri	*Dynamene bidentata*	Egg predator	Holdich (1975)
Clypeoniscus spp.	Idotheids	Egg predator	Caullery & Mesnil (1920), Sheader (1977)
Amphipoda			
Phronimid larvae	*Acanthephyra* sp.	Guts full of egg contents	Gordon (1968)
Paracyphocaris praedator	*Oplophorus novaezeelandiae*	Egg predation not confirmed, egg mimic	Bowman & Wasmer (1984)
Ischyrocerus cf. *anguipes*	*Paralithodes camtschatica*	Guts packed with eggs	Kuris et al. (in press)
Parapleustes commensalis	*Panulirus interruptus*	Guts filled with eggs, *in vitro* observations on feeding rates	Kuris, Okazaki & Chapman (unpubl. observ.)

The most studied elements of the egg mass biota are the several species of nemerteans that consume a substantial fraction of the eggs of some economically important crustaceans (*C-ancer magister, C. anthonyi, Callinectes sapidus, Homarus americanus, Paralithodes camtschatica*) (Humes 1942, Wickham 1979, 1986, Aiken et al. 1985, Campbell & Brattey 1986, Shields et al. 1990a, Kuris et al. in press) and of the shore crab, *Hemigrapsus oregonensis* (Kuris & Wickham 1987, Shields & Kuris 1988a). Their impact on crustacean hosts and other pertinent facets of their biology have recently been reviewed (Wickham & Kuris, 1985, Wickham 1986, Kuris & Wickham 1987, Roe 1988) and will not be treated in detail here. New species of symbiotic nemertean egg predators are rapidly accumulating (e.g. Shields et al. 1989, Gibson et al. 1990, Shields & Kuris 1990). Some of these have independently evolved life histories as symbiotic egg predators (Wickham & Kuris 1985). Newly recognized host associations continue to include crustacean families and suborders that had previously not been recorded as infested with nemertean egg predators (e.g. Campbell et al. 1989, Shields et al. 1989, Lafferty & Kuris in prep.).

The relative contribution of egg predators and parasites to egg mortality has been controversial (e.g. Fisher & Wickham 1976, Wild 1983, Shields & Kuris 1988b). A multifactorial in vitro experimental study involving bacteria, fungi and nemertean worms at 3 temperatures has shown that bacterial fouling had no effect on egg mortality (Shields & Kuris 1988b). Predatory nemerteans were the major cause of egg mortality *in vitro* and in nature. Nemertean feeding rates increased at elevated temperatures. There is no evidence that bacterial fouling of the egg mass reduces reproductive output in natural populations. However, under certain laboratory conditions bacterial fouling may be associated with massive destruction of the clutch (Fisher & Wickham 1976). The benign nature of bacterial growth on egg surfaces is supported by observations of Fisher (1983a) and Harper & Talbot (1984). Indeed, bacteria may sometimes be protective (Gil-Turnes et al. 1989).

Observation of certain organisms in clutches where substantial mortality is present is an insufficient demonstration that these organisms are predators or parasites. In some cases these organisms are scavengers or saprobes, feeding on remains of eggs killed by other organisms. This has been experimentally demonstrated for a chytrid fungus, *Rhiyzophydium littoreum*, by Shields (in prep., see also Sparrow 1973, Miller & Fleming 1983). Although early records implicate several species of symbiotic turbellarians as egg predators (Table 4), there is no recent convincing confirmation of egg predation. The intensity of the abundant turbellarians infesting the egg masses of red king crabs in Alaskan waters is not associated with egg mortality (Kuris et al. in press).

Some of the most highly specialized egg predators exhibit some remarkable adaptive features. For instance, several groups of egg predators are also egg mimics, including the nicothoid copepods, cryptoniscid isopods and some amphipods (Hansen 1897, Nielsen & Strömberg 1965, Bowman & Wasmer 1984, Daly & Damkaer 1986).

Interestingly, the highly modified copepod family, Nicothoidae, includes two distinctive types of symbiotic host associations. Those that infest the marsupia of ostracodes and most peracaridans are parasitic castrators that suppress development of the ovary of the host (Bowman & Kornicker 1967, Hammond 1973, Kuris 1974, Just 1978, Bishop 1982, Gotto 1984, Costello & Myers 1989). Nicothoids infesting the egg mass of decapods and mysids are egg predators, consuming host embryos (Gotto 1979, Heron & Damkaer 1986). In contrast to this, most parasitic isopods are parasitic castrators, often of decapods (Kuris 1974), whereas species of *Clypeoniscus* and *Ancyroniscus* are egg predators of peracaridans (Caullery & Mesnil 1920, Holdich 1975, Sheader 1977).

5 CONCLUSION

When heavy egg mortality has been reported, egg predation has been repeatedly demonstrated. This strongly suggests that egg predators should be sought in the many instances where brood losses are inexplicably high. For example, Chapman (1980) noted that brood losses in *Nephrops norvegicus* are generally in the range of 40-70%.

Equally of interest would be further studies of the relative roles of various mortality factors, particularly mechanical losses. There is also a need for quantitative observations on brood losses of several important malacostracan taxa, including carideans, mysids, euphausiids and stomatopods.

ACKNOWLEDGEMENTS

I thank Dan Wickham, Jeff Shields, Kevin Lafferty and John Chapman for access to their unpublished observations; Cadet Hand, Adrian Wenner, Keven Lafferty, Eric Schultz, Jeff Shields, and Elzbet Diaz de Leon for their comments on the manuscript and John Chapman for his much appreciated assistance with the peracarid literature. This work is a result of research sponsored in part by NOAA, National Sea Grant College Program, Department of Commerce, under grant number NA80AA-D00120, project numbers R/F-75 and R/F-AL-B (to AMK and D.E. Wickham), through the California and Alaska Sea Grant College Programs, and in part by the California State Resources Agency. The US Government is authorized to reproduce and distribute it for governmental purposes.

REFERENCES

Aiken, D.E. & S.L. Waddy 1986. Environmental influences on recruitment of the American lobster, *Homarus americanus*: a perspective. *Can. J. Fish. Aquatic Sci.* 43: 2258-2270.

Aiken, D.E., S.L. Waddy & L.S. Uhazy 1985. Aspects of the biology of *Pseudocarcinonemertes homari* and its association with American lobster, *Homarus americanus*. *Can. J. Fish. Aquat. Sci.* 42: 351-356.

Allen, D.M. 1984. Population dynamics of the mysid shrimp *Mysidopsis bigelowi* W.M. Tattersall in a temperate estuary. *J. Crustacean Biol.* 4: 25-34.

Atkins, D. 1954. A marine fungus *Plectospira dubia* n. sp. (Saprolegniaceae), infecting crustacean eggs and small crustaceans. *J. Mar. Biol. Assoc. UK.* 33: 721-732.

Atkins, D. 1955. *Pythium thalassium* n. sp. infecting the egg-mass of the pea-crab *Pinnotheres pisum*. *Trans. Brit. Mycol. Soc.* 38: 31-46.

Balasundaram, C. & T.J. Pandian 1982. Egg loss during incubation in *Macrobrachium nobilii* (Henderson & Mathai). *J. Exp. Mar. Biol. Ecol.* 59: 289-299.

Bauer, R.T. 1979. Antifouling adaptations of marine shrimp (Decapoda: Caridea): gill cleaning mechanisms and grooming of brooded embryos. *Zool. J. Linn. Soc.* 65: 281-303.

Betz, K.H. 1974. Phenologie, Reproduktion und Wachstum der valviferan Assel *Idotea chelipes* (Pallas, 1766) in der Schlei. *Kieler Meeres.* 30: 65-79.

Bishop, J.D.D. 1982. The growth, development and reproduction of a deep-sea cumacean (Crustacea: Peracarida). *Zool. J. Linn. Soc.* 74: 359-380.

Bland, C.E. & H.W. Amerson 1973. Observations on *Lagenidium callinectes*: isolation and sporangial development. *Mycologia* 65: 310-320.

Botsford, L.W. 1986. Population dynamics of Dungeness crab. *Can. Special Pub. Fish. Aquat. Sci.* 92: 140-153.

Botsford, L.W., R.D. Methot & J.E. Wilen 1982. Cyclic covariation in the California king salmon, *Oncorhyn-*

chus tshawytscha, silver salmon, *O. kisutch*, and Dungeness crab, *Cancer magister*, fisheries. *Fish. Bull.* 80: 791-801.

Bowman, T.E. & L.S. Kornicker 1967. Two new crustaceans; the parasitic copepod *Sphaeronellopsis monothrix* (Choniostomatidae) and its myodocopid ostracod host *Parasterope pollex* (Cylindroleberidae) from the southern New England coast. *Proc. US Nat. Mus.* 123: 128.

Bowman, T.E. & R.A. Wasmer 1984. The deep-sea amphipod *Paracyphocaris praedator* (Gammaridea: Lysianassidae) associated with the pelagic shrimp *Oplophorus novaezeelandiae* as an egg mimic. *Proc. Biol. Soc. Washington* 97: 844-848.

Bregazzi, P.K. 1972. Life cycles and seasonal movements of *Cheirimedon femoratus* (Pfeffer) and *Tryphosella kergueleni* (Miers) (Crustacea: Amphipoda). *Brit. Antarct. Surv. Bull.* 30: 1-34.

Brereton, J. LeG. 1956. A study of some factors controlling the populations of some terrestrial isopods. Ph.D. Thesis, Oxford University.

Broekhuysen, G.J. 1936. On development, growth and distribution of *Carcinides maenas*. *Arch. Neerl. Zool.* 2: 257-399.

Brunel, P. 1963. Troisième série d'observations sur la biologie et la biométrie du crabe-araignée *Chionoecetes opilio* (Fabr.), *Sta. Biol. Mar. Grande-Rivière Rapp. Ann.* 1962: 91-100.

Buchanan, J.B. 1963. The biology of *Calocaris macandreae* (Crustacea: Thalassinidea). *J. Mar. Biol. Assoc. UK.* 43: 729-747.

Burkenroad, M.D. 1947. Reproductive activities of decapod Crustacea. *Amer. Nat.* 81: 392-398.

Cachon, J. & M. Cachon 1968. Cytologie et cycle evolutif des *Chytriodinium* (Chatton). *Prostistologica* 4: 249-261.

Campbell, A. & J. Brattey 1986. Egg loss from the American lobster, *Homarus americanus*, in relation to nemertean, *Pseudocarcinonemertes homari*, infestation. *Can. J. Fish. Aquat. Sci.* 43: 772-780.

Campbell, A., R. Gibson & L.H. Evans 1989. A new species of *Carcinonemertes* (Nemertea: Carcinonemertidae) ectohabitant on *Panulirus cygnus* (Crustacea: Palinuridae) from Western Australia. *Zool. J. Linn. Soc. 95: 257-268.*

Carton, Y. 1970. Le genre *Paranicothoe*, un nouveau representant de la famille des Nicothoidae. *J. Parasit.* 56 Sect. II: 47-48.

Caullery, M. 1906. Sur un amoebien parasite de embryons de *Peltogaster curvatus*. *C. R. Soc. Biol. Paris* 61: 266-269.

Caullery, M. & F. Mesnil 1920. *Ancyroniscus bonnieri* C. & M. Epicaride parasite d'un Sphéromide *Dynamene bidentata* (Mont.). *Bull Biol. France Belg.* 54: 1-36.

Chapman, C.J. 1980. Ecology of juvenile and adult *Nephrops*. In J.S. Cobb & B.F. Phillips (eds), *The Biology and Management of Lobsters*. Vol. II. pp. 143-177. New York: Academic Press.

Cherkashina, N. Ya. 1970. Reproduction of crayfish (Astacidae) along the southeastern coast of the Caspian Sea. *Gidrobiol. Zh.* 6: 83-85.

Cheung, T.S. 1966. The development of egg-membranes and egg attachment in the shore crab, *Carcinus maenas*, and some related decapods. *J. Mar. Biol. Assoc. UK* 46: 373-400.

Clutter, R.I. & G.H. Theilacker 1971. Ecological efficiency of a pelagic mysid shrimp; estimates from growth, energy budget and mortality studies. *Fish. Bull.* 69: 93-115.

Connolly, C.J. 1929. A new copepod parasite *Choniosphaera cancrorum* gen. et sp. n., representing a new genus, and its larval development. *Proc. Zool. Soc. London* 1929: 415-427.

Costello, M.J. & A.A. Myers 1989. Observations on the parasitism of *Aora gracilis* (Bate), (Amphipoda) by *Sphaeronella leuckartii* Salensky (Copepoda), with a review of amphipod-*Sphaeronella* associations. *J. Nat. Hist.* 23: 81-91.

Couch, J.N. 1942. A new fungus on crab eggs. *J. Elisha Mitchell Sci. Soc.* 58: 158-164.

Daly, K.L. & D.M. Damkaer 1986. Population dynamics and distribution of *Neomysis mercedis* and *Alienacanthomysis* (Crustacea; Mysidacea) in relation to the parasitic copepod *Hansenulus trebax* in the Columbia River estuary. *J. Crustacean Biol.* 6: 840-857.

Dannevig, A. 1928. Beretning om Flødevigens utklekningsanstalt for 1926/1927. *Arsberetn. Norg. Fisk.* 1937: 150-156.

De Bary, H.A. 1879. *Die Erscheinung der Symbiose*. Strassburg: Cassel.

Della-Valle, A. 1893. *Gammarini des Golfes di Napoli Monografia. Fauna und Flora des Golfes van Neapel.* 289-292.

Downer, D.F. & D.H. Steele 1979. Some aspects of the biology of *Amphiporeia lawrenciana* Shoemaker (Crustacea, Amphipoda) in Newfoundland waters. *Can. J. Zool.* 57: 257-263.

Edwards, E. & R.A. Meaney 1967. Meat yields from the European edible crab in Irish waters. *Irish Sea Fish. Board Fish. Dev. Div. Resource Dev. Note* 1: 1-5.

Emlen, J.M. 1970. Age specificity and ecological theory. *Ecology* 51: 588-601.

Ennis, G.P. 1986. Stock definition, recruitment variability, and larval recruitment processes in the American lobster, *Homarus americanus*: a review. *Can. J. Fish. Aquat. Sci.* 43: 2072-2084.

Fenwick, G.D. 1985. Life-histories of four co-occurring amphipods from a shallow sand bottom at Kaikoura, New Zealand. *New Zealand J. Zool.* 12: 71-105.

Fielder, D.R. 1964. The spiny lobster, *Jasus lalandii* (H. Milne-Edwards), in South Australia. II. Reproduction. *Aust. J. Mar. Freshw. Res.* 15: 133-144.

Fish, J.D. 1975. Development, hatching and brood size in *Bathyporeia pilosa* and *B. pelagica* (Crustacea: Amphipoda). *J. Mar. Biol. Assoc. UK* 55: 357-368.

Fisher, W. 1956. Untersuchungen uber einen für die deutsch Bucht neuen parasitaren Copepoden: *Lecithomyzon maenadis* Bloch & Gallien (Familie Choniostomatidae) an *Carcinus maenadis* Pennant (Crustacea, Decapoda) *Helgolander Wiss. Meeres.* 5: 320-352.

Fisher, W.S. 1983a. Eggs of *Palaemon macrodactylus:* II. Association with aquatic bacteria. *Biol. Bull.* 164: 201-213.

Fisher, W.S. 1983b. Eggs of *Palaemon macrodactylus*: III. Infection by the fungus, *Lagenidium callinectes*. *Biol. Bull.* 164: 214-226.

Fisher, W.S. & W.H. Clark 1983. Eggs of *Palaemon macrodactylus*: I. Attachment to the pleopods and formation of the outer investment coat. *Biol. Bull.* 164: 189-200.

Fisher, W.S., E.H. Nilson, J.F. Steenbergen & D.V. Lightner 1978. Microbial diseases of cultured lobsters; a review. *Aquaculture* 14: 115-140.

Fisher, W.S. & D.E. Wickham 1976. Mortality and epibiotic fouling of eggs from wild populations of the Dungeness crab, *Cancer magister. Fish. Bull.* 74: 201-207.

Gallien, L. & F. Bloch 1936. Recherches sur *Lecithomyzon maenadis* Bloch & Gallien, copepode parasite de la ponte de *Carcinus maenas* Pennant. *Bull. Biol. France Belg.* 70: 36-53.

Gibson, R., D.E. Wickham & A.M. Kuris 1990. A new genus and species of monostiliferoidean nemertean (Nemertea, Enopla) found on an egg mass of the anomuran decapod *Paralithodes camtschatica*. *Zool. J. Linn. Soc.* 98: 185-198.

Gil-Turnes, S., M.E. Hay & W. Fenical 1989. Symbiotic marine bacteria chemically defend crustacean embryos from a pathogenic fungus. *Science* 246: 116-118.

Gnanamuthu, C.P. 1954. *Choniosphaera indica*, a copepod parasite on the crab *Neptunus* sp. *Parasitology* 44: 371-378.

Goerke, H. 1971. *Nereis fucata* (Polychaeta, Nereidae) als Kommensale von *Eupagurus bernhardus* (Crustacea, Paguridae) Entwicklung einer Population und Verhalten der Art. *Veroff. Inst. Meeres. Bremerhaven* 13: 79-118.

Gordon, I. 1966. Crustacea: general considerations. *Mem. Inst. Fond. Afrique Noire* 77: 27-86.

Gordon, I. 1968. Phronimid larvae (Amphipoda) attacking *Acanthephyra* embryos (Decapoda). *Crustaceana* 14: 217-218.

Gotto, R.V. 1979. The association of copepods with marine invertebrates. *Adv. Mar. Biol.* 15: 1-109.

Gotto, R.V. 1984. Two new species of *Rhizorhina* (Copepoda: Siphonostomatoida) from leptostracan and tanaidacean hosts. *J. Nat. Hist.* 18: 811-817.

Goudeau, M. & F. Lachaise 1983. Structure of the egg funiculus and deposition of embryonic envelopes in a crab. *Tiss. Cell* 15: 47-62.

Green, J. 1974. Parasites and epibionts of Cladocera. *Trans. Zool. Soc. London* 32: 417-515.

Gulland, J.A. 1973. Can a study of stock and recruitment aid management decisions? *Rapp. P.V. Reun. Cons. Int. Explor. Mer* 164: 368-372.

Gulland, J.A. 1974. *Management of Marine Fisheries*, Univ. Washington, Seattle.

Hammond, R. 1973. The marine and brackish-water copepods of Norfolk. Calanoida, Misophrioida, Cyclopoida, Monstrilloida, Notodelphyoida and *incertae sedis*. *Cah. Biol. Mar.* 14: 335-360.

Hansen, H.J. 1897. *The Choniostomatidae, a family of Copepoda, parasites on Crustacea Malacostraca*. Copenhagen: Andr. Fred. Host & Son.

Harper, R.E. & P. Talbot 1984. Analysis of the epibiotic bacteria of lobster (*Homarus*) eggs and their influence on the loss of eggs from the pleopods. *Aquaculture* 36: 9-26.

Hartnoll, R.G. 1968. Reproduction in the burrowing crab, *Corystes cassivelaunus* (Pennant, 1777) (Decapoda, Brachyura). *Crustaceana* 15: 165-170.

Hayes, M.L. 1983. Variation in the abundance of crab and shrimp with some hypotheses on its relationship to environmental causes. In W.S. Wooster (ed.), *From Year to Year: Interannual Variability of the Environment and Fisheries of the Eastern Bering Sea and Gulf of Alaska*. pp. 86-101. Washington Sea Grant Program, University of Washington, Seattle.

Haynes, E.B. & R.G. Wigley 1969. Biology of the northern shrimp *Pandalus borealis* in the Gulf of Maine. *Trans. Amer. Fish. Soc.* 98: 60-76.

Hepper, B.T. & C.J. Gough 1978. Fecundity and rate of embryonic development of the lobster *Homarus gammarus* (L.), off the coast of North Wales. *J. Cons. Int. Explor. Mer* 38: 54-57.

Heron, G.A. & D.M. Damkaer 1986. A new nicothoid copepod parasite on mysids from northwestern North America. *J. Crustacean Biol.* 6: 652-665.

Hilsinger, J.R. 1976. Aspects of the reproductive biology of female snow crabs, *Chionoecetes bairdi* from Prince William Sound and the adjacent Gulf of Alaska. *Mar. Sci. Commun.* 2: 201-225.

Hobbs, R.C. & L.W. Botsford 1989. Dynamics of an age-structured prey with density- and predator-dependent recruitment: the Dungeness crab and a nemertean egg predator worm. *Theor. Pop. Biol.* 36: 1-22.

Hoglund, H. 1943. On the biology and larval development of *Leander squilla* (L.). *Svenska Hydrograf. Biol. Komm. Skr. N. Ser. Biol.* 2: 1-43.

Holdich, D.N. 1968. Reproduction, growth and bionomics of *Dynamene bidentata* (Crustacea: Isopoda). *Proc. Zool. Soc. London* 156: 137-153.

Holdich, D.M. 1975. *Anyroniscus bonnieri* (Isopoda, Epicaridea) infecting British populations of *Dynamene bidentata* (Isopoda, Sphaeromatidae). *Crustaceana* 28: 145-151.

Holt, P.C. 1963. A new branchiobdellid Branchiobdellidae: *Cambarincola*. *J. Tennessee Acad. Sci.* 38: 97-100.

Hopkins, S.H. 1947. The nemertean *Carcinonemertes* as an indicator of the spawning history of the host, *Callinectes sapidus*. *J. Parasit.* 33: 146-150.

Humes, A.G. 1942. The morphology, taxonomy, and bionomics of the nemertean genus *Carcinonemertes*. *Illinois Biol. Monogr.* 18: 1-105.

Hyman, L.H. 1951. *The Invertebrates: Acanthocephala, Aschelminthes and Entoprocta. The Pseudocoelomate Bilateria*. Vol. 3. New York: McGraw-Hill.

Incze, L.S., D.A. Armstrong & S.L. Smith 1987. Abundance of larval tanner crabs (*Chionoecetes* spp.) in relation to adult female stocks and regional oceanography of the southeastern Bering Sea. *Can. J. Fish. Aquat. Sci.* 44: 1143-1156.

Incze, L.S., R.S. Otto & M.K. McDowell 1986. Recruitment variability of juvenile red king crab, *Paralithodes camtschatica* in the southeastern Bering Sea. *Can. Special Publ. Fish. Aquat. Sci.* 92: 370-378.

Ingle, R.W. & W. Thomas 1974. Mating and spawning of the crayfish *Austropotamobius pallipes* (Crustacea: Astacidea). *J. Zool.* 173: 525-538.

Jamieson, G.S. 1986. Implications of fluctuations in recruitment in selected crab populations. *Can. J. Fish. Aquat. Sci.* 43: 2085-2098.

Janzen, D.H. 1969. Seed-eaters versus seed size, number, toxicity and dispersal. *Evolution* 23: 1-27.

Jefferies, D.J. 1964. The moulting behavior of *Palaemonetes varians* (Leach) (Decapoda: Palaemonidae). *Hydrobiology* 24: 457-488.

Jewett, S.C., N.A. Sloan & D.A. Somerton 1985. Size at sexual maturity and fecundity of the fjord-dwelling golden king crab *Lithodes aequispina*. *J. Crustacean Biol.* 5: 377-385.

Johnson, D.F., L.W. Botsford, R.E. Methot, Jr & T.C. Wainwright 1986. Wind stress and cycles in Dungeness crab (*Cancer magister*) catch off California, Oregon, and Washington. *Can. J. Fish. Aquat. Sci.* 43: 838-845.

Johnson, M.W. 1957. The copepod *Choniosphaera cancrorum* parasitizing a new host, the green crab *Carcinides maenas*. *J. Parasit.* 43: 470-473.

Johnson, T.W. & R.R. Bonner 1960. *Lagenidium callinectes* Couch in barnacle ova. *J. Elisha Mitchell Sci. Soc.* 76: 147-149.

Jones, M.B. 1974. Breeding biology and seasonal population changes of *Jaera nordmanni nordica* Lemercier (Isopoda: Asellota). *J. Mar. Biol. Assoc. UK* 54: 727-736.

Jones, M.B. 1977. Breeding & seasonal population changes of *Petrolisthes elongatus* (Crustacea, Decapoda, Anomura) at Kaikoura, New Zealand. *J. Roy. Soc. New Zealand* 7: 259-272.

Jones, M.B. 1978. Aspects of the biology of the big-handed crab, *Heterozius rotundifrons* (Decapoda, Brachyura) from Kaikoura, New Zealand. *New Zealand J. Zool.* 5: 783-794.

Just, J. 1978. Taxonomy, biology and evolution of the circumarctic genus *Acanthonotozoma* (Amphipoda), with notes on *Panoploeopsis*. *Acta Arctica* 20: 1-211.

Kemp, P.F., F.A. Cole & R.C. Swartz 1985. Life history and productivity of the phoxocephalid amphipod *Rhepoxynius abronius* (Barnard). *J. Crustacean Biol.* 5: 449-464.

King, J.E. 1948. A study of the reproductive organs of the common marine shrimp, *Penaeus setiferus* (Linnaeus). *Biol. Bull.* 94: 244-268.

Kinne, O. 1954. Eidonomie, Anatomie und Lebenszyklus von *Sphaeroma hookeri* Leach (Isopoda). *Kieler Meeres.* 10: 100-120.

Kinne, O. 1959. Ecological data on the amphipod *Gammarus duebeni*. A monograph. *Veroff. Inst. Meeres. Bremerhaven* 6: 177-202.

Klapow, I.A. 1970. Ovoviviparity in the genus *Excirolana* (Crustacea: Isopoda). *J. Zool.* 2: 359-369.

Kon, T. 1974. Fisheries biology of the Japanese tanner crab-VI. On the number of ovarian eggs held in the pleopods. *Bull Jap. Soc. Sci. Fish.* 40: 465-469.

Kuris, A.M. 1971. Population interactions between a shore crab and two symbionts. Ph.D. Thesis, University of California, Berkeley, 447 pp.

Kuris, A.M. 1974. Trophic interactions: similarity of parasitic castrators to parasitoids. *Quart. Rev. Biol.* 49: 129-148.

Kuris, A.M. 1978. Life cycle, distribution and abundance of *Carcinonemertes epialti*, a nemertean egg predator of the shore crab, *Hemigrapsus oregonensis*, in relation to host size, reproduction and molt cycle. *Biol. Bull.* 154: 121-137.

Kuris, A.M., S.F. Blau, A.J. Paul, J.D. Shields & D.E. Wickham 1990. Infestation by brood symbionts and their impact on egg mortality in the red king crab, *Paralithodes camtschatica*, in Alaska: geographic and temporal variation. *Can. J. Fish. Aquat. Sci.* (in press).

Kuris, A.M., G.O. Poinar, & R. Hess 1980. Mortality of the internal isopodan parasitic castrator, *Portunion conformis* (Epicaridea, Entoniscidae), in the shore crab, *Hemigrapsus oregonensis* with a description of the host response. *Parasitology* 80: 211-232.

Kuris, A.M. & D.E. Wickham 1987. Effect of nemertean egg predators of crustaceans. *Bull. Mar. Sci.* 41: 151-164.

Lawlor, L.R. 1976. Parental investment and offspring fitness in the terrestrial isopod, *Armadillidium vulgare*. *Evolution* 30: 775-755.

Lightner, D.V. 1981. Fungal diseases of marine crustaceans. In E.W. Davidson (ed.), *Pathogenesis of Invertebrate Microbial Diseases*. pp. 451-454. New York: Allanheld & Osman, Ottawa.

Lochhead, M.S. & C.L. Newcombe 1942. Methods of hatching eggs of the blue crab. *Virginia J. Sci.* 3: 76-86.

Manier, J.-F. 1977. Cycle, ultrastructure, d'une *Catenaria* (Phycomycètes, Blastocladiales) parasites de Crustacés, Cyclopoïdes. *Ann. Parasit. Hum. Comp.* 52: 363-376.

Marsden, I.d. 1982. Population biology of the commensal asellotan, *Isais pubescens* and its sphaeromatid host, *Exosphaeroma obtusum* (Dana) (Isopoda). *J. Exp. Mar. Biol. Ecol.* 58: 233-257.

Mason, J.C. 1970. Egg-laying in the western North American crayfish, *Pacifastacus trowbridgii* (Stimpson) (Decapoda, Astacidae). *Crustaceana*, 19: 37-44.

Mathews, D.C. 1959. Observations on ova fixation in the hermit crab *Eupagurus prideauxi*. *Publ. Staz. Zool. Napoli* 31: 248-263.

McCauley, J.E. 1964. Gastropod larvae from the brood pouch of an arctic shrimp. *Trans. Amer. Micr. Soc.* 83: 290-293.

McMullen, J.C. & H.T. Yoshihara 1971. Deposition of unfertilized eggs in unmated king crabs, *Paralithodes camtschatica* (Tilesius). *Trans. Amer. Fish. Soc.* 100: 583-584.

Medawar, P.B. 1957. *The Uniqueness of the Individual*. Methuen, London.

Methot, R.D. & L.W. Botsford 1982. Estimated preseason abundance in the California Dungeness crab (*Cancer magister*) fisheries. *Can. J. Fish. Aquat. Sci.* 39: 1077-1083.

Michajłow, W. 1968. Euglenoidina (Flagellata) – parasites of Cyclopidae (Copepoda). *Acta Protozool.* 5: 181-217.

Michel, R. 1971. Champignones et protistes parasites de Copépodes Calanoïdes et Cyclopoïdes, dans la région de Montpellier et d'Alès. These 3° cycle Montpellier, 115 pp.

Milham, S. & A.M. Gittleson 1965. Parental age and malformations. *Human Biol.* 3: 13-22.

Miller, J.D. & L.C. Fleming 1983. Fungi associated with an infestation of *Pseudocarcinonemertes homari* on *Homarus americanus*. *Trans. Brit. Mycol. Soc.* 80: 9-12.

Miller, R.M. & G.N. Cameron 1983. Intraspecific variation of life history parameters in the terrestrial isopod *Armadillidium vulgare*. *Oecologia* 57: 216-226.

Moore, P.A. 1981. The life histories of the amphipods *Lembos websteri* and *Corophium bonellii* Milne Edwards in kelp holdfasts. *J. Exp. Mar. Biol. Ecol.* 49: 1-50.

Morgan, G.R. 1972. Fecundity in the western rock lobster *Panulirus longipes* (George) (Crustacea: Decapoda; Palinuridae). *Aust. J. Mar. Freshw. Res.* 23: 133-141.

Morizur, Y., G. Conan, A. Guénolé & M.H. Omnes 1981. Fécondité de *Nephrops norvegicus* dans le Golfe de Gascogne. *Mar. Biol.* 63: 319-324.

Nielsen, S.-O. & J.-O. Strömberg 1965. A new parasite of *Cirolana borealis* Lilljeborg belonging to the Cryptoniscinae (Crustacea Epicardea). *Sarsia* 18: 37-62.

Pandian, T.J. 1970. Ecophysiological studies on the developing eggs and embryos of the European lobster, *Homarus gammarus*. *Mar. Biol.* 5: 154-167.

Paris, O.H. & F.A. Pitelka 1962. Population characteristics of the terrestrial isopod *Armadillidium vulgare* in California grassland. *Ecology* 43: 229-248.

Payne, J.F. 1971. Fecundity studies on the crayfish *Procambarus hayi*. *Tulane Stud. Zool. Bot.* 17: 35-37.

Peer, D.L., L.E. Linkletter & P.W. Hicklin 1986. Life history and reproductive biology of *Corophium volutator* (Crustacea: Amphipoda) and the influence of shorebird predation on population structure in Chignecto Bay, Bay of Fundy, Canada. *Netherlands J. Sea Res.* 20: 359-373.

Penn, G.H. 1943. A study of the life history of the Louisiana red crawfish, *Cambarus clarkii* Girard. *Ecology* 24: 1-18.

Pérez, Ch. 1903. Sur un organisme nouveau, *Blastulidium paedophthorum*, parasite de embryons de Daphnies. *C. R. Soc. Biol. Paris.* 55: 715-716.

Perkins, H.C. 1971. Egg loss during incubation from offshore northern lobsters (Decapoda: Homaridae). *Fish. Bull.* 60: 451-453.

Pike, R.B. 1960. The biology and post-larval development of the bopyrid parasites *Pseudione affinis* G.O. Sars and *Hemiarthrus abdominalis* (Kröyer) (=*Phryxus abdominalis* Kröyer). *J. Linn. Soc. Zool.* 44: 239-251.

Platt, T. & N. Yamamura 1986. Prenatal mortality in a marine cladoceran, *Evadne nordmanni*. *Mar. Ecol. Prog. Ser.* 29: 127-139.

Prins, R. 1968. Comparative ecology of the crayfishes *Orconectes rusticus rusticus* and *Cambarus tenebrosus* in Doe Run, Mead County, Kentucky. *Int. Rev. Ges. Hydrobiol.* 53: 667-714.

Rasmussen, B. 1953. On the geographic variation in growth and sexual development of the deep sea prawn (*Pandalus borealis* Kr.). *Fisk. Skr. Havundersøk.* 10(3): 1-160.

Reeve, M.R. 1969. The laboratory culture of the prawn, *Palaemon serratus*. *Fish. Invest. London (Ser. 2)* 26: 1-38.

Rhodes, C.P. & D.M. Holdich 1982. Observations on the fecundity of the freshwater crayfish, *Austropotamobius pallipes* (Lereboullet) in the British Isles. *Hydrobiologia* 89: 251-286.

Ricker, W.E. 1954. Stock and recruitment. *J. Fish. Res. Bd. Can.* 11: 559-623.

Roe, P. 1984. Laboratory studies of feeding and mating in species of *Carcinonemertes* (Nemertea: Hoplonemertea). *Biol. Bull.* 167: 426-436.

Roe, P. 1988. Ecological implications of the reproductive biology of symbiotic nemerteans. *Hydrobiologia* 156: 13-22.

Rogers-Talbert, R. 1948. The fungus *Lagenidium callinectes* Couch (1942) on eggs of the blue crab in Chesapeake Bay. *Biol. Bull.* 95: 214-228.

Rothlisberg, P.C. 1982. Vertical migration and its effect on dispersal of penaeid shrimp larvae in the Gulf of Carpentaria, Australia. *Fish. Bull.* 80: 541-554.

Roughgarden, J., S. Gaines & H. Possingham 1988. Recruitment dynamics in complex life cycles. *Science* 241: 1460-1466.

Ryan, E.P. 1967. Structure and function of the reproductive system of the crab *Portunus sanguinolentus* (Herbst)

(Brachyura: Portunidae). II. The female system. *Proc. Symp. Crustacea Mar. Biol. Assoc. India Symp. (Ser.2)*2: 522-544.

Saila, S.B. & J.M. Flowers 1965. A simulation sudy of sex ratios and regulation effects with the American lobster, *Homarus americanus*. *Proc. Gulf Carib. Fish. Inst.* 18: 16-78.

Salemaa, H. 1979. Ecology of *Idotea* spp. (Isopoda) in the northern Baltic. *Ophelia* 18: 133-150.

Schlieder, R.A. 1980. Effects of dessication and autospasy on egg hatching success in stone crab, *Menippe mercenaria*. *Fish. Bull.* 77: 695-700.

Sheader, M. 1977. The breeding biology of *Idotea pelagica* (Isopoda: Valvifera) with notes on the occurrence and biology of its parasite *Clypeoniscus hanseni* (Isopoda: Epicaridea). *J. Mar. Biol. Assoc. UK* 57: 659-674.

Sheader, M. 1983. The reproductive biology and ecology of *Gammarus duebeni* (Crustacea: Amphipoda) in southern England. *J. Mar. Biol. Assoc. UK* 63: 517-540.

Sheader, M. & F.S. Chia 1970. Development, fecundity and brooding behaviour of the amphipod, *Marinogammarus obtusatus*. *J. Mar. Biol. Assoc. UK* 50: 1079-1099.

Shields, J.D. & A.M. Kuris 1988a. Temporal variation in abundance of the egg predator, *Carcinonemertes epialti* (Nemertea), and its effect on egg mortality of its host, the shore crab, *Hemigrapsus oregonensis*. *Hydrobiologia* 156: 31-38.

Shields, J.D. & A.M. Kuris 1988b. An *in vitro* analysis of egg mortality in *Cancer anthonyi:* The role of symbionts and temperature. *Biol. Bull.* 174: 267-275.

Shields, J.D. & A.M. Kuris 1990. *Carcinonemertes wickhami* n. sp. (Nemertea), a symbiotic egg predator from the spiny lobster, *Panulirus interruptus*, in southern California, with remarks on symbiont-host adaptations. *Fish. Bull.* 88: 279-287.

Shields, J.D., R.K. Okazaki, & A.M. Kuris 1990a. Brood mortality and egg predation by the nemertean, *Carcinonemertes epialti*, on the yellow rock crab, *Cancer anthonyi*, in southern California. *Can. J. Fish. Aquat. Sci.* 47: 1275-1281.

Shields, J.D., R.K. Okazaki & A.M. Kuris 1990b. Fecundity and the reproductive potential of the yellow rock crab, *Cancer anthonyi*. *Fish. Bull.* (in press).

Shields, J.D., D.E. Wickham & A.M. Kuris 1989. *Carcinonemertes regicides* n. sp. (Nemertea), a symbiotic egg predator from the red crab, *Paralithodes camtschatica* (Decapoda: Anomura), in Alaska. *Can. J. Zool.* 67: 923-930.

Shillaker, R.O. 1977. The biology of *Lembos websteri* and *Corophium bonnellii* (Crustacea: Amphipoda) in relation to turbulence and turbidity. Ph.D. Thesis. University of Glasgow. 232 pp.

Shirley, S.M., T.C. Shirley & S.D. Rice 1987. Latitudinal variation in the Dungeness crab, *Cancer magister:* zoeal morphology explained by incubation temperature. *Mar. Biol.* 95: 371-376.

Siegel, P.R. & A.M. Wenner 1984. Abnormal reproduction of the sand crab *Emerita analoga* in the vicinity of a nuclear generating station in Southern California. *Mar. Biol.* 80: 341-345.

Sindermann, C.J. & A. Rosenfield 1967. Principal diseases of commercially important marine bivalve Mollusca and Crustacea. *Fish. Bull.* 66: 335-385.

Somerton, D.A. & R.A. MacIntosh 1985. Reproductive biology of the female blue king crab *Paralithodes platypus* near the Pribilof Islands. *J. Crustacean Biol.* 5: 365-376.

Sparrow, F.F. 1973. The peculiar marine phycomycete *Atkinsiella dubia* from crab eggs. *Arch. Microbiol.* 93: 137-144.

Steele, D.H. & V.J. Steele 1969. The biology of *Gammarus* (Crustacea, Amphipoda) in the northwestern Atlantic. I. *Gammarus duebeni* Lillj. *Can. J. Zool.* 47: 235-244.

Steele, D.H. & V.J. Steele 1970. The biology of *Gammarus* (Crustacea, Amphipoda) in the northwestern Atlantic. III. *Gammarus obtusatus* Dahl. *Can. J. Zool.* 48: 989-995.

Steele, D.H. & V.J. Steele 1972a. Biology of *Gamarellus angulosus* (Crustacea, Amphipoda) in the northwestern Atlantic. *J. Fish. Res. Bd. Can.* 29: 1337-1340.

Steele, D.H. & V.J. Steele 1972b. The biology of *Gammarus* (Crustacea, Amphipoda) in the northwestern Atlantic. VI. *Gammarus tigrinus* Sexton. *Can. J. Zool.* 50: 1063-1068.

Steele, D.H. & V.J. Steele 1973. Some aspects of the biology of *Calliopius laeviusculus* (Kröyer) (Crustacea, Amphipoda) in the northwestern Atlantic. *Can. J. Zool.* 51: 723-728.

Steele, D.H. & V.J. Steele 1975a. The biology of *Gammarus wilkitzkii* Birula, *Gammarus stoerensis* Reid, and *Gammarus mucronatus* Say. *Can. J. Zool.* 53: 1105-1109.

Steele, D.H. & V.J. Steele 1975b. The biology of *Gammarus* (Crustacea, Amphipoda) in the northwestern Atlantic. X. *Gammarus finmarchicus* Dahl. *Can. J. Zool.* 53: 1110-1115.

Steele, V.J. & D.H. Steele 1972. The biology of *Gammarus* (Crustacea, Amphipoda) in the northwestern Atlantic. V. *Gammarus oceanicus* Segerstråle. *Can. J. Zool.* 50: 801-813.

Stickney, A. 1978. A previously unreported peridinian parasite in the egg of the northern shrimp, *Pandalus borealis*. *J. Invert. Pathol.* 32: 212-215.

Sundelin, B. 1983. Effects of cadmium on *Pontoporeia affinis* (Crustacea: Amphipoda) in laboratory soft-bottom microcosms. *Mar. Biol.* 74: 203-212.

Sutton, S.L. 1968. The population dynamics of *Trichoniscus pusillus* and *Philoscia muscorum* (Crustacea Oniscoidea) in a limestone grassland. *J. Anim. Ecol.* 37: 424-444.

Tack, P.I. 1941. The life history and ecology of the crayfish *Cambarus immunis* Hagen. *Amer. Midl. Nat.* 25: 420-446.

Talbot, P. & R. Harper 1984. Abnormal egg stalk morphology is correlated with clutch attrition in laboratory-maintained lobsters (*Homarus*) *Biol. Bull.* 166: 349-356.

Talbot, P., P. Wilson & C. Thaler 1984. Spawning, egg attachment and egg retention in captive lobsters (*Homarus americanus*). *Aquaculture* 37: 239-249.

Tang, Q. 1985. Modification of the Ricker stock recruitment model to account for environmentally induced variation in recruitment with particular reference to the blue crab fishery in Chesapeake Bay. *Fish. Res.* 3: 13-21.

Tharp, T.P. & C.E. Bland 1977. Biology and host range of *Haliphthoros milfordensis* Vishniac. *Can. J. Bot.* 55: 2936-2944.

Thomas, H.J. 1964. The spawning and fecundity of Norway lobsters *Nephrops norvegicus* (L.) around the Scottish coast. *J. Cons. Int. Explor. Mer.* 29: 221-229.

Threlkeld, S.T. 1979. Estimating cladoceran birth rates: the importance of egg mortality and the egg age distribution. *Limnol. Oceanogr.* 24: 601-612.

Uhazy, L.S., D.E. Aiken & A. Campbell 1985. Morphology and systematics of the nemertean *Pseudocarcinonemertes homari* (Hoplonemertea: Monostilifera) from the American lobster, *Homarus americanus*. *Can. J. Fish. Aquat. Sci.* 42: 342-350.

Underwood, A.J. & P.C. Fairweather 1989. Supply-side ecology and benthic marine assemblages *Trends Ecol. Evol.* 4: 16-20.

Vader, W. & L. de Wolf 1988. Biotope and biology of *Armadillidium album* Dollfus, a terrestrial isopod of sandy beaches, in the SW Netherlands. *Netherlands J. Sea Res.* 22: 175-183.

Van Wyk, P.M. 1982. Inhibition of the growth and reproduction of the porcellanid crab, *Pachycheles rudis* by the bopyrid isopod, *Aporobopyrus muguensis*. *Parasitology* 85: 459-473.

Vlasblom, A.G. 1969. A study of a population of *Marinogammarus marinus* (Leach) in the Oosterschelde. *Neth. J. Sea Res.* 4: 317-338.

Ward, P.I. 1985. The breeding behaviour of *Gammarus duebeni*. *Hydrobiologia* 121: 45-50.

Wenner, A.M. 1977. Food supply, feeding habits, and egg production in Pacific mole crabs (*Hippa pacifica* Dana). *Pac. Sci.* 31: 39-47.

Wenner, A.M., Y. Ricard & J. Dugan 1987. Hippid crab population structure and food availability on Pacific shorelines. *Bull. Mar. Sci.* 41: 221-233.

Wickham, D.E. 1979. Predation by the nemertean *Carcinonemertes errans* on eggs of the Dungeness crab, *Cancer magister*. *Mar. Biol.* 55: 45-53.

Wickham, D.E. 1986. Epizootic infestations by nermertean brood parasites on commercially important crustaceans. *Can. J. Fish. Aquat. Sci.* 43: 2295-2302.

Wickham, D.E. & A.M. Kuris 1985. The comparative ecology of nemertean egg predators. *Amer. Zool.* 25: 127-134.

Wickham, D.E. & A.M. Kuris 1988. Diversity among nemertean egg predators of decapod crustaceans. *Hydrobiologia* 156: 23-30.

Wickham, D.E., P. Roe & A.M. Kuris 1984. Transfer of nemertean egg predators during host molting and copulation. *Biol. Bull.* 167: 331-338.

Wild, P.W. 1983. The influence of seawater temperature on spawning, egg development, and hatching success of the Dungeness crab, *Cancer magister*. *California Fish Game Bull.* 172: 197-213.

Williams, J.A. 1978. The annual pattern of reproduction of *Talitrus saltator* (Crustacea: Amphipoda: Talitridae). *J. Zool.* 184: 231-244.

Wittmann, K.J. 1981. On the breeding biology and physiology of marsupial development in Mediterranean *Leptomysis* (Mysidacea: Crustacea) with special reference to the effects of temperature and egg size. *J. Exp. Mar. Biol. Ecol.* 53: 261-279.

Woodlock, B. & J.D. Reynolds 1988a. Laboratory breeding studies of freshwater crayfish, *Austropotamobius pallipes* (Lereboullet). *Freshw. Biol.* 19: 79-86.

Woodlock, B. & J.D. Reynolds 1988b. Reproduction in an Irish lake population of the crayfish, *Austropotamobius pallipes* (lereboullet). *Freshw. Biol.* 19: 79-86.

Egg size in squat lobsters (Galatheoidea): Constraint and freedom

Cindy Lee Van Dover
Woods Hole Oceanographic Institution, Massachusetts, USA

Austin B. Williams
National Marine Fisheries Service, Systematics Laboratory, National Museum of Natural History, Washington, DC, USA

ABSTRACT

Within the superfamily Galatheoidea, egg volumes of 52 species in the families Chirostylidae (*Chirostylus, Eumunida, Uroptychus*) and Galatheidae (*Cervimunida, Galathea, Munida, Munidopsis, Pleuroncodes*) were determined from measurements of eggs on females in museum collections. Correlations between egg size, body size, brood size, and depth of collection were examined for this data set. In general, significant, positive correlations between egg size and body size and between brood size and body size were observed. We found a strict dichotomy in egg size that corresponds to the generic classification of species examined but not to habitat. We noted a narrow range of egg sizes in species presumed to have planktotrophic larvae and a broad range of egg sizes in species presumed to have lecithotrophic larvae. These observations are discussed in the context of phylogenetic and biological constraints.

1 INTRODUCTION

Even before Thorson's (1950) influential thesis on reproduction and larval ecology of marine benthic invertebrates, biologists attempted to interpret egg size and type of larval development among these animals. Thorson, echoing an observation made more than 75 years earlier by Thomson (1874), predicted that species producing small eggs and planktotrophic larvae would be scarce in the deep sea, and that direct development would dominate there. This prediction arose from three major assumptions regarding the deep-sea environment – low food availability, low temperature, and uniform habitat – but was supported by limited data.

Although Thorson's hypothesis has had great appeal and support, accumulating information indicates that distinctions between two of his proposed modes of larval development – small eggs, planktotrophy: large eggs, lecithotrophy – should be viewed as strong tendencies subject to variation, rather than fixed pathways (fully reviewed by Hines 1982, 1986a, b). Variability and flexibility in size, concentration of stored nutrient, and biochemical pathways all may modify bimodality in egg-sizes; but evidence from diverse sources still gives Thorson's central idea utility.

Much remains to be learned concerning reproductive biology of deep-sea invertebrates, even of decapod crustaceans whose females, with the exception of penaeid shrimps, display their reproductive efforts beneath their abdomens. There are observations on mode of reproduction in deep-sea decapods, and measurements of egg size in taxonomic descriptions. For example, Bruun (1954: 86) noted that 'some abyssal animals must have abysso-pelagic larvae; females of the crab *Ethusa* carrying very small eggs... are in our collections;' Wolff (1961: 30) remarked that in the pagurid, *Probeebei mirabilis* Boone, 'as in other abyssal pagurids,... the size of the egg is very small...'. Ivanova & Vassilenko (1987), surveying number and weight of eggs per brood and female weight in selected representatives from 10 orders of Crustacea, noted no correlation between egg and female size among 13 species of caridean shrimps from the White and Barents Seas, though an apparent tendency for larger species to produce large eggs; they found only small eggs on an anomuran and three brachyuran species from the Bering and Okhotsk Seas. However, there has been no systematic survey of reproductive and distribution patterns among deep-sea decapod species. Here we attempt to reduce that deficiency with observations on the Galatheoidea.

Munidopsis and other genera within the superfamily Galatheoidea collectively include both shallow- and deep-water species that exhibit an array of body sizes. They produce either large or small eggs. We examined the relationships between egg size, body size and depth in this group, and discuss the results in terms of phylogenetic and biological constraints. Our data do not allow us to consider other important reproductive factors such as frequency of brood production or reproductive seasonality.

2 MATERIALS AND METHODS

Materials measured came from the crustacean collections of the Allan Hancock Foundation (AHF) and University of Southern California, Los Angeles; Scripps Institution of Oceanography (SIO), La Jolla, California; and the United States National Museum of Natural History (USNM), Smithsonian Institution, Washington, DC, from each of which bathymetric and locality data were also taken. Species choices were necessarily limited to samples containing ovigerous females.

Size of individuals measured was based on maximum carapace width in mm. Inasmuch as members of the genera studied have varied carapace shapes, this width was chosen as the best, easily measured indicator of size.

Eggs of all species were measured in fluid directly with the aid of a binocular dissecting microscope equipped with an ocular micrometer, or their images were projected with the aid of a camera lucida, traced, and subsequently measured. For spherical eggs, diameter alone was recorded, but for ellipsoid eggs, by far the majority, both greater and lesser diameter were recorded. Volumes were computed by conventional formulas.

Number of eggs in a brood were counted directly for *Munidopsis* and other large-egged species, but estimated for species having many small eggs by counting those on a single pleopod and extrapolating this tally to total number in brood.

Statistical tests were based on Zar (1974).

3 EGG SIZE AND INFERRED MODE OF DEVELOPMENT IN THE GALATHEOIDEA

Only a few species in the superfamily Galatheoidea have been reared in the laboratory. Of these species, *Pleuroncodes planipes* Stimpson, *Galathea rostrata* A. Milne Edwards, and an undesignated *Munida* sp. produce small eggs and planktotrophic larvae that pass through a series of zoeal stages and substages (Boyd & Johnson 1963, Fagetti & Campodonico 1971, Gore 1979). *Munidopsis tridentata* (Esmark) produces large eggs (1.65 mm diameter) and has an abbreviated number (3) of lecithotrophic zoeal stages with non-functional mouthparts (Samuelsen 1972). *Munidopsis polymorpha* Koelbel also produces large eggs (1.5 to 1.8 mm diameter) and was inferred by Fage & Monod (1936) to have direct development; however, its development may include non-feeding zoeal stages as in *M. tridentata.*

Interpretation of egg size in decapod crustaceans is not necessarily straightforward. Increases in egg size during the period between deposition on the pleopods and hatching averaged between 30 and 50% for a variety of decapod families (41% in Galatheidae, based on species with 4-5 zoeal stages; Gore 1985). Furthermore, while it may be generally true that species with large eggs have some form of abbreviated, lecithotrophic development, there are exceptions within the Decapoda in which species with large eggs pass through numerous planktotrophic zoeal stages (discussion in Rabalais & Gore 1985). Variability and flexibility in egg size within a species in response to physiological and environmental conditions are also poorly known (Hines 1986a).

We limit our interpretations to differences and trends that are sufficiently large to be 'signals' superimposed on a lesser, albeit significant, 'noise'. We assume that studies of larval development of species within the Galatheoidea outlined above are indicative of the relationship between egg size and mode of development within the superfamily as a whole.

4 RESULTS

Summaries of egg-volume, carapace-width, and sampling depth statistics for *Munidopsis* species and species in other genera within the Galatheoidea are provided in Tables 1 and 2. Standard deviations for egg volumes given in these tables reflect variability in all eggs measured from individuals of a given species.

Mean egg volumes for *Munidopsis* species ranged from 0.51 to 11.26 mm^3. Egg volumes for *Chirostylus* and *Uroptychus* species fell within this range. In *Cervimunida, Eumunida, Galathea, Munida* and *Pleuroncodes* species, mean egg volumes were consistently and considerably smaller, ranging from 0.01 to 0.18 mm^3. Coefficients of variation (V) for mean egg volumes of species in these two series ranged from 7.3 to 50%. In the species with large eggs, V ranged from 8.9 to 40.5% and in those with small eggs from 7.3 to 50%. The values were, for the most part, far above those usual for such variates in relatively unified samples (Simpson et al. 1960) and may reflect variable egg size increases accompanying embryonic development, as well as variable effects of preservation. There was a notable distortion and erosion of eggs in some lots of preserved specimens, but there was no question of distinct size difference in the two series.

Average depths of collection for non-*Munidopsis* Galatheoidea were all < 1000 m; average depths of collection for *Munidopsis* species ranged from < 1 to 4390 m, but were predominantly > 1000 m.

Average carapace width ranged from about 4 mm to a maximum of 32.9 mm in *Munidopsis* species and from 3.8 to 46.9 mm in non-*Munidopsis* species.

Variability in egg volume among individuals of a given species is reflected in data from *Munidopsis simplex* (A. Milne Edwards) (Table 3). Standard deviations of egg volume ranged from 6 to 21% of the mean egg volume in 6 individuals. Mean egg volume in *M. simplex* was

Table 1. Means (± standard deviation) egg volume (mm^3), female carapace width (mm), depth of collection (m); and number of individuals with eggs measured and counted for 36 species of *Munidopsis*; eggs for single specimens either totaled or unlisted if obviously deficient.

Species	Egg volume	(±s)	Carapace width	(±s)	Depth	(±s)	Individuals	Eggs measured	Eggs/individual
M. abbreviatus (A. Milne Edwards)	3.36	0.79	15.9	3.46	887	352	5	42	14
M. alaminos Pequegnat & Pequegnat	0.75	0.30	10.2		776		2	16	53
M. andamanica MacGilchrist	0.92	0.14	10.6		706		2	20	31
M. antonii (A. Milne Edwards)	10.00	2.61	27.5		3897		2	33	
M. armata (A. Milne Edwards)	1.35	0.37	8.1	0.58	1221	152	4	12	3
M. aspera (Henderson)	0.69	0.11	7.5	1.03	1072	324	3	33	37
M. bairdii Smith	5.01	0.80	19.8		2295		1	30	41
M. barrerai Bahamonde	0.51	0.08	11.7		800		1	30	
M. bermudezi Chace	4.64	0.53	22.0		3179		1	30	104
M. ciliata Wood Mason	4.04	0.95	10.5	4.33	2153	710	5	27	7
M. columbiana Pequegnat & Pequegnat	10.48	1.36	19.6		4152		2	28	23
M. crassa Smith	8.45	1.37	24.1		3186		1	24	24
M. curvirostra Whiteaves	1.27	0.34	8.7	0.91	1659	246	4	51	50
M. depressa Faxon	0.62	0.15	9.5		760		1	4	
M. diomedeae (Faxon)	5.41	0.96	20.0	3.75	1687	230	4	73	40
M. hystrix Faxon	1.26	0.34	11.0		915		2	60	32
M. latirostris Faxon	4.97	1.12	12.1	0.99	2800		5	32	10
M. lentigo Williams & Van Dover	4.12	0.94	10.0		2600		2	24	18
M. nitida (A. Milne Edwards)	4.40	0.50	12.8		2149		3	30	30
M. ornata Faxon	0.66	0.09	5.3		1212		1	4	
M. polymorpha Koelbel	0.62	0.17	4.4		<1		1	3	
M. quadrata Faxon	0.54	0.10	7.5		714		1	30	
M. riveroi Chace	1.18	0.20	11.3		4390		2	30	45
M. rostrata (A. Milne Edwards)	7.96	1.14	32.9		2008		1	30	
M. scabra Faxon	1.25	0.43	10.4		1083		2	23	
M. sigsbei (A. Milne Edwards)	2.06	0.23	9.2		1107		2	14	
M. similis Smith	4.35	0.39	11.6		1939		1	4	
M. simplex (A. Milne Edwards)	1.11	0.45	7.5	1.15	1697	399	7	53	
M. sinclairi McArdle	1.63	0.20	10.8		1525		1	30	

M. subsquamosa (Henderson)	8.92	2.24	29.0	3.41	2491	3	3	60
M. trachynotus (Anderson)	11.23	1.67	28.2		1375		1	30
M. tridentata (Esmark)	1.39	0.21	5.6		1472		1	2
M. valdiviae (Doflein & Bàlss)	5.14	1.14	17.7	3.36	1378	37	3	30
M. verrilli Benedict	4.19	0.80	9.4	1.89	1860	340	5	16
M. verrucosus Khodkina	11.26	1.39	19.5		4194		1	7
M. villosa Faxon	7.25	1.77	22.2		1003		2	50

Table 2. Mean egg volume (mm^3) ± standard deviation, individual or mean carapace width (mm), sampling depth, number of individuals and eggs in sample for 16 non-*Munidopsis* species.

Species	Egg volume	(±s)	Carapace width	Depth	Individuals	Eggs measured
Galatheidae						
Cervimunida johni Porter	0.10	0.016	36.0	<1	1	30
Galathea californiensis Benedict	0.05	0.010	14.1	87	1	30
G. rostrata A. Milne Edwards	0.03	0.004	5.4	51	2	60
G. strigosa (Linnaeus)	0.18	0.015	34.4	<1	1	30
Munida gracilipes Faxon	0.03	0.008	14.0	243	1	30
M. hawaiiensis Baba	0.04	0.011	11.8	<1	1	30
M. hispida Benedict	0.04	0.009	25.7	256	1	30
M. irrasa A. Milne Edwards	0.02	0.006	3.9	39	1	30
M. mexicana Benedict	0.01	0.005	4.1	59	1	30
M. propinqua Faxon	0.04	0.008	14.5	921	1	30
M. quadrispina Benedict	0.02	0.006	6.2	595	1	30
M. tenella Benedict	0.01	0.003	6.7	104	1	30
Pleuroncodes planipes Stimpson	0.03	0.006	10.5	90	2	50
Chirostylidae						
Chirostylus sp.	3.13	0.603	18.1	950	1	30
Eumunida picta Smith	0.15	0.011	46.9	512	1	30
Uroptychus nitidus (A. Milne Edwards)	0.96	0.30	3.78	592	5	26

Table 3. *Munidopsis simplex*, within species variability of egg volume at 5 depths: r_1 = correlation coefficient between carapace width (mm) and mean egg volume (mm^3); r_2 = correlation coefficient between depth and mean egg volume; r_3 = correlation coefficient between carapace width and depth; $r_1 = 0.685$; DF = 6; n.s.; $r_2 = -0.921$; DF = 6; $p < 0.01$; $r_3 = -0.839$; DF = 6; $p < 0.01$.

Carapace width	Depth (m)	Volume (±s)	Eggs measured
9.3	1110	1.70 ± 0.24	13
8.3	1281	1.51 ± 0.11	5
7.5	1564	0.96 ± 0.06	7
6.3	1825	1.21 ± 0.19	6
8.1	1825	0.82	1
6.1	2138	0.67 ± 0.14	8
7.0	2138	0.70 ± 0.08	13

Table 4. Within-species variability in 6 species of *Munidopsis* at one depth; comparisons between 2 samples of mean egg volume (mm^3); carapace width in mm; Student's t statistic.

Carapace width	Volume (±s)	Eggs measured
Munidopsis abbreviatus	(1246 m)	
13.0	3.03 ± 0.44	10
17.7	2.81 ± 0.31	4
	t = 1.07: DF = 12: n.s.	
Munidopsis simplex	(2138 m)	
6.1	0.67 ± 0.14	8
7.0	0.70 ± 0.08	13
	t = 0.553: DF = 19: n.s.	
Munidopsis valdiviae	(1399 m)	
14.0	4.02 ± 0.63	10
18.4	6.38 ± 0.62	10
	t = 8.49: DF = 18: p < 0.01	
Munidopsis columbiana	(4152 m)	
18.0	9.40 ± 0.98	13
21.2	11.41 ± 0.85	15
	t = 5.75: DF = 26: p < 0.01	
Munidopsis riveroi	(4390 m)	
11.0	1.10 ± 0.16	15
11.5	1.25 ± 0.21	15
	t = 2.25: DF = 28: n.s.	
Munidopsis lentigo	(2600 m)	
9.5	4.42 ± 0.57	12
10.4	3.82 ± 1.14	12
	t = 1.66: DF = 22: p < 0.01	

not significantly correlated with body size but was negatively correlated with depth of collection ($p < 0.01$). Egg volume decreased from 1.70 to 0.67 mm^3 over a 1000 m increase in depth. This contrasted with overall trends of increasing egg size with depth and body size among other species of *Munidopsis* and allied genera examined.

For 13 species of *Munidopsis*, we found more than one individual at a given depth, allowing us to consider variation in egg volume with size of female while depth remained constant. We compared mean egg volumes in six of these species (Table 4). The means were significantly

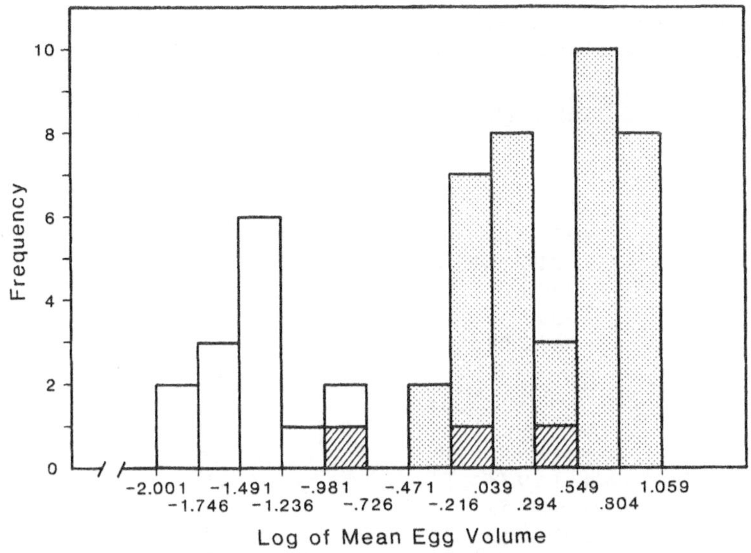

Figure 1. Histogram showing log mean egg volumes of galatheoids classed on horizontal scale and frequency of species observed within classes on vertical scale. Egg sizes of species from genera with small eggs (*Cervimunida, Eumunida, Galathea, Munida, Pleuroncodes*) and those from genera with large eggs (*Chirostylus, Munidopsis, Uroptychus*) form two distinct clusters. Bimodality in cluster of large egg sizes is attributed to sampling error. Screen shading = *Munidopsis* species, open = other galatheid species, diagonal hatching = chirostylid species.

different between two individuals in *M. valdiviae* (Doflein & Balss), *M. columbiana* Pequegnat & Pequegnat and *M. lentigo* (Williams & Van Dover) (t-test, $p < 0.01$); the larger mean egg volume corresponded to the individual with the larger carapace width in the first two of these species, but not in the last. We also calculated correlation coefficients for egg volume versus carapace width in four *Munidopsis* species obtained from a single depth (Table 5). A significant positive correlation was found in *M. ciliata* Wood Mason and *M. latirostris* Faxon. No correlations were identified in the other 2 species.

A size-frequency histogram for egg volume is given in Figure 1. This figure clearly illustrates a dichotomous distribution of egg volume. Based on egg size, we infer that planktotrophic development predominates in species with small eggs (< 0.2 mm^3), and that lecithotrophic development predominates in species with large eggs (> 0.5 mm^3).

Scatter plots of egg volume versus depth of collection and carapace width are shown in Figures 2 and 3. Data points were derived from average values listed in Tables 1 and 2, and the summary statistics accompanying these plots are given in Tables 6 and 7. Egg volume was positively correlated with carapace width and with depth in *Munidopsis* species; a similar pattern of correlation was detected when *Uroptychus* and *Chirostylus* species data were combined with those from *Munidopsis* species. Carapace width increased with depth in this combined data set. Egg volume was also positively correlated with carapace width in eight species of *Munida* combined with species of *Cervimunida, Eumunida, Galathea,* and *Pleuroncodes.* No significant correlation between carapace width and depth could be detected in this combined data set.

Within 31 species of *Munidopsis*, we examined the relationships between brood size (total

Table 5. Within species variability of egg volume and female size in 4 species of *Munidopsis*. Data for each species are from one depth; r is the correlation coefficient between carapace width (mm) and mean egg volume (mm^3).

Carapace width	Volume (±s)	Eggs measured
Munidopsis ciliata	(2666 m)	
6.7	2.44	2
7.5	3.97	1
7.9	4.12 ± 1.26	3
	$r = 0.968$; $DF = 2$: $0.01 > p < 0.05$	
Munidopsis nitida	(2149 m)	
11.9	4.20 ± 0.42	10
12.5	4.76 ± 0.50	10
13.9	4.24 ± 0.38	10
	$r = -0.164$; $DF = 2$: n.s.	
Munidopsis latirostris	(2800 m)	
10.6	3.84 ± 0.51	3
11.9	4.53 ± 0.38	11
12.3	5.49 ± 1.53	9
12.7	5.04 ± 0.89	7
13.2	6.59	2
	$r = 0.902$; $DF = 4$: $0.01 > p < 0.05$	
Munidopsis verrilli	(2012 m)	
7.3	3.78 ± 0.34	3
8.5	3.59	2
8.8	3.78	2
12.2	4.22 ± 0.48	6
	$r = 0.863$; $DF = 3$: n.s.	

Table 6. Within genus correlations using mean egg volumes (mm^3) from Tables 1 and 2.

	Depth	Carapace width
Munidopsis spp.	$DF = 34$	$DF = 35$
Egg volume	$r = 0.611$	$r = 0.853$
	$p < 0.01$	$p < 0.01$
Carapace width	$r = 0.419$	
	$p < 0.05$	
Munida spp.	$DF = 6$	$DF = 7$
Egg volume	$r = 0.589$	$r = 0.864$
	n.s.	$p < 0.01$
Carapace width	$r = 0.279$	
	n.s.	
Galathea spp.	$DF = 1$	$DF = 2$
Egg volume	$r = 1.000$	$r = 0.985$
	$p < 0.01$	n.s.
Carapace width	$r = 1.000$	
	$p < 0.01$	

number of eggs per individual), egg volume and carapace width. Figure 4 and Table 8 summarize the data. Brood size is highly variable among individuals within a species. We think that this is in part a consequence of the tendency for eggs to be lost during collection of material in

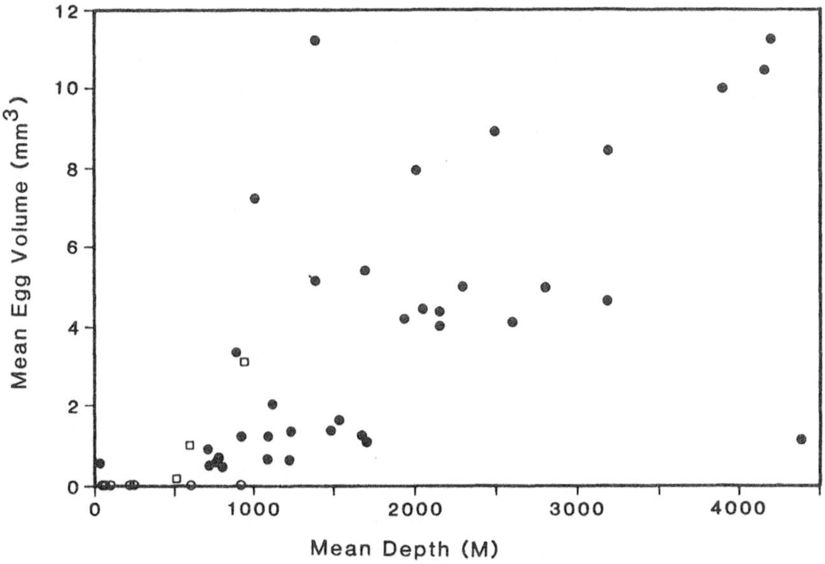

Figure 2. Scatter plot, mean depth (m) of collection per species of Galatheoidea versus mean egg volume (mm^3) for ovigerous females of species examined (Tables 1 and 2). ● = *Munidopsis* species, ○ = other galatheid species, □ = chirostylid species.

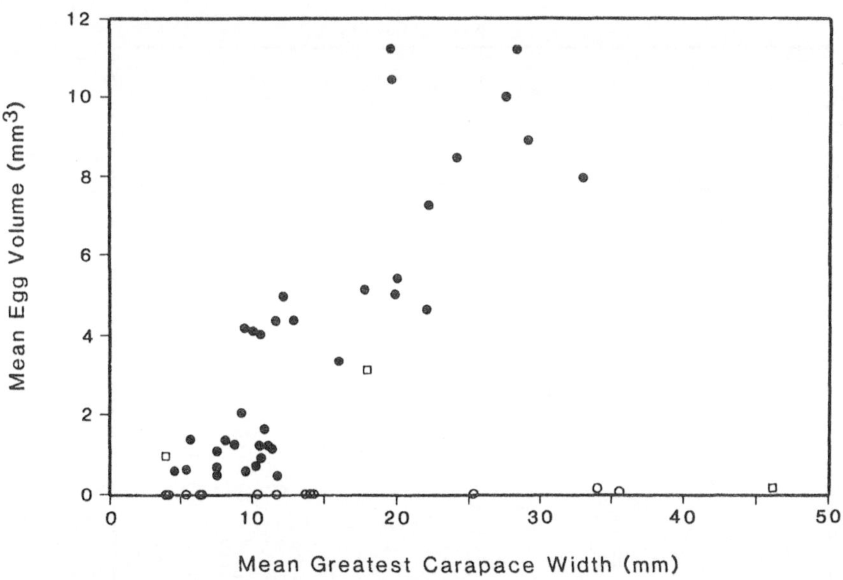

Figure 3. Scatter plot, mean greatest carapace width (mm) versus mean egg volume (mm^3) for ovigerous females of galatheoid species examined (Tables 1 and 2). ● = *Munidopsis* species, ○ = other galatheid species, □ = chirostylid species.

Table 7. Additional correlations using mean egg volumes in Tables 1 and 2.

	Depth	Carapace width
Munidopsis, Chirostylus, Uroptychus (Large eggs)	DF = 36	DF = 37
Egg volume	r = 0.621 p < 0.01	r = 0.848 p < 0.01
Carapace width	r = 0.429 p < 0.01	
Munida, Cervimunida, Eumunida, Galathea, Pleuroncodes (Small eggs)	DF = 10	DF = 13
Egg volume	r = 0.356 n.s.	r = 0.891 p < 0.01
Carapace width	r = 0.374 n.s.	

Table 8. Correlations between brood size, mean carapace width (mm) and mean egg volume (mm^3) within 31 species of *Munidopsis*.

	Egg volume	Carapace width
Number of eggs	DF = 30 r = 0.240 n.s.	 r = 0.609 p < 0.01

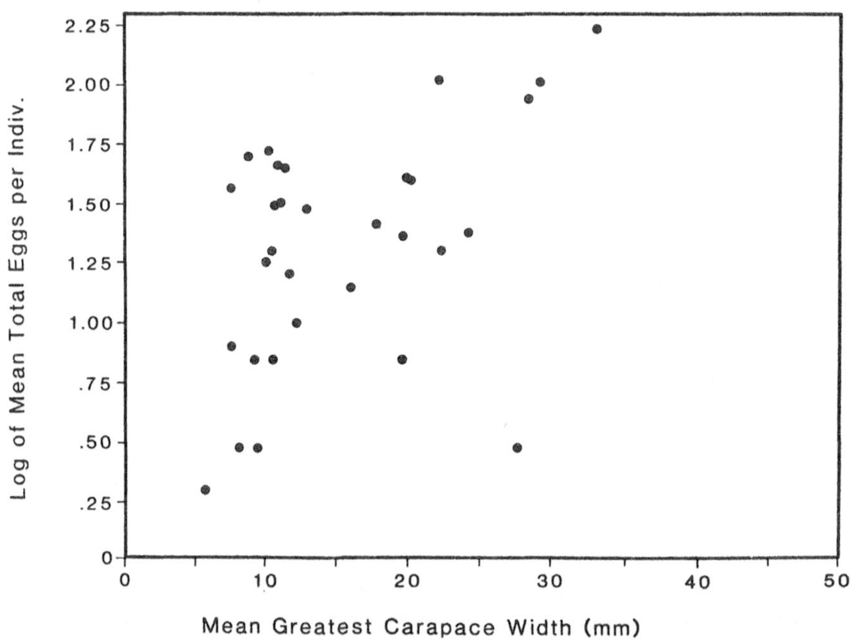

Figure 4. Scatter plot, mean greatest carapace width (mm) versus log of mean total eggs per individual for ovigerous *Munidopsis* females examined.

trawls or other sampling devices. Mean or single brood size of each species was used for correlation analyses. We found no significant correlation between brood size and egg volume but a highly significant correlation between brood size and carapace width.

Species of *Munida* and allied genera with small eggs are estimated to have approximately an order of magnitude more eggs per brood than *Munidopsis* species of comparable body sizes, based on sample estimates from single individuals of *Galathea rostrata, Munida irrasa* and *Eumunida picta.*

5 DISCUSSION

We have presented results of simple statistical analyses on egg size, body size, brood size and depth. Selected analyses, such as intraspecific variation in egg size or correlation between egg size versus body size with depth held constant, indicate variability in egg size and identify some of the correlating factors. Significant correlations such as those observed between egg size and body size or brood size and body size are correlations that have been described for other taxa and discussed at length elsewhere (see Hines 1982, 1986a, b). We narrow the focus of our discussion to two main points: 1) the observed dichotomy in egg size as a function of generic classification and 2) the narrow range of egg sizes in species with presumed planktotrophic larvae versus the broad range of egg sizes among species with presumed lecithotrophic larvae.

5.1 *Phylogenetic constraints*

Within the superfamily Galatheoidea, we examined egg sizes of 52 species in 8 genera. There is a strict dichotomy in egg size that corresponds without exception to the generic classification (Fig. 1). Species belonging to the genus *Munida* invariably have small eggs (average volume = 0.010 - 0.044 mm^3); species belonging to the genus *Munidopsis* have large eggs (average volume > 0.51 mm^3). *Cervimunida, Eumunida, Galathea* and *Pleuroncodes* species also have relatively small eggs; *Chirostylus* and *Uroptychus* have large eggs with average volumes that are within the range of egg volumes of *Munidopsis* species.

Habitats and geographical distributions of the species included in our survey are diverse: intertidal, subtidal; continental shelf, deep-sea; pelagic, benthic; soft-substrate, hard-substrate; high latitude, low latitude; Atlantic, Pacific. A few species are known to be associated with ephemeral resources such as wood islands (Turner 1973) or hydrothermal vents (Williams & Van Dover 1983), or with unusual environments such as shallow-water lava tubes (Fage & Monod 1936). No consistent, systematic pattern in habitat type is reflected in the Galatheoidea at the generic level.

A phylogenetic or evolutionary constraint that precludes a shift from lecithotrophic development with large eggs to planktotrophic development with small eggs (*sensu* Jagersten 1972) was suggested by Van Dover et al. (1985) for two species of *Munidopsis*. These authors, in examining the reproductive patterns of decapod crustaceans that occur at hydrothermal vents in the deep-sea, concluded that habitat *per se* does not dictate egg size and the mode of larval development. Our observations here, based on a larger survey within the Galatheoidea, are consistent with this hypothesis.

Hines (1986a) discussed developmental canalization and cited examples of evolutionary constraints in reproductive strategies that operate at higher taxonomic levels within the Crustacea, such as families of Brachyura (Rice 1980, Hines 1986a). Within the invertebrates in general, Strathmann (1978) discussed the biased transition to non-feeding larvae that results in large clades lacking a feeding larval stage. The *Munida/Munidopsis* dichotomy serves as an additional example of this general phenomenon. Curiously, within the genus *Munida,* we find no examples of species with large, lecithotrophic eggs, even though loss of planktotrophy is a conceivable evolutionary event.

Selection for large eggs and non-feeding larval stages in *Munidopsis* species may be linked to the success of this genus in the deep sea where planktonic food has been presumed to be rare. However, see Wiebe et al. (1988) for a modified view of the paucity of planktonic food at depths of 2000 m; zooplankton biomass at depths 100 m above the bottom in the Gulf of California is just a factor of 10 less than surface zooplankton biomass. In this context, *Munidopsis polymorpha*, a littoral species (Wilkens & Parzefall 1974, Parzefall & Wilkens 1975), could represent a reinvasion of shallow water by some deep sea *Munidopsis* ancestor that had irreversibly lost its feeding planktonic larva.

5.2 *Biological constraint and freedom*

Egg size, expressed as volume, is constrained within a narrow range of less than 0.20 mm^3 in galatheoid species that produce planktotrophic larvae. Minimum requirements for biochemical processes and structures involved in embryogenesis of a planktotrophic larva ultimately constrain the smaller end of this size range. Multiplicative physiological and ecological constraints may operate at the larger end. For example, maximum egg size may correspond to the size at which there is a balance between the maximum amount of energetic resources that can be efficiently budgeted to embryogenesis of a planktotrophic larva, and trade-offs between egg size and number of planktotrophic propagules (eggs per brood). The close constraints imposed on egg size in planktotrophic galatheoids correspond conceptually to the unimodal or optimal egg size at one extreme in theoretical models of Vance (1973a, b) and Christiansen & Fenchel (1979).

In contrast, egg size may vary over a large range (10.75 mm^3) in galatheoid species inferred to have lecithotrophic development. Minimum egg size corresponds to the size at which energetic resources budgeted to embryogenesis and development are minimized. Maximum egg size in these species is ultimately constrained by maximum allowances for biochemical and biophysical processes involved in embryogenesis and development, and trade-offs between egg size and number of lecithotrophic propagules.

The simplest way to account for tightly constrained egg sizes (s) in galatheids with planktotrophic larvae is to require that the number of propagules (n) be maximized. Then for a given reproductive output, R, where

$$R = sn$$

if n is to be maximized and s has a physiologically prescribed minimum value, s must be tightly constrained around that minimum value. Even when R is allowed to vary as a function of body size, s is tightly constrained. Obviously, if n need not be maximized, as it may not be for species with lecithotrophic development, then for a given R, s is also free to vary within the scope of physiologically feasible maximum and minimum egg sizes. When R is allowed to vary as a function of body size, variability in s may be increased.

ACKNOWLEDGEMENTS

We are indebted to the following for aid with aspects of this paper that have been in progress since 1983. J. Haig (AHF) and S. Luke (SIO) gave curatorial assistance. Critical comments were offered by Drs B.B. Collette and M. Vecchione, National Marine Fisheries Service, Systematics Laboratory (USNM), J.F. Grassle, Woods Hole Oceanographic Institution, Woods Hole, Mass., and A.H. Hines, Smithsonian Environmental Research Center, Edgewater, Md., and Dr L.-A.C. Hayek, Smithsonian Institution gave statistical counsel. R.E. Gibbons helped with measurements, entered data for computation and ran many of the statistical tests with the aid of M.S. Nizinski, V.R. Thomas processed the manuscript. C.L. Van Dover received aid from an NSF Graduate Fellowship.

REFERENCES

Boyd, C.M. & M.W. Johnson 1963. Variation in the larval stages of a decapod crustacean, *Pleuroncodes planipes* Stimpson (Galatheidae). *Biol. Bull.* 124: 141-152.

Bruun, A.F. 1954. General discussion. In: On the distribution and origin of the Deep-Sea Bottom Fauna. *Int. Union Biol. Sci.* Ser. B, 6: 86.

Christiansen, F.B. & T.M. Fenchel 1979. Evolution of marine invertebrate reproductive patterns. *Theor. Pop. Biol.* 16: 267-282.

Fage, L. & Th. Monod 1936. Biospelogica. LXIII. La faune marine du Jameo de Aqua. Lac souterrain de l'Ile de Lanzarote (Canaries). *Archs. Zool. Exp. Gen.* 78: 97-113.

Fagetti, E. & I. Campodonico 1971. Larval development of the red crab *Pleuroncodes monodon* (Decapoda Anomura: Galatheidae) under laboratory conditions. *Mar. Biol.* 8: 70-81.

Gore, R.H. 1979. Larval development of *Galathea rostrata* under laboratory conditions, with a discussion of larval development in the Galatheidae (Crustacea, Anomura). *Fish. Bull., US* 76: 781-806.

Gore, R.H. 1985. Molting and growth in decapod larvae. In A.M. Wenner (ed.), *Crustacean Issues* 2: *Larval Growth*: pp. 1-65. Rotterdam: A.A. Balkema.

Hines, A.H. 1982. Allometric constraints and variables of reproductive effort in brachyuran crabs. *Mar. Biol.* 69: 309-320.

Hines, A.H. 1986a. Larval patterns in the life histories of brachyuran crabs (Crustacea, Decapoda, Brachyura). *Bull. Mar. Sci.* 39(2): 444-466.

Hines, A.H. 1986b. Larval problems and perspectives in life histories of marine invertebrates. *Bull. Mar. Sci.* 39: 506-525.

Ivanova, M.B. & S.V. Vassilenko 1987. Relationship between number of eggs, brood weight, and female body weight in Crustacea. *Int. Revue ges. Hydrobiol.* 72: 147-169.

Jagersten, G. 1972. *Evolution of the metazoan life cycle*. New York, NY: Academic Press.

Parzefall, J. & H. Wilkens 1975. Zur Ethologie augenreduzierter Tiere. Untersuchungen an *Munidopsis polymorpha* Koelbel. *Ann. Spéléol.* 30: 325-335.

Rabalais, N.N. & R.H. Gore 1985. Abbreviated development in decapods. In A.M. Wenner (ed.), *Crustacean Issues* 2: 67-126. *Larval Growth*. Rotterdam: A.A. Balkema.

Rice, A.L. 1980. Crab zoeal morphology and its bearing on the classification of the Brachyura. *Trans. Zool. Soc. London* 35: 271-424.

Samuelsen, T.J. 1972. Larvae of *Munidopsis tridentata* (Esmark) (Decapoda, Anomura) reared in the laboratory. *Sarsia* 48: 91-98.

Simpson, G.G., A. Roe, & R.C. Lewontin 1960. *Quantitative Zoology (Rev. Ed.)* New York, NY: Harcourt, Brace & World, Inc.

Strathmann, R.R. 1978. The evolution and loss of feeding larval stages of marine invertebrates. *Evolution* 32: 894-906.

Thomson, C.W. 1874. *The Depths of the Sea*. London: McMillan.

Thorson, G. 1950. Reproductive and larval ecology of marine bottom invertebrates. *Biol. Rev.* 25: 1-45.
Turner, R.D. 1973. Wood-boring bivalves, opportunistic species in the deep-sea. *Science* 180: 1377-1379.
Vance, R.R. 1973a. On reproductive strategies in marine benthic invertebrates. *Amer. Nat.* 107: 339-352.
Vance, R.R. 1973b. More on reproductive strategies in marine benthic invertebrates. *Amer. Nat.* 107: 353-361.
Van Dover, C.L., J.R. Factor, A.B. Williams & C.J. Berg 1985. Reproductive patterns of decapod crustaceans from hydrothermal vents. *Bull. Biol. Soc. Washington.* 6: 223-227.
Wiebe, P.H., N. Copley, C. Van Dover, A. Tamse & F. Manrique 1988. Deep-water zooplankton of the Guaymas Basin hydrothermal vent field. *Deep-Sea Res.* 35: 985-1013.
Wilkens, H. & J. Parzefall 1974. Die Ökologie der Jameos del Agua (Lanzarote). Zur Entwicklung limnischer Höhlentiere aus marinen Vorfahren. *Ann. Spéléol.* 29: 419-434.
Williams, A.B. & C.L. Van Dover 1983. A new species of *Munidopsis* from submarine thermal vents of the East Pacific Rise at 21°N (Anomura: Galatheidae). *Proc. Biol. Soc. Washington.* 96: 481-488.
Wolff, T. 1961. Description of a remarkable deep-sea hermit crab, with notes on the evolution of the Paguridea. *Galathea Rep.* 4: 11-32.
Zar, J.H. 1974. *Biostatistical analysis.* Englewood Cliffs, NJ: Prentice-Hall, Inc.

Morphological and environmental restraints on egg production in amphipods

D.H. Steele & V.J. Steele
Department of Biology, Memorial University of Newfoundland, St. John's, Canada

ABSTRACT

Egg production per brood considered as brood volume equals the number of eggs times the mean volume of an egg. The upper limit for egg production is determined by the volume of the gonads. The latter is constrained by the volume of the body. Regressions of brood volume on body volume indicate that most species show restraint and produce less than their potential. In amphipods, egg size is correlated with the structure of the brood pouch. Species with narrow oöstegites produce larger eggs than those with broad oöstegites. The size of the egg is an important component of the reproductive adaptations. The size of the young when they hatch, and size at maturity are both correlated with egg size. Egg size influences the position of a species on the r-K continuum and on the timing of the reproductive cycle.

1 INTRODUCTION

Egg production in amphipods is analyzed here in terms of non-genetic variation in the sizes of eggs and the numbers and volumes of clutches or broods of eggs. Although these parameters can be altered by selection (Doyle & Hunte 1981a, b), this source of variation will not be considered here. Also, while crustacean eggs may vary in chemical composition both between species (Herring 1974, McLaren 1966, Shakuntala & Reddy 1982) and within a species (Attard & Hudon 1987), such variation in amphipods has been little studied and will not be addressed here.

Amphipods are good subjects for the study of egg production, since the embryos are held loosely in a ventral brood pouch of the female while they develop. The eggs hatch into small individuals which grow without marked morphological change into adults except in hyperiids (Laval 1968, 1980). Growth may or may not stop with the onset of sexual maturity. Living and preserved embryos can be readily removed. When the ova become pigmented in the late stages of development in the ovary they can be seen and counted in live females if the integument over the ovary is translucent. Alternatively the ovary can be dissected and the ova measured and counted. For convenience here and elsewhere we have referred to all stages of development in the ovary as oögonia.

Egg production is compared in amphipods of various gammaridean families, with a concentrated effort on *Gammarus* species. The genus *Gammarus* contains a large number of species from both fresh water and marine environments of the northern hemisphere. Although attempts are being made to subdivide these species into separate genera, all evidence to date indicates that the *Gammarus* species *sensu latu* are closely related. Here they are considered a group within the gammaridean amphipods whose egg production can be compared. The only other group for which we have much data is the Family Lyssianassidae.

We confine ourselves here to our own published and unpublished measurements, since they are comparable. Different techniques can obscure significant relationships. Amphipod eggs increase significantly in size both during development and after preservation (Sheader & Chia 1970, Steele & Steele 1970a). In our comparisons we have included only measurements of eggs in an early stage of development (Stage A, Steele & Steele 1969). Amphipod eggs are cuboidal in the ovary, almost spherical when deposited in the brood pouch, and prolately spheroidal while the embryo develops. The measurement of size is: Average diameter = Maximum length + Maximum width. Volume of an egg is calculated as $4/3\,\pi \times \text{length}/2 \times (\text{width}/2)^2$. The volume of a brood is simply: Number of eggs × Volume of an egg.

Body length, measured from the tip of the head to the distal end of the telson, is used as the descriptor of parental female body size. Size-specific fecundity is calculated as the number of eggs or fecundity/length. Size-specific brood volume has been calculated as brood volume/length^3, since Jensen (1958) concluded that length^3 'is an applicable measure of the volume of a female'.

In animals with indeterminate growth, such as amphipods, adult size of a species is somewhat difficult to specify. Many authors use the largest known specimen or the smallest mature female, but this means the size is specified by a single individual. As a measure of adult size we have used either the mid point of the range of sizes of the adult females (termed mid-sized females) or the size at which 50% of the females are mature. We consider the latter the most accurate. It can be calculated using probit analysis (Steele & Steele 1969) for large samples. Whenever only a few specimens were available, size at maturity was approximated by inspection of the data. Size at maturity can be difficult to determine if females mature in the season when ovarian diapause occurs. Mature females in ovarian diapause with small oögonia and unsetose oöstegites are difficult to distinguish from immature females with the same characteristics. Size at maturity varies with location and season (V.J. Steele & Steele 1972), but such variations have not been studied often.

2 SIZE OF EGGS

2.1 *Variation in egg size*

The range of egg sizes in the gammarideans is quite large (Fig. 1) and the potential for variation great. The smallest known gammaridean egg (0.25 mm) was reported in *Seborgia minima* (Bousfield 1970). Since gammarideans hatch fully developed, this is probably their smallest expected size. Peracarid crustaceans with incomplete development, such as the mancoids (e.g. isopods and cumaceans) and the hyperiid amphipods, could probably develop from even smaller eggs. Shallow water gammaridean eggs range upwards in size to 2.38 mm, as in the lysianassid *Socarnes bidenticulatus*. Even larger eggs occur in the very large deep sea lysianassids (Barnard & Ingram 1986).

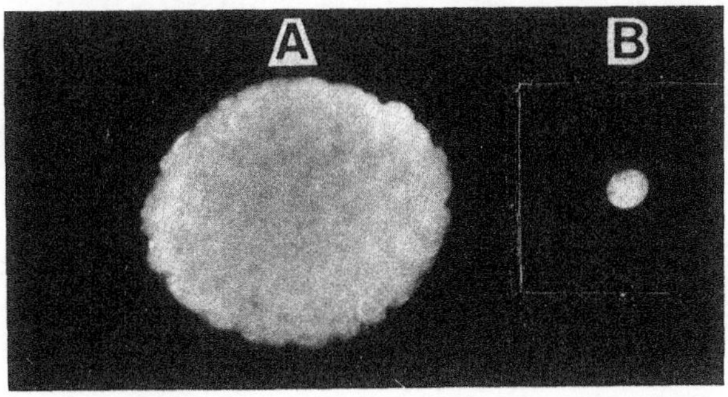

Figure 1. Representative amphipod eggs in Stage A development (× 10). A. *Socarnes bidenticulatus*. B. *Gammarus mucronatus*.

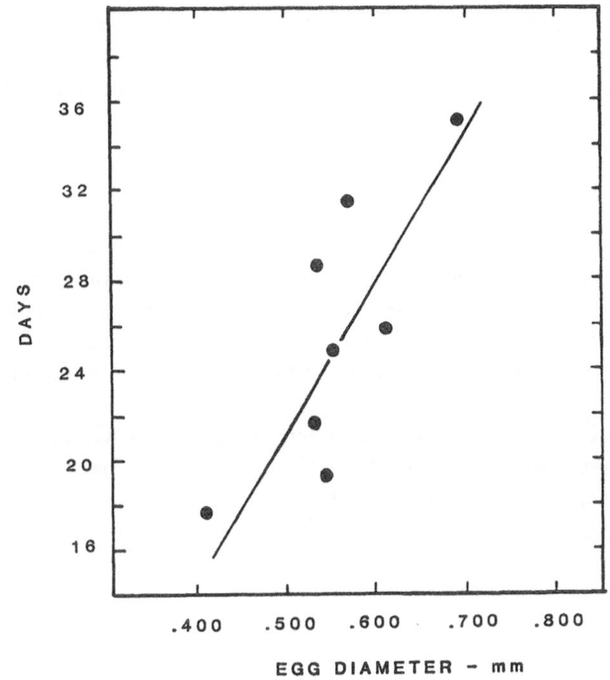

Figure 2. Duration of embryonic development for *Gammarus* eggs of varying sizes. The regression line is: Duration = 61.57 × Diameter − 9.21 (r = .82).

2.2 *Rate of embryonic development and egg size*

The rate at which crustacean embryos develop is affected by environmental conditions, particularly by temperature. In addition, different groups of crustaceans vary in their mode of development and develop at different rates (Steele & Steele 1975a). Within a group, such as *Gammarus sensu latu* (Fig. 2), development rates are correlated with egg size (Steele & Steele 1973b). Small eggs develop more rapidly than large, because of the relationship between size and metabolic rate.

Egg size, together with timing of the onset and termination of ovarian diapause, determine when the young will hatch and be released. The timing of the release is critical if there is a short season of suitable food for the young. At high latitudes with a short season, a long ovarian diapause and a large egg cause an iteroparous species (Table 1), such as *Gammarus setosus* to

Table 1. Terminology of amphipod reproduction (Steele & Steele 1986).

Term	Definition
Semelparous	Female dies after producing 1 brood
Iteroparous	Female has the potential to produce more than 1 brood after it matures
Univoltine	1 Generation/year
Multivoltine	More than 1 generation/year
Semi-annual	Each female produces more than one brood/year
Annual	Each female produces 1 brood/ year
Biennial	Each female produces 1 brood/2 years
Triennial	Each female produces 1 brood/3 years

be annual and univoltine. At lower latitudes where the growing season is longer, the ovarian diapause is short and the eggs are smaller. Iteroparous species such as *Gammarus lawrencianus* are multivoltine and semiannual. Egg size is an important component of the adaptations to the environmental conditions of different latitudes. It influences how many broods and generations are possible per year as well as the size of each brood.

2.3 *Size of hatched young*

The size of hatched young of gammaridean amphipods is correlated with the size of the egg (Fig. 3). Large young have often been considered to have better chances of survival than small (Thorson 1959, Rabalais & Gore 1985, Skadsheim 1984), although there seems to be no direct evidence from amphipods to support this hypothesis. If true, large hatched young would require large eggs.

In the lysianassids, the young (6.1 mm) hatched from the large eggs (2.38 mm) of an Arctic species such as *Socarnes bidenticulatus* may equal the adult size of other species in the same habitat and are larger than the mean size (5.8 mm) of mature lysianassids in the tropical environment of the Indo-Pacific region (Steele 1983).

2.4 *Egg size and size at maturity*

Size at maturity in the marine species of *Gammarus* from the northwestern Atlantic is correlated with egg size (Fig. 4). An exception is *Gammarus obtusatus*, which has relatively larger eggs than expected from the relationship in the other species. This observation is of interest and supports the conclusion of Stock (1969) that *G. obtusatus* belongs to *Eulimnogammarus*, a genus whose distribution is centered in the fresh waters of Lake Baikal. The large eggs of *G. obtusatus* suggest a recent fresh water origin of this species.

The eggs of small fresh water and marine species do not differ appreciably in size, but the larger species of fresh water amphipods have relatively large eggs, similar to *G. obtusatus*. These observations show that *Gammarus* species can and do produce larger eggs both in the ocean and in fresh water. The larger size of eggs of freshwater crustaceans compared to their marine relatives has often been related to the more complete embryonic development of the freshwater forms. However, this explanation cannot account for the larger size of eggs of *Gammarus* species in fresh water, since there is no known difference in the type of development of marine versus freshwater amphipods.

The relationship between egg size and size at maturity among species of *Gammarus sensu*

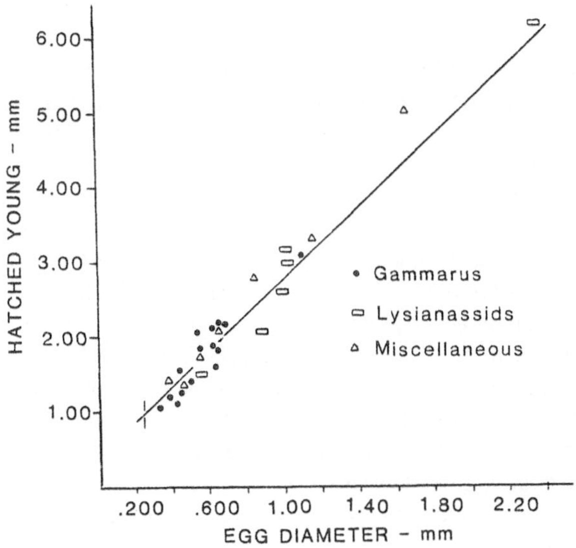

Figure 3. Relation between diameter of preserved Stage A eggs and length of hatched young in gammarid, lysianassid and miscellaneous amphipods. The regression line for all values is: Hatched young = 2.35 × Diameter + 0.43 (r = .96). The vertical line at the left indicates egg diameters and adult size of *Seborgia minima*.

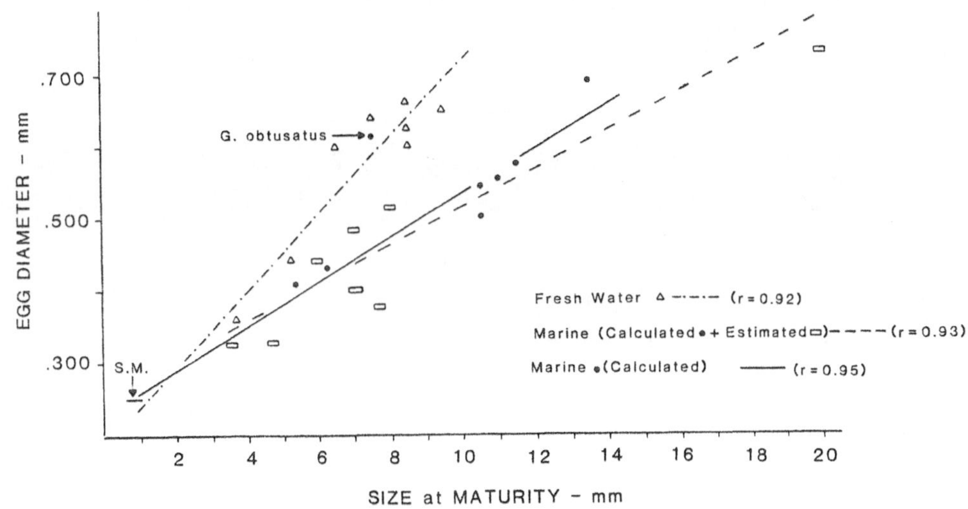

Figure 4. Relation between diameters of preserved Stage A eggs and the size at 50% maturity in marine and freshwater *Gammarus* (excluding *G. obtusatus*). Solid symbols indicate values calculated by probit analysis and open symbols values estimated from less data. The value for *G. obtusatus* is indicated and the egg diameter and adult size range of *Seborgia minima* shown by a horizontal line at left.

latu is not necessarily applicable to other amphipod groups (see next section) especially at large body sizes; but it should apply to small species. In *Seborgia minima*, with the smallest known egg size, the hatchlings (calculated from the egg size/hatchling size regression to be 1.02 mm) are already of adult size (0.8-1.0 mm) (Bousfield 1970) and could be mature when they hatch.

2.5 *Egg size and oöstegite type*

The ventral brood pouch of amphipods is formed by oöstegites attached to pereiopods 2 to 5. Observations on other related peracarids, which also have brood pouches, indicate that the primitive condition in amphipods is for the oöstegites to be broad and have moderate setae on their margins (pers. observ.). We identified two evolutionary trends in the amphipods (Fig. 5): 1) oöstegites remain short and broad with short setae, the first and last tending to become non-functional and disappear, 2) oöstegites persist on pereiopods 2 to 5, but become long and narrow with long setae; the setae, rather than the oöstegites themselves form the brood pouch.

A plot of egg size against mid-adult size of females (Fig. 6) indicates that egg size increases slightly with body size in those species with broad oöstegites, but greatly in those with narrow oöstegites. These differences may be related to the probability of an egg falling out of the brood pouch. In a small individual there would be little difference between narrow and broad oöstegites. As body length increases, the oöstegites move farther apart and lose contact if they are narrow, so that brood pouch is formed mainly by the setae. In this case it is more likely that a small egg would fall out than it is that a large egg would be lost.

However, this line of reasoning cannot explain why species with broad oöstegites do not

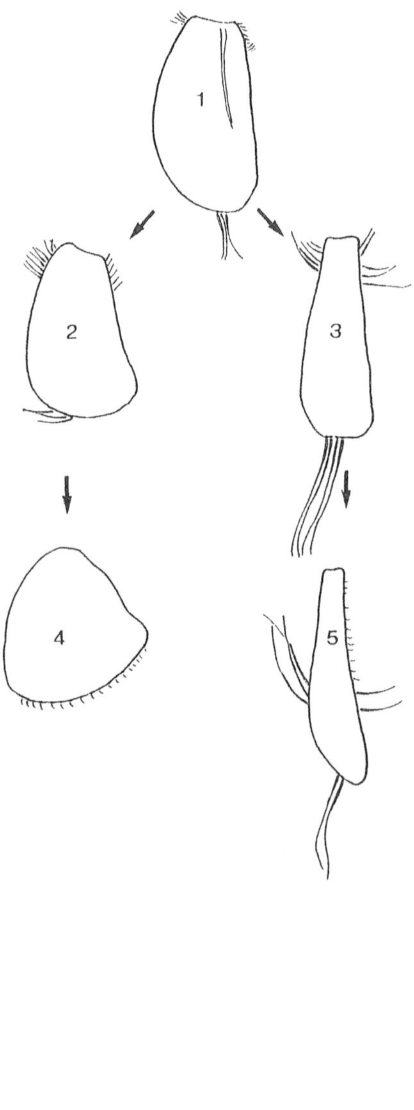

Figure 5. Postulated trends in the oöstegite of pereiopod 3. 1. *Gammaracanthus loricatus*. 2. *Pontogeneia inermis*. 3. *Gammarus oceanicus*. 4. *Caprella septentrionalis*. 5. *Pseudalibrotus litoralis*.

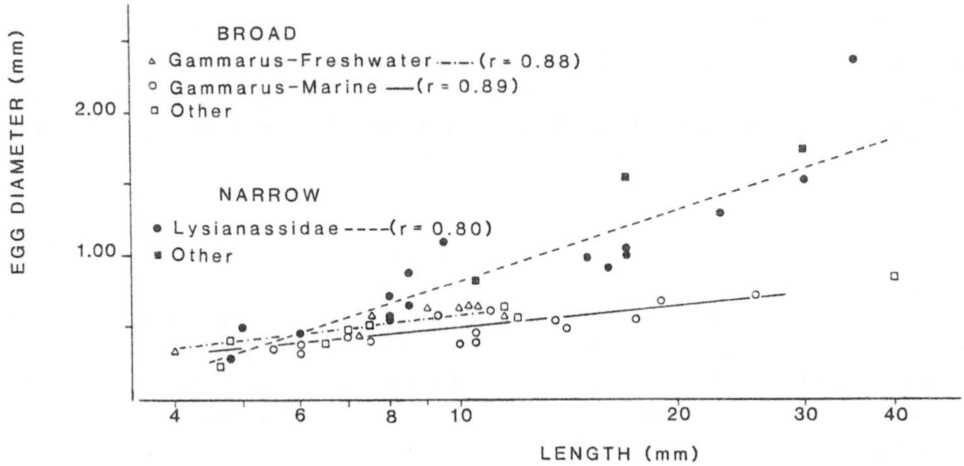

Figure 6. Mid-size of mature females and the diameters of Stage A embryos (note the log scale for body length). Marine and fresh water gammarid and lysianassid data are shown separately. Regression lines fitted for lysianassids, marine and fresh water gammarid data. Square symbols are for other amphipods and are not included in the regressions.

have large eggs. The *Gammarus* species are of particular interest, since both the relative size of the eggs and the structure of their brood pouches vary. The marine *Gammarus* species, except for *G. obtusatus*, have small eggs (Fig. 4). The eggs of fresh water species are relatively larger when related to size at maturity (Fig. 4) but there is little difference when compared to the mid-size of adults (Fig. 6).

The brood pouch of most *Gammarus* species is intermediate between the broad and narrow types of oöstegites. The two anterior oöstegites are broad [Fig. 5 and Sket (1971)], but the last two are narrow. The setae on all four are long. Cole (1985), however, found some fresh water *Gammarus* in which all the oöstegites, including the first two, are narrow. These observations suggest that in fresh water *Gammarus*, large egg size evolved before the oöstegites became narrow. This would be expected if egg size is related to the mechanical problem of maintaining eggs in the brood pouch. This also suggests that egg size and the structure of the brood pouch are adaptations to habitats and are likely to have evolved repeatedly. Their value in reconstructing phylogenies must be evaluated carefully.

3 EGG NUMBER

3.1 *Egg number and body size*

Fecundity increases with the size of the female, both within a species and between species. The relationship is described adequately by a straight line for species with a small size range of mature females. For the combined data of many species the relationship is not linear and a power function gives a better fit. Large females produce relatively more eggs than do small females and within *Gammarus lawrencianus* mean fecundity/length increases from 2.0 at 5 mm to 4.0 at 10 mm. In a large (29 mm) *Gammarus wilkitzkii* the mean size specific fecundity is 7.24, whereas in a small (5 mm) *Gammarus stoerensis* it is only 2.00.

3.2 *Restraints on egg number*

When the actual number of eggs produced by a female, measured either as enlarged oögonia or embryos (Fig. 7), is compared with the maximum numbers observed, it is evident that most individuals within a species produce less than the maximum. There is little difference between oögonia and embryos when allowance is made for the growth of the female after molting in females with embryos.

Similarly when species are compared either in the genus *Gammarus sensu latu* (Fig. 7), or in other genera (Fig. 8) many produce fewer eggs than the maximum number accommodated by their body size.

The relative reduction in egg production from the observed maximum declines with body length. In *Gammarus*, it varies from 84 to 89% and in the other amphipods from 96 to 90%. Much of the variation between species is due to differences in egg size (see next section). Compare, for example, *Stegocephalus inflatus* (large eggs) with *Gammaracanthus loricatus* (small eggs).

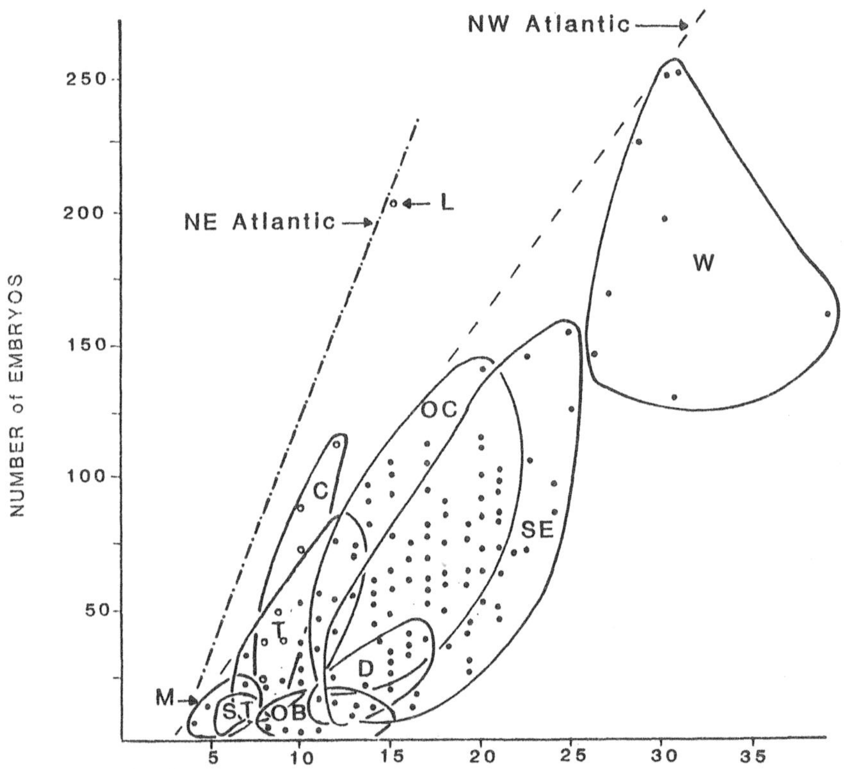

Figure 7. Numbers of early embryos carried by marine *Gammarus* species. Maximum production of species from the northwestern Atlantic to the right of the dashed line. Maximum production of species from the northeastern Atlantic to the right of the dot-dash line. W= *G. wilkitzkii*, C= *G. crinicornis*, L= *G. locusta*, OC= *G. oceanicus*, SE= *G. setosus*, D= *G. duebeni*, T= *G. tigrinus*, OB= *G. obtusatus*, St= *G. stoerensis*, M= *G. mucronatus*.

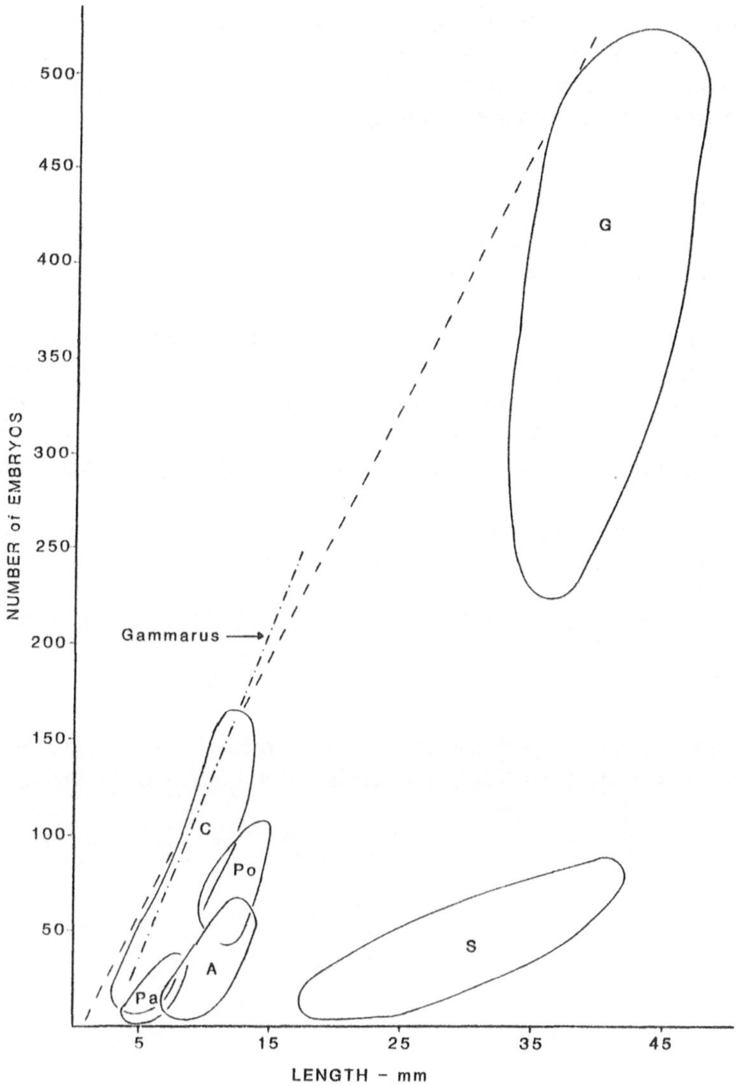

Figure 8. Numbers of early embryos carried by various amphipod species. The maximum numbers observed in all marine *Gammarus* species is shown for comparison. G= *Gammaracanthus loricatus*, C= *Calliopius laeviusculus*, Po= *Pontoporeia femorata*, A= *Amphiporeia lawrenciana*, Pa= *Parhyalella pietschmanni*.

4 BROOD VOLUME

Since the volume of a brood accounts for differences in both number and size of eggs it provides a better measure of egg production than does either egg size or egg number alone.

4.1 *Brood volume and length*[3]

Brood volume increases with length, both within a species and between species. The relationship can be described by a linear function within a species, but is non-linear when a large range of sizes is available within a species, or when many species covering a large range of sizes are analyzed. The relative volume of egg production (Brood volume/Length^3) measured for either oögonia or embryos decreases with an increase in body length.

4.2 *Constraints in brood volume*

When comparing the actual brood volumes produced with the maxima at different body sizes (Fig. 9), it is apparent that most individuals of a species do not attain maximum production in terms of volume or numbers.

This is evident whether *Gammarus* species (Fig. 9) or amphipods in general (Fig. 10) are compared. However, since egg size and number are inversely correlated there is much less variation in the volume-size relationship than there is in the number-size relationship. *Stegocephalus inflatus* exceeds *Gammaracanthus loricatus* in brood volume production because of its larger eggs. Brood volumes of *Gammarus* are also smaller than those of amphipods in general (Fig. 10).

The constraint in brood volume production is great, but generally less than that observed in egg numbers. In *Gammarus*, it declines with body size from 92 to 75% and in the other amphipods from 90 to 69%.

5 DISCUSSION

Egg production in gammaridean amphipods has been reviewed by Steele & Steele (1975d), Nelson (1980), Van Dolah & Bird (1980), Kolding & Fenchel (1981), Skadsheim (1984) and Wildish (1982).

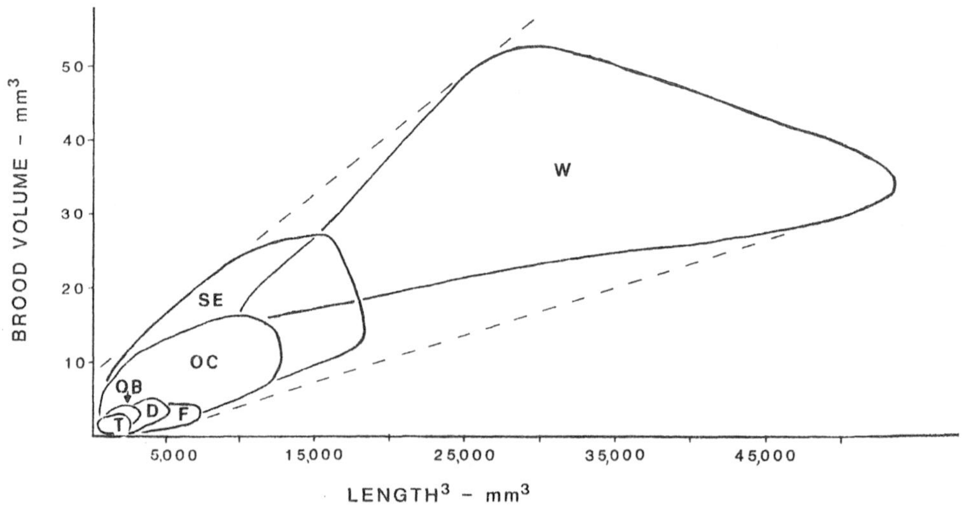

Figure 9. Brood volumes of early Stage A embryos related to length in marine *Gammarus* of the northwestern Atlantic. Maximal and minimal values enclosed by dashed lines. F=*G. finmarchicus*, others the same as Figures 7 and 8.

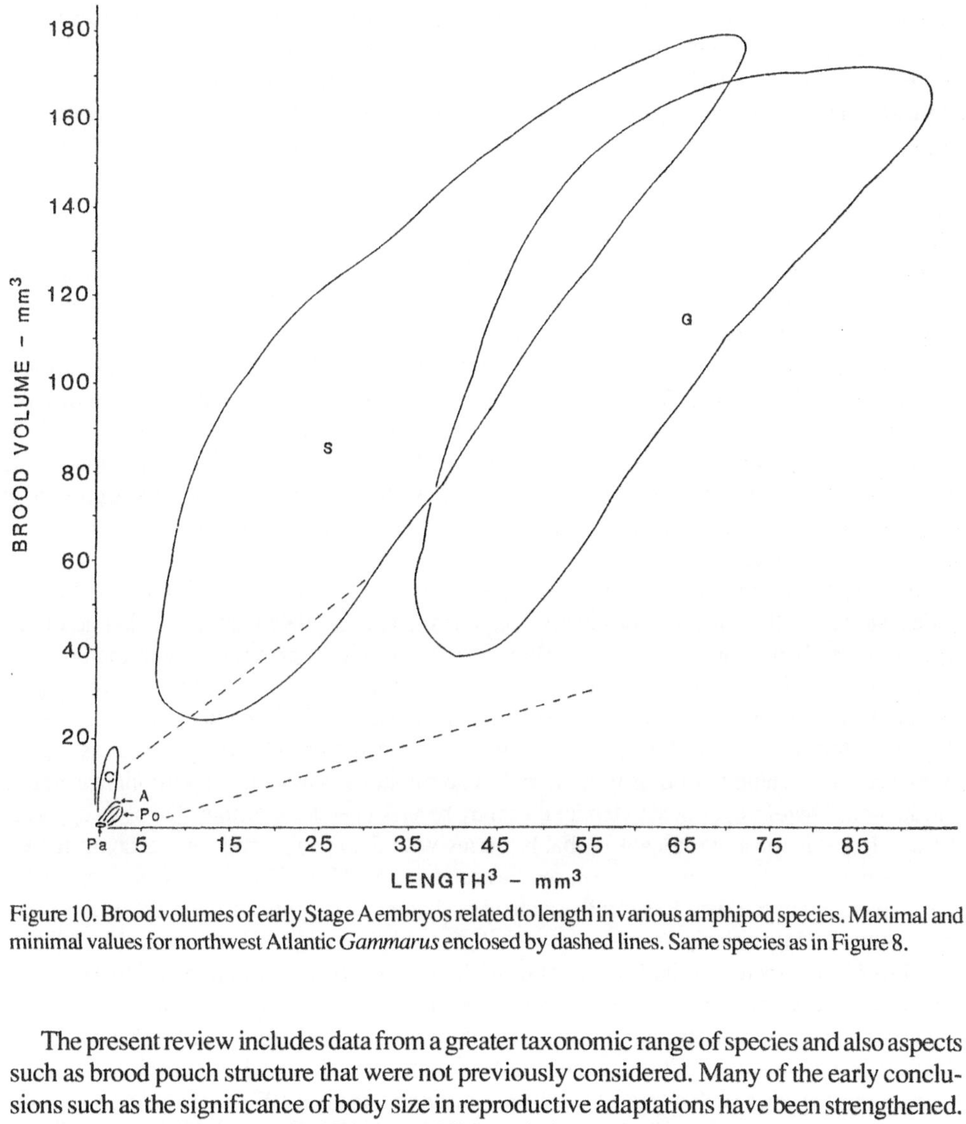

Figure 10. Brood volumes of early Stage A embryos related to length in various amphipod species. Maximal and minimal values for northwest Atlantic *Gammarus* enclosed by dashed lines. Same species as in Figure 8.

The present review includes data from a greater taxonomic range of species and also aspects such as brood pouch structure that were not previously considered. Many of the early conclusions such as the significance of body size in reproductive adaptations have been strengthened. Others have not. For example, Nelson (1980) based his conclusions on the small body size and lower reproductive output of infaunal amphipods mainly on haustoriids. However, lysianassids are primarily infaunal and include the largest amphipods known. Similarly the larger eggs of infaunal amphipods compared to the epifaunal amphipods as reported by Van Dolah & Bird (1980) could be related to the narrow oöstegites of the haustoriid amphipods that provided most of their data for infaunal amphipods.

Kolding & Fenchel (1981) interpreted the reproductive cycles of northern European *Gammarus* species quite differently from what has been proposed here. They interpreted apparent temporal differences in breeding as character displacements in order to reduce competition for females by the males of coexisting species. However, not all species will form interspecific pairs even when confined in the laboratory (Kolding 1986). Kolding & Fenchel did not detect an ovarian diapause in any species (Kolding & Fenchel 1979, Kolding 1981); they deduced the durations of the breeding seasons from the presence of ovigerous females. *Gammarus* males cannot enter precopula with an unreceptive female, but are capable of entering precopula with a receptive female. Thus a lack of ovigerous females need not mean that the males are incapable of mating and competing for females; therefore, the interpretation by Kolding & Fenchel that temporal shifts in breeding seasons represent character displacement is not conclusive. If breeding seasons were readily shifted, they should vary with the assemblage of coexisting species. However, this generally is not the case. For example, when allowance is made for differences in environmental temperature, *Gammarus setosus* has a similar cycle whether it is found alone on an Arctic beach or in the Bay of Fundy, where it lives with seven other species (V.J. Steele & Steele 1970). In addition, *Gammarus oceanicus* and *Gammarus duebeni* have similar cycles on the two sides of the Atlantic.

Kolding & Fenchel (1981) and Skadsheim (1984) have also developed a model of egg production in *Gammarus*. That model is based on the assumptions that the survival of large young from large eggs is better than that of small young and that there is a fixed reproductive effort (egg size × egg number = constant). However, there is no direct evidence to support these assumptions and it does not agree with evidence shown here, namely that egg size is correlated with a number of other parameters. Their model is undoubtedly too simple (see also Clarke, Skadsheim & Holmes 1985).

Despite their morphological similarities, gammaridean amphipods exhibit diverse life histories. However, the various reproductive parameters are correlated, so that a change in one brings about changes in many others. Hence selection of one parameter over another may ensure survival. Thus, with an increase in egg size, the duration of embryonic development, the size of the hatched young, and the size of adults increases while fecundity decreases. Larger eggs therefore result in lower egg production, since fewer broods and generations can be produced in the same period of time. Larger size of adults is associated with increased egg production. Therefore, opposing tendencies may be produced by a simple change in egg size. Which factor is most significant probably varies with the species and may account for the impressive diversity of gammaridean species accompanied by so little morphological difference. Species can coexist successfully in the same habitat with quite different life histories and reproductive adaptations.

Nevertheless there are significant trends in the life histories of amphipods. The latitudinal variation in food production in the intertidal and shallow water marine habitats, with a short season in the Arctic and a prolonged or continuous season in the tropics must be accomodated. With a continuous supply of food, young can be produced continuously. Semiannual, multivoltine, tropical species such as *Parahyalella pietschmanni* (Steele 1973) are small in size, with small eggs, small young, small adults and low fecundity. Their primary adaptation is small body size. The low egg production per female is more than compensated for by the short brood interval and the multivoltine life cycle which allow a geometric rate of population increase. This must be balanced by a high mortality rate. In part at least, a high mortality will result from reproduction itself (Steele & Steele 1986).

In the Arctic, only a single brood can be produced in the short season. Amphipods are annual or biennial univoltine species, large in size, high in fecundity and with large eggs and young. Their low rate of egg production is partially compensated for by their larger adult size and the greater size range of adults. The primary adaptation is a large adult size. However, this allows only an arithmetic increase by production of additional young compared to the geometric increase by multiplication correlated with a small adult size. By surviving longer the females produce larger broods.

REFERENCES

Attard, J. & C. Hudon 1987. Embryonic development and energetic investment in egg production in relation to size of female lobster (*Homarus americanus*). *Can. J. Fish. Aquat. Sci.* 44: 1157-1164.

Barnard, J.L. & C.L. Ingram 1986. The supergiant amphipod *Alicella gigantea* Chevreux from the North Pacific gyre. *J. Crust. Biol.* 6: 825-839.

Bousfield, E.L. 1970. Terrestrial and aquatic amphipod Crustacea from Rennell Island. In *The Natural History of Rennell Island, British Solomon Islands* 6: 155-168.

Clarke, A., A. Skadsheim & L.J. Holmes 1985. Lipid biochemistry and reproductive biology in two species of Gammaridae (Crustacea: Amphipoda). *Mar. Biol.* 88: 247-263.

Cole, G.A. 1985. Analysis of the *Gammarus-pecos* complex (Crustacea: Amphipoda) in Texas and New Mexico, USA. *J. Arizona-Nevada Acad. Sci.* 20: 93-103.

Downer, D.F. & D.H. Steele 1979. Some aspects of the biology of *Amphiporeia lawrenciana* Shoemaker (Crustacea, Amphipoda) in Newfoundland waters. *Can. J. Zool.* 57: 257-263.

Doyle, R.W. & W. Hunte 1081a. Genetic changes in 'fitness' and yield in a crustacean population in a controlled environment. *J. Exp. Mar. Biol. Ecol.* 52: 147-156.

Doyle, R.W. & W. Hunte 1981b. Demography of an estuarine amphipod (*Gammarus lawrencianus*) experimentally selected for high 'r': A model of the genetic effects of environmental change. *Can. J. Fish. Aquat. Sci.* 28: 1120-1127.

Herring, P.J. 1974. Size, density and lipid content of some decapod eggs. *Deep-Sea Res.* 21: 91-94.

Jensen, J.P. 1958. The relation between body size and number of eggs in marine malacostrakes. *Medd. Dan. Havunders.* 2: 1-125.

Kolding, S. 1981. Habitat selection and life cycle characteristics of five species of the amphipod genus *Gammarus* in the Baltic. *Oikos* 37: 173-178.

Kolding, S. 1986. Interspecific competition for mates and habitat selection in five species of *Gammarus* (Amphipoda: Crustacea). *Mar. Biol.* 91: 491-495.

Kolding, S. & T.M. Fenchel 1979. Coexistence and life cycle characteristics of five species of the amphipod genus *Gammarus*. *Oikos* 33: 323-327.

Kolding, S. & T.M. Fenchel 1981. Patterns of reproduction in different populations of five species of the amphipod genus *Gammarus*. *Oikos* 37: 167-172.

Laval, P. 1968. Devéloppement en élevage et systématique d'*Hyperia schizogeneios* Stebb. (Amphipode Hypéride) *Arch. Zool. Exp. Gén.* 109: 25-67.

Laval, P. 1980. Hyperiid amphipods as crustacean parasitoids associated with gelatinous zooplankton. *Oceanogr. Mar. Biol. Ann. Rev.* 18: 11-56.

McLaren, I.A. 1966. Predicting development rate of copepod eggs. *Biol. Bull.* 131: 457-469.

Nelson, W.G. 1980. Reproductive patterns of gammaridean amphipods. *Sarsia* 65: 61-71.

Rabalais, N.N. & R.H. Gore 1985. Abbreviated development in decapods. In A.M. Wenner (ed.), *Larval Growth. Crustacean Issues* 2: 67-127. Rotterdam: Balkema.

Shakuntala, K. & S.R. Reddy 1982. Crustacean egg sizes as an indicator of egg/fat protein reserves. *Internat. J. Invert. Reprod.* 4: 381-384.

Sheader, M. & F.-S. Chia 1970. Development, fecundity and brooding behaviour of the amphipod *Marinogammarus obtusatus*. *J. Mar. Biol. Ass. UK.* 50: 1079-1099.

Skadsheim, A. 1984. Coexistence and reproductive adaptations of amphipods: the role of environmental heterogeneity. *Oikos* 43: 94-103.

Sket, B. 1971. Zur systematik und phylogenie der Gammarini (Amphipoda). *Bulletin Scientifique*, Yugoslavia (A)16: 6.

Steele, D.H. 1967. The life cycle of the marine amphipod *Stegocephalus inflatus* Krøyer in the northwest Atlantic. *Can. J. Zool.* 45: 623-628.

Steele, D.H. 1973. The biology of *Parhyalella pietschmanni* Schellenberg, 1938 (Amphipoda, Hyalellidae) at Nosy Bé, Madagascar. *Crustaceana* 25: 276-280.

Steele, D.H. 1983. Size compositions of lysianassid amphipods in cold and warm water habitats. *Austral. Mus. Mem.* 18: 113-119.

Steele, D.H. & V.J. Steele 1969. The biology of *Gammarus* (Crustacea, Amphipoda) in the northwestern Atlantic. I. *Gammarus duebeni* Lillj. *Can. J. Zool.* 47: 235-244.

Steele, D.H. & V.J. Steele 1970a. The biology of *Gammarus* (Crustacea, Amphipoda) in the northwestern Atlantic. III. *Gammarus obtusatus* Dahl. *Can. J. Zool.* 48: 899-995.

Steele, D.H. & V.J. Steele 1970b. The biology of *Gammarus* (Crustacea, Amphipoda) in the northwestern Atlantic. IV. *Gammarus lawrencianus* Bousfield. *Can. J. Zool.* 48: 1261-1267.

Steele, D.H. & V.J. Steele 1972. The biology of *Gammarus* (Crustacea, Amphipoda) in the northwestern Atlantic. VI. *Gammarus tigrinus* Sexton. *Can. J. Zool.* 50: 1063-1068.

Steele, D.H. & V.J. Steele 1973a. Some aspects of the biology of *Calliopius laeviusculus* (Krøyer) (Crustacea, Amphipoda) in the northwestern Atlantic. *Can. J. Zool.* 51: 723-728.

Steele, D.H. & V.J. Steele 1973b. The biology of *Gammarus* (Crustacea, Amphipoda) in the northwestern Atlantic. VII. The duration of embryonic development in five species at various temperatures. *Can. J. Zool.* 51: 995-999.

Steele, D.H. & V.J. Steele 1974. The biology of *Gammarus* (Crustacea, Amphipoda) in the northwestern Atlantic. VIII. Geographic distribution of the northern species. *Can. J. Zool.* 52: 1115-1120.

Steele, D.H. & V.J. Steele 1975a. Egg size and duration of embryonic development in Crustacea. *Int. Revue. ges. Hydrobiol.* 60: 711-715.

Steele, D.H. & V.J. Steele 1975b. The biology of *Gammarus* (Crustacea, Amphipoda) in the northwestern Atlantic. IX. *Gammarus wilkitzkii* Birula, *Gammarus stoerensis* Reid, and *Gammarus mucronatus* Say. *Can. J. Zool.* 53: 1105-1109.

Steele, D.H. & V.J. Steele 1975c. The biology of *Gammarus* (Crustacea, Amphipoda) in the northwestern Atlantic. X. *Gammarus finmarchicus* Dahl. *Can. J. Zool.* 53: 1110-1115.

Steele, D.H. & V.J. Steele 1975d. The biology of *Gammarus* (Crustacea, Amphipoda) in the northwestern Atlantic. XI. Comparison and discussion. *Can. J. Zool.* 53: 1116-1126.

Steele, D.H. & V.J. Steele 1976. Some aspects of the biology of *Gammaracanthus loricatus* (Sabine) (Crustacea, Amphipoda). *Astarte* 8 (1975): 69-72.

Steele, D.H. & V.J. Steele 1978. Some aspects of the biology of *Pontoporeia femorata* and *Pontoporeia affinis* (Crustacea, Amphipoda) in the northwestern Atlantic. *Astarte* 11: 61-66.

Steele, D.H. & V.J. Steele 1986. The cost of reproduction in the amphipod *Gammarus lawrencianus* Bousfield, 1956. *Crustaceana* 51: 176-182.

Steele, V.J. & D.H. Steele 1970. The biolody of *Gammarus* (Crustacea, Amphipoda) in the northwestern Atlantic. II. *Gammarus setosus* Dementieva. *Can. J. Zool.* 48: 659-671.

Steele, V.J. & D.H. Steele 1972. The biology of *Gammarus* (Crustacea, Amphipoda) in the northwestern Atlantic. V. *Gammarus oceanicus* Segerstrale. *Can. J. Zool.* 50: 801-813.

Steele, V.J. & D.H. Steele 1986. The influence of photoperiod on the timing of reproductive cycles in *Gammarus* species (Crustacea, Amphipoda). *Amer. Zool.* 26: 459-467.

Stock, J.H. 1969. Members of Baikal amphipod genera in European waters, with description of a new species, *Eulimnogammarus macrocarpus* from Spain. *Proc. Kon. Ned. Akad. Wetensch.* (C)72: 66-75.

Thorson, G. 1959. Reproductive and larval ecology of marine bottom invertebrates. *Biol. Rev.* 25: 1-45.

Van Dolah, R.F. & E. Bird 1980. A comparison of reproductive patterns in epifaunal and infaunal gammaridean amphipods. *Est. Coast. Mar. Sci.* 11: 593-604.

Wildish, D.J. 1982. Evolutionary ecology of reproduction in gammaridean Amphipoda. *Internat. J. Invert. Reprod.* 5: 1-19.

Variation in reproductive output of the isopod, *Armadillidium vulgare*

Michael S. Brody
US Environmental Protection Agency, Washington, DC, USA

ABSTRACT

Two size classes of the female isopod, *Armadillidium vulgare*, were kept in the laboratory to compare growth against reproduction. A variable food supply simulated natural variation in food availability to field populations of these isopods. During the course of a molt cycle, the supply of food determined the biomass allocated to growth or reproduction. When food was abundant, both sizes of females grew and reproduced. When food was scarce, both small and large females reduced their growth, but still allocated energy to reproduction. Thus, reproductive effort, measured as the portion of biomass allocated to reproduction during one molt cycle (after physiological maintenance requirements), was higher during periods of food shortages.

1 INTRODUCTION

During their lifetimes, many animals and plants experience shortages of energy or resources. Fluctuations in supply of critical resources or energy may cause changes in an organism's pattern of growth and reproduction. Specifically, 'tradeoffs' between the reproductive and nonreproductive activities and their associated costs have been treated in life history theory (Gadgil & Bossert 1970, Taylor et al. 1974, Pianka & Parker 1975, Reznick 1985).

Reproductive effort, the portion of an organism's energy allocated to reproduction, has received considerable attention in theoretical and empirical life history literature (Fisher 1930, Williams 1966, Tinkle 1969, Schaffer 1974a, Hirshfield & Tinkle 1975), and may be optimized by natural selection (Schaffer 1974a, b, Schaffer & Gadgil 1975). According to theory, optimal reproductive effort maximizes the lifetime reproductive success of an organism (Pianka & Parker 1975).

Energetics of life history functions were investigated in the terrestrial isopod, *Armadillidium vulgare*, an isopod of indeterminate growth. Field research has shown that *A. vulgare* inhabits areas in which the availability of its food supply fluctuates significantly, both between years and within reproductive seasons (Brody 1980, Lawlor 1976a, Paris 1963). Field and laboratory studies indicated that these fluctuations have had ecological or evolutionary effects on this organism's reproductive capabilities (Lawlor 1976b). The interval between molts provides a clearly defined time period for measuring both resource acquisition and energy allocation. Growth and reproduction resulting from tissue development during a molt cycle are easily distinguished from growth and reproduction of previous and subsequent molt cycles. Also, available food within a given molt cycle mainly affects growth and reproduction during that molt cycle. Thus, periods of resource acquisition and energy allocation overlap.

A. vulgare feeds predominantly on leaf litter, detritus, some living green vegetation (Paris 1963, Paris & Sikora 1965), and may occasionally be predacious (Edney et al. 1974). This organism's food supply is normally very abundant, although availability fluctuates with weather changes. Terrestrial crustaceans, lacking some adaptations by which insects prevent desiccation, often maintain their body water content by behavioral regulation of activities. During periods of high summer temperatures during low humidity, sowbugs remain inactive below the surface layers of the leaf litter and do not feed. Thus, access to their otherwise abundant food source becomes weather dependent. The experimental laboratory treatments described below were designed to reflect natural fluctuation of food availability. These experiments examined life history attributes of *A. vulgare* provided two levels of available food. These food levels reflected changes in field food availability during the course of a reproductive season. One level provided a very abundant food supply. The second, reduced level represented a biologically significant food shortage that provided some energy in excess of minimum maintenance levels. These food levels, abundant and reduced, were supplied to two size classes of this isopod: a group of small adult females and a group of large adult females. This laboratory study thus tested the effects of two levels of food availability on the growth, reproductive output and reproductive effort in two size classes of *A. vulgare* during two complete molt cycles.

2 METHODS

Isopods were collected from an area within the Brackenridge Field Laboratory of the University of Texas, Austin, Texas. The sampled populations lived in an area of grasses and mixed species woodland consisting mainly of live oaks and hackberries. Isopods for the experiments were collected by two methods in mid-May, early June, and early July 1978. (1) Litter was cleared away from areas one foot square. Water was poured on these areas of soil, which were then covered with foot-square wooden boards; this procedure maintained a moist, preferred microhabitat from which isopods were easily gathered. (2) Active isopods walking the base of concrete walls enclosing the study areas were easily caught by hand. Though opportunistic, the collection methods ensured a fairly representative sample of the population.

After collection from the field the isopods were kept in the laboratory prior to the experiments. Selection from these isopods of the individuals actually to be used in the experimental treatments was designed to minimize any differences among them due to molt or reproductive condition, size, or food supply. Females nearing a molt were examined daily. When these females molted, and if they extruded a brood (resulting from conditions during the molt cycle prior to the experimental treatments) then they were used in the experiments. Since collection of isopods for the experiments occurred in spring, females producing broods were far more abundant than those failing to produce broods. In order to minimize any residual effects from microhabitat variation in food quality or availability in the field, all isopods had been provided with a constantly available excess of the same food during the laboratory holding period prior to the experimental treatments.

Two size classes were selected: small (S), 30 to 45 mg; and large (L), 65 to 125 mg. The wide interval between size classes prevented the smaller size class from growing into the larger class during the course of the experiment. Preliminary studies indicated that an intermediate class would merely show intermediate, not qualitatively different, behavior.

There were four experimental treatments for both molt cycles. The two size classes of females (small [S] and large [L]) were each subdivided and subjected to two levels of food availability (abundant [A] and reduced [R]). The two available food levels were chosen on the basis of a pilot study, such that (1) the high level of available food (abundant, or A) provided a great excess above bare maintenance needs, and (2) the low food level (reduced, or R) provided a small excess of energy above bare maintenance needs. A total of 50 females were assigned to each of the four treatments for the first molt cycle: large females, abundant food (LA); small females, abundant food (SA); large females, reduced food (LR); and small females, reduced food (SR).

For the high level of food availability (A), food was present 7 days per week. At the low level treatment (R), food was available 2 fixed consecutive days each week (always Tuesday and Wednesday) and either present or absent. When present, it was there in excess to prevent any possibility of competition for food. The diet composition of both levels of food availability was identical, so there was no difference in food quality. The laboratory diet included natural food items: soil, litter, and dead plants that occurred at the collection site. The diet was supplemented with a small portion of fresh carrots. Previous studies on male *Armadillidium* growth rates (Lawlor, unpubl.) were used to choose the specific food items in the diet; the carrots ensured high food quality. Each item in the diet, including the carrots, was present in excess so that its availability to the isopods would be constant.

The laboratory treatments of food availability (A and R) were imposed for a full molt cycle on both small and large females, making a total of four experimental treatment groups. The assignment of the 200 females to the four groups was by a stratified random method. Within each size class, females were randomly chosen for either level of available food.

During the treatments, the isopods were kept in plastic kitchen dishpans, with a maximum of 25 animals in each pan. The number of animals per pan would vary slightly during the course of the experiment, but even the maximum number represented a very low level of density compared to some of the aggregations observed in the field. High humidity was maintained by keeping moist paper towels in the dishpans and enclosing the entire dishpan within a plastic bag. Temperature ranged from 22-24°C. Photoperiod was automatically maintained at 16L:8D. All females were marked on their posterior dorsal surface with glue-on plastic bee numbers. Although females store sperm, possibly making male isopods unnecessary, 6 male isopods were always present to ensure fertilization of all eggs produced. Without sperm, females will ovulate unfertilized eggs (Lawlor 1973).

Molt cycle 1 neared completion as females entered the premolt stage. Females were examined on a daily basis, and if identified as premolt females, they were isolated into 100 × 15 mm petri dishes. A dampened piece of filter paper was placed inside the dish to maintain humidity. This isolation into petri dishes did not interfere with normal activity. *Armadillidium* becomes quiescent immediately preceding ecdysis and continues to be inactive following ecdysis until the new cuticle has hardened. Ecdysis and hardening of the postmolt cuticle were completed within the petri dish. During this period of isolation, the isopods were not given food. They do not normally feed during these three phases of the molt cycle (Lawlor 1973). At this point females were ready for the experimental treatments of molt cycle 2.

The same two treatments of food availability were then applied during molt cycle 2 (MC2). Half of the small females and half of the large females remained on the same level of available food; the remaining females were changed to the other level of food availability. The selection of the females to be switched to the different level of food was similar to the method used when assigning them to food level for MC1. Females completing MC2 were isolated, with the same procedures followed as used in MC1. Females producing broods at this point were kept for the following 3 weeks into their next cycle, in order to retrieve the offspring.

Somatic growth during the molt cycle and biomass of broods were measured. Growth was measured as the difference between the postmolt live weight before the experimental food regime and the postmolt live weight after the food regime. Preliminary studies showed that this procedure of live weight measurement provides the most accurate estimate of an isopod's dry weight. Dry weight was preferred for calculation of reproductive effort, which compared weights of reproductive tissues to somatic tissues. Weights were measured using an analytical balance to the nearest 0.1 mg. At the end of the molt cycle, eggs were extruded into the brood pouch at ecdysis. Reproduction can be measured when offspring hatch out from the brood pouch after about the third week of the subsequent molt cycle. Females with broods nearing completion of development were isolated into petri dishes in order to collect all the newly released young. The females were further isolated in standard 38 × 10 mm plastic capsules. These capsules contained perforations permitting the newly born young to escape while retaining the mother within the enclosed circumstances of the petri dish; the mother might have otherwise eaten her young.

Offspring were first counted and then vacuum dried in an oven. A total of 10 representative individuals were selected and weighed together on a Cahn 21 automatic electrobalance to the nearest .001 mg. Pilot studies demonstrated that the weight of 10, corrected for total clutch size, provided an accurate estimate of the weight of the entire clutch.

During the experiment, significant mortality occurred due to some form of infection, with all treatments affected. Ultraviolet light treatment of the litter coincided with a decrease in mortality. A chi-square test comparing the proportion of isopods that died within each treatment during MC1 was not significant. Because differential mortality among treatments due to the infection did not occur, it is unlikely that the infection biased the results. Finally, 2-way ANOVAs for unequal sample sizes were calculated according to Steel & Torrie (1960).

3 RESULTS

3.1 *Reproductive frequency*

In three treatments during MC1, nearly all females reproduced. Only the treatment of small females with reduced food supply (SR1) had a significant number of females that did not produce broods. Approximately half of the SR1 group molted without producing progeny. This reduction was statistically significant when compared to both the treatment of small females with the abundant level of food (SA1) and the treatment of large females with the reduced level of food availability (LR1) (SA1: chi-square = 8.36, $p < .005$; LR1: chi-square = 13.3, $p < .005$).

Similar results occurred in MC2. Fewer small females on the low food supply (SR2) reproduced. This was statistically significant when compared to small females with high food availability (SA2); but not when compared to large females with low food availability (LR2) (SA2: chi-square = 4.40, $p < .05$; LR2: chi-square = 1.64, n.s.). Data showing both the percent and actual number of females producing broods during MC1 were reported in Brody & Lawlor (1984).

3.2 *Growth in reproductive females*

Growth was very similar for small and large reproductive females kept at identical food availabilities (Fig. 1). Growth was substantial for females on the abundant food level, but near zero on the reduced level. Differences in growth due to food level were statistically significant for small females in each molt cycle (MC1: $t = 14.7$, $p < .001$; MC2: $t = 9.92$, $p < .001$) and for large females (MC1: $t = 10.1$, $p < .001$; MC2: $t = 24.1$, $p < .001$).

3.3 *Brood biomass*

Within either level of food availability, large females produced broods that weighed about twice as much as did those of smaller females (Fig. 2). Figure 2 shows biomass of broods produced during MC 2. Data for MC 1 were similar to data for MC2 (Brody & Lawlor 1984). Female size thus affected reproductive biomass, regardless of food supply.

Food level had a small effect on brood weight, unlike its large effect on somatic growth.

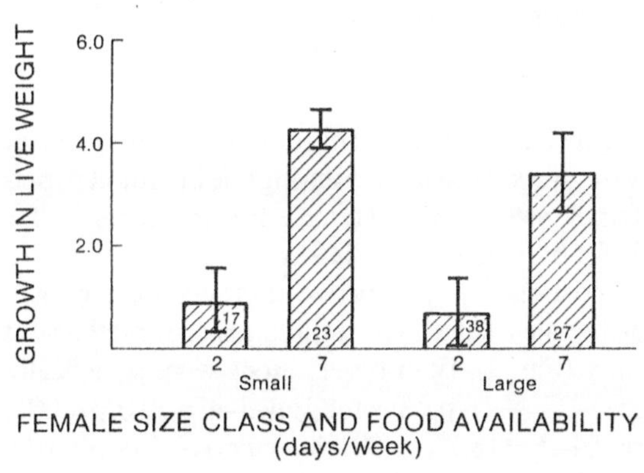

Figure 1. Growth (mgs live weight + SE) of reproductive females during molt cycle 1. Values are means ± Standard error; numbers within bars are sample sizes.

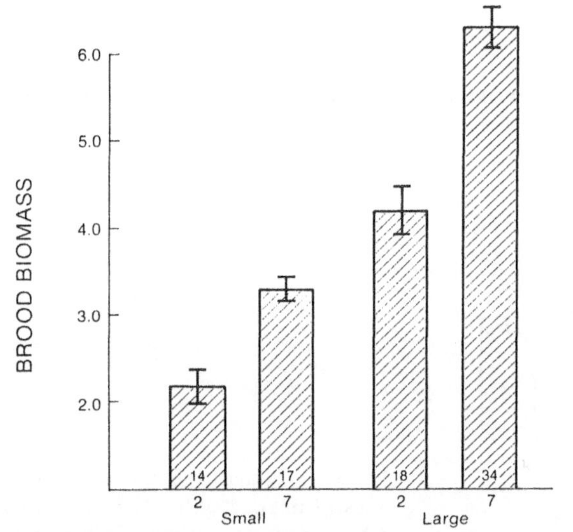

Figure 2. Biomass (mgs dry weight + SE) of broods produced during molt cycle 2. Values are means ± Standard error; numbers within bars are sample sizes.

Reproductive biomass was directly correlated to food availability: it usually decreased with decreased food availability for both large and small females during both molt cycles (Fig. 2). Differences in brood biomass caused by food availability were statistically significant for small females in each molt cycle (MC1: $t = 1.66$, n.s.; MC2: $t = 4.80$, $p < .001$) and for large females (MC1: $t = 2.08$, $p < 00\,1$; 2: $t = 5.58$, $p < .001$). Brood biomass, not egg number, was reported here since individual egg size also varies with food availability. Under conditions of reduced food availability, egg size increases even though brood biomass decreases (Brody & Lawlor 1984). So for the purposes of calculating reproductive effort brood biomass was a more appropriate measure than clutch size.

3.4 *Reproductive effort*

In this experiment, reproductive effort was defined as the ratio of reproduction to the sum of reproduction and growth:

$$\frac{\text{(mgs dry weight of brood)}}{\text{(mgs dry weight of brood)} + (0.25 \times \text{mgs change in live net weight})}$$

where 0.25 converts live weight to dry weight (Lawlor 1973). This definition represents an allocation to reproduction of a fraction of the dry weight production during one intermolt period and does not directly contain a measure of female size. Effects of female size are accounted for by calculating reproductive effort separately for the discrete size classes.

In both molt cycles, for both small and large females, reproductive effort was higher when food availability was reduced. Further, large females had a higher reproductive effort than did small females, regardless of food availability (Fig. 3). Both observations were statistically significant (2-way ANOVA, MC1: female size – $F = 8.71$, $p < .05$; food level – $F = 20.0$, $p < 001$; MC2: female size – $F = 16.1$, $p < .001$, food level – $F = 18.7$, $p < .001$, interaction effect, $p < .01$).

4 DISCUSSION

In these experiments most female isopods reproduced during both molt cycles, apparently responding to energy levels above maintenance threshold. During conditions of relative food

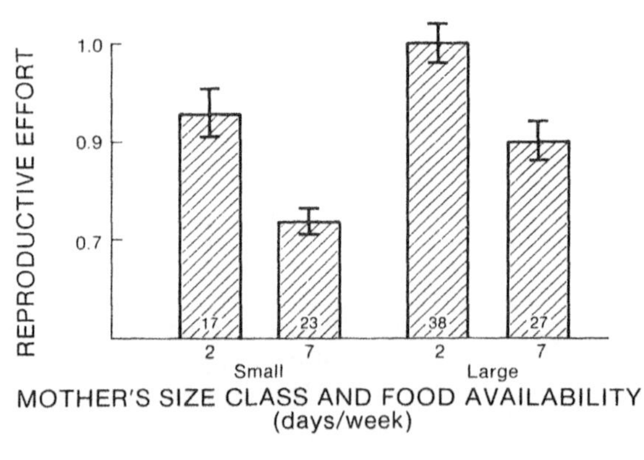

Figure 3. Reproductive effort (see definition in text) during molt cycle 1. Values are means ± Standard error; numbers within bars are sample sizes.

shortages most large females managed to reproduce, but a significant portion of the small females did not. In terms of total offspring produced, females delaying reproduction did not display reproductive success equal to those females that did reproduce. The group of females that did not reproduce during molt cycle 1 did not reproduce with greater frequency or with larger broods in molt cycle 2. Further, growth of nonreproductives was never statistically different from the growth of reproductives (Brody 1980). In a parallel experiment, males grew in the same treatments of reduced food, which is further evidence that the reduced food treatment supplied energy above maintenance threshold levels. Perhaps large females have sufficient stored reserves, but small females lack such reserves. Food level during molt cycle 1 might have affected the likelihood of reproduction during molt cycle 2. There was a trend for females on the reduced food level during molt cycle 1 to reproduce with less frequency during molt cycle 2, when compared to females who had the abundant food supply during molt cycle 1 (not statistically significant, $p = 0.19$).

The growth of females that reproduced during either molt cycle was determined by the level of available food. In general, females kept on the reduced food level showed virtually no change in their live weight from one molt to the next, regardless of size. In marked contrast, females with abundant food showed significant increases in live weight. Compared to the responsiveness of growth, brood weight among reproductive females was relatively unaffected by different levels of food supply. For both small and large females, the mean dry weight of a brood was always slightly larger for better-fed females. Females subjected to food shortages thus nearly cease somatic growth, but maintain their reproductive output at fairly high levels.

Reproductive effort represents the fraction of available resources that an organism assimilates and then allocates to reproduction. Reproductive effort has frequently been estimated for various organisms as a ratio of clutch weight to body weight. This estimate does represent an allocation of standing crop biomass, but it is inadequate as a measure of the allocation of recently assimilated biomass. Whereas body weight represents an allocation process that has been going on for the lifetime of the organism, clutch weight represents an allocation process that has been proceeding for only a fraction of that lifetime. The definition of reproductive effort used in these experiments minimizes this problem of different time scales. All measurements account for events within one molt cycle.

For *Armadillidium*, reproductive effort was higher for females subjected to food shortages. Starved females had less biomass to allocate, but because they sometimes stopped growing nearly all of this biomass was allocated to reproduction. Metabolic differences based upon body size or food availability could affect the interpretation of these results. Standard metabolic rate is the minimum fasting metabolism at a given temperature (Gordon 1968). At 20°C, SMR is approximately 140 mls of oxygen consumed/gram/hour for *A. vulgare* (Edney 1964, Reichle 1967). Basal metabolism for active animals, however, shows a substantial increase above this rate (Hubbell unpubl., Reichle 1967). The females in the reduced food treatment were far less active than those on the abundant food level, and they tended to aggregate in one area of the dishpan on days without food. A factor that probably raised the metabolic rate of well-fed females was specific dynamic action,the increase in metabolic rate due to ingestion of food (Vander et al. 1970). This had a substantial effect on metabolic rate of *Armadillidium* (Hubbell 1971); females that feed more often expend far more energy on specific dynamic action. On average, females on the high abundance level have much higher metabolic energy expenditures, as well as higher overall expenditures of biomass on growth and reproduction. Therefore, metabolic considerations would amplify the differences that have already been measured.

Reproductive effort may increase with age of an organism (Gadgil & Bossert 1970), but the evidence from these experiments is equivocal. Large females are older than small females. Among starved females, larger (older) ones were more likely to reproduce. Field studies (Brody 1980, Lawlor 1976a) show that large females are also more likely to produce a second brood during a single reproductive season. Further, within either food level, large females had higher reproductive efforts than small ones, although this may have simply been due to the lower per unit body weight respiratory costs of larger individuals. However, reproductive effort should have been higher during the second molt cycle compared to the first. For three of the four treatments, it was actually lower.

According to Pianka & Parker (1975), adverse conditions for reproduction should result in lower current reproductive effort and an allocation of resources to somatic growth. This behavior should increase survival and future reproductive efforts, but an assumption of an optimal life history model requires that such life history variables can evolve without physiological constraints (Stearns 1980).

The data presented here describe the relationship between current reproduction and growth. Increased body size is highly correlated to increased brood size; large body size also greatly increases survivorship (Brody et al. 1983). Therefore, it is reasonable to compare these data to models that predict optimal balances between current and future levels of reproductive effort. Some optimality models predict that reproductive effort should increase with age (e.g. Pianka & Parker 1975, Gadgil & Bossert 1970), however, young food stressed females have much higher levels of effort than older well fed females. However, the allocation to current levels of effort does result in reduced future size and therefore future fecundity, a key assumption in many theoretical treatments of life history (Williams 1966).

Half of the small females on the reduced food supply grew without reproducing. Yet they did not grow an extra amount. This group's overall reproductive success was simply less than those that did reproduce. There is currently no satisfactory explanation for the loss off energy in this group. A somewhat similar situation occurs in guppies; non-reproductive females appeared to have less overall energy to partition to either somatic or reproductive tissues than did sibling females which reproduced (Reznick 1983).

Brood biomass and survivorship are both size dependent, and size is growth dependent. Perhaps the optimal reproductive tactic is constrained by the interrelationships of these developmental variables in the sense of Stearns (1980).

Further, it is possible that reproductive effort is not really being regulated at all. The relatively constant reproductive biomass may actually be regulated, leading to apparent changes in reproductive effort. Reznick (1983) pointed out that either growth or reproductive effort may have evolved in guppies. Or in this case, what has been fixed may be a form of phenotypic plasticity; as these experiments do not provide data on any underlying genetic variability (Reznick 1985). These experiments did not identify any popultions of this species which exhibit different tradeoffs and therefore different costs and benefits. In either case, regulation of its energy budget may be reflected in these measurements of reproductive effort. Maintenance of high current reproductive effort, whether achieved directly or indirectly, may have contributed to its success in environments that fluctuate in food supply.

ACKNOWLEDGEMENTS

I am especially grateful to my wife, Betty, for her editorial comments and for typing this manuscript. Drs Edie Marsh and Jim Palmer provided valuable reviews. This research was supported in part by NSF Grant DEB 78-10656 to Dr Lawrence R. Lawlor, Department of Zoology, of the University of Texas at Austin. The figures were prepared with the support of The Sea Grant College Program of Texas A & M University.

REFERENCES

Brody, M.S. 1980. Reproductive adaptations to food shortages in the terrestrial isopod, *Armadillidium vulgare*. Ph.D. Dissertation. The University of Texas at Austin.

Brody, M.S. & L.R. Lawlor 1984. Adaptive variation in offspring size in the terrestrial isopod, *Armadillidium vulgare*. *Oecologia* 61: 55-59.

Brody, M.S., L.R. Lawlor & M. Edgar 1983. A cost of reproduction in a terrestrial isopod. *Evol.* 37: 653-655.

Edney, E.B. 1964. Acclimation to temperature in terrestrial isopods I. Lethal temperatures. *Physiol. Zool.* 37: 378-394.

Edney, E.B., W. Allen & J. McFarlane 1974. Predation by terrestrial isopods. *Ecol.* 55: 428-433.

Fisher, R.A. 1930. *The Genetical Theory of Natural Selection*. Oxford University Press, London.

Gadgil, M. & W.H. Bossert 1970. Life historical consequences of natural selection. *Amer. Natur.* 104: 1-24.

Gordon, M. 1968. *Animal Function: Principles and Adaptations*. Macmillan Company, London.

Hirshfield, M.F. & D.W. Tinkle 1975. Natural selection and the evolution of reproductive effort. *Proc. Natl. Acad. Sci. USA* 72: 2227-2231.

Hubbell, S.P. 1971. Of sowbugs and systems: The ecological bioenergetics of a terrestrial isopod. In B.C. Patten (ed.), *Systems Analysis and Simulation in ecology*. Vol. 1. pp. 269-324. Academic Press, New York.

Lawlor, L.R. 1973. The ecology of reproduction in the terrestrial isopod, *Armadillidium vulgare*. Ph.D. Dissertation. The University of California, Berkeley.

Lawlor, L.R. 1976a. Molting, growth, and reproductive strategies in the terrestrial isopod, *Armadillidium vulgare*. *Ecol.* 57: 1179-1194.

Lawlor, L.R. 1976b. Parental investment and offspring fitness in the terrestrial isopod, *Armadillidium vulgare*. *Evol.* 30: 775-785.

Paris, O.H. 1963. The ecology of *Armadillidium vulgare* in California grassland: Food, enemies, and weather. *Ecol. Monogr.* 33: 1-22.

Paris, O.H. & A. Sikora 1965. Radiotracer demonstration of isopod herbivory. *Ecol.* 46: 729-734.

Pianka, E.R. & W. Parker 1975. Age-specific reproductive tactics. *Amer. Natur.* 109: 453-464.

Reichle, D.E. 1967. Radioisotope turnover and energy flow in terrestrial isopod populations. *Ecol.* 48: 351-366.

Reznick, D. 1983. The structure of guppy life histories: the tradeoff between growth and reproduction. *Ecol.* 64: 862-873.

Reznick, D. 1985. Costs of reproduction: an evaluation of the empirical evidence. *Oikos* 44: 257-267.

Schaffer, W.M. 1974a. Selection for optimal life histories: The effects of age structure. *Ecol.* 55: 291-303.

Schaffer, W.M. 1974b. Optimal reproductive effort in fluctuating environments. *Amer. Natur.* 108: 783-790.

Schaffer, W.M. & M.D. Gadgil 1975. Selection for optimal life histories in plants. In M. Cody & J. Diamond (eds), *Ecology and Evolution of Communities*, pp. 142-157. Harvard University Press, Cambridge.

Stearns, S.C. 1980. A new view of life-history evolution. *Oikos* 35: 266-281.

Steel, R.G.D. & J.H. Torrie 1960. *Principles and Procedures in Statistics*. McGraw-Hill, New York.

Taylor, H.M., R. Gourley, C.E. Lawrence & R. Kaplan 1974. Natural selection of life history attributes: An analytical approach. *Theor. Popul. Biol.* 5: 104-122.

Tinkle, D.W. 1969. The concept of reproductive effort and its relation to the evolution of life histories of lizards. *Amer. Natur.* 103: 501-516.

Vander, A.J., J.H. Sherman & D. Luciano 1970. *Human Physiology: The Mechanisms of Body Function*. McGraw-Hill, New York.

Williams, G.C. 1966. Natural selection, the costs of reproduction, and a refinement of Lack's principle. *Amer. Natur.* 100: 687-690.

Analysis of embryo production in a caridean shrimp guild from a tropical seagrass meadow

Raymond T. Bauer
Center for Crustacean Research, University of Southwestern Louisiana, Lafayette, USA

ABSTRACT

Embryo production was compared among nine caridean shrimp species inhabiting seagrass meadows in a tropical location, the north coast of Puerto Rico. Reproduction occurred throughout the year in all species. During a short life span (less than one year), females reproduced continuously after reaching maturity, producing successive broods of embryos at short intervals. Analysis of allometry in brood size (number of embryos/brood) as a function of female size (carapace length) was done by calculating slopes of brood size/female size regressions with log transformed variables. Increase in brood size with female size was isometric in seven species and showed positive allometry in two others. However, the relationship of species mean log brood size on species mean log female size was negatively allometric, indicating that females of smaller species produced relatively larger broods for their size than did females of larger species. Analysis of covariance corroborated this conclusion. Log brood size means adjusted for differences in female size were significantly different among species, and adjusted species brood means were negatively correlated with species female size. The mechanism for enhanced reproductive output (in terms of brood size) in smaller species is unclear. In some crustacean groups, reduced embryo size has been selected for as a means of increasing brood size. Additionally, since embryo size and incubation period (duration of embryonic development) showed a significant positive correlation in this group of caridean species, smaller embryo size would also result in shorter incubation periods and thus larger numbers of broods per reproductive period. However, there was no relationship between species embryo size and species female size in this group, indicating that differential selection among species on embryo size as a means of increasing reproductive output has not occurred.

1 INTRODUCTION

In a study on the temporal pattern of reproduction and recruitment in nine caridean shrimp species inhabiting a tropical seagrass meadow (Bauer 1989), detailed observations on brood size, embryo size and female size were recorded on a large number of breeding females. These and supplemental observations on embryo incubation periods permitted the present analysis of the interrelationships of brood size, embryo size, female size and incubation period in these shrimps. Since the members of this suite of caridean species inhabit the same meadows and were collected during the same time period, confounding variation in reproductive output due to latitudinal or habitat differences among species is reduced, i.e. breeding females of the guild were exposed to very similar environmental conditions. The goal of such an analysis is to propose and test hypotheses on the selective forces acting on reproductive output in these and, ultimately, other crustacean species.

2 NATURAL HISTORY AND REPRODUCTIVE PATTERNS

The nine most abundant carideans in *Thalassia testudinum-Syringodium filiforme* seagrass meadows at a site on the north coast of Puerto Rico (Dorado; 18°29'N, 66°15'W), sampled in 1981-82, were: *Thor manningi, Hippolyte curacaoensis, Latreutes fucorum, L. parvulus* (Hippolytidae); *Leander tenuicornis, Periclimenes americanus* (Palaemonidae); *Processa bermudensis, P. riveroi* (Processidae); and *Alpheus normanni* (Alpheidae) (Bauer 1985). *Hippolyte curacaoensis, Latreutes fucorum* and *L. parvulus* cling to seagrass blades, while *Thor manningi, Periclimenes americanus*,and *Leander tenuicornis* range over both the substrate and seagrass blades. *Alpheus normanni* constructs burrows in which adult male-female pairs occur, and the processid species burrow directly into the sediment by day. Laboratory observations show that burrowing species emerge at night and also that all other species appear most active in swimming and foraging at night. These laboratory observations confirmed the hypothesis, suggested by results of day-night sampling, that these species are primarily nocturnal. Densities estimated from night sampling were significantly higher than those based on day collecting in all species (Bauer 1985).

Reproduction (embryo production) was continuous throughout the year in all species at the population level (Bauer 1989). The percentage of mature females carrying embryos was never less than 20% in any species in any month and generally was much higher. The monthly variations in breeding intensity (% ovigerous females) were not concordant among the nine species, i.e. reproduction was continuous with no common seasonal pattern in the highs and lows of breeding intensity. Three environmental variables (temperature, salinity and qualitive observations on seagrass cover) were relatively constant throughout the year (Bauer 1985). The tropical seagrass meadows sampled were a classic example of a physically benign, stable tropical nearshore marine habitat (cf. Orton 1920). Breeding patterns observed for these seagrass shrimp populations generally fit predictions (Orton 1920, Thorson 1950, Sastry 1983 about continuous reproduction in littoral and neritic tropical marine invertebrates).

Embryo production was continuous at the individual as well as at the population level in these shrimp species. Laboratory observations revealed that females carrying embryos near to hatching also had ovaries filled with large vitellogenic oöcytes. After embryos hatched, the female molted within a day or two. Spawning of a new brood of embryos then took place within a day if no male was present, sooner if a male was available for mating (similar to the carideans *Heptacarpus pictus* and *H. paludicola*, Bauer 1976, 1979). Statistical analyses on data for more than three thousand embryo-carrying females confirmed that females carrying embryos near hatching had ovaries ripe for a new spawning (Bauer 1989). Embryo incubation periods were short (5-10 days); once females reached maturity, they apparently produced uninterrupted successive spawns at short intervals until death. These small seagrass shrimp species probably have short life spans [less than one year; estimated from cohort analysis at 5 months in *Thor manningi* (Bauer 1986) corroborated by measurements on development and growth in laboratory-reared *Thor* spp. (Dobkin 1968); estimated at 2-5 months in *Latreutes* spp. from cohort analysis (Salva 1984)]. Thus, females of these seagrass shrimp species appear to mature rapidly and to produce several successive spawns within a single short lifetime reproductive period.

3 DATA BASE

More than 3200 embryo-carrying females were examined from preserved collections taken in the year-long sampling program (Bauer 1985) in the Dorado, Puerto Rico, seagrass meadows. The number of embryos carried by females (= brood size) was counted only in females carrying embryos in early developmental stages. These stages (1 & 2) are modified (Bauer 1986) from a subjective scale of 1-4 for embryo development given in Allen (1966). Embryos observed at 6-25× under a stereomicroscope were staged as follows: Stage 1, embryo entirely of yolk, no tissue cap (blastodisc) visible; Stage 2, blastodisc present, but no eye or eye pigment developed; Stage 3, eye developed or developing, cephalothorax and abdomen of embryo not separated; Stage 4, embryo near hatching, with little yolk, cephalothorax and abdomen visibly separated. Broods containing Stage 1 & 2 embryos were rather tightly arranged under the female abdomen. However, broods with later-stage embryos were distended away from the abdomen and pleopods due to an increase in embryo size during development. Such advanced (Stage 3 & 4) embryos were easily dislodged from females during sampling and especially in handling after preservation; females carrying advanced embryos were not used in the analyses. The number of females examined with Stage 1 or 2 embryos was 661 (median number/species = 78, minimum = 24, maximum = 131).

Measurements of the greater and lesser diameters of Stage 1 and 2 embryos were taken using an ocular micrometer in a stereomicroscope at 25 or 50× magnifications. Five embryos were chosen haphazardly from a group of embryos separated from a given female, and the median greater and lesser diameters were used to represent embryo size from that female (median less affected by extremes than the mean in a small sample size). Embryo volume was calculated from the formula for an oblate spheroid (cf. Turner & Lawrence 1979):

$$v = \pi \times (d_1)^2 \times (d_2)/6$$

where v = volume (mm^3), d_1 = lesser embryo diameter and d_2 = greater diameter. Carapace length, a measure of female size, was the distance from the posterior edge of the eye orbit to the posterior mid-dorsal margin of the carapace.

4 BROOD SIZE AS A FUNCTION OF FEMALE SIZE

Table 1 gives mean brood size and mean breeding female size for the nine caridean species. The number of embryos incubated (brood size) by a female is a direct expression of its reproductive output. Intraspecific analyses and interspecific comparisons of the brood size/female size relationship often reveal possible selective forces or constraints acting on reproductive output in species from related taxa (e.g. Reaka 1979, Hines 1982). Allometric relationships can best be examined by linear regressions obtained from log transformation of the power curve $Y = aX^b$ (Gould 1966, Hartnoll 1982, Hines 1982) so that

$$\log Y = b (\log X) + \log a$$

Here, the allometry of brood size is examined within and among the nine caridean species. In

Table 1. Mean ($\overline{X}$) brood size (number of embryos/brood) and mean size (carapace length, mm) (± standard deviation) of breeding females in the nine species of tropical seagrass carideans. n: number of specimens.

Species	n	$\overline{X}$ brood size	$\overline{X}$ female size
A. normanni	55	95 ± 72	4.5 ± 0.7
H. curacaoensis	131	153 ± 70	2.4 ± 0.3
L. fucorum	98	57 ± 24	2.3 ± 0.3
L. parvulus	78	59 ± 23	2.0 ± 0.2
L. tenuicornis	40	593 ± 311	6.7 ± 1.0
P. americanus	99	56 ± 27	2.1 ± 0.2
P. bermudensis	84	129 ± 82	3.5 ± 0.5
P. riveroi	24	384 ± 186	5.3 ± 0.7
T. manningi	52	66 ± 27	1.9 ± 0.2

regression equations dealing with brood size and female size (Table 2), Y = number of embryos carried by a female (= brood size) and X is carapace length, a standard measure of body size in carideans (e.g. Bauer 1986, Kuris et al. 1987).

4.1 *Intraspecific variation in brood size*

Table 2 presents the regression equations of log brood size on log female size for all 9 caridean species. The slope indicating isometry in allometric analyses depends on the dimensions of the dependent and independent variables (Gould 1966). Thus, if Y and X have the same dimensions, e.g. are both linear measurements, then a slope of b = 1 indicates isometry. If, on the other hand, X is a linear measure (L) and Y is a volume (L^3), then a slope of 3 indicates isometry. The measure of brood size used here, i.e. number of embryos, has no dimension *per se*. However, it is proposed that a slope of 3 indicates isometry in the brood size on female size regressions because the number of embryos approximates a volume, i.e. the volume of the collective mass of embryos. The slopes (b) of regressions in Table 2 vary around a median of 3.05 from 2.38 to 3.90. The hypothesis that b = 3 can be accepted in those seven species is which the 95% confidence limits on b include b = 3 (all species except *Latreutes fucorum* and *Alpheus normanni*).

Negative allometry in brood size might indicate senescence of reproductive output in larger (older) females, while positive allometry could indicate selection for morphological or physiological changes in larger females increasing this measure of reproductive output. However, in 7 of 9 species, brood size increased isometrically with female size. The two species that showed a positive allometry in brood size, *Latreutes fucorum* and *Alpheus normanni*, differ greatly in general morphology, behavior and ecology. It is likely that the morphological and/or physiological adaptations accounting for the positive allometry are quite different in the two species. In *Alpheus normanni*, brood size in smaller females begins at a very low level (for their size) relative to other species. This is shown by the Y-intercept of the *A. normanni* (Table 2; significantly different from Y = 0, t-test, $p < 0.05$), lowest of any species. The differential increase in brood size (b = 3.90) in A. normanni may reflect some compensatory morphological or physiological adaptation lifting embryo production with increasing female size from a relatively low level in the smaller females.

4.2 *Interspecific comparisons of brood size*

It is obvious (Table 1, Fig. 1) that females from larger-sized species can carry greater absolute numbers of embryos than those from smaller species. The question that will be asked in this section is if females of smaller-sized species compensate for their smaller absolute body size by a relative increase in reproductive output, here measured by number of embryos in a brood.

Figure 1 shows a plot of log mean number of embryos (brood size) on log mean carapace length (female size) for all species. The linear regression for these data is:

$$\log(\bar{Y}) = 1.60 (\log \bar{X}) + 1.30$$

As discussed above, a slope of 3 indicates isometry in allometric equations of number of embryos on female carapace length. In this regression, $b = 1.60$ and the null hypothesis of $b = 3$ is rejected (95% confidence limits on b are 0.78, 2.42). Clearly, there is negative allometry in brood size when these species of different body size are compared. This analysis indicates that females of smaller species produced relatively more embryos for their size than did those of larger species.

The hypothesis that smaller species have a relatively higher reproductive output (in terms of brood size) than larger species was examined with analysis of covariance (ANCOVA) of log number of embryos (brood size) on the covariate log female carapace length (female size). The assumption of homogeneity of slopes could not be met for all nine species ($p = 0.01$). *Alpheus normanni* and *Latreutes fucorum* were eliminated from the analysis since the slopes of the brood size/female size regressions (Table 2) were significantly different from 3 while those of the other 7 species were not. The hypothesis of homogeneity of slopes in these seven species was accepted ($p = 0.53$). In the seven species, the hypothesis that log brood size means (adjusted for female body size) are not different (equivalent to testing for homogeneity of the Y-intercepts, Sokal & Rohlf 1981: 521) is rejected ($p < 0.001$). The ANCOVA involved extrapolation of

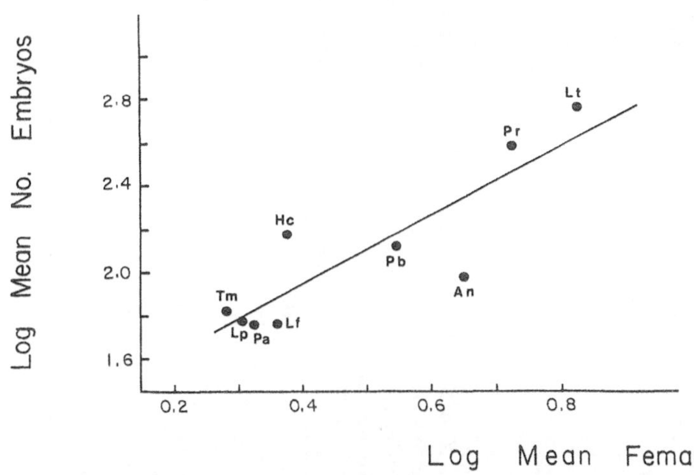

Figure 1. Interspecific comparisons of brood size as a function of female size in the nine seagrass shrimp species. Log mean number of embryos (brood size) is plotted against log mean breeding female carapace length (CL, in mm). The regression equation is: $\log(\bar{Y}) = 1.60 (\log \bar{X}) + 1.30$. Species symbols are: An – *Alpheus normanni*; Hc – *Hippolyte curacaoensis*; Lf – *Latreutes fucorum*; Lp – *L. parvulus*; Lt – *Leander tenuicornis*; Pa – *Periclimenes americanus*; Pb – *Processa bermudensis*; Pr – *P. riveroi*; Tm – *Thor manningi*.

Table 2. Regression equations of log number of embryos (brood size) on log CL (carapace length in mm, the measure of female size). Y: number of embryos, X: female carapace length, a: Y-intercept, b: slope, n: number of specimens, r^2: coefficient of determination (squared Pearson correlation coefficient). In all regressions, $p (H_0: b = 0) < 0.001$. 95% confidence limits on b are given.

Species	n	log (number of embryos) = b (log CL) + log a	r^2	95% limits
A. normanni	55	log (Y) = 3.90 (log X) – 0.65	0.78	3.32, 4.48
H. curacaoensis	131	log (Y) = 2.94 (log X) + 1.03	0.75	2.64, 3.23
L. fucorum	98	log (Y) = 3.66 (log X) + 0.38	0.59	3.04, 4.27
L. parvulus	78	log (Y) = 2.67 (log X) + 0.96	0.52	2.07, 3.26
L. tenuicornis	40	log (Y) = 3.17 (log X) + 0.12	0.86	2.74, 3.60
P. americanus	99	log (Y) = 3.28 (log X) + 0.62	0.64	2.78, 3.77
P. bermudensis	84	log (Y) = 2.88 (log X) + 0.50	0.54	2.30, 3.46
P. riveroi	24	log (Y) = 3.05 (log X) + 0.34	0.66	2.08, 4.02
T. manningi	52	log (Y) = 2.38 (log X) + 1.10	0.31	1.37, 3.39

Table 3. Means ($\bar{X}$) of Stage 1 and 2 embryo volumes (mm^3) for the nine caridean species. The greater (d_2) and lesser (d_1) diameters (mm) corresponding to those of a measured embryo with a volume closest to the mean volume are given for Stage 1 embryos for each species. A single asterisk by the species name indicates those species in which the mean volume of Stage 2 embryos was significantly different from that of Stage 1 embryos (t-test, $p < 0.05$). n: number of female broods examined, **no observations.

Species	$\bar{X}$ volume	Stage 1 embryos (n)	d_2	d_1	$\bar{X}$ volume	Stage 2 embryos (n)
*A. normanni**	0.030	35	0.44	0.36	0.033	20
*H. curacaoensis**	0.005	66	0.24	0.20	0.007	65
L. fucorum	0.012	46	0.40	0.24	0.012	50
L. parvulus	0.012	30	0.36	0.24	0.013	48
*L. tenuicornis**	0.028	22	0.42	0.36	0.033	18
*P. americanus**	0.013	59	0.36	0.26	0.016	37
*P. bermudensis**	0.008	45	0.32	0.22	0.012	39
*P. riveroi**	0.010	10	0.30	0.26	0.015	14
*T. manningi**	0.026	39	0.42	0.34	**	**

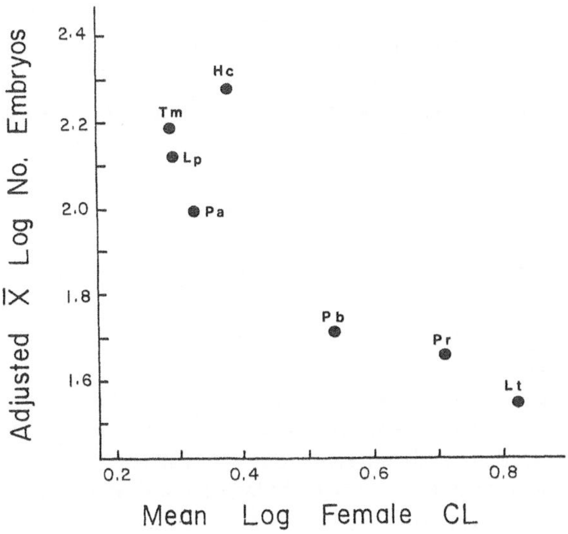

Figure 2. Mean log brood size (number of embryos/brood) for seven species adjusted by ANCOVA for body size plotted against mean log female carapace length (CL). r = - 0.90. Species symbols as in Figure 1.

brood size/female size regressions in order to compare species in which female sizes did not overlap. This biased the analysis in favor of accepting the hypothesis of no difference in adjusted log brood size means (Snedecor & Cochran 1967: 430). In spite of this possible bias, adjusted log brood size means were significantly different among the seven species. *Latreutes fucorum* and *A. normanni* were similarly analysed, and for these two species, adjusted log brood size means were also significantly different ($p < 0.001$).

In Figure 2, adjusted log brood size means are plotted as a function of mean log female carapace length (size) for the group of seven species. There is a significant negative correlation between adjusted log brood size means and log female size means ($r = -0.90$; $p < 0.05$), i.e. in this group, females of smaller-sized species produced relatively more embryos (for their size) than did females of larger species. Similarly, the adjusted log brood size mean of *Latreutes fucorum* (2 102) is significantly larger ($p < 0.001$) than that of *Alpheus normanni* (1 182), the species with larger female size (Table 1). Females of smaller species are limited by absolute size in the number of embryos they can carry, and there has apparently been selection for a relatively higher reproductive output (in terms of numbers of embryos produced and carried by females) to compensate for this.

5 EMBRYO SIZE

The size of the newly fertilized egg (embryo) may reflect the action of various, sometimes opposing, selective pressures acting on reproductive output and larval development. Small embryo size permits a larger brood size for a given energetic allocation to reproduction, thus increasing the number of larvae hatched (Steele & Steele 1975). On the other hand, large embryo size may allow the larva to hatch and later settle at a larger size, with increased probability of survival (Strathmann 1977, Hines 1986). In comparisons of embryo size among species within a taxon, larger embryo size may indicate hatching at an advanced stage of development. Such an advanced larva is exposed for a shorter time to the intense mortality presumed characteristic of the planktonic environment (Thorson 1950, Grahame & Branch 1985), although this advantage is counterbalanced by the disadvantage of reduced larval dispersal.

Duration of embryonic development is positively correlated with embryo size in some crustacean groups (Wear 1974, Steele & Steele 1975), but not in others (reviewed in Hines 1986). In the seagrass caridean species under consideration, the time that embryos are incubated (= developmental time from spawning and fertilization to hatching) could be an important factor in the number of broods produced per reproductive period, since females of these species produce broods successively throughout their short lifetimes (Bauer 1989). If incubation period varies positively with embryo size, reproductive output could be increased by a reduction in embryo size, with the concomitant shortening of incubation period and the resultant increase in the number of broods per reproductive period.

Wear (1974), both with his own data and that reviewed in the literature, showed that embryo size increased during development in several decapod species. In the seagrass caridean species, the mean volume of Stage 2 embryos was significantly different (larger than that of Stage 1 to 6 of 8 species) (Table 3). Only observations on broods composed of Stage 1 embryos were used the subsequent analyses to avoid confounding variation due to differences in embryo developmental stage and for consistency among species.

5.1 *Interspecific comparisons in embryo size*

It was shown earlier that the smaller species produced, for their size, relatively higher numbers of embryos per brood than did larger species. Increased brood size could be accomplished by a reduction in embryo size, here measured as embryo volume. The question was asked: do smaller species tend to have smaller embryo sizes than larger species? The answer is negative: there was no correlation between species mean log embryo volume and species mean log female carapace length in this group of shrimps ($r = 0.30$, $p > 0.05$).

5.2 *Duration of embryonic development and embryo size*

The duration of embryonic development (period incubated by females) is positively correlated with embryo size (volume) ($r = 0.75$; $p < 0.05$) (Fig. 3) in the eight of these nine species for which incubation periods have been estimated (Bauer 1989). Species could conceivably increase

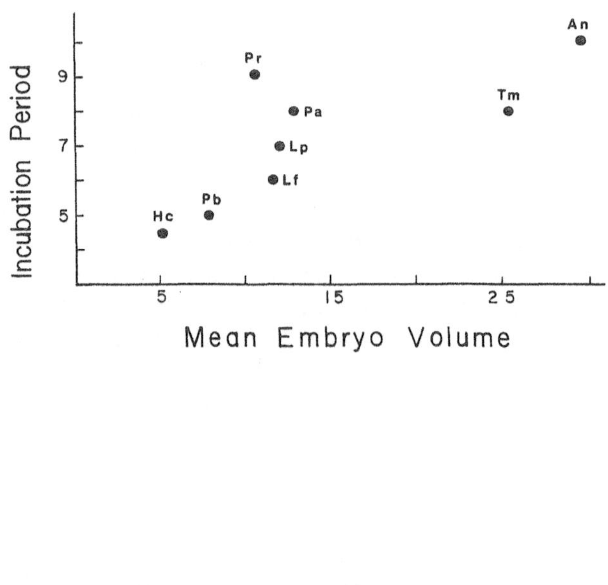

Figure 3. Embryo incubation period (days) of 8 seagrass species plotted against their mean embryo volumes ($mm^3 \times 10^3$). $r = 0.75$. Species symbols as in Figure 1.

reproductive output by increasing the number of broods per unit time through a decrease in embryo developmental time by means of smaller embryo size as discussed above. The question was asked: do smaller species also have greater reproductive output in terms of increased number of broods per unit time as a result of shortened incubation periods? There is no significant correlation between species measures of incubation period and mean breeding female body size (carapace length) ($r = 0.48$, $p > 0.05$). It appears that there is no tendency towards shortened incubation periods in smaller species in this group of seagrass shrimps.

Wear (1974) suggested that developmental time of incubated embryos varied with the quantity of yolk (embryo volume) in the newly spawned embryo, i.e. more yolk to metabolize, longer developmental times at the same temperature. Steele & Steele (1975) and Steele (1977) proposed that faster developmental times of smaller embryos simply reflects the well-known inverse relationship between size and metabolic rate in biological systems. The embryo incubation periods of these tropical seagrass species were quite short compared to temperate or boreal species (cf. Wear 1974, Allen 1966). Wear (1974) found, as one might suspect, that incubation periods decreased with increasing temperature in the British species studied. Incubation periods were extrapolated to temperatures (27-30°C) similar to those (26-28°C) at which the tropical seagrass species in this study were maintained. Although the embryos of British species did not survive at these temperatures, it is interesting to note that the predicted incubation periods were similar (6-11 days) to those observed in the tropical carideans. Short embryo incubation periods in these tropical species may be a life history adaptation for increasing the number of broods spawned per unit time. However, the Wear (1974) study suggests the alternative hypothesis that short developmental periods in tropical carideans may be a simple metabolic response to elevated water temperatures in the tropical habitat.

6 LIFE HISTORY CONSIDERATIONS

Reproductive characteristics of these nine tropical caridean species do not clearly fit into contrasting predictions of life history models (Pianka 1970, Stearns 1976, Grahame & Branch 1985). They do approach the *r* end of the *r-K* continuum of deterministic models. Life spans of these shrimps are probably short (a few to several months), and growth to reproductive maturity appears rapid (Bauer 1986, Salva 1984). Once maturity is reached, embryos are produced in successive spawns at short intervals (1-2 weeks) between spawns (Bauer 1989). Although the term iteroparous is usually associated with multiple spawnings, these species might be considered semelparous in the sense that there is a brief but continuous burst of reproduction during a single reproductive period in a relatively short lifetime (cf. Sastry 1983), an *r* life history characteristic of fluctuating, unstable environments. The adult habitat (tropical seagrass meadows on the north coast of Puerto Rico) appears relatively stable in terms of basic environmental parameters such as temperature, salinity, and substrate (seagrass cover) (Bauer 1985, 1989). The stability of these species' larval planktonic habitat is unknown. Although tropical planktonic environments are often considered to be predictable relative to temperate and boreal ones, the episodic recruitment, concordant among all species, observed for this group of shrimps (Bauer 1989) may indicate otherwise. Thus the reproductive and other life history traits observed for these species may better fit stochastic ('bet-hedging') models for species, in which adults live in stable environments but in which juvenile (= larval) mortality is variable (Stearns 1976). In the physically stable tropical seagrass meadows in which these species live, predation pressure by fish appears intense (Randall 1967, Heck 1979, Weinstein & Heck 1979, Heck &

Orth 1980, Roblee & Zieman 1984). This source of mortality (stable throughout the year?) is probably responsible for the short (less than one year) life spans of these shrimps and may be the major selective force favoring their rapid maturity and high reproductive effort (Bauer 1989).

These species are relatively small for caridean shrimp, and body size clearly limits absolute embryo production in these as in most other crustaceans. In terms of embryo production as a life history trait, the negative allometry in brood size exhibited by these species suggest that there has been selection such that small species put more resources into reproduction than do large species. Smaller species might increase the number of progeny produced per brood by decreasing embryo size (as in brachyuran crabs, Hines 1982) which would also have the effect of decreasing embryo incubation period and thus increasing the number of broods produced per reproductive life span. However, there was no trend toward smaller embryo size in the smaller species of this group. The mechanism that allows smaller species to produce larger broods (relative to their size) remains unknown, but it might be a morphological feature such as a relatively high allocation of cephalothoracic space for ovarian development and egg accumulation.

ACKNOWLEDGEMENTS

The field work for this study was done when I was a faculty member of the Department of Biology, University of Puerto Rico (UPR), Rio Piedras. I wish to acknowledge the UPR FIPI and OCEGI grants whose funds aided in field sampling and laboratory collection of data. I am grateful to student assistants for their valuable help. Brood size and embryo size observations on *Latreutes* spp. were done by Y. Salva as part of a master's thesis at UPR, and I gratefully acknowledge her contribution concerning these species. This is Contribution No. 14 of the University of Southwestern Louisiana's Center for Crustacean Research.

REFERENCES

Allen, J.A. 1966. The dynamics and interrelationships of mixed populations of Caridea found off the north-east coast of England. In H. Barnes (ed.), *Some Contemporary Studies in Marine Science* pp. 45-66. London: Allen & Unwin.

Bauer, R.T. 1976. Mating behaviour and spematophore transfer in the shrimp *Heptacarpus pictus* (Stimpson) (Decapoda: Caridea: Hippolytidae). *J. Nat. Hist.* 10: 415-440.

Bauer, R.T. 1979. Sex attraction and recognition in the caridean shrimp *Heptacarpus paludicola* Holmes (Decapoda: Hippolytidae). *Mar. Behav. Physiol.* 6: 157-174.

Bauer, R.T. 1985. Diel and seasonal variation in species composition and abundance of caridean shrimps (Crustacea, Decapoda) from seagrass meadows on the north coast of Puerto Rico. *Bull. Mar. Sci.* 36: 150-162.

Bauer, R.T. 1986. Sex change and life history pattern in the shrimp *Thor manningi* (Decapoda: Caridea): a novel case of partial protandric hermaphroditism. *Biol. Bull.* 170: 11-31.

Bauer, R.T. 1989. Continuous reproduction and episodic recruitment in nine shrimp species inhabiting a tropical seagrass meadow. *J. Exp. Mar. Biol. Ecol.* 127: 175-187.

Dobkin, S. 1968. The larval development of a species of *Thor* (Caridea, Hippolytidae) from south Florida. *Crustaceana Suppl.* 2: 1-18.

Gould, S.J. 1966. Allometry and size in ontogeny and development. *Biol. Rev. Camb. Philos. Soc.* 41: 587-640.

Grahame, J. & G.M. Branch 1985. Reproductive patterns of marine invertebrates. *Oceanogr. Mar. Biol. Ann. Rev.* 23: 373-398.

Hartnoll, R.G. 1982. Growth. In L.G. Abele (ed.), *The Biology of Crustacea*, Vol. 2 pp. 111-196. New York: Academic Press.

Heck Jr, K.L. 1979. Some determinants of the composition and abundances of motile invertebrate species in tropical and temperate turtlegrass (*Thalassia testudinum*) meadows. *J. Biogeogr.* 6: 183-200.

Heck Jr, K.L. & R.J. Orth 1980. Seagrass habitats: the roles of habitat complexity, competition, and predation in structuring associated fish and motile macroinvertebrates assemblages. In V.S. Kennedy (ed.), *Estuarine Perspectives* pp. 449-464. New York: Academic Press.

Hines, A.H. 1982. Allometric constraints and variables of reproductive effort in brachyuran crabs. *Mar. Biol. (Berl.)* 69: 309-320.

Hines, A.H. 1986. Larval patterns in the life histories of brachyuran crabs (Crustacea, Decapoda, Brachyura). *Bull. Mar. Sci.* 39: 444-466.

Kuris, A.M., Z. Ra'anan, A. Sagi & D. Cohen 1987. Morphotypic differentiation of male Malaysian giant prawns, *Macrobrachium rosenbergii*. *J. Crust. Biol.* 7: 219-237.

Orton, J.H. 1920. Sea temperatures, breeding and distribution of marine animals. *J. Mar. Biol. Assc. UK* 12: 339-366.

Pianka, E.R. 1970. On 'r' and 'K' selection. *Amer. Nat.* 104: 592-597.

Randall, J.E. 1967. Food habits of reef fishes of the West Indies. *Stud. Trop. Oceanogr. (Miami)* 5: 665-847.

Reaka, M.L. 1979. The evolutionary ecology of life history patterns in stomatopod Crustacea. In S. Stancyk (ed.), *Reproductive Ecology of Marine Invertebrates* pp. 235-260. Columbia: Univ. South Carolina Press.

Roblee, M.B. & J.C. Zieman 1984. Diel variation in the fish fauna of a tropical seagrass feeding ground. *Bull. Mar. Sci.* 43: 335-345.

Salva, Y.M. 1984. Life history patterns of the shrimps *Latreutes fucorum* and *L. parvulus* Stimpson (Caridea: Hippolytidae) in Puerto Rican seagrass (*Thalassia*) meadows. Master's thesis, Univ. Puerto Rico, Rio Piedras, 69 pp.

Sastry, A.N. 1983. Ecological aspects of reproduction. In F.J. & W.B. Vernberg (eds), *The Biology of Crustacea*. Vol. 8 pp. 179-270. New York: Academic Press.

Snedecor, G.W. & W.G. Cochran 1967. *Statistical Methods*. Ames: Iowa State Univ. Press.

Sokal, R.R. & F.J. Rohlf 1981. *Biometry*. New York: W.H. Freeman.

Stearns, S.C. 1976. Life-history tactics: a review of the ideas. *Quart. Rev. Biol.* 51: 3-47.

Steele, D.H. 1977. Correlation between egg size and developmental period. *Amer. Nat. 111: 371-372*.

Steele, D.H. & V.J. Steele 1975. Egg size and duration of embryonic development in Crustacea. *Int. Rev. gesamten Hydrobiol.* 60: 711-715.

Strathmann, R.R. 1977. Egg size, larval development, and juvenile size in benthic marine invertebrates. *Amer. Nat.* 111: 373-376.

Thorson, G. 1950. Reproductive and larval ecology of marine bottom invertebrates. *Biol. Rev. Camb. Philos. Soc.* 25: '-45.

Turner, R.L. & J.M. Lawrence 1979. Volume and composition of echinoderm eggs: implications for use of egg size in life-history models. In S.E. Stancky (ed.), *Reproductive Ecology of Marine Invertebrates* pp. 25-40. Columbia: Univ. South Carolina Press.

Wear, R.G. 1974. Incubation in British decapod Crustacea, and the effects of temperature on the rate and success of embryonic development. *J. Mar. Biol. Assc. UK* 54: 745-762.

Weinstein, M.P. & K.L. Heck Jr 1979. Icthyofauna of seagrass meadows along the Caribbean coast of Panama and in the Gulf of Mexico: composition, structure, and community ecology. *Mar. Biol. Berlin* 50: 97-107.

The reproductive ecology and fecundity of *Cancer* crabs

Jeffrey D. Shields
Department of Biological Sciences & Marine Science Institute, University of California, Santa Barbara, USA
(Present address: Department of Parasitology, University of Queensland, St. Lucia, Brisbane, Australia)

ABSTRACT

Patterns in *Cancer* reproduction varied within the general themes of life history, behavior, and body size. Patterns existed in the maturation and mating of these crabs, in the processes of extrusion and oviposition of their eggs, and in their size-fecundity relationships. Fecundity and reproductive potential were defined in terms of the number of eggs per brood and the potential number of broods over the life span of an individual. The patterns in egg production and fecundity were correlated with aspects of maturation, adult size and seasonality. Egg production appeared constrained by the number of juvenile and adult instars, crab longevity, body size and the habitat/climatic regime. These constraints may be surmounted at the expense of crab growth or adult size; changes in egg production may result from precocious maturation or from modifications in the energetics of production.

1 INTRODUCTION

Efficient management of commercial crab fisheries requires knowledge of the reproduction and life history of the exploited species. Considering the commercial importance of some *Cancer* crab stocks, it is surprising how little is known about the reproductive biology and reproductive ecology of most species. While the physiology of spermatogenesis, oögenesis and fertilization have been reported for several species of Brachyura, an analysis of fecundity and reproductive ecology of most species is lacking.

Here I review and synthesize the reproductive ecology of the genus *Cancer*. Aspects of maturation, mating and reproductive behavior, egg extrusion, oviposition, fecundity and brood mortality are presented. Factors affecting reproduction, such as seasonality and temperature, are described, and the temporal patterns of egg production are presented. In addition, the reproductive potential is estimated for each species.

2 MATURATION AND MATING

Spermatogenesis, oögenesis and fertilization are integral components of maturation and mating. These processes have received wide attention in the brachyurans (e.g. Adiyodi & Subramoniam 1983, Pochon-Masson 1983), including *Cancer* (Langreth 1969, Eurenius 1973). Spermatogenesis and oögenesis are closely tied to circadian rhythms and neurosecretory hormone cycles (e.g. Adiyodi & Subramoniam 1983, Charniaux-Cotton & Payen 1988). Crab reproduction, therefore, has a strong temporal component which may be regulated by seasonal and biochemical cues.

2.1 *Maturation*

Several studies have examined sexual maturation in male and female *Cancer* crabs. A female crab capable of mating may possess unripe gonads. Hence, to determine maturity, the following should be examined: completeness and maturation of gonads, presence of secondary sexual characteristics (including pre-pubertal and pubertal molt characters), minimum and average size at first copulation, molt increment vs. size for female crabs, and the obvious presence of sperm plugs or eggs.

Maturation, as defined by the ability to copulate, in *Cancer* occurs at the pubertal molt. Considerable variation exists in the number of juvenile instars that occur prior to maturation (Table 1). For most species of *Cancer*, 9 to 10 juvenile instars precede maturity which occurs

Table 1. Maturity and reported longevity of various female *Cancer* crabs. Unknown data are indicated by (–); incomplete data, (?); carapace width, CW.

Species	References	Number of juvenile instars	Years to maturity (from megalopa)	Minimum size at maturity (CW in mm)	Maximum number of instars	Maximum age (years)
C. antennarius	Carroll (1982)	10-12	1.5-2	73	16	7
C. anthonyi	Anderson & Ford (1976)	11-13?	2?	89	16?	–
C. borealis	Carpenter (1978)	–	2?	89	–	–
C. gracilis	Orensanz & Gallucci (1988)	9-10	2	54	12	4?
C. irroratus	Krouse (1976)	5-9	2	14	14	7-8
C. magister	Orensanz & Gallucci (1988)	9-10	2	95	13	8?
C. oregonensis	Orensanz & Gallucci (1988)	4-5	0.5	10	11	–
C. pagurus	Bennett (1974)	–	–	91	–	–
C. productus	Orensanz & Gallucci (1988)	9-10	1	70	13	4
C. setosus	Gutierrez & Zuniga (1976)	7-8	2	83	12	5

Table 2. Variation in the minimum and maximum size of ovigerous *Cancer irroratus* with latitude; sites listed from north to south. *Mature non-ovigerous crabs.

Location	References	Minimum size	Maximum size	Sample size (N)
Northumberland Strait	Scaratt & Lowe (1972)	60	99	53*
Bay of Fundy	Campbell & Eagle (1983)	41	100	50
Northern Main	Krouse (1972)	55	99	6635
Rhode Island	Reilly & Saila (1978)	14	82	31
Mid-Atlantic Bight	Haefner (1976)	21	100	130*
Chesapeake Bay	Shotton (1973)	28	106	32

approximately two years after settlement of the larval megalopa. These species attain maturity at a minimum size of about 90 mm carapace width (CW) (e.g. *C. anthonyi, C. borealis, C. magister* and *C. pagurus*). The notable exception is *C. gracilis* which attains but 54 mm CW after a similar number of juvenile instars over two years (Table 1). *Cancer oregonensis* has the fewest juvenile instars; it achieves maturity after 4 to 5 juvenile instars within approximately 6 months of settlement (Orensanz & Gallucci 1988). *Cancer productus* retains the average number of juvenile instars, but has a short maturation period of approximately one year (Orensanz & Gallucci 1988).

The minimum size at maturity is partially dependent upon the number of juvenile instars (based on the minimum number of juvenile instars, $R = 0.773$, $P < 0.05$, $N = 8$; Table 1). The number of adult instars may also be dependent upon the number of juvenile instars. However, the number of juvenile instars of *C. irroratus* may increase with latitude (Table 2). Northern populations of *C. irroratus* attain sexual maturity at a relatively large size after 13 to 14 juvenile instars (Krouse 1976), while populations off Rhode Island attain maturity after 5 to 9 juvenile instars (Reilly & Saila 1978).

Individual crabs mature at different rates and sizes. Hence, the size at which 50% of the population matures is often used to indicate the 'typical' size at which crabs are considered adults (Wenner et al. 1974, Campbell & Eagle 1983, LeFoll 1986). While this estimate may be indicative of maturity for a local population, considerable variation is reported between populations. For example, *C. irroratus*, matures at a larger adult size in the boreal waters of the Bay of Fundy and Nova Scotia (Campbell & Eagle 1983) than it does off Rhode Island and in waters around the Chesapeake Bay (Table 2) (Haefner 1976, Reilly & Saila 1978). Similarly, *C. pagurus* matures at a larger size in the northern waters of England than it does off southern Brittany (Brown & Bennett 1980, LeFoll 1986). Further, the contribution of smaller crabs to the overall production of eggs may be small (Gutierrez & Zuniga 1976), but if small mature crabs are abundant they may account for substantial egg production (Hankin et al. 1985).

2.2 *Mating*

The two most important factors in the mating of crabs are the maturity of the individuals and their stages in the molt cycle (Hartnoll 1969). In addition, behavioral aspects that affect the locating of mates and the probability of successful matings are important features of recently described mating systems.

Chemical cues are used by some *Cancer* species to attract mates. Prior to ecdysis and mating, female *C. antennarius* and *C. anthonyi* secrete a mating pheromone into the surrounding water that attracts male crabs at distances of up to 30 cm (Kittredge et al. 1971). Male crabs are both attracted to and placated by the pheromone (Kittredge et al. 1971). Upon contact with the pheromone, a male crab approaches and gently clambers over the female, positioning himself dorsally over her. Although an attraction to female crabs was noted in *C. pagurus* (Edwards 1966), a mating pheromone has not been detected in the urine or surrounding seawater of 'precopula' females (Hartnoll & Smith 1979). Work with *C. pagurus* is inconclusive; pheromones may not necessarily be released over the entire course of 'precopula.' Rather, they may be secreted over a much shorter time and have a short period of biological activity, as in *Portunus sanguinolentus* (Christofferson 1970).

The male and female crab remain in the mating embrace for some time. The precopulatory embrace or 'precopula' lasts from 3 to 21 days in *C. pagurus* (Edwards 1966), at least 8 days in *C. magister* (Snow & Nielsen 1966), and at least 13 days in *C. borealis* (Elner et al. 1985). The male and female are positioned sternum to sternum (Snow & Nielsen 1966), or sternum to carapace (Edwards 1966); or they switch between these positions (Williamson 1904, Elner et al. 1985). When ecdysis is imminent, the female signals her impending molt to the male, via increased antennal and chelar activity (Snow & Nielsen 1966). The male crab assists ecdysis of the female crab by helping to push off her exuvium (Edwards 1966, Snow & Nielsen 1966). Upon ecdysis, the male carefully positions the female sternum to sternum and copulation takes place.

Copulation was first described for *Cancer pagurus* (Williamson 1904). Additional observations have been provided by Pearson (1908), Butler (1960), Edwards (1966), Snow & Nielsen (1966), Hartnoll (1969), Shotton (1973) and Elner et al. (1985). Copulation occurs immediately or shortly after the female crab molts and lasts longer than 2 hours (Edwards 1966, Snow & Nielsen 1966, Elner et al. 1985). Successful mating only occurs when the female crab is immediately post-molt (Edwards 1966, Hartnoll 1969, Elner et al. 1985).

Spermatophores transferred at copulation break open during intercourse and release spermatozoa into the oviducts (Williamson 1904, Shotton 1973). The released spermatozoa travel up the oviducts into the internal spermathecae. Sperm plugs seal the oviducts after copulation. Sperm plugs may function to keep spermatozoa in the spermathecae, to lubricate the oviduct, and/or to stop forced mating by dominant male crabs (Williamson 1904, Edwards 1966, Hartnoll 1969, Elner et al. 1985). Sperm plugs in *C. pagurus* are thought to arise from male secretions (Hartnoll 1969), although Williamson (1904) suggests that they may originate from the glands lining the spermathecae of the female.

Some male *Cancer* crabs are polygynous. In laboratory studies, a single male *C. pagurus* mated with 10 female crabs (Edwards 1966). Large male *C. pagurus* force smaller male crabs off and away from 'precopula' females. Female crabs do not appear to select larger males and, in the absence of large males, successfully mate with smaller males (Edwards 1966). Males of *C. magister* also mate with more than one female in the laboratory (Butler 1960). The presence of mating marks on the carapace of trapped male crabs is used to indicate their mating history (Cleaver 1949, Butler 1960).

After copulation, the male and female crab remain in a postcopulatory embrace. The embrace lasts from 1 to 12 days or until the new carapace of the female is sufficiently hardened for her protection (Edwards 1966). Males removed from females during the postcopulatory embrace return to embrace the females (Edwards 1966, Elner et al. 1985). The use of chemical cues has not been investigated during this period. After mating and the postcopuolatory embrace, the males of many species molt (Edwards 1966, Hartnoll 1969). In extreme cases, the male molts several months after the female (Haefner & Van Engel 1975).

2.2.1 *Mating seasonality*

Molting and mating of mature *Cancer* crabs exhibit distinct seasonal patterns (Fig. 1). The majority of the species molt and mate from late winter (February-March) to mid autumn (October-November), with a mating peak in summer (Fig. 1). Little if any mating occurs in late autumn and early winter. Many females approach oviposition or were ovigerous at that time (Fig. 2). Exceptions to the seasonal pattern are those crabs that breed year-round (*C. antennarius* and *C. anthonyi*), those with a relatively long and irregular brooding season (*C. irroratus* and *C. productus*), and those found in the southern hemisphere, where the austral summer is from December to February (*C. setosus*).

Variations within a species in the seasonal patterns of molting and mating are numerous and occur between distinct stocks or latitudes (see Conan 1985). For example, *C. pagurus* molts during summer off northwest England, during autumn off Scotland (Broekhuysen 1936), year round off southwest England (Bennett 1974), and during spring and autumn off Brittany (Drach 1949). Similarly, *C. irroratus* molts during summer and autumn in the northern portion of its range, and during winter and spring in the southern portion of its range (Scaratt & Lowe 1972, Haefner & Van Engel 1975, Krouse 1980). These patterns may be related to seasonal changes in temperature, photoperiod or primary productivity (Conan 1985, see also 4.2).

2.2.2 *Mating systems*

Distinct mating systems have recently been proposed for three species of *Cancer* crabs (Oren-

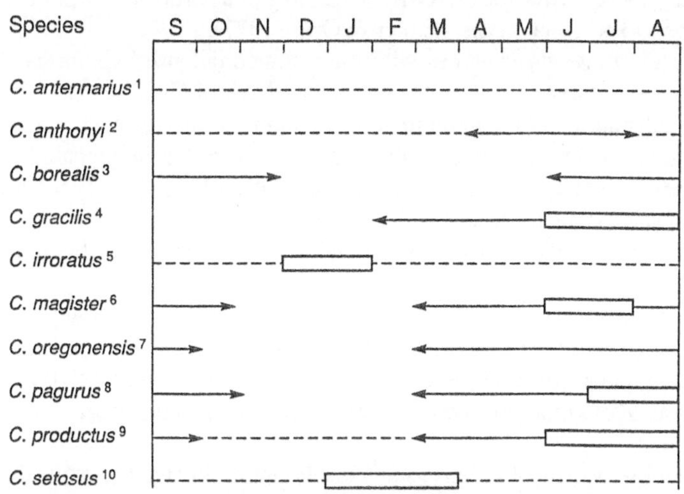

Figure 1. Molting and mating seasons for females of various species of *Cancer* crabs; (box) peak (5% of female population) in the number of molting females, (solid line) moderate levels (5%) of molting and mating, (dashed line) low levels (1%) of molting and mating. Note in some cases the proportion of molting or mating females in a population has been determined subjectively on the basis of descriptions in references. References: [1] Carroll (1982), Carroll & Winn (1988); [2] Winn (1985), Carroll & Winn (1987); [3] Haefner (1977), Carpenter (1978); [4, 7, 9] Knudsen (1964); [5] Krouse (1972, 1976), Terretta (1973), Reilly & Saila (1978); [6] MacKay (1942), Cleaver (1949), Butler (1961); [8] Bennett (1974), Brown & Bennett (1980); [10] Gutierrez & Zuniga (1976).

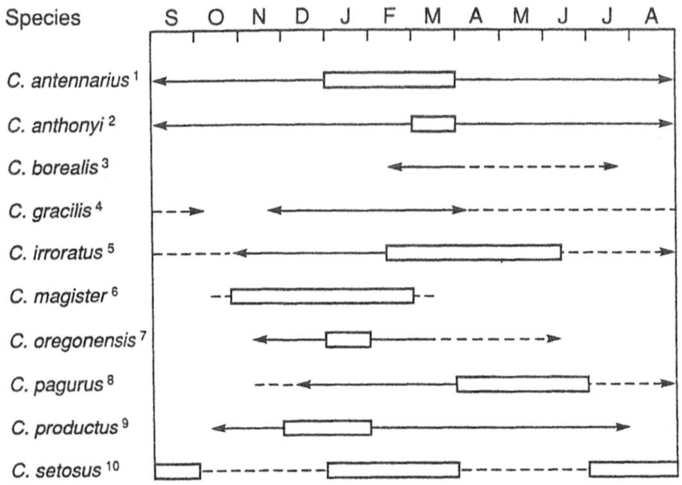

Figure 2. Ovigerous or brooding seasons of various species of *Cancer* crabs; (box) peak in the number of ovigerous females, (solid line) many (5%) ovigerous females in population, (dashed line) few (1%) ovigerous females in population. Note in some cases the proportion of ovigerous females in a population has been determined subjectively on the basis of descriptions in references. See Table 1 for references unless noted: [2] Reilly (1987); [4,6,7,9] Orensanz & Gallucci (1988).

sanz & Gallucci 1988). These mating systems are based on the polygynous behavior of male crabs, the behavioral dynamics of the species and the structure of the populations. Directed by sexual selection, these mating systems may be driven by 'the ability of a portion of the population to control access of others to potential mates' (Emlen & Oring 1977).

The mating behavior of *C. oregonensis* agrees with the resource defense hypothesis of polygyny (Emlen & Oring 1977, Orensanz & Gallucci 1988). Males of *C. oregonensis* indirectly control access to small 'harems' of female crabs by dominating critical resources, in this case, small refuges. The ratio of males to females in the population is low, as large males are outnumbered by their 'harem' of smaller females (Orensanz & Gallucci 1988). In effect, the male crabs defend or dominate separate refuges to which females are attracted.

The structure of the mating system of *C. gracilis* corresponds to the female defense hypothesis of polygyny (Emlen & Oring 1977, Orensanz & Gallucci 1988). The males of *C. gracilis* directly control access to females by contests with other males and by an increased activity over that of females. Females of *C. gracilis* form large aggregates, and males actively staying within the aggregate increase their chance for multiple copulations (Orensanz & Gallucci 1988). Aggregating females are assured of a mate and increase their likelihood of mating with a highly competitive, dominant male.

The structure of the mating system of *C. magister* closely approximates the explosive breeding assemblage of the male dominance hypothesis of polygyny (Emlen & Oring 1977, Orensanz & Gallucci 1988). Males and females of *C. magister* converge during a short but highly synchronized mating period (Butler 1961, Diamond & Hankin 1985). A male-dominated sex ratio ensures that females find mates, but the highly synchronous mating period drives the sex ratio of mating crabs towards unity. Consequently, searching behavior to find a mate increases in importance. Sequestering of females or the dominating of subordinates appears less important during this synchronized mating period (Orensanz & Gallucci 1988).

3 EXTRUSION AND OVIPOSITION

3.1 *Extrusion*

Fertilization is internal in *Cancer* and presumably occurs in the oviduct. Contractions of the surrounding muscles squeeze ova out of the ovary, through the oviduct and out onto the pleopods (Binford 1913). Spermathecal secretions lubricate the oviduct for the passage of eggs during extrusion (Anilkumar & Adiyodi 1977). The eggs then become attached to setae on the pleopods.

Oviposition does not occur immediately after copulation. In *Cancer antennarius*, oviposition occurs approximately 11 weeks after mating (Carroll 1982). Extrusion in *C. magister* occurs 3 months after mating; and in *C. pagurus*, extrusion occurs more than 4 months after mating (Figs. 1 and 2). Viable sperm remain in the spermathecae for over six months (Shields pers. observ.) and indirect evidence suggests that sperm may remain viable for over one to two years (Pearson 1908, Knudsen 1964, Ebert et al. 1983, Hankin et al. 1985, Orensanz & Gallucci 1988).

3.2 *Egg attachment*

The process of egg attachment to the setae of the pleopods among the Decapoda has been much debated. In shrimps and lobsters, an attachment cement is secreted at oviposition by glands in the pleopods (see Fisher & Clark 1983). Secretions from the pleopod have not been observed in crabs (Cheung 1966). Glandular openings, however, have been observed on the exopods of the pleopods of *C. anthonyi* (Okazaki 1988). Further work may elucidate the role of the exopods in egg attachment.

The outer egg coat of the crab egg has been shown to effect the attachment of eggs to the maternal pleopods (Goudeau & Lachaise 1980, Goudeau & Lachaise 1983). The attachment of eggs occurs when the outer coat contacts the seta of an endopod; the external face of the outermost vitelline membrane sticks to the seta, elongates and hardens to form a funiculus, stalk, or connectant (Goudeau & Lachaise 1983). Curiously, eggs do not adhere readily to each other, rather, the strong adhesion of eggs occurs only to the setae of the endopod of the pleopod. Perhaps an epoxy-type 'resin' is secreted onto the setae prior to or during extrusion. The 'resin' could interact with its counterpart on the surface of the chorion to form a strong bond. Alternatively, aeration by the maternal pleopods, which ensures that eggs come into contact with the setae, could rupture the chorion, and the perivitelline cement first observed by Williamson (1904) then effects the attachment of eggs. In any case, the attachment of eggs to the setae remains unclear for *Cancer* crabs as well as for most brachyurans.

3.3 *Oviposition*

3.3.1 *Ovipositional behavior*

A female *Cancer* crab precedes oviposition by selecting a suitable substrate for extrusion. She typically burrows into sand or a 'soft' bottom substrate. There, she extrudes eggs into the incubatory chamber formed by her abdomen and the surrounding substrate. A soft substrate was required for the proper retention of eggs of *C. anthonyi, C. irroratus, C. magister,* and *C. pagurus* (Krouse 1980, Howard 1982, Ebert et al. 1983, Shields pers. obs.).

Changes in behavior occur during oviposition and brooding. Ovipositing crabs are quiescent during and immediately after extrusion. Female *C. pagurus* remain quiescent throughout the ovigerous period; they do not actively enter traps (Brown & Bennet 1980) nor do they feed (Howard 1982). However, ovigerous females of *C. antennarius* and *C. anthonyi* are not quiescent and actively enter commercial traps (Shields pers. obs.). Further, ovigerous females of *C. magister* enter traps in their early breeding period and later become less abundant in traps (Hankin et al. 1985).

3.3.2 *Reproductive migrations*

The patterns of migration in relation to reproduction for *Cancer* crabs were recently reviewed (Orensanz & Gallucci 1988). The females of *C. magister, C. pagurus* and *C. productus* migrate several kilometers to offshore breeding grounds. A distance of 230 miles was recorded for one female *C. pagurus* (Edwards 1979). In addition, the early zoeal larvae of these species may often be found several kilometers offshore (Nichols et al. 1982). *Cancer setosus* migrates shorter distances to deeper waters (Gutierrez & Zuniga 1976).

Migrations are not necessarily to offshore locations. The females of *C. borealis, C. irroratus* and *C. pagurus* appear to migrate inshore and offshore; their migrations appear related to habitat (i.e. substrate type for suitable oviposition) (Edwards 1979, Krouse 1980), or thermal preferences (Jeffries 1966, Haefner & Van Engel 1975). Some crabs do not migrate and remain in the same general area, for example, *C. antennarius* and *C. anthonyi* (Carroll 1982, Winn 1985). Further, the migratory patterns of male crabs may differ from conspecific females (Orensanz & Gallucci 1988).

4 EMBRYOGENESIS

4.1 *Brooding time and temperature*

The effect of temperature on crustacean egg development has received much attention (e.g. Perkins 1972, Wear 1974). Low temperatures decrease and high temperatures increase the rate of embryogenesis. Wide variations in development times are notable between species of *Cancer* and these variations may be, in part, the result of climatic conditions. In general, those species inhabiting cold waters (< 10°C) have much longer embryogenic periods than did those inhabiting warmer waters (> 10°C) (Table 3).

Four studies have elucidated the effect of temperature on the embryos of *Cancer* crabs. Mayer (1973) and Wild (1983b) found that temperatures above 13°C hasten embryogenesis *in vitro* and cause significant mortality to the embryos of *C. magister* (see 5.4.1). Lower temperatures increase egg survival and decrease the rate of embryogenesis. Shirley et al. (1987) found that temperatures at or above 10°C hasten embryogenesis *in vivo* but do not cause significant mortality to the embryos of *C. magister*. Embryogeneis of *C. magister* took 160 d at 5°C, 60 d at 10°C, and 42 d at 15°C. Development ceases at 1°C (Shirley et al. 1987). The embryos of *C. anthonyi*, a warmer water species, follow a similar pattern, but at higher temperatures (Shields & Kuris 1988). Embryogenesis of *C. anthonyi* takes approximately 55 d at 10°C, 42 d at 15°C, and 27 d at 20°C. Embryos exposed to 4°C for 10 days cease development, but resume normal development upon being returned to 15°C (Shields & Kuris 1988).

Diapause, wherein embryogenesis slows or stops, was observed in two species of *Cancer* (Table 3). The embryos of *C. pagurus* undergo a uniform period of diapause that remains

Table 3. Egg development time, number of broods per year, range in size of ovigerous females, and reported range in fecundity per brood in various *Cancer* crabs. Unknown data are indicated (?), and embryonic diapause. D: Ambient seawater temperature during the brood period is for the location/area where the study occurred (Knudsen 1964, Rathbun 1930, Carroll 1982, Wild et al. 1983, Le Foll 1986, Shields pers. observ.).

Species	References	Embryogenic development (in days)	Female size (mm CW)	Reported range in fecundity per brood	Range in temperature while brooding
C. antennarius	Carroll (1982), Shields et al. (unpubl. data)	50- 60	116-143	$\approx 1.0 \times 10^6$	10-16
C. anthonyi	Shields et al. (submitted)	40- 45	90-153	$0.7\text{-}3.3 \times 10^6$	10-20
C. borealis	Haefner (1977), Carpenter (1978)	?	105-139	$0.3\text{-}1.6 \times 10^6$	8-14
C. gracilis	Knudsen (1964), Orensanz & Gallucci (1988)	?	54-100	?	7-11
C. irroratus	Krouse (1976), Haefner (1976), Reilly & Saila (1978), Campbell & Eagle (1983)	176-235 D	14-106	$4.4\text{-}567.7 \times 10^3$	8-21
C. magister	Orensanz & Gallucci (1988), Wild (1983a, b, c), Hankin et al. (1985)	90-120	110-171	$0.5\text{-}1.5 \times 10^6$	9-14
C. oregonensis	Knudsen (1964), Orensanz & Gallucci (1988)	103-118	10- 43	$1.0\text{-}3.3 \times 10^4$	7-11
G. pagurus	Edwards (1979), Brown & Bennett (1980), Le Foll (1986)	235-265 D	115-205	$0.5\text{-}3.0 \times 10^6$	7-18
C. productus	Toole (1985)	≈ 120	70-129	?	7-14
C. setosus	Gutierrez & Zuniga (1976)	?	83-151	$0.6\text{-}1.7 \times 10^6$	–

unaffected by changes in water temperature (Wear 1974). However, the embryos of *C. irroratus* have a variable period of diapause that is affected by changes in water temperature (Haefner 1976, Reilly & Saila 1978). The variable timing of diapause in the embryos of *C. irroratus* may synchronize hatching of larvae over a one to two month summer season (see 4.2).

The effect of temperature on embryogenesis was noted in field studies of *C. irroratus* and *C. magister*. Embryos of *C. irroratus* oviposited early in the reproductive season (late fall-early winter) hatch at a similar time as those oviposited later in the season when water temperature is increasing (Haefner 1976). Embryos of C. magister hatch after 90 days in central California (Point Lobos-Point Arena) vs 120 days in the colder climes of Washington and further north (MacKay 1942, Butler 1960, Wild 1983b).

4.2 *Eclosion*

Hatching in species of *Cancer* occurs when the embryo reaches the prezoea stage (Pearson 1908, Wear 1974). Prezoeae of *C. anthonyi* appear to imbibe water and enlarge in size before hatching (Shields pers. observ., cf. Davis 1965). Hatching of *Cancer anthonyi* prezoeae typically occurs at night or in relative darkness.

During the process of eclosion, female *Cancer anthonyi* actively aid the hatching process (Shields et al. in press). Females stand upon their walking legs and aerate the clutch vigorously by agitating the pleopods. In addition, the branchiae appear to direct water currents anteriorly through the brood, facilitating zoeal eclosion and escape from the egg mass. After eclosion, female crabs either immediately strip the hatched remnants and detritus from their pleopods (*C. antennarius*, *C. anthonyi*, Kuris & Wickham 1987; Shields pers. obs.; *C. pagurus*, Williamson 1904) or lose the adhering detritus at their next molt (*C. magister*, Kuris & Wickham 1987). Since female *C. magister* may oviposit a second brood prior to molting (Ebert et al. 1983), they may strip their pleopods of detritus before extrusion.

Hatching of *Cancer* larvae may be synchronized with seasonal primary productivity. Hatching typically occurs in the spring for *C. anthonyi* and *C. antennarius*. Along the southern California coast primary production is at its highest at this time (Mullin 1986). The larvae of *C. borealis*, *C. irroratus* and *C. pagurus* also hatch when nutrient levels are high, i.e. in the summer months (from comparison with Harvey 1955, Dunbar 1979).

5 FECUNDITY

5.1 *Brood size*

Brood size of *Cancer* crabs shows a conservative relationship with body size between species (Table 3). For example, the brood size reported for the smallest *C. magister* (450 000 eggs at 110 mm CW) is equivalent to that of the largest ovigerous *C. irroratus* (440 000 eggs at 100 mm CW). In addition, temperature may indirectly affect brood size by directly affecting the size at which crabs mature (e.g. *C. irroratus*, *C. pagurus*; see 2.1).

In general, the smaller *Cancer* species (*C. irroratus* and *C. oregonensis*) produce relatively fewer eggs at a given size than do the larger species. There is, however, one notable exception. For its size, *C. anthonyi* produces relatively and absolutely more eggs per brood than any other *Cancer* crab, including both of the larger species, *C. magister* and *C. pagurus* (Table 3). Given the strong size-fecundity relationship found in brachyuran families (Hines 1982, Hartnoll 1985), it is surprising that *C. magister* and *C. pagurus* carry fewer eggs than the smaller *C. anthonyi*. Differences in water temperature cannot explain the differences in the fecundities between these species, because *C. antennarius* and *C. productus* – two species that are sympatric with *C. anthonyi* (Nations 1975) – do not have an increased brood size in relation to their species size (Table 3).

5.2 *Size-fecundity relationships*

Reproductive output per brood in brachyurans is strongly correlated with body size (carapace width) and weight within a species (Hines 1982, Hartnoll 1985). The body size–clutch size (fecundity) relationship was investigated in six species of *Cancer*. Fecundity was linearly or logarithmically correlated with body size, or not correlated with body size (Table 4).

Table 4. Size-fecundity relationships for different species of *Cancer*. Key to symbols: [a]Fecundity estimated from total number of eggs per brood. [b]Fecundity estimated from total number of eggs per second pleopod. [c]Carapace length used to calculate values. [d]Relationship calculated from mean carapace length and mean fecundity. NR: Not reported. [e]Fecundity estimated from internal ova and external eggs. $^{+}0.05 < P < 0.10$, $^{*}P < 0.05$, $^{**}P < 0.01$.

Species	References	Slope	Intercept	r	N
Linear					
C. anthonyi	Shields et al. (in press), Shields (unpubl. data)	5551[b]	-371040	0.559*	220
Logarithmic					
C. anthonyi	Shields et al. (in press), Shields (unpubl. data)	2.308[b]	0.646	0.554**	220
C. irroratus	Reilly & Saila (1978)	3.010[a]	-0.335	0.986**	33
C. irroratus	Campbell & Eagle (1983)	2.553[a]	0.583	0.857**	50
C. magister	Hankin et al. (1985)	NR	NR	≈0.245*	85
C. pagurus	Le Foll (1986)	4.174[ac]	-2.417	0.720**	29
C. pagurus	Edwards (1979)	3.633[a]	-1.854	0.708**	16
No significant logarithmic relationship					
C. anthonyi	Shields et al. (in press), Shields (unpubl. data)	1.020[a]	4.267	0.449	12
C. borealis	Shields et al. (in press), Shields (unpubl. data)	0.977[ae]	3.869	0.273	14
C. magister	Hankin et al. (1985)	NR	NR	NR	35
C. setosus	Gutierrez & Zuniga (1976)	1.449[ad]	3.101	0.719+	6

Size-fecundity data for *Cancer* are highly variable. Thus, logarithmic tranformations of fecundity (log Fecundity = b log CW ± log a) are used to reduce heteroscedasticity, stabilize the variance, and provide a better fit to regression models. The regression coefficients (slopes) of those relationships that fit logarithmic regression models vary from 2.308 to 4.174 (Table 4).

The lack of relationship between body size and fecundity for *C. borealis*, *C. magister*, and *C. setosus* may have several causes (Table 4). (1) Large female crabs (155 mm CW) may be senescent and produce fewer eggs than smaller, more virile females (Gutierrez & Zuniga 1976, Hankin et al. 1985, Orensanz & Gallucci 1988). (2) Large female crabs do not molt annually (Pearson 1908, Ebert et al. 1983, Hankin et al. 1985), hence they may possess less stored sperm, and thus fertilize fewer eggs than smaller crabs (Pearson 1908). (3) The narrow range in size of adult females, coupled with high variation in fecundity, may reduce the ability to detect a significant correlation (see below).

The senescence hypothesis states that large adult female crabs are senescent because of their decreased molting probability, the decline in their number of eggs per brood, and their low annual survival rates. The data supporting the senescence hypothesis have potentially confounding variables. First, the proposed mortality rate (hence, senescence) of older females of *C. magister* is based on the collection of crabs in shallow, inshore waters during spring (Diamond & Hankin 1985). At that time, younger, smaller females migrate inshore, molt and mate. Since older females did not molt annually (Hankin et al. 1985), their molting and mating pattern differs from that of younger crabs. Such older females may remain offshore. Hence, their mortality rate, based on inshore trapping in the spring (Hankin et al. 1985), may be exaggerated. Second, adults of *C. gracilis* and *C. productus* which are thought to have died of senescence (Orensanz & Gallucci 1988) may have died from disease, changes in temperature, or from the severity of winter storms (El Niño in 1983). The range in sizes of these dying or dead crabs and the fact that most crab mortality occurred in a single year (1983) argues against senescence as a physiological process in these crabs. Further, with lab-held *C. anthonyi*, second or third broods in an instar are significantly smaller than the first brood in an instar even if the crab were to molt again (Shields et al. in press). Factors producing such effects may include reduction in the quality of sperm stored from copulation at the previous molt. Hence, such broods cannot be used as evidence for senescent processes. Third, egg predation in broods of *C. magister* by a nemertean, *Carcinonemertes errans*, was sufficiently intense at that time (1983) to reduce the fecundity of the large, late instar females, further reducing the number of eggs present in their broods (Kuris & Wickham 1987) (see also 5.4.2).

Sampling procedures may also account for the lack of a statistically significant relationship between body size and fecundity in some of these species. For *C. borealis*, the lack of relationship between size and fecundity may be due to the narrow range of the female sizes examined, the variation inherent in fecundity, and the small sample size (size range = 100.5-119.7, mean fecundity = 763 793 ± 132112 eggs per brood, N = 14; from Carpenter 1978). For *C. anthonyi*, log crab size (CW) was positively correlated with log fecundity per pleopod ($R = 0.554$) when the sample size was large. Yet, total fecundity from 12 crabs revealed no significant relationship with body size ($R = 0.449$; Shields et al. in press). The partial correlation of log crab size with log total fecundity was, however, significant when log fecundity per pleopod was held constant (Partial correlation – $R = 0.635$, $P < 0.05$, $N = 12$; Shields et al. in press).

5.3 *Broods per instar and seasonality*

As with fecundity, the number of broods per instar may be size-dependent in *Cancer* crabs. Recent adult instars (i.e. A1, A2) appear to molt annually (*C. magister*) or semi-annually (*C. oregonensis*) (Hankin et al. 1985, Orensanz & Gallucci 1988). These young adults generally produce one brood per instar which corresponds to one brood per reproductive season. Older adult crabs (e.g. *C. magister* 155 mm CW; *C. pagurus* 170 mm CW) which do not molt annually (Pearson 1908, Hankin et al. 1985) produce more than one brood per instar (Table 5) (Hankin et al. 1985, Orensanz & Gallucci 1988), and infrequently produce more than one brood per reproductive season (Knudsen 1964, Orensanz & Gallucci 1988).

Generally, one brood per instar typifies annual reproduction for most species of *Cancer* (Table 5). However, two species, *C. antennarius* and *C. anthonyi*, regularly oviposit two or more broods per instar (Shields et al. in press, Shields pers. observ.). Since both species have short embryogenic periods (Table 3), it is apparent that they produce more than one brood per reproductive season. The mating and reproductive seasons of *C. setosus* suggest that it, too, may regularly produce more than one brood per instar (Gutierrez & Zuniga 1976).

The breeding season of most species of *Cancer* appears to be winter (Fig. 2). Of the ten species examined, all bore eggs between January and March. For at least seven of these species,

Table 5. Broods per instar (potential number of broods per instar in late instars), maximum number of adult instars, mean fecundity per brood (in millions of eggs, mean size in parentheses), and the reproductive potential of individual female *Cancer* crabs. Unknown data is indicated by (–), incomplete data by (?). Note that the potential number of broods per instar is not equivalent to the number of broods per reproductive season which in most cases is one brood per year. Reproductive potential calculated by: Estimated number of adult instars × mean fecundity at mean size × number of broods per instar (when capable of producing more than one brood per year).

Species	Broods per instar	Maximum number of adult instars	Mean fecundity at mean size	Individual reproductive potential (millions of eggs)
C. antennarius	1-2	5	1.0 (120)	5.0-10.0
C. anthonyi	3-4	2-4	2.6 (130)	15.6-41.6
C. borealis	1 (?)	–	0.8 (120)	–
C. gracilis	1 (2)	2	– (65)	–
C. irroratus	1	1-6?	0.3 (83)	0.3-1.8
C. magister	1 (2)	3-4	1.0 (131)	3.0-4.0
C. oregonensis	1 (2)	6	0.02 (21)	0.1-0.2
C. pagurus	1 (3)	3-4	1.2 (153)	3.6-4.8
C. productus	1 (2)	3	– (107)	–
C. setosus	1 (?)	2-3?	1.0 (105)	2.0-3.0

the period between January and March represents a peak in their reproductive season. The females of three species (*Cancer anthonyi*, *C. antennarius*, and *C. setosus*) oviposit year round, with peak periods in the winter. *Cancer setosus* had two peaks in reproduction (Gutierrez & Zuniga 1976). One peak occurs in January through March, which corresponds with the peak breeding season of crabs in the northern hemisphere; the other occurs in late June through September, the austral winter season. Ovigerous *C. irroratus*, *C. pagurus* and *C. productus* are found throughout much of the year but there is a distinctive peak in reproduction in the winter or spring.

Most species of *Cancer* do not have a well defined reproductive season (Fig. 2). In contrast, *C. magister* has a synchronized reproductive season. Virtually all of the females present in a population of *C. magister* mate and lay eggs within a month or two of each other (Wild 1983a, b, Orensanz & Gallucci 1988). Northern populations (e.g. Washington) of *C. magister* bred later in the season compared to southern populations (e.g. central California) and in northern stocks synchrony is reduced (MacKay 1942). Oviposition in *C. magister* occurs in October-November in central and northern California, but it occurs progressively later at more northern latitudes (MacKay 1942, Cleaver 1949, Butler 1961, Wild 1983a, b). Similar observations were made with *C. irroratus* in relation to water temperature (Terretta 1973).

5.4 *Brood mortality*

Brood mortality may affect fecundity and natality of crabs. Physical, chemical and biological factors may cause brood mortality in *Cancer*. *Cancer* crabs brood their eggs on pleopods in an open clutch; hence, the distal protions of the clutch may be exposed to a variety of potentially harmful agents.

5.4.1 *Physical factors*

Mechanical damage may be incurred on the periphery of the clutch of *C. anthonyi* (Shields

1987, Shields et al. in press). As ovigerous crabs forage, the outermost eggs of their clutches contact the substrate. This may weaken or abrade the egg coats. Egg mortality due to mechanical damage is estimated to be 4.0-6.0% of the embryos of the margins of the clutch (Shields 1987, Shields et al. in press). Negligible mortality due to abrasion occurs in the medial portion of the clutch.

Water temperature may also contribute to egg mortality and under extreme conditions may lead to massive mortality *in vitro* and *in vivo*. Wild (1983b) found that the embryos of *C. magister* experienced higher mortality at 17°C than did those held at lower temperatures. Hatching success was inversely correlated with temperature. Curiously, mortality was notable at the ambient temperature (13°C). Mayer (1973) observed that embryos of *C. magister* experienced massive mortalities at 15° and 20°C after only 3-6 days. Shirley et al. (1987) found negligible mortality *in vivo* at temperatures between 5° and 15°C, but substantial mortality at 1°C. *Cancer anthonyi* had a wider range of thermal tolerance; embryos survived for over 10 days at 4° and 20°C (Shields & Kuris 1988). *Cancer* crabs appear to have a thermal limit in their distribution of 4° to 23°C (MacKay 1943). Since crab embryos are sensitive to warm temperatures, the thermal limit could directly affect egg and larval survival in the transition zones.

5.4.2 *Chemical factors*

Pollutants have been cited as causal agents in egg mortality and in the decline of certain crab stocks (Wild 1983c). The role of pollution in the collapse and non-recovery of the *Cancer magister* fishery off central California has received little attention. An increase in pollutants (chlorine) temporally matched the decline in the fishery (Horne et al. 1983). Several pesticides affected the ability of embryos to hatch *in vitro* (Buchanan et al. 1970, Caldwell 1977). Concentrations of pesticides required to kill embryos were higher than those killing larvae.

Other organic pollutants affect *Cancer* egg production and fecundity. Ovarian necrosis in *C. magister* results from an increase in organic solutes leached from timber at log transfer facilities (O'Clair & Freese 1988). Ovigerous crabs found in contact with the bark deposits have fewer eggs, reduced survivorship, and inhibited feeding responses when compared with crabs from sites adjacent to log transfer facilities and crabs from control sites. Similarly, crabs exposed to oiled sediments produce fewer eggs and less hardy larvae than do unexposed crabs (Karinen et al. 1985).

The effects of several metals in embryo survivorship, hatching success, and prezoeal survivorship have been documented for *Cancer anthonyi* (Macdonald et al. 1988). Metal toxicity levels for *C. anthonyi* embryos are similar to those causing toxicities in zoeae of *C. magister* (Glickstein 1978, Martin et al. 1981). Various metals, particularly iron and copper, delay hatching of *C. anthonyi* embryos at concentrations 100-times lower than those causing mortalities. Further, manganese and iron are selectively adsorbed to the vitelline membrane of embryos of *C. pagurus* following extrusion (Martin 1976). Apparently, embryo mortality is delayed due to the added protection of the vitelline membrane, which reduces the rate of metal absorption. Since crab embryos imbibe water at or near hatching (Davis 1965), hatching success may have declined due to an increased exposure to the various metals (Macdonald et al. 1988).

5.4.3 *Biological factors*

Several biological agents cause egg mortality in *Cancer* crabs. Bacteria, fungi, nemertean worms, and amphipods have all been observed parasitizing or eating crab eggs and embryos. Bacteria may cause substantial mortality in the broods of *C. magister* (Fisher 1976, Fisher & Wickham 1976). In controlled experiments, bacterial filaments and egg mortality increase in the presence of nutrients and in the absence of antibiotics (Fisher 1976). These results, however, may have been confounded by the presence of the egg predator *Carcinonemertes errans* (Wickham 1979b). In an *in vitro* analysis, bacteria do not contribute significantly to egg mortality of *Cancer anthonyi* (Shields & Kuris 1988). In the same analysis, the fungus *Lagenidium callinectes* leads to significant mortality only at high temperatures and high zoospore densities. The nemertean worm, *Carcinonemertes epialti*, contributes most to the egg mortality of *Cancer anthonyi* (Shields & Kuris 1988, Shields et al. in press).

Nemertean worms have been found at high densities on several different species in several crab populations (Wickham 1986, Kuris & Wickham 1987, Shields & Kuris 1987). Brood mortality is high in these crab populations and reaches levels where virtually all of the eggs in a local population are eaten (Wickham 1986, Shields & Kuris 1987). These worms have been implicated in the non-recovery of the central California stock of *Cancer magister* (Wickham 1979a, 1986, Kuris & Wickham 1987). Their role in the collapse and non-recovery of this crab stock remains in contention (see Kuris & Wickham 1987).

The impact of brood mortality on certain populations of *C. magister* has led to the suggestion that the causal factors may contribute, in part, to the cycling of crab populations (Wickham 1979a; but see Hankin 1985, Jamieson 1986). While the roles of these factors and agents are difficult to determine in nature, their study remains important to our understanding of the biology and environmental sensitivity of these commerically important crabs.

6 REPRODUCTIVE POTENTIAL AND REPRODUCTIVE PATTERNS

6.1 *Reproductive potential*

Reproductive potential has been previously quantified in terms of fecundity, size at maturity, fishing mortality, proportion of females in each size class, and growth of individuals in a population (Campbell & Robinson 1983). Populations chosen for examination in this manner typically reproduce but once per year. A developing model may provide a comprehensive analysis of the impact of these variables on reproductive potential (Hartnoll & Gould 1988). Information on size at maturity, growth, proportions in size classes, and fishing mortality are lacking for most of the species of *Cancer* examined.

Here, I quantify reproductive potential in terms of the individual crab. Reproductive potential is the total number of eggs produced throughout the *maximum* adult life span (lifetime reproductive success) of an individual of the species (Shields et al. in press). The maximum adult life span was chosen for four reasons: late instar females may contribute most to the overall egg production of their cohort; hence, they may contribute most to the fitness of the species; estimates of mortality rates are unknown or unreported for most species; and confounding correlations between fecundity and mortality rate are eliminated. Estimates of the reproductive potential of an individual crab are based on (1) the mean fecundity at the mean size, (2) the minimum and maximum number of mature instars, (3) the minimum and maximum number of broods per instar, and (4) in relation to (3), the number of broods oviposited per year (see below and Table 5). This estimate represents a crude but effective means of examining the influence of life history constraints on the reproductive potential of *Cancer* crabs.

For example, the reproductive potential, R_p, of *Cancer magister* is derived by:

$$R_p = Fec_{A1} + Fec_{A2} + Fec_{A3''} + \mathrm{Fec}_{A3''}$$

Table 6. Patterns in the reproduction of *Cancer* crabs. See text for further explanation.

Pattern	Mature size	Number of juvenile instars	Number of adult instars	Broods per instar	Relative reproductive potential	Examples
General	Small-large	9-11	2-4	1 (2)	Moderate	*C. pagurus* *C. gracilis* *C. magister*
Precocious maturation	Small	4-5	4-6	1 (2)	Low	*C. irroratus* *C. oregonensis*
Maximal production	Moderate	9-11	3-5	2+	High	*C. antennarius* *C. anthonyi*

where, R_p = mean number of eggs per brood at mean size × number of broods per instar (per year) × maximum number of adult instars. Thus, for *C. magister:* $R_p = 1.0$ million $\times 4 = 4.0$ million eggs per crab, or where Fec_{A1} is the fecundity of the first adult instar; and $Fec_{A3'}$ and $Fec_{A3''}$ represent the fecundity of the first and second brood of the third adult instar. This estimate is similar to that proposed by MacKay (1942) – 3 to 5 million eggs per crab – and may represent an upper limit.

The reproductive potential of *Cancer anthonyi* is greater than that of any other known *Cancer* species (Table 5). The ability of *C. anthonyi* to bear multiple broods per instar, coupled with its high size-specific fecundity yields this enormous reproductive potential. The effect of longevity or the maximum number of adult instars is observed with *C. irroratus,* which has up to 6 adult instars (varying in size from 28-100 mm CW). A large number of adult instars, coupled with the positive size-fecundity relationship, produces for *C. irroratus* a reproductive potential approaching that of *C. magister*, or in the case of *C. antennarius*, exceeding that of *C. magister*.

6.2 *Patterns in reproduction*

Cancer crabs appear to have three distinct patterns in egg production (Table 6). The general pattern of egg production was typified by those species having 9 to 11 juvenile instars and 2 to 4 adult instars (Table 6). There is usually one brood per instar and per season. These species mature at either small or large sizes. Egg production may be scaled with size such that the size-fecundity relationship is relatively similar between species. Production of a single brood per instar by such species may optimize the use of available food resources.

Abbreviating parts of the life history may lead to a precocious maturation pattern of egg production. Such species have small adult sizes, maximize their reproductive effort by maturing after a few juvenile instars, and increase the number of adult instars (Table 6). The increased number of adult instars, coupled with the potential of producing more than one brood per season (Knudsen 1964, Orensanz & Gallucci 1988) renders a high reproductive potential over the life span of the species. Species characterized by the precocious maturation pattern have a reproductive potential that approaches that of those species with a general pattern of egg production. An advantage of precocious maturation may be an increase in fitness relatively early in the life history of the species.

The maximal production pattern derives from a modification in the energetics of egg production. This pattern conserves the life history constraints of the general pattern (i.e. number of juvenile and adult instars – Table 5), but production energetics are substantially increased (see below). Species with the maximal production pattern produce more than one brood per instar, have a relatively short embryogenic period, and have extremely high fecundity. Egg production appears to be optimized at the expense of growth as multiple broods within an instar effectively preclude intervening molts. Species having this pattern are of moderate size (90-150 mm CW). Interestingly, species with a maximal egg production are not quiescent when ovigerous. They forage, acquire nutrient resources, and undergo ovarian growth for further egg production while bearing eggs.

7 CONCLUSIONS

Species of *Cancer* crab show variations in their reproductive ecology at intraspecific and interspecific levels. Interspecific variations may be controlled by many factors, including climatic regime, habitat, and biological constraints. The variations in reproductive ecology tend to follow latitudinal or temperature gradients. Temperature influences maturation, molting and mating, synchronized oviposition (*C. magister*), and the duration of embryogenesis (brooding). Lastly, features of the embryonic and larval stages of these crabs lend themselves to studies of environmental sensitivity. An increase in pollution in the habitats used by *Cancer* warrants the use of life history stages as bioassays to monitor the healthiness of crab stocks.

ACKNOWLEDGEMENTS

I thank Julie Shields for her unswerving support of this project. Drs. Armand Kuris and Adrian Wenner receive special thanks for providing me with the opportunity to contribute to this series.

REFERENCES

Adiyodi, R.G. & T. Subramoniam 1983. Arthropoda – Crustacea. In K.G. Adiyodi & R.G. Adiyodi (eds), *Reproductive Biology of Invertebrates, Vol. II: Spermatogenesis and Sperm Function.* pp. 443-495. Chichester: Wiley & Sons.

Anderson, W.R. & R.F. Ford 1976. Early development, growth and survival of the yellow crab *Cancer anthonyi* Rathbun (Decapoda, Brachyura) in the laboratory. *Aquaculture* 7: 267-279.

Anilkumar, G. & K.G. Adiyodi 1977. Spermatheca of the freshwater crab, *Paratelphusa hydrodromus* (Herbst) in relation to the ovarian cycle. In K.G. Adiyodi & R.G. Adiyodi (eds), *Advances in Invertebrate Reproduction, Vol. 1.* pp. 269-274. Karivellur: Peralaam-Kenoth.

Bennett, D.B. 1974. Growth of the edible crab (*Cancer pagurus* L.) off south-west England. *J. Mar. Biol. Assoc. UK* 54: 803-823.

Bennett, D.B. & C.G. Brown 1976. Crab migrations in the English Channel, 1968-75. *Fish. Not., MAFF, Direct Fish. Res., Lowestoft* 44: 1-12.

Binford, R. 1913. The germ cells and the process of fertilization in the crab *Menippe mercenaria. J. Morphol.* 24: 147-202.

Broekhuysen, G.J. 1936. On development, growth and distribution of *Carcinides maenas* (L.). *Arch. Néerl. Zool.* 2: 257-399.

Brown, C.G. & D.B. Bennett 1980. Population and catch structure of the edible crab (*Cancer pagurus*) in the English Channel. *J. Cons. Inst. Explor. Mer.* 39: 88-100.

Buchanan, D.V., R.E. Millemann & N.E. Stewart 1970. Effects of the insecticide Sevin on various stages of the Dungeness crab, *Cancer magister. J. Fish. Res. Bd. Canada* 27: 93-104.

Butler, T.H. 1960. Maturity and breeding of the Pacific edible crab, *Cancer magister* Dana. *J. Fish. Res. Bd. Canada* 17: 641-646.

Butler, T.H. 1961. Growth and age determination of the Pacific edible crab, *Cancer magister* Dana. *J. Fish. Res. Bd. Canada* 18: 873-891.

Caldwell, R.S. 1977. Biological effects of pesticides on the Dungeness crab. *Environmental Portection Agency, USA, EPA-600/3-77-131, PB 276 978: 1-123.*

Campbell, A. & M.D. Eagle 1983. Size at maturity and fecundity of rock crabs, *Cancer irroratus*, from the Bay of Fundy and southwestern Nova Scotia. *Fish. Bull.* 81: 357-362.

Campbell, A. & D.G. Robinson 1983. Reproductive potential of three American lobster (*Homarus americanus*) stocks in the Canadian maritimes. *Can. J. Fish. Aquat. Sci.* 40: 1958-1967.

Carpenter, R.K. 1978. Aspects of growth, reproduction, distribution and abundance of the jonah crab, (*Cancer borealis*) Stimpson, in Norfolk Canyon and the adjacent slope. MA Thesis, Univ. Virginia, Charlottesville.

Carroll, J.C. 1982. Seasonal abundance, size composition, and growth of rock crab, *Cancer antennarius* Stimpson, off central California. *J. Crust. Biol.* 2: 549-561.

Carroll, J.C. & R.N. Winn 1988. Species profiles: life histories and environmental requirements of coastal fishes and invertebrates (Pacific Southwest) – Rock crabs: brown, red, and yellow crab. US *Fish Wildl. Serv. Biol. Rep. 82, US Army Corps of Engineers, Tr EL-82-4.*

Charniaux-Cotton, H. & G. Payen 1988. Crustacean reproduction, In H. Laufer & R.G.H. Downer (eds), *Endocrinology of selected Invertebrate Types* pp. 279-303. New York: Liss.

Cheung, T.S. 1966. The development of egg-membranes and egg attachment in the shore crab, *Carcinus maenas*, and some related decapods. *J. Mar. Biol. Assoc. UK* 46: 373-400.

Christofferson, J.P. 1970. An electrophysiological and chemical investigation of the female sex pheromone of the crab *Portunus sanguinolentus*. Ph.D Thesis, Univ. Hawaii, Hilo.

Cleaver, F.C. 1949. Preliminary results of the coastal crab (*Cancer magister*) investigation. *Washington State Dept. of Fisheries, Biol. Rept.*, 49A: 47-82.

Conan, G.Y. 1985. Periodicity and phasing of molting, in A.M. Wenner (ed.), *Factors in adult growth, Crustacean Issues* 3: 75-99. Rotterdam: Balkema.

Davis, C.C. 1965. A study of the hatching process in aquatic invertebrates. XX. The blue crab *Callinectes sapidus* Rathbun, XXI. The nemertean, *Carcinonemertes carcinophila* (Kolliker). *Chesapeake Sci.* 6: 201-208.

Diamond, N. & D.G. Hankin 1985. Biases in crab tag recovery data. In *Proceedings of the Symposium on Dungeness crab biology and management. Lowell Wakefield Fisheries Symposia Series, Alaska Sea Grant Report* 85-3: 341-355.

Drach, P. 1949. Périodes de mue chez un crustacé à croissance lente *Xantho floridus* Montagu dans le district septentrional de son aire de répartition. *C.R. Acad. Sci. Paris* 228: 128-130.

Dunbar, M.J. 1979. Biological production in the Gulf of St Lawrence. In M.J. Dunbar (ed.), *Marine production mechanisms*, pp. 151-171. Cambridge: Cambridge Univ. Press.

Ebert, E.E., A.W. Hazeltine, J.L. Houk & R.O. Kelly 1983. Laboratory cultivation of the Dungeness crab, *Cancer magister. Calif. Fish & Game Fish Bull.* 172: 259-309.

Edwards, E. 1966. Mating behavior in the European edible crab (*Cancer pagurus* L.). *Crustaceana* 10: 23-30.

Edwards, E. 1979. *The Edible Crab.* Farnham, Surrey: Fishing News Books Ltd.

Elner, R.W., S.A. Gass & A. Campbell 1985. Mating behavior of the Jonah crab, *Cancer borealis* Stimpson (Decapoda, Brachyura). *Crustaceana* 48: 34-39.

Emlen, S.T. & L.W. Oring 1977. Ecology, sexual selection, and the evolution of mating systems. *Science* 197: 215-223.

Eurenius, L. 1973. An electron microscope study on the developing oöcytes of the crab *Cancer pagurus* L. with special reference to yolk formation (Crustacea). *Z. Morph. Tiere* 75: 243-254.

Fisher, W.S. 1976. Relationships of epibiotic fouling and mortalities of eggs of the Dungeness crab (*Cancer magister*). *J. Fish. Res. Bd. Canada* 33: 2849-2853.

Fisher, W.S. & W.H. Clark, Jr 1983. Eggs of *Palaemon macrodactylus:* I. Attachment to the pleopods and formation of the outer investment coat. *Biol. Bull.* 164: 189-200.

Fisher, W.S. & D.E. Wickham 1976. Mortality and epibiotic fouling of eggs from wild populations of the Dungeness crab, *Cancer magister. Fish. Bull.* 74: 201-207.

Glickstein, N. 1978. Acute toxicity of mercury and selenium to *Crassostrea gigas* embryos and *Cancer magister* larvae. *Mar. Biol.* 49: 113-117.

Goudeau, M. & F. Lachaise 1980. Fine structure and secretion of the capsule enclosing the embryo in a crab (*Carcinus maenas (L.)). Tissue & Cell* 12: 287-308.

Goudeau, M. & F. Lachaise 1983. Structure of the egg funiculus and deposition of embryonic envelopes in a crab. *Tissue & Cell* 15: 47-62.

Gutierrez, J.M. & O.R. Zuniga 1976. *Cancer setosus* Molina en la Bahia de Mejillones L sur (Crustacea, Decapoda, Brachyura). *Rev. Biol. Mar.* 16: 1-25.

Haefner Jr, P.A. 1976. Distribution, reproduction, and moulting of the rock crab, *Cancer irroratus* Say, 1917, in the mid-Atlantic Bight. *J. Nat. Hist.* 10: 377-379.

Haefner Jr, P.A. 1977. Aspects of the biology of the jonah crab, *Cancer borealis* Stimpson, 1959 in the mid-Atlantic Bight. *J. Nat. Hist.* 11: 303-320.

Haefner Jr, P.A. 1985. The biology and exploitation of crabs. In A.J. Provenzano, Jr (ed.), *The Biology of the Crustacea, Vol. 10*, pp. 111-166. Orlando: Academic Press.

Haefner Jr, P.A. & W.A. Van Engel 1975. Aspects of molting, growth and survival of male rock crabs *Cancer irroratus*, in Chesapeake Bay. *Chesapeake Sci.* 16: 253-265.

Hankin, D.G. 1985. Proposed explanations for fluctuations in abundance of Dungeness crabs: A review and critique. In *Proceedings of the Sumposium on Dungeness crab biology and management. Lowell Wakefield Fisheries Symposia Series, Alaska Sea Grant Report* 85-3: 305-326.

Hankin, D.G., N. Diamond, M. Mohr & J. Ianelli 1985. Molt increments, annual molting probabilities, fecundity and survival rates of adult female Dungeness crabs in northern California. In *Proceedings of the Symposium on Dungeness crab biology and management. Lowell Wakefield Fisheries Symposia Series, Alaska Sea Grant Report* 85-3: 189-206.

Hartnoll, R.G. 1969. Mating in the Brachyura. *Crustaceana* 16: 161-181.

Hartnoll, R.G. 1985. Growth, sexual maturity and reproductive output. In A.M. Wenner (ed.), *Factors in adult growth, Crustacean Issues* 3: 101-128. Rotterdam: Balkema.

Hartnoll, R.G. & P. Gould 1988. Brachyuran life history strategies and the optimization of egg production. *Symp. Zool. Soc. London* 59: 1-9.

Hartnoll, R.G. & S.M. Smith 1979. Pair formation in the edible crab (Decapoda, Brachyura). *Crustaceana* 36: 23-28.

Harvey, H.W. 1955. *The chemistry and fertility of sea waters.* Cambridge: Cambridge Univ. Press.

Hines, A.H. 1982. Allometric constraints and variables of reproductive effort in brachyuran crabs. *Mar. Biol.* 69: 309-320.

Horne, A.J., M. Bennett, R. Valentine, R.E. Selleck, P.P. Russell & P.W. Wild 1983. The effects of chlorination of wastewater on juvenile Dungeness crabs in San Francisco Bay waters. *Calif. Fish & Game Fish Bull.* 172: 215-226.

Howard, A.E. 1982. The distribution and behavior of ovigerous edible crabs (*Cancer pagurus*), and consequent sampling bias. *J. Cons. Int. Explor. Mer.* 40: 259-261.

Jamieson, G.S. 1986. Implications of fluctuations in recruitment in selected crab populations. *Can. J. Fish. Aquat. Sci.* 43: 2085-2098.

Jeffries, H.P. 1966. Partitioning of the estuarine environment by two species of *Cancer. Ecology* 47: 477-481.

Karinen, J.F., S.D. Rice & M.M. Babcock 1985. Reproductive success in Dungeness crabs (*Cancer magister*) during long-term exposures to oil-contaminated sediments. *Final Report – OCSEAP P-Unit 3008, 28 pp. Anchorage, Alaska.*

Kittredge, J.S., M. Terry & F.T. Takahashi 1971. Sex pheromone activity of the moulting hormone, crustecdysone, on male crabs (*Pachygrapsus crassipes, Cancer antennarius, Cancer anthonyi*). *Fish. Bull.* 69: 337-343.

Knudsen, J.W. 1964. Observations of the reproductive cycles and ecology of the common Brachyura and crablike Anomura of Puget Sound, Washington. *Pac. Sci.* 18: 3-33.

Krouse, J.S. 1972. Some life history aspects of the rock crab, *Cancer irroratus*, in the Gulf of Maine. *J. Fish. Res. Bd. Canada* 29: 1479-1482.

Krouse, J.S. 1976. Size composition and growth of young rock crab, *Cancer irroratus*, on a rocky beach in Maine. *Fish. Bull.* 74: 949-954.

Krouse, J.S. 1980. Distribution and catch composition of Jonah crab, *Cancer borealis*, and rock crab, *Cancer irroratus*, near Boothbay Harbor, Maine. *Fish. Bull.* 77: 685-693.

Kuris, A.M. & D.E. Wickham 1987. Effect of nemertean egg predators on crustaceans. *Bull. Mar. Sci.* 41: 151-164.

Langreth, S.G. 1969. Spermiogenesis in *Cancer* crabs. *J. Cell Biol.* 43: 575-603.

LeFoll, A. 1984 (1986). Contribution á l'étude de la biologie du crabe-tourteau *Cancer pagurus* sur les côtes de Bretagne Sud. *Rev. Trav. Inst. Pêches marit.* 48: 5-22.

Macdonald, J.M., J.D. Shields & R.K. Zimmer-Faust 1988. Acute toxicities of eleven metals to early life history stages of the yellow crab *Cancer anthonyi*. *Mar. Biol.* 98: 201-207.

MacKay, D.C.G. 1942. The Pacific edible crab, *Cancer magister*. *Bull. Fish. Res. Bd. Canada* 62: 1-32.

MacKay, D.C.G. 1943. Temperature and the world distribution of crabs of the genus *Cancer*. *Ecology* 24: 113-115.

Martin, J.L. 1976. Métabolism de Fe, Cu, Zn, Mg, Mn, et Co dans les oeufs de *Cancer irroratus* (Crustacé Décapode). *C.R. Soc. Biol. Fil.* 170: 157-162.

Martin, M., K.E. Osborn, P. Billig & N. Glickstein 1981. Toxicities of ten metals to *Crassostrea gigas* and *Mytilus edulis* embryos and *Cancer magister* larvae. *Mar. Pollut. Bull.* 12: 305-308.

Mayer, D.L. 1973. The ecology and thermal sensitivity of the Dungeness crab, *Cancer magister* and related species of its benthic community on Similk Bay, Washington. PhD Thesis, Univ. Wash., Seattle.

Mullin, M.M. 1986. Spatial and temporal scales and patterns. In R.W. Eppley (ed.), *Plankton dynamics of the Southern California Bight, Vol. 15*, pp. 216-273. Berlin: Springer-Verlag.

Nations, J.D. 1975. The genus *Cancer* (Crustacea: Brachyura): Systematics, biogeography, and fossil record. *Nat. Hist. Mus. Los Angeles County Bull.* 23: 1-104.

Nichols, J.H., B.M. Thompson & M. Cryer 1982. Production, drift and mortality of the planktonic larvae of the edible crab (*Cancer pagurus*) off the north-east coast of England. *Neth. J. Sea Res.* 16: 173-184.

O'Clair, C.E. & J.L. Freese 1988. Reproductive condition of Dungeness crabs, *Cancer magister*, at or near log transfer facilities in southeastern Alaska. *Mar. Environ. Res.* 26: 57-81.

Okazaki, R.K. 1988. Factors regulating predation on crab embryos by the nemertean *Carcinonemertes:* the role of host ecdysteroids. PhD Thesis, Univ. Calif., Santa Barbara.

Orensanz, J.M. & V.F. Gallucci 1988. Comparative study of postlarval life-history schedules in four sympatric species of *Cancer* (Decapoda: Brachyura: Cancridae). *J. Crust. Biol.* 8: 187-220.

Pearson, J. 1908. *Cancer* (the edible crab). L.M.B.C. Memoirs, No. XVI. *Proc. Liverpool Biol. Soc.* 22: 291-498.

Perkins, H.C. 1972. Developmental rates at various temperatures of embryos of the northern lobster (*Homarus americanus* Milne Edwards). *Fish. Bull.* 70: 95-99.

Pochon-Masson, J. 1983. Arthropoda–Crustacea. In K.G. Adiyodi & R.G. Adiyodi (eds), *Reproductive Biology of Invertebrates, Vol. II*, pp. 407-449. Chichester: Wiley & Sons.

Rathbun, M.J. 1930. The cancroid crabs of America of the families Euryalidae, Portunidae, Atelecyclidae, Cancridae, and Xanthidae. *Bull. US Natl. Mus.* 152: 1-609.

Reilly, P.N. 1987. Popultions studies of rock crab, *Cancer antennarius*, yellow crab, *C. anthonyi*, and Kellet's whelk, *Kelletia kelletii*, in the vicinity of Little Cojo Bay, Santa Barbara County, California. *Calif. Fish Game* 73: 88-98.

Reilly, P.N. & S.B. Saila 1978. Biology and ecology of the rock crab, *Cancer irroratus* Say, 1817, in southern New England waters (Decapoda, Brachyura). *Crustaceana* 34: 121-140.

Sandoz, M. & R. Rogers 1942. The effect of environmental factors on hatching, moulting, and survival of zoea larvae of the blue crab *Callinectes sapidus* Rathbun. *Ecology* 25: 216-228.

Scarratt, D.J. & R. Lowe 1972. Biology of rock crab (*Cancer irroratus*) in Northumberland Strait. *J. Fish. Res. Bd. Canada* 29: 161-166.

Shields, J.D. 1987. The dynamics and mechanisms of egg mortality in *Cancer anthonyi* (Decapoda). PhD Thesis, Univ. Calif., Santa Barbara.

Shields, J.D. & A.M. Kuris 1987. Temporal variation in abundance of the egg predator *Carcinonemertes epialti* (Nemertea) and its effect on egg mortality of its host, the shore crab, *Hemigrapsus oregonensis*. *Hydrobiol.* 156: 31-38.

Shields, J.D. & A.M. Kuris 1988. An *in vitro* analysis of egg morality in *Cancer anthonyi:* The role of symbionts and temperature. *Biol. Bull.* 174: 267-275.

Shields, J.D., R.K. Okazaki & A.M. Kuris (in press). Brood mortality and egg predation by the nemertean, *Carcinonemertes epialti*, on the yellow rock crab, *Cancer anthonyi*, in southern California. *Can. J. Fish. Aquat. Sci.*

Shields, J.D., R.K. Okazaki & A.M. Kuris (in press). The fecundity and reproductive potential of *Cancer anthonyi. Fish. Bull.*

Shirley, S.M., T.C. Shirley & S.D. Rice 1987. Latitudinal variation in the Dungeness crab, *Cancer magister*: zoeal morphology explained by incubation temperature. *Mar. Biol.* 95: 371-376.

Shotton, L.R. 1973. Biology of the rock crab, *Cancer irroratus* Say, in the coastal waters of Virginia. MA Thesis, Univ. Virginia., Charlottesville.

Snow, C.D. & J.R. Nielsen 1966. Premating and mating behavior of the Dungeness crab (*Cancer magister* Dana). *J. Fish. Res. Bd. Canada* 23: 1319-1323.

Terretta, R.T. 1973. Relative growth, reproduction and distribution of the rock crab, *Cancer irroratus*, in Chesapeake Bay during the winter. MS thesis, College of William and Mary, Williamsburg.

Toole, C.L. 1985. *Rock crab survey of Humboldt Bay – Interim Report*. California Sea Grant Advisory Program, Eureka. 14 pp.

Wear, R.G. 1974. Incubation in British decapod Crustacea and the effects of temperature on the rate and success of embryonic development. *J. Mar. Biol. Assoc. UK* 54: 745-762.

Wenner, A.M., C. Fusaro & A. Oaten 1974. Size at onset of sexual maturity and growth rate in crustacean populations. *Can. J. Zool.* 52: 1095-1106.

Wickham, D.E. 1979a. Predation by *Carcinonemertes errans* on eggs of the Dungeness crab, *Cancer magister. Mar. Biol.* 55: 45-53.

Wickham, D.E. 1979b. *Carcinonemertes errans* and the fouling and mortality of eggs of the Dungeness crab, *Cancer magister. J. Fish. Res. Bd. Canada* 36: 1319-1324.

Wickham, D.E. 1986. Epizootic infestations by nemertean brood parasites on commercially important crustaceans. *Can. J. Fish. Aquat. Sci.* 43: 2295-2302.

Wild, P.W. 1983a. Comparisons of ovary development in Dungeness crabs, *Cancer magister*, in central and northern California. *Calif. Fish & Game Fish Bull. 172: 189-196.*

Wild, P.W. 1983b. The influence of seawater temperature on spawning, egg development, and hatching success of the Dungeness crab, *Cancer magister. Calif. Fish & Game Fish Bull.* 172: 197-214.

Wild, P.W. 1983c. Dungeness crab environment project studies: objectives and approach. *Calif. Fish & Game Fish Bull.* 172: 171-174.

Wild, P.W., P.M.W. Law & D.R. McLain 1983. Variations in ocean climate and the Dungeness crab fishery in California. *Calif. Fish & Game Fish Bull.* 172: 175-188.

Williamson, H.C. 1904. Contributions to the life-histories of the edible crab (*Cancer pagurus*) and of other decapod Crustacea: impregnations: spawning: casting: distribution: rate of growth. *22nd Ann. Rept. Fish. Bd. Scotland* 3: 100-140.

Winn, R.N. 1985. Comparative ecology of three cancrid crab species (*Cancer anthonyi, C. antennarius, C. productus*) in marine subtidal habitats in southern California. PhD Thesis, Univ. Southern California, Los Angeles.

3. Population and fisheries aspects

Egg production in crabs with abbreviated development

Nancy Nash Rabalais
Louisiana Universities Marine Consortium, Chauvin, USA

ABSTRACT

Crabs with abbreviated development generally bear fewer and larger eggs than species with longer periods of development. The larger, lecithotrophic eggs contain more yolk, undergo extended embryonic development, and produce morphologically advanced larvae or post-larvae. Among congeners, those that hatch many larvae produce an order of magnitude or greater number of planktotrophic eggs per brood than those with abbreviated development. Among crabs with abbreviated development, the number of eggs per brood ranges over three orders of magnitude. The number and size of eggs carried by any female is quite variable, but within a species the brood size is related primarily to the size of the female. Additional factors may include limitations in lipid and protein reserves and size limitations of internal anatomy and the abdomen.

Within a population of *Uca subcylindrica,* adult females produced one brood per year with the population split into those that were preparing for egg production and larval release either in mid- to late spring or others in mid- to late summer. Literature accounts indicate a single brood per year for other species with abbreviated development, but those data are sparse. The number of broods per year for female *U. subcylindrica* are limited by the time required to replenish lipid and protein reserves and to develop ovaries with many, large yolk-filled eggs. If the life span of *U. subcylindrica* is similar to that of other fiddler crabs, then two or three opportunities exist within a lifetime for an adult female to produce a brood of larvae. Many unknown factors, however, including size at metamorphosis, growth rate of females, and yearly reproductive output, could influence the lifetime reproductive output.

1 INTRODUCTION

The seasonal, yearly, and lifetime reproductive output of crabs with abbreviated development is unknown for any species, since data on egg production is usually incidental to faunal surveys or morphological studies of larvae. Those data, consisting primarily of brood and egg size, are compiled and compared. Information on the natural history and reproductive biology of those crabs is even more limited, and a few examples are given from the literature. Data on egg production from a population of the fiddler crab, *U. subcylindrica*, are presented here and are used to predict seasonal and yearly egg production.

2 ABBREVIATED DEVELOPMENT

2.1 *Definitions*

Abbreviated development occupies one end of a continuum of developmental types that proceeds through regular staged larval development and into extended larval development at the other extreme (Gore 1985, Rabalais & Gore 1985), but the point at which any one sequence fails to be 'regular' and becomes 'abbreviated' is not precisely definable. Abbreviated development is a sequence of shorter duration than that seen in the majority of related species in a taxon and results in fewer morphologically discrete instars. The shortened developmental times result in the more rapid production of a juvenile form possessing the general morphological characteristics of the adult.

The incidence of abbreviated development within the decapods is frequent and widespread across phyletic lines. Examples of abbreviated development occur within larval and postlarval stages. These examples have been thoroughly reviewed by Rabalais & Gore (1985), but are summarized here for easy reference within this chapter. Two types are described here – direct and advanced.

No free-swimming larval stages exist in direct development. The young hatch in the form of the adult, but often have vestiges of larval characteristics and are retained on the pleopods of the parental female. Examples of direct development in marine brachyuran crabs are *Cryptodromia octodentata* and *Petalomera lateralis*, in which the adult females retain young crabs morphologically similar to the adults but with pleopods having natatory setae, as in the megalopae of some other crabs (Montgomery 1922, Hale 1925). In some cases, late embryos, zoeae or megalopae, held under the abdomen of the adult females, do not swim, even though they possess some typically larval characteristics. The xanthid crab, *Pilumnus lumpinus*, hatches as an advanced zoea with rudimentary natatory setae on the maxillipedes (Wear 1967). These setae are not functional, and the zoeae do not swim but lie on the bottom, rapidly flexing and extending the abdomen, while molting to the megalopal stage. The megalopae are active swimmers which, although not attached to the parental female, remain in contact with her. The young of *Halicarcinus lacustris*, an Australian freshwater hymenosomatid, hatch on the female pleopods in a form that Lucas (1971) terms a 'late embryo.' These retain rudimentary natatory setae, fall to the bottom if separated from the female's pleopods and lie there until they subsequently molt to the juvenile crab stage.

In abbreviated development of the advanced type, the young hatch as zoeae but in a growth state considerably more complete than that seen in their congeneric relatives. These are functional, swimming zoeae, often in the ultimate or penultimate zoeal stage, and may thus undergo additional ecdyses prior to metamorphosis. The larval duration and ecdysial events are thus reduced, so that fewer morphological changes occur before the metamorphic molt.

Anomuran crustaceans are well known for having abbreviated larval development. The zoea I of the lithodid *Paralomis granulosa* hatches in an advanced condition, passes through a second zoeal stage with slight refinement of characters, and then metamorphoses to the megalopa (Campodonico & Guzman 1981). Other lithodids develop through three or four zoeal stages.

Advanced development occurs in several families of the Brachyura. For example, in the families Grapsidae and Pinnotheridae, where the majority of developmental sequences are extended (four or five zoeal stages), several species undergo one, two or three zoeal stages before they molt to the postlarval form. In Jamaica, there are five endemic species of grapsid crabs which never enter sea water (Hartnoll 1964, 1971). At least three of these species exhibit abbreviated development of the advanced type (Hartnoll 1964, Abele & Means 1977). Only one member of the Ocypodidae has been recorded with abbreviated development (Rabalais & Cameron 1983). Two advanced zoeal stages occur in *U. subcylindrica*, as compared to the five zoeal stages typical of other ocypodids.

2.2 *Features associated with abbreviated development*

Certain characteristics associated with abbreviated development can be recognized across phylogenetic lines. In general, species with abbreviated development bear fewer and larger eggs than do species with longer periods of development (see Section 3.1). Larger eggs contain more yolk and have a greater internal food supply. Embryonic development is extended, especially where a more advanced stage is produced at hatching. The period of incubation for eggs produced by *U. subcylindrica* is 1 to 1.5 months (Rabalais & Cameron 1983), compared to other *Uca* spp. with extended development and an incubation period of 12-15 days (Feest 1969, von Hagen 1970, Ringold 1979, Christy 1982). The periods of incubation for *Pilumnus novaezealandiae* and *P. lumpinus* (Wear 1967) are similar but are considerably longer for other crabs with abbreviated development. Approximately 2.5 to 3 months are required for *Metopaulius depressus* (see Hartnoll 1964) and 3 to 5 months for *Heterozius rotundifrons* (see Wear 1968, Jones 1978).

Associated with the larger eggs, longer incubation period, and advanced morphology of these larvae is their lecithotrophic pattern of embryogenesis. These larvae hatch from large yolky eggs, upon which they depend for nutrition. Larvae from planktotrophic eggs, by comparison, hatch from small eggs without a yolk reserve and depend on planktonic prey for energy. Larvae from lecithotrophic eggs usually hatch with substantial amounts of yolk still available and may be characterized by relatively undeveloped mouthparts and reduced feeding ability (Hartnoll 1964, Provenzano 1968, Soh 1969, Abele & Means 1977, Rabalais & Cameron 1983). Stored yolk reserves would be advantageous in certain ecological settings where suitable planktonic prey may be in short supply, such as in the deep sea (Provenzano 1968, Wear & Batham 1975), in collected water of large tank bromeliads (Hartnoll 1964), or in ephemeral puddles formed after rainfall (Rabalais & Cameron 1983, 1985a).

Species with lengthened incubation periods and advanced larvae often have stored energy reserves, and there are obvious differences in maturation rates and survivorship as compared to those species with shorter incubation periods and longer planktonic duration. In populations of *U. subcylindrica*, megalopae are present within two days of hatching and crab I stages within seven to nine days of hatching (Rabalais & Cameron 1983, 1985a). Other species of *Uca* require an average of 17-28 days to reach megalopa and 27-40 days to reach crab I (Hyman 1920, Herrnkind 1968, Feest 1969, Kurata 1970, Montague 1980). Survival to the megalopal and first crab stages is considerably higher in *U. subcylindrica* than in other members of the genus (Christiansen & Yang 1976, Christy 1982, Rabalais & Cameron 1983, 1985a, M. Flynn pers. comm.). Increased survival may be related to decreased nutritional vulnerability, reduced number of critical premetamorphic molt periods, and morphological and physiological adaptations that reduce osmotic stress in environments of dilute media.

3 EGG PRODUCTION IN CRABS WITH ABBREVIATED DEVELOPMENT

There are few data on egg production in crabs with abbreviated development, limited mostly to brood sizes of females collected during faunal surveys or of females from which larvae were described for morphological features and phylogenetic classifications. Data on reproductive biology and life history patterns of crabs with abbreviated development are even more scarce than egg production data. In the following sections I provide examples from the literature and some original data.

3.1 *Comparative information*

Differences between size and number of eggs in species with abbreviated development and in those species with longer developmental sequences are most evident among close relatives. For example, most members of the genus *Uca* produce smaller eggs (0.21-0.38 mm, diameter, *n* = 5) than *U. subcylindrica* (1.06 mm diameter). Along with larger egg size, females produce fewer eggs. The number of eggs produced by *U. subcylindrica* is considerably smaller than the number produced by congeners of similar size (Fig. 1).

The number of eggs per female in different species of brachyuran crabs with abbreviated development patterns ranges over three orders of magnitude, varies within species, and is proportional to size of the female (Fig. 2). In Figures 1 and 2, similarly-sized females with

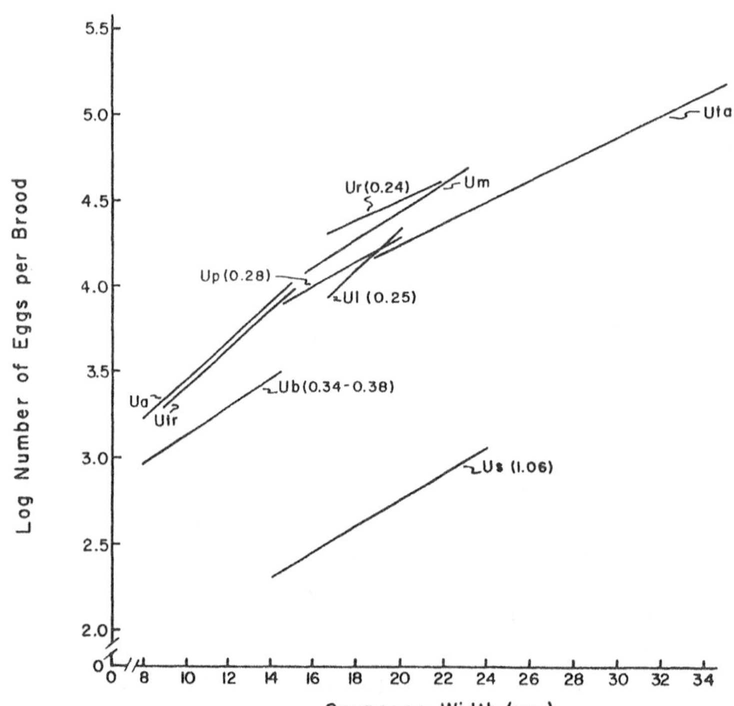

Figure 1. Log number of eggs per brood versus carapace width for several species of fiddler crabs. Egg diameter (mm), given next to species where known. Initials correspond to Ua, Uta and Utr – *Uca annulipes, U. tangeri* and *U. triangularis* (Feest 1969); Ub – *U. burgersi* (Gibbs 1974); Ul, Up, Ur and Us – *U. longisignalis, U. panacea, U. rapax* and *U. subcylindrica* (Rabalais & Cameron 1983); Um – *U. minax* (Gray 1942, Hines 1982). Egg size for *Uca* given by Crane (1941) as 0.22-0.27 mm, diameter. (After Rabalais & Gore 1985.)

similar body form have been compared because of the allometric constraints of female body size on reproductive 'effort' (Hines 1982).

The number of eggs lost or potentially lost during incubation is not known for most of the species figured. The potential for egg loss in a female with a small brood size may be an important factor in the number of individuals successfully hatched. Jones (1978) reported egg mortality for *Heterozius rotundifrons* to be low, fluctuating between 3.8 and 16.4% per brood, with no apparent relationship to size of the incubating females. The data for brood size in *U. subcylindrica* as a function of female size were analyzed for differences, according to four stages of egg development, from newly deposited to near hatching. There were no differences in the slopes of the relationships, which indicated no substantial loss of eggs during incubation.

Behavioral adaptations of ovigerous females in this and other species with abbreviated development may reduce the potential for loss of eggs during incubation. If ovigerous females of *U. subcylindrica* were collected outside their burrows, they were usually releasing larvae. Otherwise, the ovigerous females were most often found in burrows with plugged entrances. The burrow shaft was often plugged as far as 30 cm below the sediment surface. Burrow plugging activity intensified during the summer as sediments became drier.

No obvious relationship was apparent between size of female and size of eggs for crabs with abbreviated development (Fig. 2). A similar conclusion was reached for a study of brachyurans with regular development patterns (Hines 1986). [When available, data for egg size at a time near hatching are presented preferentially, because there is usually an increase in the size of the egg during embryonic development (Wear 1970), Wear & Batham 1975, Abele & Means 1977,

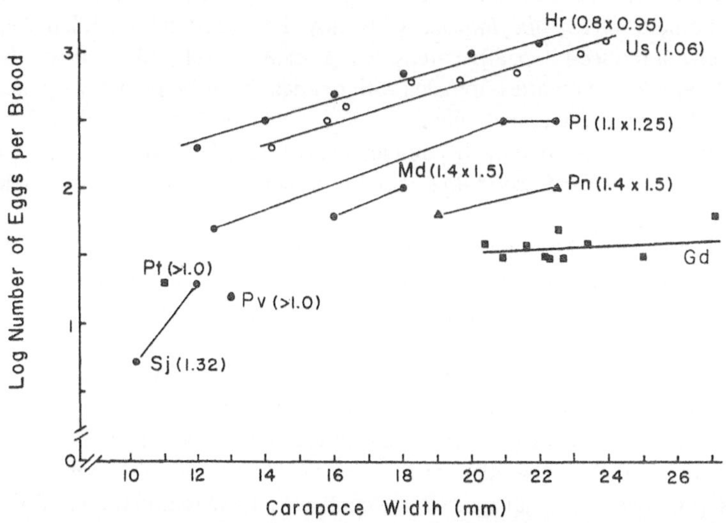

Figure 2. Log number of eggs per brood versus carapace width for several species of brachyuran crabs with abbreviated development. Egg size in mm (either diameter, or length by width), where known, given next to species designation. Species marked with open or closed circles have advanced development; species marked with squares or triangles have direct development (different types). Initials correspond to Hr – *Heterozius rotundifrons* (Jones 1978); Us – *Uca subcylindrica* (Rabalais & Cameron 1983); Pl and Pn – *Pilumnus lumpinus* and *P. novaezealandiae* (Wear 1967); Pv – *Pilumnus vestitus* (Hale 1931); Md – *Metopaulius depressus* (Hartnoll 1964); Pt – *Platydromia thomsoni* (Hale 1925); Sj – *Sesarma jarvisi* (Abele & Means 1977); Gd – *Geothelphusa dehaani* (Yamaguchi & Takamatsu 1980). (Modified from Rabalais & Gore 1985.)

Jones 1978). Very often, however, stage of embryonic development of the egg associated with the stated size was not recorded.]

Additional data for egg size in crabs with abbreviated development were available but were not complete enough for inclusion in Figure 2, or the females were not within the size range considered. Examples follow: Eggs were exceptionally large (3.0 to 3.6 mm, or 6.0 mm, diameter) in the freshwater crab family Potamidae (Koba 1936, Hines 1986, Tyagi 1972). Eggs of *Dromidia australis* were 2.0 mm diameter (Hale 1927). The number of eggs carried by *Cymonomus bathamae* was quite small (11 to 26), but the adults were also small (4.0 to 5.1 mm carapace width) (Wear & Batham 1975). Twenty individuals were carried on the pleopods of an adult (carapace width not reported) *Petalomera lateralis*, a dromiid with direct development (Montgomery 1922). Ovigerous females (size not recorded) of *Sesarma (Geosesarma) perracae* carried 25-70 eggs that measured 1.6 mm diameter (Soh 1969). Dissection of an adult *Sesarma cookei* (19.0 mm carapace width) revealed 21 ova measuring 1.25 mm in fully-developed gonads (Abele & Means 1977).

3.2 *Limitations to brood size*

There are morphological and physiological limitations to the number of large, yolk-rich eggs a female of a species with abbreviated development can produce. During ovarian development, substantial quantities of protein and lipid that are synthesized from ingested food are required. During peak breeding activity, or when successive broods are produced, stored lipid reserves may be borrowed from the hepatopancreas and transferred to the ovaries, and protein may be mobilized from muscle (Wear 1974). Because successive batches of eggs became smaller in number, Wear suggested that females may have increasing difficulty in mobilizing sufficient protein-lipid reserves. Females of *U. subcylindrica* with fully-developed gonads filled with large, yolk-rich eggs have a reduced hepatopancreas. It is sometimes difficult to locate the hepatopancreas in such females. In contrast, females with ovaries that have not fully ripened have a relatively larger hepatopancreas that fills most of the cephalothoracic region. This suggests that lipid reserves are depleted in the hepatopancreas of *U. subcylindrica* during the production of a brood. The size of the brood thus may be influenced by the available lipid reserves.

Size of the branchial chamber and cephalothoracic region where large, yolky eggs are held prior to deposition on the pleopods may limit the number of eggs produced in a brood. In *U. subcylindrica*, shape of the carapace is much more globular and enlarged than in other members of the genus (hence, the specific name). This modification may be related to respiratory physiology (Rabalais & Cameron 1985b), but, in any event, it also provides additional internal space for eggs. If the shape of the carapace in *U. subcylindrica* were similar to that of its congeners, i.e. more flattened, there would be less space available for the large developing ova and thus fewer ova. The reduced size of the hepatopancreas in female *U. subcylindrica* with fully-developed ovaries also provides additional space in the cephalothoracic region for additional ova.

A large egg mass might limit the activity of a female by restricting her movement. The average volume of an egg mass in *U. subcylindrica* is more than twice that of the *Uca* spp. (Rabalais & Gore 1985). Ovigerous female *U. subcylindrica* appear heavily encumbered and move with difficulty. The legs are extended and the abdomen touches the sediment. By comparison, other ovigerous *Uca* spp. have better mobility. The limited mobility of *U. subcylindrica* when carrying eggs is consistent with their inactivity in plugged burrows during egg incubation.

Where abbreviated development occurs in species for which habitat conditions may cause desiccation of the egg mass (i.e. terrestrial or semi-terrestrial), the number of eggs in a brood may be limited by size of the abdomen. Abele & Means (1977) noted that the abdomen of ovigerous *Sesarma cookei* was not as large in proportion to the body size as in other grapsid crabs and was held tightly against the sternum, preventing desiccation of the eggs. These crabs occupy rock rubble habitats in montane environments (i.e. where the potential for desiccation is high) and produce few eggs (see Section 3.1). The number of eggs produced may be, in part, a function of the size of the abdomen that can cover them. On the other hand, the abdomen of *U. subcylindrica* does not cover the egg clutch. These ovigerous females offset desiccation by digging burrows deep to the water table, plugging them, and limiting activity outside the burrow (see above).

3.3 *Egg production*

Little is known about seasonal, yearly, or lifetime production of eggs in crabs with abbreviated development. Summaries of reproductive biology for a few species with abbreviated development follow with a more detailed discussion of *U. subcylindrica*.

Metopaulias depressus is a grapsid crab endemic to Jamaica (Hartnoll 1964). Individuals inhabit tanks (leaf axils) of large bromeliads. Water contained in the bromeliads provides nursery areas for the two zoeal and one megalopal stages. Egg production occurs from March through November, with peaks in May and October. The incubation period is 2.5 to 3 months. Hartnoll (1964) reported newly ovigerous females with flaccid gonads and recently ovigerous females with developing ovaries. Based on this, he suggested that a spent female could be prepared to produce another brood by the next breeding season. It was not clear in his discussion if the next brood could be hatched in the next peak or in the next year. With intervals of 6 months between periods of predicted hatching, the possibility exists that females may be able to hatch larvae during each of these. Otherwise, the females would be limited to a single brood per year. Lifetime reproductive output was not discussed.

Heterozius rotundifrons is a xanthid crab (Belliidae, in Guinot 1976) endemic to New Zealand (Jones 1978). Individuals typically bury in sand and gravel under stones and pebbles in the intertidal region. There are two zoeal stages in the development sequence (Wear 1968). Ovigerous females are found in all months except February with production peaks in May-July and October-November (Jones 1978). Incubation periods are 3 months in summer and 5 months in winter. Based on size frequency distribution of females with eggs in various stages of development, Jones (1978) concluded that there was no evidence to suggest that successive broods are produced, indicating a single brood per year. Gonads of reproductively active females were not examined, nor was lifetime reproductive output discussed.

The freshwater crab, *Barythelphusa guerini*, which exhibits abbreviated development of the direct type, has a single annual reproductive cycle (Gangotri et al. 1971). Gonadal indices of adult females are high during April and May; eggs and young occur on female abdomens in June and July, during which time the gonadal index decreased. During January to April gonads are in a 'preparatory' period. Female crabs are capable of producing young in the second and third years of life, but do so primarily in the second. Similarly, the freshwater crab, *Geothelphusa dehaani*, with direct development, has a single reproductive period from mid-July to mid-September. The adult female population is divided into two groups with gonadal development and egg laying timed for production of young twice during the reproductive period (Yamaguchi & Takamatsu 1980).

4 REPRODUCTIVE BIOLOGY OF *UCA SUBCYLINDRICA*

Egg production and reproductive biology has been studied in detail for *U. subcylindrica*, a fiddler crab which inhabits semi-arid coastal and inland environments of the extreme western Gulf of Mexico (Rabalais 1983, Rabalais & Cameron 1983, 1985a). Biweekly or more frequent samples were taken throughout the year from a well-defined area along a creek bank in southern Texas, USA. Information on ovigerous females was most difficult to obtain, because of their seclusive nature. However, adequate samples of the adult, larval, and early postlarval populations supplemented egg production data with information on seasonal and yearly reproductive activities. This information permits an assessment of seasonal, yearly, and lifetime reproductive output.

4.1 *Natural history*

The preferred habitats for *U. subcylindrica* are banks of ephemeral ponds and intermittently flowing streams, vegetated areas of extensive wind tidal flats, and shorelines of bays and lagoons subject to hypersalinity and/or freshwater flooding. The areas are characterized by low rainfall (20-24 in/yr) of extreme variability (Brown et al. 1976, Norwine et al. 1977). Conditions are typically more arid than indicated by average rainfall values, because infrequent, heavy rains distort the rainfall figures (Hedgpeth 1967). There is some rain in spring and summer, but most is concentrated in late summer and early fall during the tropical storm season (see Fig. 7). Prolonged trends of moist periods and droughts last 10 to 25 years (Brown et al. 1977, Norwine et al. 1977).

The abbreviated development sequence for *U. subcylindrica* consists of two morphologically advanced zoeae and an equally advanced megalopa. Metamorphosis to the megalopal stage occurs generally within two days of hatch, at which time the postlarvae are capable of a benthic existence similar to that of the crab I stage. The larval and early postlarval stages have been found only in, or associated with, ephemeral pools formed from rainfall. These pools may dry up within a few days, depending on weather conditions. Female crabs release larvae in these temporary pools, even though a more permanent and apparently suitable source of water may be available. Larval release coincides with sporadic yet seasonally predictable rainfall; this indicates a timing of reproductive activities through the year among the adult crabs of the population permitting exploitation of more frequent rainfall periods.

4.2 *Adult female population*

Assessment of the adult female population was determined by a timed search of active burrows along a well-defined area of the creek bank, additional collections outside of the area, and searches of known nursery areas where larval and early postlarval stages had been collected previously. The eggs of ovigerous females were categorized as (1) newly-deposited: filled with dark yellowish-orange yolk, (2) early: eye pigment, incipient carapace and cephalothoracic appendages, (3) mid: well-developed eyes, formed telson, beginning of chromatophores, reduced yolk except for that contained within the carapace, and beating heart, or (4) late: well-developed chromatophores and reduced yolk, overall appearance pale brownish-yellow, an embryo actively moving within the egg membrane. Most ovigerous females were measured after their larvae were released. Unhatched eggs were measured, and larvae and unhatched eggs were counted. Presence of unhatched eggs or egg membranes on the pleopods of an adult

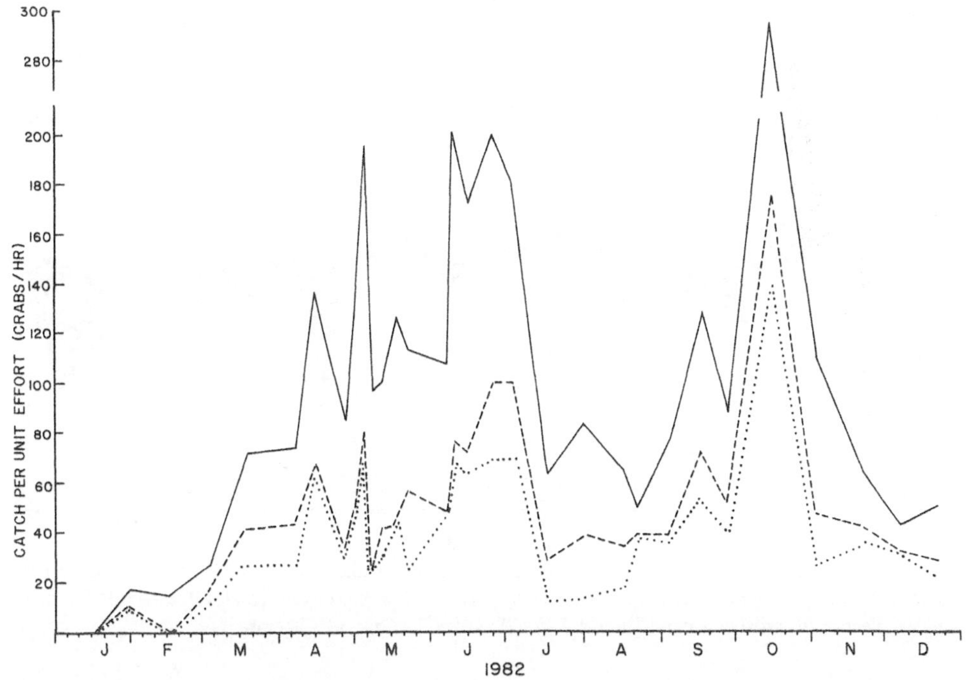

Figure 3. Catch per unit effort for *Uca subcylindrica* in population subsamples at Santa Gertrudis Creek, Texas, USA in 1982. Solid line represents all crabs (male and female) both mature and immature, dashed line represents all females, and dotted line represents adult females (≥ 12 mm, carapace width).

female indicated larval release by that female within the last 24 hours (= recently ovigerous).

Assessing the adult female population of *U. subcylindrica* required biweekly or more frequent samples during the year, because the number of collectable females was often low and the collection of ovigerous females in the timed searches was infrequent. Crabs were active through most of the year except during the coldest months (January through early March) and during the hotter, drier months of July and early August (Fig. 3). If a sex ratio of 0.5 [proportion of males in the population, M/(M+F)] can be assumed for fiddler crabs (Montague 1980), then during the entire sample period the sex ratio = 0.512 was close to the expected value of equal males and females. Yet the ratio was seldom 0.5 for any sample (Fig. 4). Thus, the number of adult females was considerably lower or higher than expected during certain times of the year. The catches of ovigerous and recently ovigerous females were not similar (see Fig. 7), and information from ovigerous females alone could not be relied upon for determining the reproductive activity of the population.

Collections of ovigerous females were supplemented by data obtained from dissections of adult females to provide information on gonadal condition, cycles of reproductive activity, and seasonal and yearly reproductive output. Gonads were categorized as (1) fully-developed with large (0.56 mm diameter), yolky eggs, (2) developed with a medium number of small (0.22 mm diameter) or medium-sized (0.38 mm diameter) eggs, (3) developed but with a small number of small eggs, or (4) flaccid and empty. The hepatopancreas was reduced in size or not obvious in crabs characterized by fully-developed gonads.

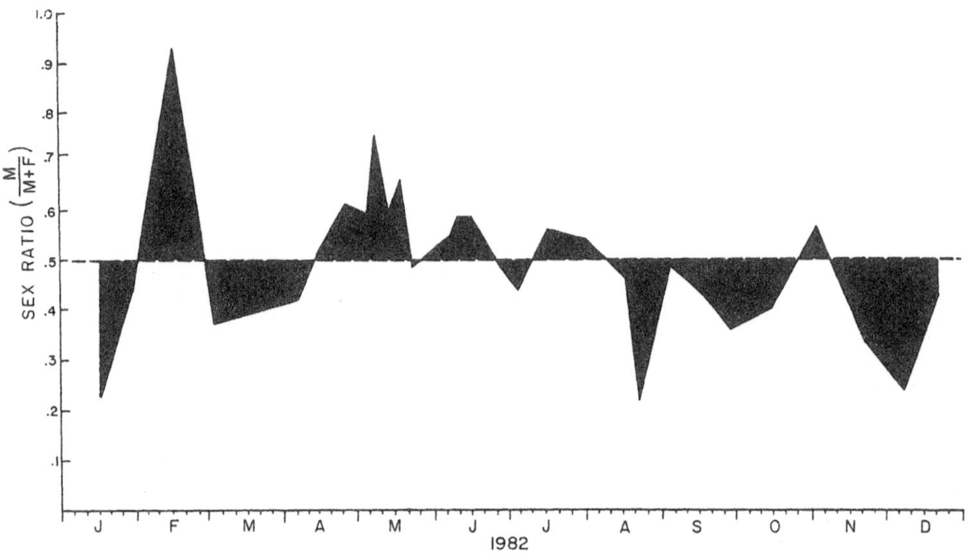

Figure 4. Temporal changes in sex ratio (proportion of males in population) at Santa Gertrudis Creek in 1982. Ratios greater than 0.5 represent a majority of males; less than 0.5, a majority of females. Total $n = 3166$; for any single collection, n varied between 15 and 294. For 78% of collections, $n > 50$ crabs.

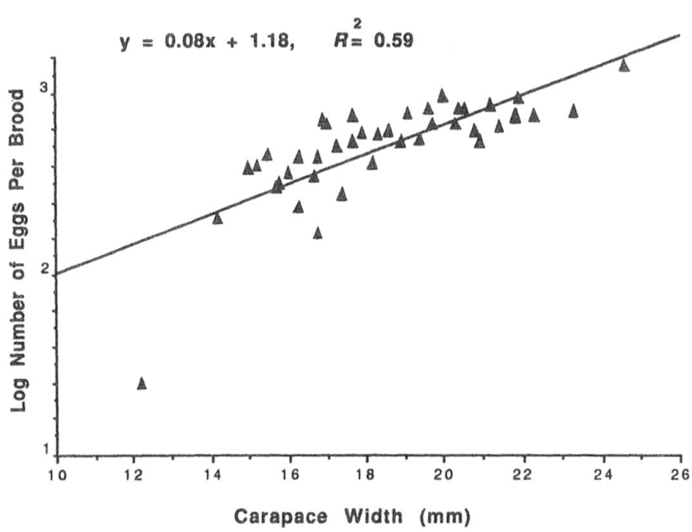

Figure 5. Relationship for *Uca subcylindrica* between log number of eggs per brood and carapace width of egg-bearing females. Number of eggs per brood = total of unhatched eggs and/or live hatched zoeae and/or dead hatched zoeae. R^2 = coefficient of determination. $n = 41$.

4.2.1 *Egg production*

Females bearing eggs and recently ovigerous females ranged in size from 12.2 to 24.6 mm carapace width (Fig. 5) and averaged 17.8 ± 0.3 mm ($n = 75$). Females as small as 10 mm carapace width were found to contain developing ovaries with small eggs, and this size may be the lower limit for ovigerous females. The size frequency of ovigerous females was normally

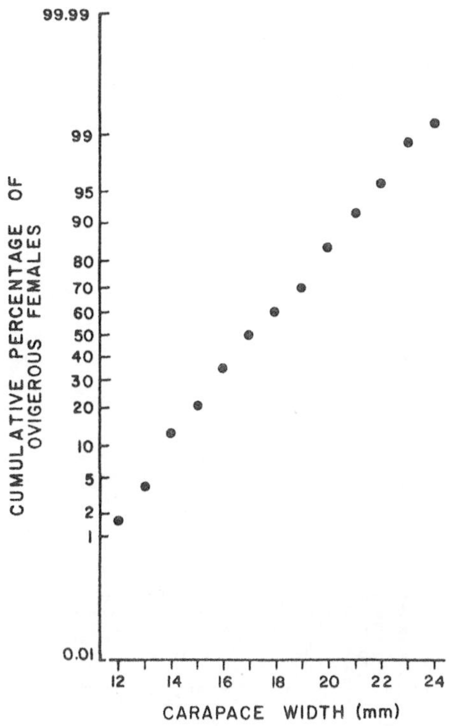

Figure 6. Size distribution of ovigerous females of *Uca subcylindrica*, shown as cumulative percentage ($n = 75$).

distributed across the adult female population (Fig. 6); the mean size did not differ significantly between spring and summer (ANOVA, $P < 0.05$).

Females produced from 25 to 1412 eggs per brood with a mean of 580 ($n = 41$) (Fig. 5). Fecundity varied linearly with the size of the female and did not differ between those females that had broods in spring and in summer (ANCOVA, $P < 0.05$).

Ovigerous and recently ovigerous females were collected in March through early June and in late July through September (Fig. 7). Evidence for the presence of ovigerous females in the population was also found in early July and early October, as determined by the occurrence of zoeae and megalopae in nursery areas. Ovigerous females were not collected between early October and mid-March, nor were there any larvae in nursery areas during this period. Although there were ovigerous females in the population from mid-March through early October, there were peaks in numbers in mid- to late spring and in mid- to late summer, with a greater proportion occurring in the latter season. These peaks corresponded to the ascending limb of average rainfall peaks during the warmer months of the year – May and early June and again in August and early September (Fig. 7).

4.2.2 *Incubation period and seasonal egg production*

Females bearing newly deposited eggs were collected in late March through early April, and in late July, with the exception of a single individual in late August (Fig. 7). Most ovigerous females bore late stage eggs which hatched within a few days, if not immediately. A series of ovigerous females with a progression of various stage eggs was collected from late July through late August. From these data, laboratory incubation of eggs in various stages of development,

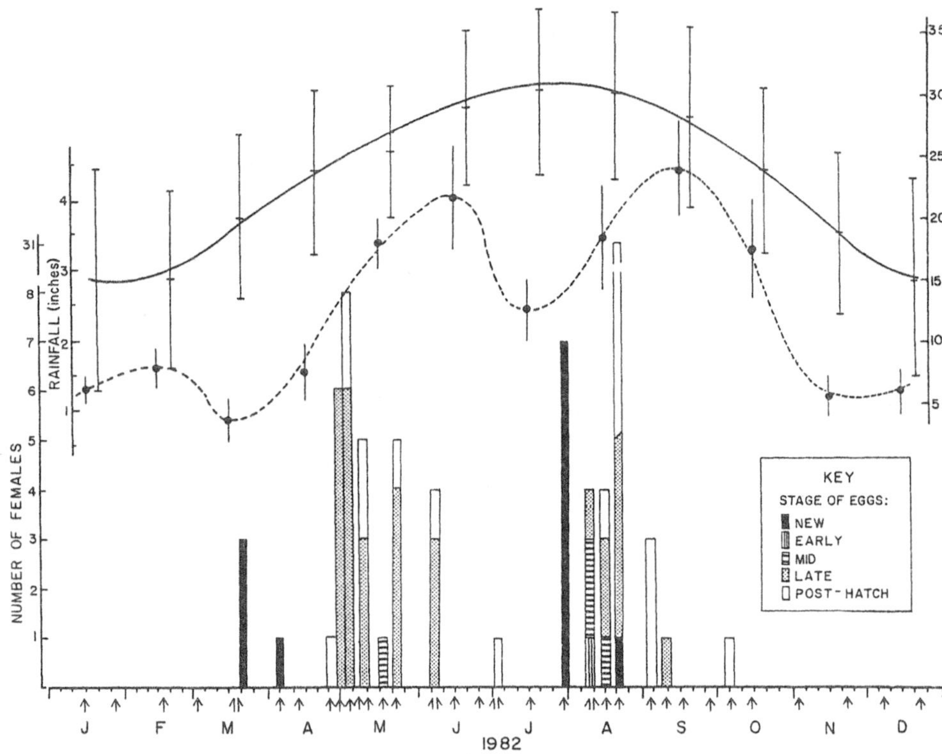

Figure 7. Occurrence of females of *Uca subcylindrica* at Santa Gertrudis Creek in 1982 with eggs in various stages of development (see Key; post-hatch females indicated by females bearing unhatched eggs or egg membranes on pleopods, or by presence of larvae or early postlarvae in nursery areas). Air temperatures (mean, mean maximum and mean minimum for 1982; solid line) and rainfall (23-yr monthly means and ranges; dotted line) superimposed on figure. Collection dates (arrows) marked across time scale.

and the occurrence of zoeae and megalopae in nursery areas, I estimated that the incubation period during the late summer peak of reproductive activity was approximately 1 month. Based on the relatively few collections of females with newly-deposited or early stage eggs during the spring peak, and laboratory experiments with larval development, I estimated that the incubation period for this season was 1.5 months. Mean air temperatures varied from 20°C for spring to near 30°C for summer.

The period of incubation for each of the periods of peak reproductive activity corresponded with the periods during the warmer months, when rainfall was likely to occur. Ovigerous females were absent from the population in the intervening one to one and a half months. Based solely on this information, I postulate that the females are limited, at a maximum, to a single brood in each reproductive season, one brood per spring and one brood per late summer.

4.3 *Yearly egg production*

Rabalais & Gore (1985) proposed that, even though abbreviated development produced a greater drain on the energy reserves of the parental female than would production of many

Table 1. Gonadal condition of ovigerous and recently ovigerous femals of *Uca subcylindrica* (n = 25).

Condition of ovaries	Ovigerous females	Recently ovigerous females	Total
Flaccid, empty	24%	7%	31%
Developed, but with small number of small eggs	10%	38%	48%
Developed with medium number of small eggs or medium-sized eggs	0%	21%	21%
Total	34%	76%	100%

small, yolk-poor eggs (because of the substantial quantities of protein and lipid required), the benefit was that embryonic development would be long; during that time the female could replenish her lipid reserves in preparation for the next deposition of eggs. Based on peaks in breeding activity and occurrence of ovigerous females spaced temporally to allow for mating, egg deposition and incubation of the large eggs, Rabalais (1983) and Rabalais & Gore (1985) also proposed that females of *U. subcylindrica* may be able to produce multiple (two) broods per year. Temperate zone *Uca* spp. produce multiple (4 to 5) broods per year; it seemed reasonable that *U. subcylindrica* could produce two per year.

Dissections of 25 adult female *U. subcylindrica* provided information relative to this hypothesis. In 31% of the ovigerous and recently ovigerous females, the ovaries were in the flaccid, empty condition with the hepatopancreas reduced or not evident (Table 1). An additional 48% were developed but with a small number of small eggs; these individuals also had reduced hepatopancreases. Only a few (20.7%) were in the medium stage of development, and none were well-developed and full of large eggs. Most of the females dissected were recently ovigerous and thus had oviposited 1 to 1.5 months previously. The ovigerous females which were dissected carried eggs in various stages of development, but most were in the late stage of development and represented a similar period since deposition of eggs. During this time, the hepatopancreas had not recovered, lipid reserves were still reduced, and gonads were not becoming fully redeveloped with large ova. These observations suggest that females were not redeveloping a subsequent brood shortly after release of larvae and are thus limited to a single brood per year.

Dissections of gonads of non-ovigerous adult females further supported this conclusion. The sample size of ovigerous or recently ovigerous females was small compared to the total adult female population (5.5 to 25%). The percentage of non-ovigerous females was larger and more indicative of the adult female population and thus a better indicator of egg production sequences in the population. The percentage of females with 'small' or flaccid ovaries closely paralleled the ovigerous or recently ovigerous population (Fig. 8). Sample size for the 'large' category was smaller than those of the 'small' or 'medium' (but still 'large'), indicating that these individuals may have become less active (i.e. less catchable) immediately prior to egg deposition.

Peaks within each gonadal category can be followed in the population through time (Fig. 9). Climatic events, such as average air temperature, may influence life cycles and thus modify the trend from year to year. Individuals with small or flaccid ovaries in winter and early spring progress to medium-developed ovaries in the late winter and spring, to large ovaries in spring and early summer, and to egg deposition that results in larval hatching in mid- to late summer. Individuals with small or flaccid ovaries in late summer and fall progress to medium-developed

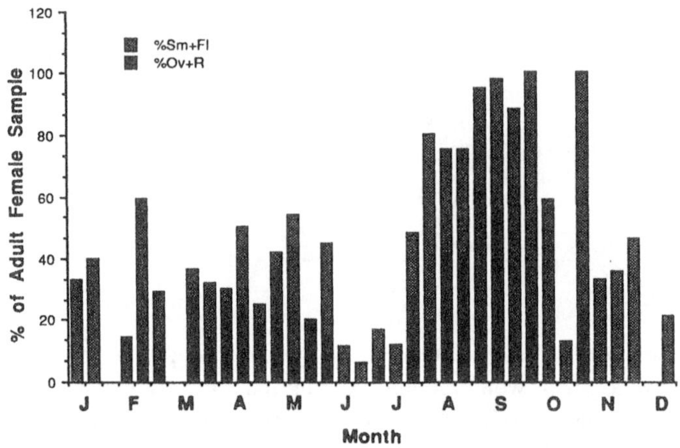

Figure 8. Percentage of adult females that were ovigerous or recently ovigerous (%Ov + R, black shading) or that had ovaries either flaccid or developed but small with small number of small eggs (%Sm + Fl, stippled shading).

ovaries in fall, to large ovaries in late fall and early winter, and to egg deposition that results in larval hatching in mid- to late spring. As about one-third (29%) of the recently ovigerous females have 'medium' ovaries (Table 1), it is possible that they may be able to produce two broods per year (Fig. 9); but as a proportion of the population, these individuals were few.

Adult females produce one brood per year; the population is thus divided into those that are preparing for egg production and larval release in mid- to late spring or in mid- to late summer. The peaks are coincident with periods of higher probability of rain. It would be advantageous for portions of the population to be primed for larval release at these times because of the likely occurrence of suitable nursery areas.

4.4 *Lifetime reproductive output*

Flexibility in timing of reproductive events would be advantageous within populations of *U. subcylindrica* living in a region of seasonal though sporadic rainfall. Thus, some individuals could release larvae during either of the periods with the most rain. What would happen to the reproductive output of females in a season when there is no rain for a long period or none at all? In the case of the former, larval release might be delayed. In the case of the latter, the reproductive output of those females would probably be lost. Even with other aquatic habitats available, such as creeks, bays or lagoons, ephemeral rain puddles are the only nursery areas selected. Samples of burrow water never contained any of the early larval or postlarval stages. Successful larval release depends on the occurrence of rain and the formation of ephemeral ponds or puddles. The probability is low that there will be no rain during both of the periods of expected rain. Hence, at least a portion of the population will successfully produce larvae during one of the two periods of higher rainfall during the year.

Length of adult life is not known for *U. subcylindrica*. If it is similar to that of other fiddler crabs, then the females usually live 2 or 3 years (Colby & Fonseca 1984). There would therefore be 2 or 3 opportunities, depending on the growth rate of the females, to successfully produce a brood of larvae. The success in producing larvae during any one year would also be a factor and would be mostly dependent on suitable rainfall. There is no indication that females of *U. subcylindrica* enter the reproductive population at a smaller size than do females of its congeners (Colby & Fonseca 1984). The megalopal and first crab stages of *U. subcylindrica* are

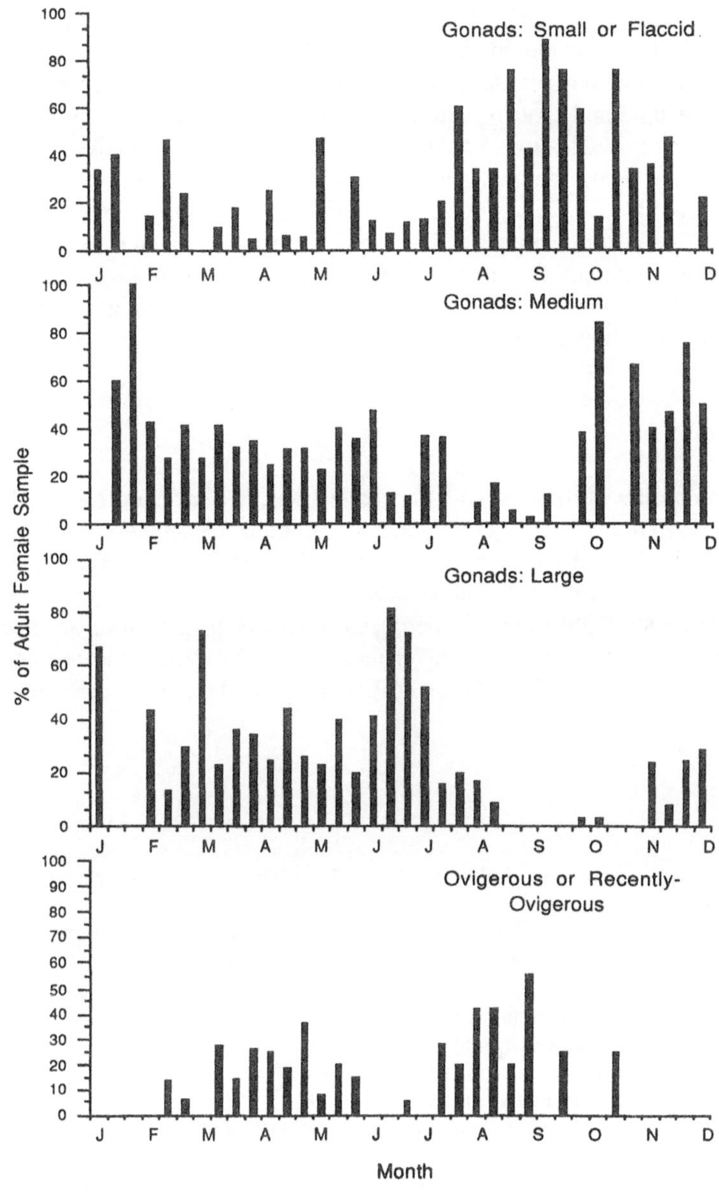

Figure 9. Percentage of adult females that had gonads either flaccid or developed but small with a small number of small eggs, gonads that were developed with a medium number of medium-sized or small eggs, gonads that were well-developed and full of large eggs, or that were ovigerous or recently ovigerous.

larger than those of other *Uca* (Rabalais & Gore 1985). Whether these initial differences provide a mechanism for faster attainment of size of sexual maturity is also unknown. Knowing these factors would be necessary to determine the lifetime reproductive output for *U. subcylindrica*.

4.5 *A model for crabs with abbreviated development?*

Can what is known of the reproductive biology of *U. subcylindrica* be used to develop a reasonable model of egg production in crabs with abbreviated development? Based on the detailed data for *U. subcylindrica* and the limited published information available on other species, it appears that egg production is limited to a single brood per year. Egg production is limited by the quantities of lipid and protein required to produce an egg mass with large, yolk-filled eggs and the time required to replenish depleted reserves. Many unknowns, such as growth rate of females, size at sexual maturity and yearly success, would influence lifetime reproductive output. The reproductive biology data for *U. subcylindrica*, while more complete than for most other species, are still not adequate to develop a general model for egg production in crabs with abbreviated development. The life history pattern of *U. subcylindrica*, however, may be similar to those of other crabs as patterns of egg size, egg number, and larval development sequence converge across taxonomic lines.

ACKNOWLEDGEMENTS

I am indebted to the many people who helped with field collections: A. Gracia, R. Rogers, O. Rabalais, S.C. Rabalais, R. Henry, T. Garza, D.L. Felder, R. Schmidt, T. Van Ness and A.H. Chaney. L. Smith helped with dissections and measurements. The manuscript benefited from the reviews of A.B. Williams, M. Salmon, R.E. Turner, A.M. Wenner and A. Kuris. Support was provided by NSF Grant PCM 80-24358 to J.N. Cameron, University of Texas Marine Science Institute, for data collection and by the Louisiana Universities Marine Consortium and the State of Louisiana Education Quality Support Fund Grant 86-LUM(2)-084-09 for manuscript preparation.

REFERENCES

Abele, L.G. & D.B. Means 1977. *Sesarma jarvisi* and *Sesarma cookei*: Montane, terrestrial grapsid crabs in Jamaica (Decapoda). *Crustaceana* 32: 91-93.

Brown Jr, L.F., J.L. Brewton, J.H. McGowen, T.J. Evans, W.L. Fisher & C.G. Groat 1976. *Environmental Geologic Atlas of the Texas Coastal Zone – Corpus Christis Area*. Bureau of Economic Geology, The Univ. of Texas at Austin, 123 pp.

Campodonico, I. & L. Guzman 1981. Larval development of *Paralomis granulosa* (Jacquinot) under laboratory conditions (Decapoda, Anomura, Lithodidae). *Crustaceana* 40: 272-285.

Christiansen, M.E. & W.T. Yang 1976. Feeding experiments on the larvae of the fiddler crab *Uca pugilator* (Brachyura, Ocypodidae), reared in the laboratory. *Aquaculture* 8: 91-98.

Christy, J.H. 1982. Adaptive significance of semilunar cycles of larval release in fiddler crabs (genus *Uca*): Test of an hypothesis. *Biol. Bull.* 116: 373-396.

Colby, D.R. & M.S. Fonseca 1984. Population dynamics, spatial dispersion and somatic growth of the sand fiddler crab *Uca pugilator*. *Mar. Ecol. Prog. Ser.* 16: 269-279.

Crane, J. 1941. Crabs of the genus *Uca* from the west coast of Central America. *Zoologica* 26: 145-207.

Feest, J. 1969. Morphophysiologische Untersuchungen zur Ontogenese und Fortpflazungsbiologie von *Uca annulipes* und *Uca triangularis* mit Vergleischbefunden an *Ilyoplax gangetica*. *Forma et Functio* 1: 159-225.

Gangotri, M.S., N. Vasantha & S.A.T. Venkatachari 1971. Sexual maturity and breeding behavior in the freshwater crab, *Barythelphusa guerini* Milne Edwards. *Proc. Indian Acad. Sci.* 87B: 195-201.

Gibbs, P.E. 1974. Notes on *Uca burgersi* Holthuis (Decapoda, Ocypodidae) from Barbuda, Leeward Islands. *Crustaceana* 27: 84-91.

Gore, R.H. 1985. Molting and growth in decapod larvae. In A.M. Wenner (ed.), *Larval Growth. Crustacean Issues 2:* pp. 1-65. Rotterdam: Balkema.

Gray, E.H. 1942. Ecological and life history aspects of the red-jointed fiddler crab, *Uca minax* (Le Conte), region of Solomons Island, Maryland. *Chesapeake Biol. Lab. Publ.* 51: 3-20.

Guinot, D. 1976. Constitution de quelques groupes naturels chez les Crustacés Décapodes Brachyoures. 1. La superfamille des Bellioidea et tres sous-familles de Xanthidae (Polydectinae Dana, Trichiinae de Haan, Actaeinae Alcock). *Mémoires du Muséum National d'Histoire Naturelle Paris (N.S.) Série A,* 97: 1-308, 19 pls.

Hale, H.M. 1925. The development of two Australian sponge crabs. *Proc. Linn. Soc. New. South Wales* 50: 405-413.

Hale, H.M. 1927. *The Crustaceans of South Australia, Part I.* British Science Guild, Adelaide.

Hale, H.M. 1931. The post-embryonic development of an Australian xanthid crab (*Pilumnus vestitus* Haswell). *Rec. South Aust. Mus.* 4: 321-331.

Hartnoll, R.G. 1964. The freshwater grapsid crabs of Jamaica. *Proc. Linn. Soc. London* 176: 113-147.

Hartnoll, R.G. 1971. *Sesarma cookei* n.sp., a grapsid crab from Jamaica (Decapoda, Brachyura). *Crustaceana* 20: 257-262.

Hedgpeth, J.W. 1967. Ecological aspects of the Laguna Madre, a hypersaline estuary. In G.A. Lauff (ed.), *Estuaries.* pp. 408-419. Amer. Assoc. Advmt. of Science Publ. No. 83, Washington, DC.

Herrnkind, W.F. 1968. The breeding of *Uca pugilator* (Bosc) and mass rearing of the larvae with comments on the behavior of the larval and early crab stages (Brachyura, Ocypodidae). *Crustaceana* suppl. 2: 214-224.

Hines, A.H. 1982. Allometric constraints and variables of reproductive effort in brachyuran crabs. *Mar. Biol.* 69: 309-320.

Hines, A.H. 1986. Larval patterns in the life histories of brachyuran crabs (Crustacea, Decapoda, Brachyura). *Bull. Mar. Sci.* 39: 444-466.

Hyman, O.W. 1920. The development of *Gelasimus* after hatching. *J. Morphol.* 33: 487-501.

Jones, M.B. 1978. Aspects of the biology of the big-handed crab, *Heterozius rotundifrons* (Decapoda: Brachyura), from Kaikoura, New Zealand. *New Zealand J. Zool.* 5: 783-794.

Koba, K. 1936. Preliminary notes on the development of *Geotelphusa dehaani* (White). *Proc. Imp. Acad. Japan* 12: 105-107.

Kurata, J. 1970. *Studies on the Life Histories of Decapod Crustacea of Georgia: part III. Larvae of Decapod Crustacea of Georgia.* Final Rep. Sapelo Island, Georgia: Univ. of Georgia Mar. Sci. Inst.

Lucas, J.S. 1971. The larval stages of some Australian species of *Halicarcinus* (Crustacea, Brachyura, Hymenosomatidae). I. Morphology. *Bull. Mar. Sci.* 22: 824-840.

Montague, C.L. 1980. A natural history of temperate western Atlantic fiddler crabs (genus *Uca*) with reference to their impact on the salt marsh. *Contrib. Mar. Sci.* 23: 25-55.

Montgomery, S.K. 1922. Direct development in a dromiid crab. *Proc. Zool. Soc. London* 1922: 193-196.

Norwine, J., R. Bingham & R.V. Zepeda 1977. Twentieth-century semi-arid and subhumid climates and climatic fluctuations in Texas and northeastern Mexico: Some lessons and implications for the future. *Intl. Conf. on the Meteorology of Semi-Arid Zones,* American and Israeli Meteorological Societies, Tel Aviv, Israel, Oct. 31-Nov. 4, 1977.

Provenzano Jr, A.J. 1968. *Lithopagurus yucatanicus,* a new genus and species of hermit crab with a distinctive larva. *Bull. Mar. Sci.* 18: 627-644.

Rabalais, N.N. 1983. Adaptations of fiddler crabs *Uca subcylindrica* (Stimpson, 1859) to semi-arid environments. Ph.D. Dissertation, The Univ. of Texas at Austin, 335 pp.

Rabalais, N.N. & J.N. Cameron 1983. Abbreviated development in *Uca subcylindrica* (Stimpson 1859) (Crustacea, Decapoda, Ocypodidae) reared in the laboratory. *J. Crust. Biol.* 3: 519-541.

Rabalais, N.N. & J.N. Cameron 1985a. The effects of factors important in semi-arid environments on the early development of *Uca subcylindrica. Biol. Bull.* 168: 147-160.

Rabalais, N.N. & J.N. Cameron 1985b. Physiological and morphological adaptations of adult *Uca subcylindrica* to semi-arid environments. *Biol. Bull.* 168: 135-146.

Rabalais, N.N. & R.H. Gore 1985. Abbreviated development in decapods. In A.M. Wenner (ed.), *Crustacean Issues 2: Larval Growth.* pp. 67-126. Rotterdam: Balkema.

Ringold, P.L. 1979. Costs to reproduction in an egg carrying mobile crustacean. *Amer. Zool.* 19: 928.

Soh, C.L. 1969. Abbreviated development in a non-marine crab, *Sesarma (Geosesarma) perracae* (Brachyura; Grapsidae), from Singapore. *J. Zool., London* 158: 357-370.

Tyagi, A.P. 1972. Life history of the freshwater crab *Parathelphusa masoniana* (Henderson). *Zoologica Poloniae* 22: 171-176.

von Hagen, H.O. 1070. Anpassungen an das spezielle Gezeitenzonen-Niveau bei Ocypodiden (Decapoda, Brachyura). *Forma et Functio* 2: 361-413.

Wear R.G. 1967. Life-history studies on New Zealand Brachyura. 1. Embryonic and post-embryonic development of *Pilumnus novaezealandiae* Filhol, 1886, and of *P. lumpinus* Bennett, 1964 (Xanthidae, Pilumninae). *New Zealand J. Mar. Freshwat. Res.* 1: 482-535.

Wear, R.G. 1968. Life-history studies on New Zealand Brachyura. 2. Family Xanthidae. Larvae of *Heterozius rotundifrons* A. Milne Edwards, 1967, *Ozius truncatus* H. Milne Edwards, 1834, and *Heteropanope (Pilumnopeus) serratifrons* (Kinahan, 1856). *New Zealand J. Mar. Freshwat. Res.* 2: 293-332.

Wear, R.G. 1970. Some larval stages of *Petalomera wilsoni* (Fulton & Grant, 1902) (Decapoda, Dromiidae). *Crustaceana* 18: 1-12.

Wear, R.G. 1974. Incubation in British decapod Crustacea, and the effects of temperature on the rate and success of embryonic development. *J. Mar. Biol. Assc., UK* 54: 745-762.

Wear, R.G. & E.J. Batham 1975. Larvae of the deep sea crab *Cymonomus bathamae* Dell, 1971 (Decapoda, Dorippidae) with observations on larval affinities of the Tymolinae. *Crustaceana* 28: 113-120.

Yamaguchi, T. & Y. Takamatsu 1980. Ecological and morphological studies on the Japanese freshwater crab, *Geothelphusa dehaani*. *Kumamoto J. Sci., Biol.* 15: 1-27.

Special problems in freshwater crayfish egg production

Jay V. Huner
Crayfish Center, University of Southwestern Louisiana, Lafayette, USA

Ossi V. Lindqvist
University of Kuopio, Finland

ABSTRACT

Fisheries managers and aquaculturists are concerned about crayfish egg production as it relates to the management of natural populations and culture of captive populations. However, there is a paucity of information about the subject, despite the economic and ecological importance of astacid and cambarid crayfishes in the Northern Hemisphere and parastacid crayfishes in the Southern Hemisphere. We review here effects of population structure on fecundity, egg production as a function of size, variation in egg size, effects of natural cues (seasonality) on egg production, effects of maternal nutrition on egg production and effects of handling on egg production.

1 INTRODUCTION

1.1 *Natural fisheries and culture*

Crayfishes are the dominant benthic invertebrates in many temperate regions of North America, Europe, and Australia (Momot 1984). In addition to their ecological significance, such populations are exploited commercially for food and profit, especially in Scandinavia, Spain and Turkey in Europe; in California, Oregon, Washington and Louisiana in the United States; and in eastern and southwestern Australia (Huner 1978, 1983). Over 60 000 ha of earthen culture ponds occur in the southern USA, with more than 90% of these in Louisiana (Roberts pers. comm.). Young crayfish are cultivated on a very small scale, comparatively speaking, in most of Europe and parts of Australia. European efforts however, have led to re-establishment of crayfish fisheries in southern Sweden, where native crayfish had been largely eliminated by a virulent crayfish pathogen, the crayfish fungus plague, *Aphanomyces astaci* (Fjälling & Fürst 1988).

1.2 *European, North American and Australian crayfishes*

There are five native crayfish species (family Astacidae) in Europe, the most important being *Astacus astacus, Astacus leptodactylus* and *Austropotamobius pallipes* (Kossakowski 1966, Mancini 1986). In contrast to the situation in Europe, there are over 300 species of North

American crayfishes in the Family Cambaridae (Hobbs 1972). Two of these species have been successfully introduced into Europe including: *Procambarus clarkii*, most successful in warm areas; and *Orconectes limosus*, most successful in cooler areas where water quality is poor, habitat has been disturbed, and/or the crayfish fungus plague is epidemic (Huner 1977, Kossakowski 1966). There is only one important astacid crayfish in North America, the West Coast species *Pacifastacus leniusculus* (Huner 1983). All North American crayfishes are apparently highly resistant to the crayfish fungus plague, but are, unfortunately, also carriers of the malady (Unestam 1972, Persson & Soderhall 1983, Vey et al. 1983).

There are approximately 100 species of crayfishes in Australia (family Parastacidae) (Morrissy 1983). These include the largest of the freshwater crayfishes. The species belonging to the genera *Cherax* and *Euastacus* are among the most important from both ecological and economic viewpoints.

1.3 *Significance of crayfish egg studies*

The topic of crayfish egg production is one of significance to fisheries managers and aquaculturists as well as environmentalists concerned with perpetuating native crayfish stocks. This brief review discusses the state of knowledge about factors affecting crayfish egg production. It concludes with suggestions for research topics to fill gaps in our knowledge about such factors.

2 OVERALL EGG OUTPUT OF A POPULATION

2.1 *Potential versus net fecundity*

Ovarian egg counts are typically used to describe potential fecundity of many aquatic species (Snyder 1983). Some may argue that because net fecundity, the number of attached eggs or potentially free living second/third instar crayfish, is always lower than potential fecundity (Corey this volume, Somers this volume), net fecundity is a more desirable population characteristic to measure. However, female crayfishes move into hiding when they lay and incubate eggs. Thus, it is difficult to collect them and, in collecting them, it is very easy to dislodge eggs and/or young crayfish. This disturbance biases the data gathered. Available literature presents ovarian and/or pleopod egg counts, although they are frequently not reported simultaneously. Summaries of cambarid and astacid crayfish ovarian and pleopod egg numbers are presented by Momot (1984) and Corey (this volume).

2.2 *Population sampling bias*

Most crayfish population studies are based on samples captured with traps. There may be great differences in the crayfish caught in terms of sex, nutritional status and stage of ovarian development depending on the time of year that the samples are taken (Abrahamsson 1971, Huner & Lindqvist 1986). Trap sampling generates a hidden 'bias' that can be difficult to avoid unless frequent samples are taken during the year or non-selective sampling techniques (such as hand collection using SCUBA gear) are used to collect samples.

3 EGG NUMBERS AS A FUNCTION OF SIZE

3.1 *Prediction of egg production based on size*

3.1.1 *Regression equations*

Although pleopod egg numbers may be as much as 50% lower than ovarian egg numbers in various crayfishes (Corey this volume), there is a direct relationship between egg numbers and size. Somers (this volume) has devoted considerable effort to evaluate methods for mathematically describing the relationship between ovarian and pleopod egg numbers versus size in crustaceans, including crayfishes. Methods use linear regression models applied to raw or logarithmically transformed data. Coefficients of correlation are remarkably similar regardless of method used to predict egg numbers.

3.1.2 *Examples of predicted egg numbers in commercially important species*

In general crayfishes that have a maximum size exceeding 10 cm in total length (TL) are regarded as suitable for human food, but they usually mature at sizes around 7 cm TL. (Total length is approximately twice the cephalothorax length.) Such large crayfishes invariably dominate the benthic communities where they occur. Therefore, larger forms are usually of more interest to fisheries managers than smaller species. Table 1 presents some selected ovarian and pleopod egg numbers for 7 cm and 10 cm TL crayfishes that have potential or realized commercial importance. [Readers are referred to Corey (this volume) and Momot (1984) for additional references dealing with smaller species.]

3.1.3 *Problems of obtaining and interpreting egg production data*

Few life history studies report egg production as some function of body length. It is very easy to

Table 1. Ovarian and pleopodal egg production in several species of large astacid and cambarid crayfishes. TL = total length.

Species	Ovarian (TL)		Reference	Pleopodal (TL)		Reference
	7 cm	10 cm		7 cm	10 cm	
Astacids						
Astacus astacus	85	280	Lindqvist & Louekari 1975	20	120	Abrahamsson 1971
Astacus leptodactylus	60	270	Kossakowski 1966	–	–	
Pacifastacus leniusculus	125	350	Mason 1975, McGriff 1983	50	170	Abrahamsson 1971
Cambarids						
Cambarus robustus	–	–		95	170	Corey 1987
Procambarus clarkii	325	605	Delong (pers. comm.)	100	535	Penn 1943
Procambarus acutus acutus	280	420	Delong (pers. comm.)	–	–	
Procambarus hayi	285	400	Payne 1971	170	260	Payne 1971
Orconectes causeyi	375	750	Dean 1969	270	465	Dean 1969
Orconectes limosus	250	400	Kossakowski 1966	110	220	Kossakowski 1966
Orconectes rusticus	270	460	Prins 1968	190	320	Prins 1968
Orconectes virilis	–	–		280	480	Weagle & Ozburn 1972

count ovarian eggs, even though it is often difficult to collect ovigerous females. Thus, it would seem that life history studies should provide ovarian egg numbers as a matter of course. This is especially critical when one considers the intraspecific variability described by Corey (this volume).

An additional problem for assessment of egg number data was that papers providing regression equations or related data often neglected to specify if carapace length were 'postorbital' cephalothorax length or cephalothorax length.

3.2 *Effects of size variation and age class on egg production*

Size at maturity for crayfishes varies both interspecifically and intraspecifically. The dwarf cambarids, *Cambarellus* spp., rarely exceed 3-4 cm TL, while some of the large cambarids, *Procambarus* spp., may exceed 13-14 cm TL. As a rule, most species mature at one-third to one-half their maximum sizes (Black 1966, Huner & Barr 1984, Huner & Romaire 1979).

Procambarus clarkii, the most cosmopolitan of all crayfishes, is native to northwestern Mexico and the southcentral USA (Huner 1977). It matures at sizes as small as 4.5 cm TL but may, under favorable conditions, reach maturity at sizes in excess to 12 cm TL (Huner & Barr 1984). Both Penn (1943) in Louisiana and Suko (1956) in Japan presented extensive observations on ovarian egg numbers in *P. clarkii*. The direct dependence of ovarian egg numbers and size is very clear, with 5-6 cm TL females producing 50-100 ovarian eggs while 11-12 cm TL females produced 500-600 ovarian eggs. In general, for crayfishes larger/older females appear to lay more eggs than smaller/younger ones. However, it is not clear if very old females start showing increased infertility. This may be a problem with long lived species such as *Astacus astacus*. See Momot (1984) for further discussion of this matter.

3.3 *Some reasons for intraspecific size variation*

In uncropped crayfish populations, old, dominant males inhibit growth and survival of young crayfish. Females in such populations are smaller and produce fewer eggs than those in closely cropped populations. For example, Morgan & Momot (1988) found that an uncropped *O. virilis* population in northwestern Ontario exhibited more erratic egg production per female than did a nearby cropped population.

Momot and co-workers (Momot & Gowing 1977, Jones & Momot 1981, Momot 1986) studied *O. virillis* populations in Michigan. Females matured at 7-8 cm TL. Hazlett & Rittschof (1985) also studied an *O. virilis* population in Michigan, but females routinely matured at sizes over 10 cm TL. Environmental conditions (food, water quality and refugia) were clearly more favorable at the latter location and contributed to greater potential egg production.

Huner & Romaire (1979) and Taylor (1983) documented responses of *P. clarkii* and *Procambarus spiculifer* populations to variation in habitat. Those in favorable habitats responded by reaching near maximum size for the species while those in unfavorable habitats matured but were considerably smaller and less fecund.

4 VARIATION IN EGG SIZE

4.1 *r-Selected versus K-selected species*

Egg size will effect the number of eggs that two crayfishes of equal size can produce. The *r*-selected crayfishes are typically short lived, grow rapidly, and produce high numbers of 'small' eggs. *Procambarus clarkii* can be considered a *r*-selected species whose mature, extruded eggs are about 2 mm in diameter. A large female of 13 cm TL may produce 600+ ovarian eggs (Suko 1958). Conversely, a *K*-selected species is long lived, grows slowly, and produces a low number of 'large' eggs. The cambarid, *Cambarus robusus*, is a *K*-selected species with an egg diameter near 3 mm. Based on data presented by Corey (1987) for *C. robustus* and Suko (1958) for *P. clarkii*, the numbers of ovarian eggs for females of approximately 7.5 cm TL would be 85 and 325, respectively.

4.2 *Value of large eggs in K-selected species*

K-selected crayfish species have been favored in harsher, colder climates or regions where incubation takes a significant amount of time, especially in astacids which oviposit in mid-autumn. Hatching occurs in midspring-midsummer, depending on latitude (Momot 1984). Apparently, strong seasonality introduces an element of predictability in the environment, which favors longer length of life, larger size at maturity with repeated spawnings, and relatively slow development (cf. Pianka 1970). Thus, eggs of such species carry more organic reserves, permitting development of viable young during the lengthy incubation period.

4.3 *Importance of r-K continuum?*

Morgan & Momot (1988) suggested that the *r-K* continuum is not especially important in describing differences related to egg size and variability in egg size. These differences, they noted, were not constant, but changed in response to exploitation, as seen in two *O. virilis* populations studied over 12 years in northwestern Ontario. An exploited population produced greater numbers of smaller eggs with less size variability (2 to 4 times smaller in diameter than in the unexploited population). The production of young in an unexploited population was related to the 'grow or die' syndrome, i.e., females produced very large eggs. The resultant young had improved survivorship in a variable, austere environment with predation and intimidation from large males. Exploitation broke this mechanism down, and females produced more and smaller eggs. Those results suggest that large males had a greater influence on egg production than did the environment. Egg size and variability in egg size, thus, appear to be measures of the intensity of population regulation.

Orconectes virilis oviposits in the spring. Thus the incubation period of about 2 months is far shorter than that observed in other North Temperate species such as *C. robustus* and *A. astacus*, which oviposit in the autumn with late spring or early summer hatching. Therefore, because less organic reserves are required for the short incubation species, species like *O. virilis* produce more and smaller eggs.

Mason (1979) noticed variation in egg size of *P. leniusculus*, with larger and presumably older females producing the largest eggs. However, he found no difference in viability associated with egg size. Skurdal & Qvenild (1986) found no correlation between female size (age) and egg diameter in a Norwegian population of *A. astacus*. Neither study addressed survival of young in the environment. Based on Morgan & Momot's data, one should expect that egg size does make a difference in whether or not the young would subsequently survive in an austere environment.

5 EFFECTS OF NATURAL CUES (SEASONALITY) ON EGG PRODUCTION

Light and temperature are natural cues that affect the reproductive cycles of organisms. Several studies have reported the effects of changes in these cues (photoperiod and/or temperature) on the ovarian cycle in various crayfishes. These studies provide a better understanding of endocrine systems in arthropods. They also generate information that permits aquaculturists to manipulate ovarian cycles, producing young crayfish on a predictable basis.

5.1 *Significance of photoperiod*

When temperature is held constant and photoperiod is varied, different results are observed. In general, ovaries in crayfish that are held on short day and normal day photoperiods mature quickly, but eventually recycle. Those held in constant darkness tend to mature and stabilize. those held on long day or constant light photoperiods recycle continuously. For specific studies see Stephens (1952) on *O. virilis*, Lowe (1961) on *Cambarellus shufeldtii*, Perryman (1969) on *Procambarus simulans*, Armitage et al. (1973) on *Orconectes nais*, and Rice & Armitage (1974) on *O. nais*.

There are two problems, however, with interpretation of results of these studies. First, there were no mating opportunities to determine if the viable offspring could be produced from the so-called mature ovaries. Second, reference was made to cyclic maturation/resorption of oöcytes in presumed immature females. By definition, ovarian maturation is not possible if animals are immature. Further, the questionably immature females were within the size ranges at which the species can be mature. In future studies, maturity should be determined by external examination (Huner & Barr 1984).

5.2 *Significance of environmental cues at lower latitudes*

Payne (1972) noted that the southern species, *Procambarus hayi*, spawned in early summer in his laboratory, 2-3 months before it would in the field. He speculated that early oviposition in his laboratory animals was due to exposure to longer photoperiods. These laboratory observations might be expected in mated females based on the recycling of ovaries noted above for female crayfishes held without males but on a long day photoperiod.

One study included mating under laboratory conditions, with temperature held constant, but with a decreasing photoperiod. Dendy (1978) noted that reduction of day length by 5 minutes per day, beginning in winter from 12h Light:12h Dark, led to early summer spawning in *P. clarkii* when the light phase declined to 2-3 hours. Again, based on earlier studies, short day caused maturation of ovaries; in a mated crayfish one would expect spawning once the ovary matured.

An understanding of the natural cycle of *P. clarkii* and *C. shufeldtii* will help interpret responses to variation in photoperiod and temperature. Both species mature in 21/2 to 6 months, generally live less than 18 months, and may spawn twice in the same year (Black 1966, Suko 1958, Huner 1975, Black & Huner 1980). Both species appear most frequently in temporary habitats, such as seasonally flooded marshes and swamps. As long as surface water is present, newly released young may be found year-round, although peak reproduction takes place in autumn with a secondary peak in spring (Penn 1943, Suko 1958, Black 1966, de la Bretonne & Avault 1977). Mature females retire to plugged earthen burrows to lay and brood eggs. Therefore, they may to a degree control photoperiod and temperature. Hence, the availability of suitable aquatic habitat may be more important for control of the ovarian cycle than are environmental cues.

5.3 *Significance of environmental cues at higher latitudes*

Aiken (1969) first demonstrated the significance of temperature on ovarian development and oviposition in high latitude crayfishes. He found that ovarian maturation in Canadian (Alberta) populations of *O. virilis* requires 4-5 months in total darkness and low temperatures circa 4°C. Temperatures had to rise to threshold levels of 10-11°C before oviposition could occur. Aiken noted that 'complete maturation is assured only when both temperature and photoperiod are in proper proportion for a prescribed period.' Berrill & Arsenault (1982) reported similar results for *O. rusticus*, also from a Canadian (Ontario) population. They concluded that while a rise in water temperature initiated breeding events, the absolute temperature increase might be inversely correlated with the duration of lengthening photoperiod. They reported that the crayfish bred in mass when water temperatures rose above 4°C. Capelli & Magnuson (1975) observed a similar pattern in Wisconsin during the spring after water temperatures reached 7°C.

The North Temperate astacids mate and spawn in the autumn, when water temperatures and day length decrease rapidly. Westin & Gydemo (1986) reported that temperature was the dominant cue controlling oviposition in *A. astacus*. Oviposition took place when light phases were increasing, decreasing and artificial (constant and 5h Light:7h Dark), as long as temperature was decreasing from a summer high of 20°C to a winter low of 4°C. Huner & Lindqvist (1985) noted that *A. astacus* held at 20°C and 12h Light:12h Dark mated and spawned when conditions were changed to 4°C and 8h Light:16h Dark. Lack of appropriate control treatments prevented separation of temperature from photoperiod effects.

5.4 *Effects of eyestalk ablation on the ovarian cycle*

Stephens (1952) observed that bilateral eyestalk ablation induced precocious spawning in *O. virilis* when eyestalks were removed in the autumn and winter. Eyestalk ablation reduced titers of ovarian inhibitory hormones produced in the eyestalks. Problems with survival of the females were encountered. Zipf (1959) stated that bilateral eyestalk ablation induced precocious oviposition in over 100 *Cambarus bartoni sciotensis* with little mortality. Huner (unpubl. data) performed unilateral eyestalk ablation on female *P. clarkii* with well-developed glair glands. No differences in time and percentage of those ovipositing when compared to intact control animals were noted. However, all females had well-developed cement glands and were probably committed to spawning as the cement glands produce the mucus-like glair material necessary for successful oviposition. Gydemo (pers. comm.) reported no oviposition in mature *A. astacus* subjected to bilateral eyestalk ablation.

Eyestalk ablation to induce oviposition in crayfish may have some aquaculture applications. This technique is used routinely to induce spawning in penaeid shrimps (Lawrence et al. 1985). However, one eyestalk rather than both is removed.

6 EFFECTS OF NUTRITION ON EGG PRODUCTION

6.1 *General observations*

Studies of the effects of nutrition on egg production in crustaceans are largely limited to penaeid shrimps, *Penaeus* spp. and homarid lobsters, *Homarus americanus* (Castell & Kean 1986). Nutrient reserves at the time of hatching are functions of yolk lipid deposited during oögenesis, and the temperature regime to which the eggs have been subjected. Larval lobsters have little, if any, yolk lipid remaining when they hatch. Diminished lipid reserves theoretically, reduce viability (Sasaki et al. 1986). The amount of energy reserves in lobster eggs is greater in eggs produced by older, larger females (Attard & Hudon 1987).

6.2 *Significance of nutrition in Astacus astacus*

6.2.1 *Annual cycle*

Accumulation of organic reserves apparently influences ovarian development in populations of *A. astacus* at high latitudes. Mature females molt once annually in the summer. They lay and incubate eggs in alternate years (Kossakowski 1966, Abrahamsson 1972, Huner & Lindqvist 1986). Molting requires accumulation of organic reserves. Because this cannot be done until midsummer, when young leave the female, there is not enough time to permit ovarian development while accumulating organic reserves for the lengthy boreal winter. Such resting females molt very early in the following summer, and ovarian development/accumulation of organic reserves for the winter commences and continues through the summer and into the autumn. For contrast, *A. astacus* females reproduce annually at lower latitudes (Kossakowski 1966). They are able to spawn twice annually under artificial conditions, as previously discussed (Westin & Gydemo 1986).

6.2.2 *Resorption of eggs*

Ovarian development does not, however, mean that the female will oviposit. Huner & Lindqvist (1986, 1988) studied a slow growing *A. astacus* population in central Finland for several years. Data suggest that only 50-60% of the reproductively active females are actually laying eggs. Food supplies appear limited in the pond. It appears, based on resorption of eggs in reproduction studies (Huner & Lindqvist 1985), that oviposition is nutritionally dependent.

6.2.3 *Interpretation of data*

Kossakowski (1966) noted that expanding *A. astacus* populations in northern Europe are characterized by absence of resting females, i.e. those that do not develop mature ovaries and spawn in the year following spawning. Lindqvist & Louekari (1975) reported a high percentage of such reproductively active females which were spawning annually in an expanding *A. astacus* population in a lake in central Finland. This seems to contradict earlier descriptions of alternating periods of ovarian development and nutrition (Kossakowski 1966, Abrahamsson 1972, Huner & Lindqvist 1986). Perhaps more high quality food is available for crayfish in an expanding population. Therefore, females are able to accumulate enough organic reserves to both molt and regenerate their ovaries. Alternatively, there may be a problem involving interpretation of data. That is, if females are examined for evidence of ovarian development early in the summer season, the catch should be dominated by newly molted females with developing ovaries. Later in the summer, the catch will be dominated by newly molted and very aggressive females that have released young but who show no evidence of ovarian development (Huner & Lindqvist 1986).

6.3 *Significance of nutrition in Orconectes virilis*

Momot (1986) summarized population studies of *O. virilis* at high latitudes. Two points related to female nutrition and egg production. First, egg production was higher in high nutrient lakes than in low nutrient lakes. Second, above a certain threshold of food availability, seasonal limits on food intake seemed to restrict food intake and subsequent bioenergetic translation of excess food into egg production. Further, Momot & Gowing (1977) correlated resorption of oöcytes in *O. virilis* populations in nutrient deficient environments.

6.4 *Significance of nutrition in Cherax tenuimanus*

Morrissy (1975) studied egg production of cultured and wild stocks of the Australian freshwater crayfish, *Cherax tenuimanus*. He noted that egg production of pond cultured stocks was negatively effected by poor nutrition, with eggs being smaller and less numerous in nutrient poor environments. He emphasized that the most productive females were those cultured in a system with an aged, well composted detritus base supplemented with commercial poultry feed.

6.5 *Nutrition studies and carotenoid deficiencies*

Crayfish nutrition is a little studied subject. Studies treat assimilation of nutrients (Moshiri & Goldman 1969, Wiernicki 1984, Brown et al. 1986, Davis & Robinson 1987, Hubbard et al. 1986) and growth responses (Huner & Meyers 1979, Huner 1984, D'Abramo et al. 1985). Increased protein level invariably enhanced growth (Huner & Meyers 1979, Huner 1984); however, most laboratory diets are deficient in carotenoids, so crayfish are pale. When such diets are supplemented with palatable vascular plant material, coloration is normal and growth is increased (Huner 1984). Because carotenoids are deposited in high concentrations in crustacean eggs (Castell & Kean 1986), it can be inferred that diets deficient in carotenoids could adversely affect viability of embryos and larvae. There is limited evidence for this condition in lobsters, *H. americanus* (Castell & Kean 1986), but none for crayfishes.

7 EFFECTS OF HANDLING ON EGG PRODUCTION

Adult crayfish are widely used to stock new crayfish culture ponds and to re-establish crayfish populations in natural waters (Huner & Barr 1984, Huner & Lindqvist 1986). Such crayfishes are captured and held out of water for periods that may extend for several days. The effects on egg production of handling, and transporting the crayfishes are little known. Huner (unpubl. data) found that harvesting female *P. clarkii* with an electrofishing device (high voltage, low amperage, direct current) had no apparent effect on egg production when compared to crayfish captured with traps (the usual method of collecting for stocking purposes). However, *P. clarkii* is a very hardy species which survives long aerial exposures as long as the branchial cavity is wet. In contrast, *A. astacus* is very sensitive to even short aerial exposures and might not survive such exposure.

8 FURTHER RESEARCH

Five areas of research into problems of freshwater crayfish egg production seem especially worthwile and timely. First, more basic intraspecific and interspecific fecundity data are needed. Second, more information is needed on the relative importance of egg size in determining the viability of incubating eggs. Third, more applied research is needed to ascertain the usefulness of varying environmental conditions (photoperiod and temperature) and manipulating hormonal levels through use of eyestalk ablation, etc. on egg production. Fourth, data on the impact that capturing, handling and transporting crayfish have on egg production are needed. Finally, more basic research is needed to determine the effects of maternal nutrition, both qualitatively and quantitatively, on egg production.

ACKNOWLEDGEMENTS

The authors of this article would like to express their appreciation to Drs. Adrian Wenner and Arman Kuris for their patience and assistance in preparing it for publication.

REFERENCES

Abrahamsson, S.A.A. 1971. Density, growth and reproduction in populations of *Astacus astacus* and *Pacifastacus leniusculus* in an isolated pond. *Oikos* 22: 373-380.

Abrahamsson, S.A.A. 1972. Fecundity and growth of some populations of *Astacus astacus* (Linn) in Sweden with special regard to introduction in northern Sweden. *Inst. Freshwater Res. Drottningholm* 52: 23-37.

Aiken, D.A. 1969. Ovarian maturation and egg laying in the crayfish *Orconectes virilis*: influence of temperature and photoperiod. *Can. J. Zool.* 60: 264102645.

Armitage, K.B., A.L. Buikema & N.J. Willems 1973. The effect of photoperiod on organic constituents and molting of the crayfish *Orconectes nais* (Faxon). *Comp. Biochem. Physiol.* 44A: 431-456.

Attard, J. & C. Hudon 1987. Embryonic development and energetic investment in egg production in relation to size of female lobster (*Homarus americanus*). *Can. J. Fish. Aquat. Sci.* 44: 1157-1164.

Berrill, M. & M. Arsenault 1982. Spring breeding of a northern temperate crayfish *Orconectes rusticus*. *Can. J. Zool.* 60: 2641-2645.

Black, J.B. 1966. Cyclic male reproductive activities in the dwarf crawfishes *Cambarellus shufeldti* (Faxon) and *Cambarellus puer* Hobbs. *Trans. Amer. Micros. Soc.* 85: 214-232.

Black, J.B. & J.V. Huner 1980. Genetics of the red swamp crawfish, *Procambarus clarkii* (Girard) (Crustacea: Decapoda) from different habitats. *Proc. World Mariculture Soc.* 11: 535-543.

Brown, P.B., C.D. Williams, E.H. Robinson, D.M. Akiyama & A.L. Lawrence 1986. Evaluation of methods of determining in vivo digestion coefficients for adult red swamp crayfish, *Procambarus clarkii*. *J. World Aquaculture Soc.* 17: 19-24.

Capelli, G.M. & J.J. Magnuson 1975. Reproduction, molting and distribution of *Orconectes propinquus* (Girard) in relation to temperature in a northern mesotrophic lake. *Freshwater Crayfish* 2: 415-428.

Castell, J.D. & J.C. Kean 1986. Evaluation of the role of nutrition in lobster recruitment. *Can. J. Fish. Aquat. Sci.* 43: 2320-2327.

Corey, S. 1987. Comparative fecundity of four species of crayfish in southwestern Ontario. *Crustaceana* 52: 276-286.

D'Abramo, L.R., J.S. Wright, K.H. Wright, C.E. Bordner & D.E. Conklin 1985. Sterol requirements of cultured juvenile crayfish, *Pacifastacus leniusculus*. *Aquaculture* 49: 245-255.

Davis, D.A. & E.H. Robinson 1987. Estimation of the dietary lipid requirement level of the white crayfish *Procambarus actus actus*. *J. World Aquaculture Soc.* 17: 19-24.

Dean, J.L. 1969. Biology of the crayfish *Orconectes causeyi* and its control of aquatic weeds in trout lakes. *Tech. Pap US Bur. Sport Fish. Wild.* 24: 1-15.

de la Bretonne Jr, L. & J.W. Avault Jr 1977. Egg development and management of *Procambarus clarkii* (Girard) in a south Louisiana commercial crawfish pond. *Freshwater Crayfish* 3: 133-140.

Dendy, J.S. 1978. Preliminary experiment with photoperiod to influence crawfish spawning. *Aquaculture* 15: 379-382.

Fjälling, A. & M. Fürst 1988. The development of a fishery for the crayfish *Pacifastacus leniusculus* in Sweden 1969-86. *Freshwater Crayfish* 7: 223-230.

Hazlett, B.A. & D. Rittschof 1985. Variation in rate of growth in the crayfish *Orconectes virilis*. *J. Crust. Biol.* 5: 341-346.

Hobbs Jr, H.H. 1972. *Crayfishes (Astacidae) of North and Middle America.* Biota of Freshwater Ecosystems. US Environ. Prot. Agency, Water Pollution Control Res. Ser. Ident. Manual 9, Washington, DC.

Hubbard, D.M., F.H. Robinson, P.B. Brown & W.H. Daniels 1986. Optimum ratio of dietary protein to energy for red swamp crayfish (*Procambarus clarkii*). *Prog. Fish-Cult.* 48: 233-237.

Huner, J.V. 1975. Observations on the life histories of recreationally important crawfishes in temporary habitats in Louisiana. *Proc. La. Acad. Sci.* 38: 20-24.

Huner, J.V. 1977. Introductions of *Procambarus clarkii* (Girard) – An update. *Freshwater Crayfish* 3: 193-202.

Huner, J.V. 1978. Exploitation of freshwater crayfishes in North America. *Fish. Bull.* 3(6): 2-5.

Huner, J.V. 1983. Crayfish production-achievements and potential. In *Second Int. Fish Farm. Conf. UK, Brighton, UK:* pp. 145-173. Janssen Services, Surrey, Kent, UK.

Huner, J.V. 1984. Growth responses of juvenile male crawfish, *Procambarus clarkii* fed artificial diets supplemented with elodea, *Egera densa*, a vascular aquatic plant. *J. World Aquaculture Soc.* 15: 129-131.

Huner, J.V. & J.E. Barr 1984 (revised edition). *Red Swamp Crawfish: Biology and Exploitation.* Louisiana Sea Grant Program, Louisiana State University, Baton Rouge, Louisiana.

Huner, J.V. & O.V. Lindqvist 1985. Effects of temperature and photoperiod on mating and spawning activities of wild caught noble crayfish, *Astacus astacus* (Linn) (Astacidae, Decapoda). *J. World Aquaculture Soc.* 16: 225-226.

Huner, J.V. & O.V. Lindqvist 1986. A stunted crayfish *Astacus astacus* population in central Finland. *Freshwater Crayfish* 6: 156-165.

Huner, J.V. & O.V. Lindqvist 1988. Response of a slow growing noble crayfish, *Astacus astacus* L., population from a pond in central Finland to several years of exploitation. *Freshwater Crayfish* 7: 81-89.

Huner, J.V. & S.P. Meyers 1979. Dietary protein requirements of the red swamp crawfish, *Procambarus clarkii* (Girard) (Decapoda, Cambaridae), grown in a closed system. *Proc. World Mariculture Soc.* 10: 751-760.

Huner, J.V. & R.P. Romaire 1979. Size at maturity as a means of comparing populations of *Procambarus clarkii* (Girard) (Crustacea, Decapoda) from different habitats. *Freshwater Crayfish* 4: 53-64.

Jones, P.D. & W.T. Momot 1981. Crayfish productivity, allochthony, and basin morphometry. *Can. J. Fish. Aquat. Sci.* 38: 175-183.

Kossakowski, J. 1966. *Crayfish (Raki).* Panstwowe Wydawnictwo Rolnicze i Lesne. Translated from Polish for US Nat. Mar. Fish. Serv., Washington, DC. Translation TT 70-55114, 1971.

Lawrence, A.L., J.P. McVey, & J.V. Huner 1985. Penaeid shrimp culture. In J.V. Huner & E.E. Brown (eds), *Crustacean and Mollusk Aquaculture in the United States:* pp. 127-158. Westport, Conn: AVI Publ. Co.

Lindqvist, O.V. & K. Louekari 1975. Muscle and hepatopancreas weight in *Astacus astacus* L. (Crustacea, Astacidae) in the trapping season in Finland. *Ann. Zool. Fennici.* 12: 237-243.

Lowe, M.E. 1961. The female reproductive cycle of the crayfish *Cambarellus shufeldti*: the influence of environmental factors. *Tulane Stud. Zool.* 13: 157-176.

Mancini. 1986. *Astacicoltura. Allevamento e pesca dei gamberi d' acqua dolce.* Edagricole, Bologna, Italy.

Mason, J.C. 1975. Crayfish production in a small woodland stream. *Freshwater Crayfish* 2: 449-480.

Mason, J.C. 1979. Significance of egg size in the freshwater crayfish, *Pacifastacus leniusculus* (Dana). *Freshwater Crayfish* 4: 83-92.

McGriff, D. 1983. Growth, maturity, and fecundity of the crayfish, *Pacifastacus leniusculus*, from the Sacramento-San Joaquin Delta. *Calif. Fish and Game* 69: 227-242.

Momot, W.T. 1984. Crayfish production: a reflection of community energetics. *J. Crustacean Biol.* 4: 35-54.

Momot, W.T. 1986. Production and exploitation of the crayfish, *Orconectes virilis*, in northern climates. *Can. Spec. Publ. fish. Aquat. Sci.* 92: 154-167.

Momot, W.T. & H. Gowing 1977. Production and population dynamics of the crayfish *Orconectes virilis* in three Michigan lakes. *J. Fish. Res. Bd. Can. 34: 2041-2055.*

Morgan, G.E. & W.T. Momot 1988. Exploitation of *Orconectes virilis* in northern climates: complimentarity of management options with self-regulatory life history strategies. *Freshwater Crayfish* 7: 69-80.

Morrissy, N.M. 1975. Spawning variation and its relationship to growth rate and density in marron, *Cherax tenuimanus* (Smith). *Fish. Bull. West. Aust.* 16: 1-32.

Morrissy, N.M. 1983. Crayfish research and industry activities in Australia. *Freshwater Crayfish* 5: 534-544.

Moshiri, G.A. & C.R. Goldman 1969. Estimation of assimilation efficiency in the crayfish *Pacifastacus leniusculus* (Dana) (Crustacea: Decapoda). *Arch. Hydrobiol.* 66: 298-306.

Payne, J.F. 1971. Fecundity studies on the crawfish *Procambarus hayi*. *Stud. Zool. Bot.* 17: 35-37.

Payne, J.F. 1972. Life history of *Procambarus hayi*. *Amer. Midl. Natur.* 87: 25-35.

Penn Jr, G.H. 1943. A study of the life history of the Louisiana red-crawfish, *Cambarus clarkii* Girard. *Ecology* 24: 1-18.

Perryman, E.K. 1969. *Procambarus simulans*: light-induced changes in neurosecretory cells in ovarian cycle. *Trans. Amer. Microsc. Soc.* 88: 514-524.

Persson, M.K. & K. Soderhall 1983. *Pacifastacus leniusculus* (Dana) and its resistance to the parasitic fungus *Aphanomyces astaci* Schikora. *Freshwater Crayfish* 5: 292-298.

Pianka, E.R. 1970. On *r*- and *K*-selection. *Amer. Midl. Natur.* 104: 592-597.

Prins, R. 1968. Comparative ecology of the crayfishes *Orconectes rusticus rusticus* and *Cambarus tenebrosus* in Doe Run, Meade County, Kentucky. *Int. Revue ges. Hydrobiol.* 53: 667-714.

Rice, P.A. & K.B. Armitage 1974. The influence of photoperiod on processes associated with molting and reproduction in the crayfish *Orconectes nais* (Faxon). *Comp. Biochem. Physiol.* 47A: 243-259.

Sasaki, G.C., J.M. Capuzzo & P. Biesiot 1986. Nutritional and bioenergetic considerations in the development of the American lobster, *Homarus americanus*. *Can. J. Fish. Aquat. Sci.* 43: 2311-2319.

Skurdal, J. & T. Qvenild 1986. Growth, maturity, and fecundity of *Astacus astacus* in lake Steinsfjorden, S.E. Norway. *Freshwater Crayfish* 6: 182-185.

Snyder, D.E. 1983. Fish eggs and larvae. In L.A. Nielsen & D.L. Johnson (eds), *Fisheries Techniques*. Bethesda: American Fisheries Society.

Stephens, G.J. 1952. Mechanisms regulating the reproductive cycle in the crayfish, *Cambarus*. I. the female cycle. *Physiol. Zool.* 25: 70-84.

Suko, T. 1956. Studies on the development of crayfish. IV. the development of winter eggs. *Sci. Rept. Saitama Univ. (Japan) Ser. B.* 2: 213-219.

Suko, T. 1958. Studies on the development of the crayfish. VI. the reproductive cycle. *Sci. Rept. Saitama Univ. (Japan) Ser. B.* 3: 79-91.

Taylor, R.C. 1983. Drought-induced changes in crayfish populations along a stream continuum. *Amer. Midl. Natur.* 110: 286-298.

Unestam, T. 1972. On the host range and origin of the crayfish plague fungus. *Inst. Freshwater Res. Drottningholm* 52: 192-198.

Vey, A., K. Soderhall, & R. Ajaxon 1983. Susceptibility of *Orconectes limosus* Raff. to the crayfish plague, *Aphanomyces astaci* Schikora. *Freshwater Crayfish* 5: 284-291.

Weagle, K.V. & G.W. Ozburn 1972. Observations on aspects of the life history of the crayfish, *Orconectes virilis* (Hagen) in northwestern Ontario. *Can. J. Zool.* 50: 366-370.

Westin, L. & R. Gydemo 1986. Influence of light and temperature on reproduction and moulting frequency of the crayfish, *Astacus astacus* L. *Aquaculture* 52: 43-50.

Wiernicki, C. 1984. Assimilation efficiency by *Procambarus clarkii* fed elodea (*Egera densa*) and its products of decomposition. *Aquaculture* 36: 203-215.

Zipf, E.M. 1959. A contribution toward knowledge of the embryological development of the crayfish, *Cambarus bartoni sciotensis*. Ph.D. Disseration, University of Virginia, Charlottesville, Virginia.

Population regulation and stock-recruitment relationships in some crayfish and lobster populations

David Edward Pollock
Sea Fisheries Research Institute, Rogge Bay, Cape Town, South Africa

ABSTRACT

Life histories and reproductive dynamics of the crayfish *Orconectes virilis*, the langoustine *Nephrops andamanicus*, and spiny lobsters of the genus *Jasus* are reviewed and compared. Fecundity per recruit values (the average number of eggs produced per female during a lifetime) were calculated to provide comparisons of mortality rates of larval and/or juvenile stages, both among the three lobster groups and between different populations of the same species. In relatively crowded populations of *Nephrops* density-dependent rates of growth and survival are reduced and sizes at maturity of females are small. Modeling of *Jasus* populations shows that fecundity per recruit is regulated by size at maturity of females, at both inter- and intraspecific levels. Egg size, phyllosoma larval size and puerulus size are correlated, and there are interspecific differences within the genus *Jasus*. Mean size at maturity is determined largely by average molt increments of juveniles, because the modal number of pre-puberty molts is not random but appears to be species-specific in these groups. Thus density-dependent growth of juveniles regulates size at maturity and egg production in these crustaceans, suggesting an indirect feed-back relationship between adult egg production and post-larval settling densities. Despite the apparent resilience afforded to lobster stocks by compensatory (density-dependent) mechanisms, overfishing is to be avoided-firstly because of the adverse economic impact, and also because reduced egg and larval production must diminish the capacity of a stock to resist environmental change.

1 INTRODUCTION

Freshwater crayfish and certain marine clawed and spiny lobsters have been much studied, largely because of their economic importance. Although many lobster populations have been both fished and researched continuously since the early part of the present century, little information exists even today about the dynamics of the important larval and juvenile stages. Lack of knowledge about relationships between adult stock and subsequent recruitment remains an important barrier to management of lobster fisheries throughout the world.

The degree of protection of brood-stock required to guarantee adequate recruitment has long been a central issue of concern for fisheries managers. Many restrictions such as closed seasons, minimum size limits, prohibition of berried females, etc. have been imposed in attempts to maintain an adult stock large enough to ensure 'sufficient' egg and larval production. The choice of an optimum minimum size limit is, however, complicated by the fact that female somatic growth of several lobster species slows down with the onset of sexual maturity (as energy reserves are channeled towards oögenesis), to the extent that optimum minimum size (from a yield per recruit viewpoint) and size at first maturity tend to coincide (Davis 1980). There is thus little motivation to introduce a female minimum size limit much larger than the size at first maturity, although male yield per recruit may well be optimized at a comparatively larger size limit. However, egg per recruit analyses (Campbell & Robinson 1982) have made it clear that, under high exploitation rates, egg production can only be maintained at desired (high) levels by enforcing relatively large size limits. Thus, if egg and larval production is to be used as an argument in favour of any particular size limit, it becomes doubly important to understand stock/recruitment relationships. Lack of knowledge in this area has hindered the development of plans that attempt to incorporate stock and recruitment considerations as well as simplistic constant recruitment 'yield per recruit' estimates into management strategies. For example, Acheson & Reidman (1982) writing on Maine lobsters, stated: '...all serious attemps by biologists to demonstrate a positive relationship between numbers of eggs and recruitment have failed.'

Below I compare egg production in crayfish, clawed lobster, and spiny lobster populations from high latitudes, which have rather similar life histories and annual breeding cycles. The use of annual spawner simplifies modeling fecundity per recruit, which is used as a tool to discern natural mortality rates of both larval and juvenile lobsters. Three species (groups) were chosen for detailed comparison: *Orconectes virilis*, a crayfish inhabiting mid-North American lakes; *Nephrops norvegicus*, a burrowing lobster of the NE Atlantic and Mediterranean Sea; and the spiny lobster genus *Jasus*, group *lalandii*, 6 closely-related species with a discontinuous circumglobal distribution in the southern hemisphere between latitudes of approx. 30°-40°S. I attempted to determine factors responsible for both regulating population size and maintaining continuous recruitment. Factors which emerged as being vitally important were: density-dependent growth and survival of juveniles in limited nursery habitats; differences in size at maturity caused mainly by density-dependent effects on molt increments of juveniles; and control over per capita and total egg production by the environment.

2 SOURCES OF DATA

Considerable emphasis is placed herein upon interspecific differences in egg size and brood size. The weight of eggs within a brood in crustaceans is often related to the logarithm or cube of body length (Sastry 1983); in mysids and in crabs brood volumes and weights are consistently about 10% of body volumes and weights (Hartnoll 1985). In other crustaceans, the relationships may be negatively allometric, and brood weight may form a decreasing proportion of female weight with increasing size (Hartnoll 1985).

Data on egg sizes in crayfish and lobster species are often lacking, as are brood weights. If weights are available, they unfortunately often refer to wet or partially drained egg masses with unknown amounts of water. However, egg counts have been made for most species so that brood size/body size relationships are usually available in the literature.

For closely-related species with very similar body shapes and morphology, such as most of the clawed and spiny lobsters, the number of eggs per unit body weight provides a rough index of egg size, i.e. egg size is inversely related to the number of eggs per gram body weight within a group of closely-related species. This is because the volume of eggs that can be accommodated within the ovaries is limited by the morphometry of the carapace and anterior abdomen. Since this volume forms a relatively constant proportion of the total body volume of mature female homarids and palinurids, egg volume or weight bears a relationship to total body volume or weight. Hence, egg size and number are dependent variables within the constraints of brood weight, and differences in egg size between species are reflected by differences in the average number of eggs per gram body weight. Similar findings have been reported by Mauchline (1980) and Hines (1982) for euphausids and crabs, respectively.

The close similarity in morphometry of some lobster and spiny lobster species is demonstrated by the similarity of body weight/carapace length relationships listed in Table 1 and shown in Figure 1. Also shown in Table 1 is the mean number of eggs per gram body weight for

Table 1. Weight/length relationships and mean number of eggs per g body weight in *Nephrops norvegicus* and *Jasus* species. W = weight (g), CL = carapace length (mm) except where indicated, CL* (cm).

Species	Weight	Eggs/g	Range of sexually mature size classes
N. norvegicus	$W = 0.0006840CL^{2.963}$ (Farmer 1975)	60	20-40 mm
J. lalandii	$W = 0.5869CL^{*2.9729}$ (Heydorn 1969)	500	70-100 mm
J. novaehollandiae	$W = 0.00225CL^{2.7}$ (Hickman 1946)	400	80-110 mm
J. edwardsii	Assumed as above	300	80-110 mm
J. frontalis	$W = 0.00225CL^{2.694}$ (Arana & Pizarro 1971)	400	80-110 mm

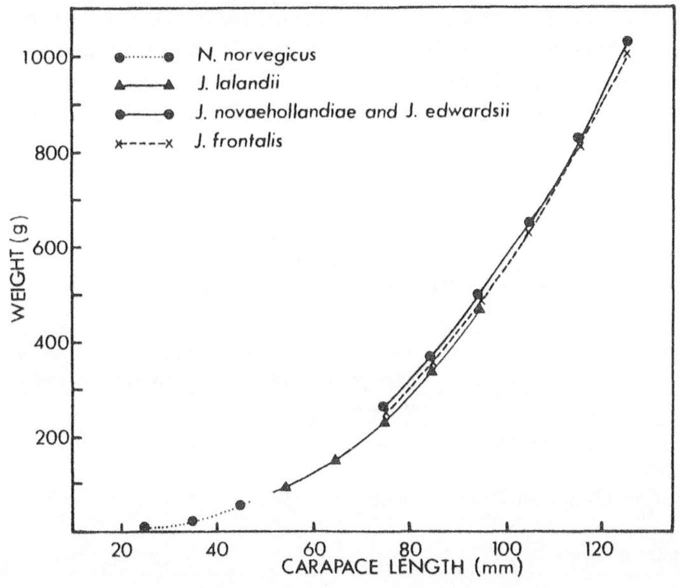

Figure 1. Weight-length relationships for *Nephrops norvegicus* and four *Jasus* species.

Nephrops norvegicus and four *Jasus* species, calculated from brood size/body size relationships, numbers which illustrate fairly large differences, even between congeners.

3 LIFE HISTORY

Cobb & Wang (1985), in their review of crayfish and lobsters, described common features in the biology, ecology, and population dynamics of these groups. The most obvious morphological difference between crayfish and lobsters on one hand and spiny lobsters on the other, is the lack of claws (pincers) in the latter group. Far more important differences between these groups lie in their reproductive patterns, brood sizes, size and type of eggs, and larval type and duration.

Freshwater crayfish no longer have a larval stage, presumably in response to evolution in riverine systems where unidirectional water flow would constantly remove larvae from the stock (Cobb & Wang 1985). After eggs hatch, the young crayfish remain attached to the female for the first two molts, so some degree of parental care is evident.

Eggs are relatively large and few in number in *Nephrops norvegicus*, the pelagic larval cycle is short, with a prelarva, only 3 larval stages, and a postlarva (Cobb & Wang 1985). Larval dispersal is fairly limited in duration (Farmer 1975), taking place in waters overlying the continental shelf.

Spiny lobsters produce large numbers of small eggs with limited yolk reserves, and the larval size at hatching is correspondingly small. A large number of larval molts (usually 11 or more) is associated with extensive larval growth so that the size at metamorphosis of the final phyllosoma larval stage into the swimming puerulus larval stage is relatively large. The size of the puerulus varies between about 6 and 13 mm carapace length at settlement, depending on species. The teleplanic phyllosoma larvae tend to be widely dispersed over broad oceanic regions (Pollock 1990).

Some of the differences between the species under study are contrasted in Table 2. The shift from crayfish (direct development) towards spiny lobsters (long-lived planktotrophic larvae) is accompanied by a reduction in egg size, an increase in the number of eggs per gram body weight, and by an increasing trend in the number of larval instars. Each molt stage in spiny lobsters does not necessarily represent a single instar, so that many more molts than stages may take place before attainment of the final phyllosoma stage, e.g. 25 molts or more for *Panulirus inflatus*, despite only 11 described stages (Phillips & Sastry 1980). This type of 'mark-time' molting (Gore 1985) allows the pelagic larval life of spiny lobsters to be greatly extended for indeterminate periods (Pollock 1990).

Table 2. Comparison of some life-history parameters for three crayfish, lobster and spiny lobster species.

Species	Area	Mean no. eggs per g body wt	No. larval molt stages	Larval duration	Lifespan (years)
Orconectes virilis	Northern Ontario lakes, Canada, at 49°N	~7	0	0	4
Nephrops norvegicus	Northern Europe 35-70°N	~60	3	2-3 weeks	~11
Jasus lalandii	Southern Africa 25-35°S	~500	11+	> 1 year	~30

4 FECUNDITY PER RECRUIT

Computations of fecundity per recruit both within and between species allow comparisons to be made of mortality rates, providing insight into the dynamics of early life history stages. Comparisons are possible because of the premise that, on average, out of all eggs produced by a female during her lifetime, only two offspring (one male, one female) survive to maturity (Royce 1984). For every species, the average fecundity of females is an evolved trait which maintains populations even during harsh climatic periods. Because the environment often differs throughout the range occupied by a species, life history parameters, including fecundity per recruit, may vary widely.

Determination of fecundity per recruit for many species has been hindered by lack of information, especially about mortality rates and growth of adult females. However, data on brood size in relation to body size are often available. Therefore, when some information on adult growth is available, it is often only necessary to make assumptions about natural mortality rates (M) in order for rough calculations of fecundity per recruit to be made. In most crustaceans, natural mortality rates usually appear to be relatively low for species living at high latitudes and in the deep-sea where temperatures are cooler. When exact values of natural mortality are unknown, it is still possible to model fecundity per recruit by using a range of likely values of M.

4.1 *Comparison by species and group*

Attempts are made here to model fecundity per recruit for the species under study. The calculations are not intended to represent highly accurate models, but are intended to serve as a broad basis for comparison among different species and groups living in different regions.

4.1.1 *Orconectes virilis*

Momot & Gowing (1977a, b, c) and Momot (1967, 1978, 1986) produced a series of papers on the population dynamics of *O. virilis* in Michigan and Ontario lakes. Size-specific brood sizes of female *O. virilis* were about twice as large in the eutrophic Ontario lakes as in the oligotrophic Michigan lakes studied by Momot (1986), suggesting that fecundity and presumably fecundity per recruit was controlled by food availability for adult females.

France (1983) found that size-specific brood sizes of *O. virilis* in four Manitoba lakes varied directly with lake productivity, as measured by phytoplankton production. In addition, increased fecundity correlated with faster somatic growth in the more productive lakes. France (1983) noted that faster growing animals in more productive lakes attained larger mean and maximum sizes and reached first maturity at larger mean sizes than those in less productive lakes.

Since *O. virilis* is a short-lived species (Table 2) with high natural mortality rates, fecundity per recruit is low. A calculated value of approx. 120 eggs per female is shown in Table 3 for females in Dock Lake, using data from Momot (1986). The results would suggest that only about 2% of progeny (one female out of 62 females born) would have to survive to reach maturity if the population is to maintain itself. This is borne out by actual measurements of survival rates, for Momot (1986) found that even large cohorts were reduced to less than 2% of the initial size of the brood by age 2.

The value for fecundity per recruit (Table 3) is an overestimate, because it was assumed, incorrectly, that all females bear eggs at age 1.5 years.

Momot (1967, 1986) found that mortality rates were high in adult females, about 80% for 2

Table 3. Fecundity per recruit of *O. virilis* in Dock Lake, Ontario. *Mean values calculated from actual data 1976 to 1980 in Momot (1986), **Brood size: pleopodal egg counts (Momot 1986).

Age (years)	Approx. mean CL (mm)	No. of survivors*	Mean brood size**	No. of eggs produced × 10^{-3}
1.5	29	1000	53	53.0
2.5	35	672	83	55.8
3.5	41	133	113	15.0
4.5	–	0	–	–
Total				123.8 (i.e. 124 eggs per female recruit)

to 3 year olds, and increased rather dramatically in the fourth growing season (i.e. 3 year olds), so that virtually no animals survived into a fifth season. Momot (1967) suggested that the high mortalities could be related to mechanical and physiological problems during molting. Crayfish die if they are unable to completely withdraw from their exoskeletons; this has been suggested as a possible source of increased mortality in large, old animals (Aiken, pers. comm. in Momot 1967; personal observations on *J. lalandii*).

4.1.2 *Nephrops norvegicus*

The larvae of *Nephrops norvegicus* in the Irish Sea are found mainly over the adult beds (Hillis 1968 in Farmer 1975), confirming that the short larval life of only 2 to 3 weeks' duration effectively restricts larval dispersal, especially in semi-enclosed waters such as the Irish Sea. Because both juveniles and adults live only in muddy habitats (mainly reworked glacial clay beds (Farmer 1975), a large retention of larvae in local waters maintains recruitment, especially where populations are relatively isolated, as in the Adriatic Sea. In keeping with the short larval duration, eggs are relatively large in size, brood sizes relatively small, and only 3 larval stages exist (Table 2).

Throughout the range occupied by *Nephrops norvegicus*, which stretches from northern Norway to Iceland to the Mediterranean and the coast of North Africa (Fig. 2), considerable variations in mean sizes, maximum sizes and sizes at maturity are evident. Figueiredo & Thomas (1967) noted that these variations are unrelated to latitude (and hence to temperatures), as illustrated in Table 4. Also evident is the correlation between mean size, maximum size and size at maturity of females.

Fecundity of *Nephrops* also varied with location, for example, Figueiredo & Thomas (1967) described steeper fecundity curves from Portugal than from Scotland. As well, Chapman (1980) noted that brood sizes increased much more rapidly with increasing body size in females from Portugal compared with those from the Irish Sea. The latter relationships are plotted in Figure 3.

In most areas adult females molt only once per year (in spring) and spawn annually (although in alternate years in far northern regions, e.g. Faroes, Scandinavia), becoming ovigerous from about August onwards. The average incubation period varies with latitude (i.e. with temperature), varying from about 6 months off Portugal to maximum of about 12-13 months off Iceland (Chapman 1980).

When computing fecundity per recruit, it is useful to compare two populations differing widely in mean and maximum sizes and in sizes at maturity. Detailed growth information was

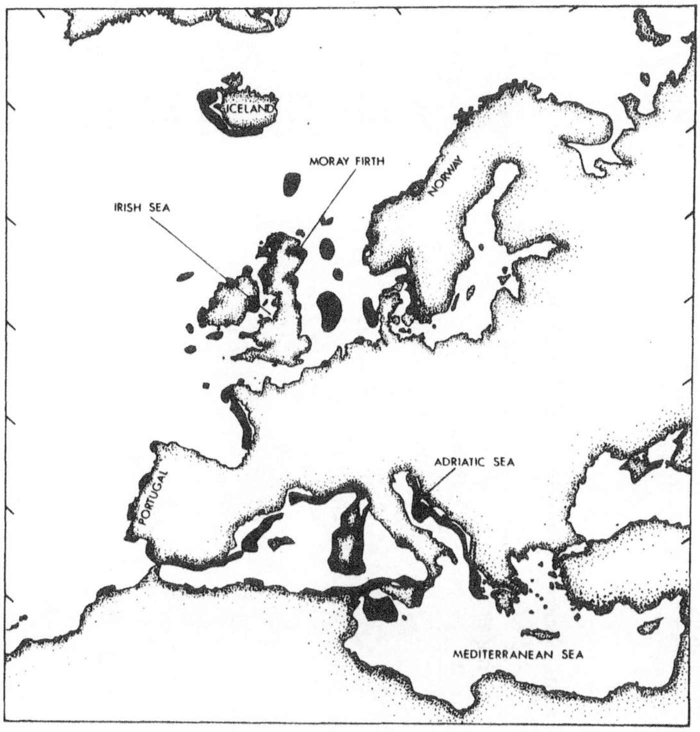

Figure 2. Distribution of *Nephrops norvegicus* (adapted from Farmer 1975).

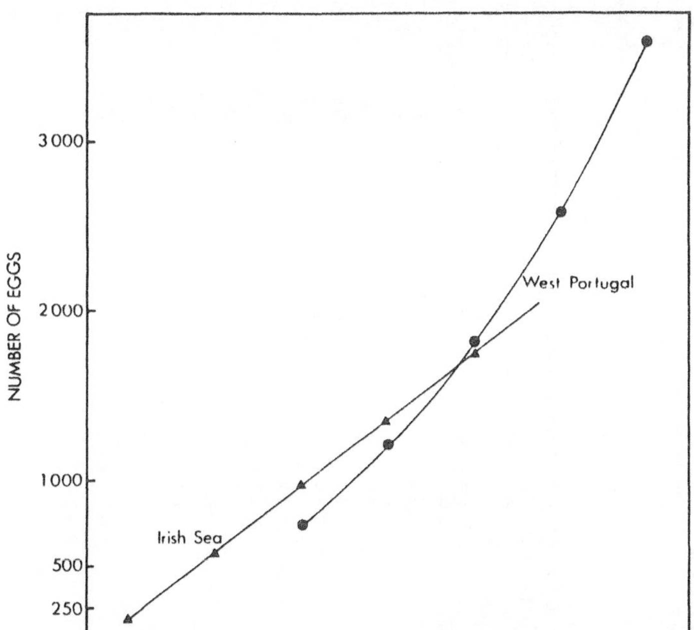

Figure 3. Brood size/body size relationships for two populations of *Nephrops norvegicus*.

Table 4. Some variations in sizes of *Nephrops* with locality and latitude. From Figueiredo & Thomas (1967) and Farmer (1975).

Area	Latitude	Carapace length (mm)				Bottom temp. range (°C)
		Mean size (males)	Maximum size		Size at 50%	
			Males	Females	maturity of females	
Moray Firth	57°N	32	57	50	22	–
Kattegat and Skagerrak	57°N	49	80	58	–	–
Irish Sea	54°N	30	50+	42	21	7-13
High Adriatic	44°N	35	72	55	27	10-15
West Portugal	38°N	48	92	74	33	–

Table 5. Fecundity per recruit calculated from simulated female *Nephrops* populations, assuming M = 0.16. %ΔS: % molt increment, using the notation of Botsford (1985).

Irish Sea %ΔS	CL (mm)	Frequency	No. of eggs × 10^{-3}	Portugal %ΔS	CL (mm)	Frequency	No. eggs × 10^{-3}
10	21	1000	266	12	33	1000	100
10	23	850	359	11	37	850	1210
9	25	723	419	11	41	723	1408
9	28	614	500	10	46	614	1700
8	30	522	506	10	50	522	1864
7	33	444	535	9	55	444	2122
6	35	377	513	8	60	377	2233
5	37	321	487	7	65	321	2323
4	39	160	268	5	69	160	1398
3	40	64	112	3	73	64	639
	42	10	19		75	10	114
Total			3984	Total			16014

only available for females from the Irish Sea, where Farmer (1975) determined that juveniles molted approximately 15 times before reaching sexual maturity at about 21 mm. Adult females increased in carapace length by about 10% per annual molt, the percentage increment decreasing gradually with increasing size and age.

Natural mortality rates for adult *Nephrops* are likely to be fairly low in comparison with rates of fishing mortality, at least for males. Natural deaths are mainly the result of fish (especially cod) predation (Rae 1967, 1968 in Farmer 1975), despite the shelter afforded by burrows. Unfortunately, no detailed information on natural mortality rates of *Nephrops* is available, although Jones (1979) suggested that values probably do not exceed M = 0.2.

Chapman (1980) suggested that heavy postlarval settlement leads to high densities and thus increased competition for food, which retards growth and results in small mean and maximum sizes in the population. In such areas of high population density, animals would presumably have to spend more time foraging out of burrows and therefore be exposed to a greater risk for predation. In this way, areas of high population density are likely to have both reduced growth (food – limited) and reduced survival (increased predation by fish) in comparison with low density populations.

In the ensuing computations, female populations from the Irish Sea and West Portugal were simulated and compared for fecundity per recruit. Maximum sizes attained by females in these areas are approx. 42 mm and 74 mm, respectively; mean sizes at 50% maturity are approx. 21 mm and 33 mm, respectively (Farmer 1973, 1975, Figueiredo & Thomas 1967). *Nephrops* in West Portugal appears to attain the largest mean and maximum sizes, as well as the largest size at maturity, out of 21 sites for which data are cited by Farmer (1975). Because females are less vulnerable to fishing than males (Farmer 1975), the consistent regional differences in size at maturity and maximum size of females are more likely to be natural environmentally-induced characteristics rather than being induced by regional differences in fishing intensity. For this reason it is assumed here, in the absence of actual growth data for *Nephrops* in Portuguese waters, that the large sizes recorded off Portugal are indicative of optimal growth rates. If the average number of juvenile molts in *Nephrops* throughout its range is approx. 15 (Farmer 1975), then the large mean size at maturity off Portugal is likely the result of larger juvenile growth increments. Pollock (1987) showed, for *Jasus lalandii*, that size at maturity of females is directly related to the mean molt increments of juvenile females which were mainly influenced by feeding. High densities resulted in reduced per capita food availability, small molt increments and a reduced size at maturity of female *J. lalandii*.

If the same causative factors apply in the case of *N. norvegicus*, then it follows that molt increments of adult *Nephrops* would also be larger off Portugal than in the Irish Sea, which would explain the larger maximum sizes attained by both sexes off Portugal (Table 4).

Calculations were made of egg production by 1,000 adult female recruits of the mean size at maturity in each area (Table 5). Numbers of females were reduced by means of natural mortality only, i.e. the models simulate *unexploited* female populations, in which a batch of eggs is produced each year throughout the lifespan of the females. The following assumptions were incorporated in the simulations:

1) For Irish Sea females, the first five annual molt increments were obtained from growth data provided by Farmer (1975). Ensuing molt increments were obtained by extrapolation; increments declined with increasing size and age, a characteristic of most macrurans (Hartnoll 1982). Approximately 10 molts occurred before the maximum size of 32 mm was attained, which suggests a maximum adult lifespan of about 10 years.

2) For females from Portugal, slightly larger percentage molt increments were assumed,

with the first increment at 12%, decreasing gradually over 10 molts until a maximum size of 75 mm was reached (close to the recorded maximum of 74 mm).

3) Annual survival rates of 85% were incorporated (corresponding to M = 0.16), but mortality was increased over the last 4 molts as maximum sizes were approached. As mentioned earlier, rather large increases in natural mortality rate amongst the largest size classes (i.e. oldest age groups) have been observed by Momot (1967) in *O. virilis*, and by Pollock (1987) in *Jasus lalandii*. This may be related to an increased frequency of deaths occurring during moulting in the oldest animals in the population. In the case of the *Nephrops* populations simulated here, an increase in natural mortality rate is required in order to 'make the population virtually disappear' at the maximum sizes recorded.

In Table 5 the numbers surviving in each age class (instar-class) are multiplied by the mean brood size applicable to that particular size. Fecundity data were available for Irish Sea and Portuguese females from the relationships:

Irish Sea: $F = 78.204\,CL - 1376.421$ (Chapman 1980)

Portugal: $\log F = 3.0557 \log CL - 1.6387$ (Chapman 1980)

where F = no. of eggs, CL = carapace length (mm).

Both relationships apply to newly laid eggs, i.e. to absolute brood sizes, and do not incorporate egg losses during incubation.

The computations indicate that fecundity per recruit of females off Portugal (approximately 16000 eggs) is some 4 times greater than that of females from the Irish Sea (approximately 4000 eggs). Additional simulations were performed using different mortality rates to gauge the sensitivity of the results to differences in M. Values ranged from about 9000 eggs (M = 0.22) to 20 000 eggs (M = 0.1) for females off Portugal.

In general, it is apparent that the large size at maturity of females in Portuguese waters permits increased egg production and higher fecundity per recruit. Even allowing for a 60% egg loss during incubation [Chapman (1980) cites estimates varying from 40% to 70% loss during development], realistic relative values of fecundity per recruit would vary maximally from about 1600 to 7800 eggs between the two areas – illustrating an approximate 5-fold difference.

These results agree closely with findings of Pollock (1987) for *Jasus lalandii* and confirm that large size at maturity is related to rapid juvenile growth (large molt increments) and this is associated with high egg production. As shown earlier, regional differences in mean and maturity sizes in *Nephrops* are not related to temperature, and it seems likely that Chapman (1980) was wholly correct in ascribing such differences to density-dependent feeding and growth.

In general, individual *Nephrops* populations differ considerably in mean growth rates (and possibly mortality rates also), which are reflected by differences in mean and maximum sizes and sizes at maturity. Where population densities are relatively high [e.g. in the Irish Sea and in the Sound of Jura (Chapman & Rice 1971)] mean growth increments are small, as are mean, maximum and maturation sizes. *Per capita* egg production is also comparatively reduced. This implies that in high density populations, many small females each produce relatively few eggs, while in low density populations the converse applies. The net effect is that egg production per unit area of seabed probably tends towards a value which is roughly correlated with the carrying capacity (food production) of the habitat.

4.1.3 *Jasus lalandii group*

In species of the genus *Jasus*, growth rates are known to vary both intra- and interspecifically.

Examples of intraspecific variations in growth are given by Pollock & Beyers (1981) and Pollock (1987) for *J. lalandii*, by McKoy & Esterman (1981), Annala (1983), and Annala & Bycroft (1985) for *J. edwardsii*, and by Pollock & Roscoe (1977) and Pollock (1981) for *J. tristani*. Street (1969) was one of the first authors to note the correlation between maximum size and size at maturity of females in *J. edwardsii* of New Zealand; similar correlations have been since noted for a variety of species.

Published information available for the 6 *Jasus* species is given in Table 6. For the two Atlantic Ocean species, *J. tristani* and *J. lalandii*, sizes at maturity vary with locality, but 50% maturity values range from about 57 to 67 mm carapace length. Maximum sizes of male *J. lalandii* do not appear to exceed about 190 mm. That was even true in the early 1900's, when fishing had had little impact on the size composition of the stock (Gilchrist 1913).

In the Indian Ocean, female *J. paulensis* at the mid-ocean islands of St. Paul and New Amsterdam appear to attain sexual maturity at sizes similar to those of *J. tristani* and *J. lalandii*, possibly at about 64 mm (Grua 1960, 1963 in Arana 1985). Fecundity has not yet been determined for *J. paulensis*. Maximum sizes recorded in Table 6 are probably underestimates, because population size composition could have been considerably altered by fishing by the time these observations were recorded.

In the South Pacific, wide regional variations in size at maturity have been recorded for both *J. novaehollandiae* and *J. edwardsii*, with mean 50% maturity values ranging from 75 mm to 114 mm in Australia and from 72 mm to 121 mm in New Zealand. At the archipelago of Juan Fernandez off the coast of Chile, 50% maturity is reached at a mean size of about 84 mm carapace length (Arana 1985). In general, both sizes at maturity and maximum sizes are much larger in Pacific *Jasus* species than elsewhere. Another example is *Jasus verreauxi* from the warm waters off east Australia and the northern tip of North Island, New Zealand. Females reach maturity at a mean size of 155 mm (Kensler 1967) and grow to a size exceeding 220 mm carapace length (Booth 1979).

Another interesting feature in Table 6 is the apparent relationship between egg size (as reflected by the mean number of eggs per gram body weight over the range of sexually mature size classes in each species) and the size of the puerulus stage. It is evident that final stage phyllosoma larvae of *J. edwardsii* are some 25 to 30% larger than those of *J. lalandii* and *J. tristani*, and the size of the puerulus stage is correspondingly larger. This appears to be correlated with a larger egg size in *J. edwardsii*. Despite a small egg size and puerulus size for *J. frontalis* at the archipelago of Juan Fernandez, two final phyllosoma stages of 28-30 mm cephalic width have been captured off the islands and described by Baez (1973). This is a surprizingly large larval size for a species with a relatively small settling stage of only 10.6 mm (Baez 1979).

The curves in Figure 4 represent brood size versus carapace length over the sexually mature size range for 4 *Jasus* species. Unfortunately no data are available for *J. tristani* or *J. paulensis*. The influence of egg size on size of brood is apparent, since brood sizes are clearly larger for any given carapace length for *J. lalandii*, similar for *J. frontalis* and *J. novaehollandiae*, and smallest for *J. edwardsii*. This correlates with the observed egg size indices in Table 6.

Probably because the different species all occupy fairly similar environments with similar temperature ranges (from about 10° to about 18°C), mainly at latitudes between about 35°S and 40°S, both molting and breeding cycles are similar among the various species. *Jasus* females spawn annually (Annala 1983, Arana 1985, Winstanley et al. 1983, Roscoe 1979, Pollock 1986), with oviposition occurring shortly after molting and mating, usually during late autumn or winter. Egg incubation takes 4 to 6 months depending on temperature. Larvae generally

Table 6. Comparison of some life history parameters for *Jasus* species of three ocean basins.

Ocean	Species	Maximum size (mm) Male Female	Size at maturity of females (mm)	Approx. no. of eggs/g body wt	Size of final stage phyllosoma (mm)	Mean CL of puerulus (mm)
Atlantic	*J. lalandii* Southern Africa	190 140 (Gilchrist 1913)	57-67 (Pollock 1987)	500	22 (Silberbauer 1971)	10 (Pollock 1973)
	J. tristani S.E.Atlantic	175 115 (Roscoe 1979, Gough)	57-59 (Pollock in prep.)	–	22 (36 TL) (Miller 1985)	9-11 (Holthuis & Sivertsen 1967)
Indian	*J. paulensis*	155 ~95 (Paulian 1957 in Silas 1967)	62-64 (Grua 1960, 1963 in Arana 1985)	–	–	~9 (Gruvel 1911, in Holthuis & Sivertsen 1967)
Pacific	*J. novaehollandiae*	215 180 (Fielder & Olsen 1967)	75-80 (S.E. of S. Australia) 112-114 (W. coast region of S. Australia) (Lewis 1981)	400	23-26 (40 TL) (McWilliam & Phillips 1987)	10.3 (Lewis 1977)
	J. edwardsii New Zealand	210 170 (Street 1969)	72-92 (N. Island) 80-121 (S. Island) (Annala et al. 1980)	300	29 (45 TL) (Lesser 1978)	11.4-12.3 (Booth pers. comm.)
	J. frontalis E. Pacific)	217 185 (Arana & Pizarro 1971)	82-85 (Arana 1985)	400	28-30 (43-45 TL) (Baez 1973)	10.6 (Baez 1979)

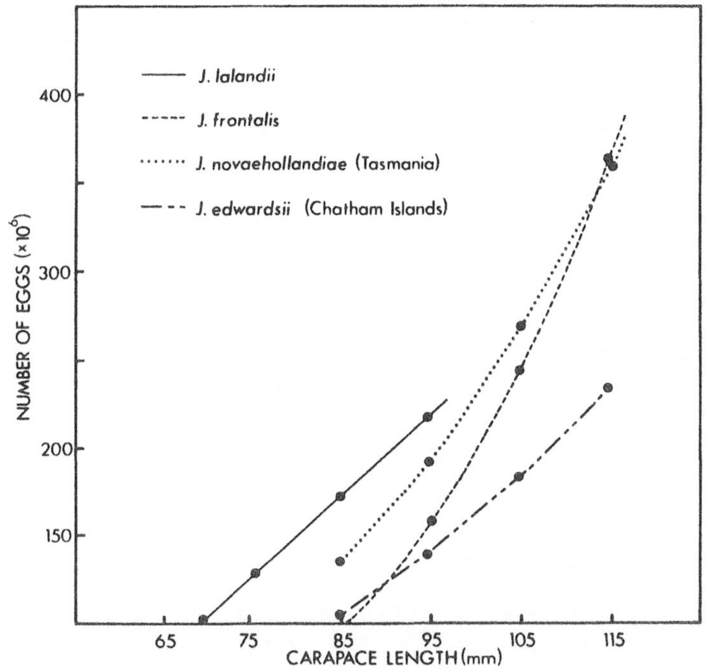

Figure 4. Brood size/body size relationships for four *Jasus* species.

hatch in late spring and early summer (Pollock 1986, Winstanley et al. 1983, Roscoe 1979).

Adult males usually molt annually, in summer. When mating takes place in April, May or June, males are hard-shelled while females are in soft-shelled condition. In certain areas and seasons, some males molt twice annually. This appears to be a characteristic feature in the warmer waters of North Island, New Zealand (Annala et al. 1980, McKoy & Esterman 1981). As a rule, however, females only molt and spawn once per annum.

Pollock (1987) calculated fecundity per recruit for female *J. lalandii* by simulating unexploited female populations from two areas on the Cape west coast of South Africa, where sizes at maturity and mean molt increments differed considerably. Using known molt increments for males and females, and sex ratios from diver-collected samples, he modeled adult female populations and calculated that in 'normal' areas where growth and survival rates were not unduly affected by overcrowding, fecundity per recruit was approximately 1×10^6 eggs for females that reached a size at 50% maturity of 66 mm. Using a similar procedure, fecundity per recruit was calculated for *J. edwardsii*. Growth data for adult females were adopted from Annala (1983) and McKoy (1985); brood sizes were calculated from the relationship: $E = 0.6921\ CL^{2.6852}$ [data from Kensler (1968) for the Chatham Islands, expressed as a regression by Arana (1985)]. A mean size at maturity of 90 mm was adopted in the model. This value lies between extremes of 72 to 121 mm quoted for New Zealand, (Annala et al. 1980) and is fairly typical of South Island females. Natural mortality rates (M) of 0.1 were used (Annala 1980). For modeling purposes fishing mortality was ignored. In effect therefore, the model simulates an unexploited female population of *J. edwardsii*, in which females reach maturity at 90 mm. As in the case of *J. lalandii*, an increase in natural mortality rate is incorporated in the model for large, old females.

Calculations of fecundity per recruit of *J. edwardsii* are shown in Table 7 together with

Table 7. Fecundity per recruit, calculations from simulated populations of female *J. lalandii* and *J. edwardsii*.

J. lalandii (S.W. Cape)				*J. edwardsii* (S. Island, N.Z.)			
CL	%ΔS	Frequency	No. eggs × 10^{-6}	CL	%ΔS	Frequency	No. eggs × 10^{-6}
66	3	1000	87.2	90	4	1000	122.4
68	3	900	86.6	94	4	900	123.8
70	3	810	85.2	98	3	810	124.6
71.5	2	729	81.6	101	3	729	121.6
73	2	656	77.9	104	3	656	118.4
74.5	2	590	74.1	107	3	590	115.1
76	2	531	70.2	110	2	531	111.4
77.5	2	478	66.5	112	2	478	105.2
79	2	430	62.7	114	2	430	99.5
80.5	1	353	53.9	116	2	387	93.6
81.5	1	289	45.4	118	1	349	88.4
82.5	1	237	38.3	119	1	279	72.3
83.5	1	195	32.4	120	1	223	59.1
84.5	1	160	27.3	121	1	179	48.5
85.5	1	131	22.9	122	1	143	39.6
86.5	1	107	19.2	123	1	114	32.3
87.5	1	88	16.2	124	1	80	23.1
88.5	1	72	13.6	125	1	56	16.6
89.5	1	59	11.4	126	1	39	11.8
90.5	1	47	9.3	127	1	27	8.3
			981.9 + 3% = 1 × 10^6 eggs per female				1535.6 + 3% = 1.6 × 10^6 eggs per female

computations for *J. lalandii* from Pollock (1987). The simulations are run for a total of 20 adult instars corresponding to 20 years. Mean molt increments of adult *J. edwardsii* are larger than those of *J. lalandii*, such that females attain max. sizes of 127 mm and 91 mm respectively after about 20 years of adult growth. Differences in instar-specific molt increments (% ΔS) appear to be largely responsible for the differences in maximum sizes attained by the two species.

From the results it is evident that fecundity per recruit of *J. edwardsii* females is considerably higher than that of *J. lalandii* females in South Africa. This is so despite the larger mean egg size of *J. edwardsii* and is mainly the result of the larger size at maturity in the latter species. Once again, it is clear that size at maturity is the most important determinant of fecundity, an observation which has also been noted by Hines (1982) for a wide variety of crab species.

Pollock (1987) determined that *J. lalandii* females larger than about 91 mm (>20 years after maturity) accounted for approximately another 3% of eggs produced, bringing the total fecundity per recruit in Table 7 to a figure of approximately 1×10^6 eggs. If the same is assumed for *J. edwardsii*, females of the latter species would produce about 1.6×10^6 eggs during a lifetime. At the maximum recorded size at maturity of about 120 mm for females from Stewart Island to the south of New Zealand, fecundity per recruit would be much greater than this, while females from North Island, which generally have smaller sizes at maturity (Table 6) would have lower values, assuming that similar brood size/body size relationships apply throughout New Zealand waters.

Unfortunately it is not yet possible to accurately model examples of fecundity per recruit of the remaining species, owing to a lack of information on growth, mortality, and/or fecundity. In general, however, the calculations illustrate that the larger sizes at maturity, and the generally larger maximum sizes attained by *Jasus* species in the Pacific Ocean (Table 6), suggest that fecundity per recruit is greater for Pacific species than for species of the Atlantic and Indian Ocean basins.

A possible explanation for this finding is discussed by Pollock (1990), who suggested that larval mortality (largely the result of non-return of larvae advected away from parent populations) may be relatively greater in the Pacific region, where larger and more variable ocean gyres and eddies exist than in the smaller Atlantic and Indian Ocean basins.

Egg production (fecundity per recruit) in the three study groups is compared in Table 8. Mortality rates (between hatching and attainment of sexual maturity) are calculated on the premise that only 2 offspring survive to reach maturity out of the average number of eggs produced by each female during a lifetime. It is self-evident that mortality rates (mainly during the early life history stages) are inversely related to adult egg production, but the real significance of the data is that even when no larval phase is present (as in freshwater crayfish),

Table 8. Fecundity per recruit and prepuberty mortality rates in some crayfish, lobster and spiny lobster species.

Species	Fecundity per recruit	Female survival to adulthood	% juvenile mortality
O. virilis	124	1 ex 62	98.4
N. norvegicus (Portugal)	7800	1 ex 3900	99.97
N. norvegicus (Irish Sea)	1300	1 ex 650	99.85
J. lalandii (S.W. Cape)	1,000,000	1 ex 500,000	99.9998
J. edwardsii (S. Island, N.Z.)	1,600,000	1 ex 800,000	99.9999

mortality rates are still so high that some form of density-dependent regulation is likely. As discussed earlier, at low settling densities (i.e. high *per capita* food availability) juvenile growth increments are large, resulting in large mean size at maturity, and high egg production per recruit. The mechanism is suggestive of a feedback regulatory system, but a detailed examination of this concept will be the subject of a separate paper in preparation.

5 DISCUSSION

Studies on the crayfish *O. virilis* and the langoustine *N. norvegicus* have shown that growth rates and mean size at maturity are primarily regulated by food availability, and that, certainly in the case of *N. norvegicus*, water temperature was not related to the geographic variations observed in this parameter. Studies conducted by Pollock (1987), and Beyers & Goosen (1987) have also shown that geographic variations is growth, mean size at maturity and egg production per recruit of *J. lalandii* are governed by density-dependent processes on the west coast of Southern Africa.

Besides these observations, other studies have suggested a possible relationship between mean size at maturity and water temperature. For instance, the following studies found that small mean size at maturity was correlated with higher water temperatures: *Homarus americanus* in Northumberland Strait (Aiken & Waddy 1980), *Homarus gammarus* from North Wales (Aiken & Waddy 1980), *J. edwardsii* from North Island, New Zealand (Bradstock 1950, Street 1969, Annala et al. 1980), *P. argus* in parts of the Caribbean (Sutcliffe 1957). On the other hand, the following studies show the reverse, namely higher water temperatures correlated with larger size at maturity: *J. novaehollandiae* from parts of South Australia (Lewis 1981), *Panulirus homarus rubellus* from Natal and Transkei (Heydorn 1969, Berry 1971) and *P. argus* from Antigua (Davis 1975).

It is therefore apparent that precise reasons for intraspecific variations in mean size at maturity are not always readily distinguishable, although it is quite clear that egg production is directly related to mean size at maturity. In many cases, the latter parameter has been shown to depend on juvenile growth rates (actually molt increments) and these in turn appear to be strongly affected by density-dependent processes, although the possible additional influence of temperature cannot be discounted.

As Chittleborough (1979) and Pollock (1987) have discussed, intensive fishing on spiny lobster stocks may cause a growth and fecundity response due to lowered adult population densities, and this may be considered as a primary regulating mechanism, allowing relatively high levels of egg production to be maintained in the face of declining adult biomass. More important, however, is the secondary response which is likely to take place after larval settlement, within the nursery habitat, if larval recruitment begins to decline, either as a result of environmental changes (e.g. changes in surface circulation patterns) or as a direct result of recruitment overfishing. If fewer larvae settle, the relative increase in *per capita* availability of food and space is likely to result in increased rates of growth and survival of juveniles. The larger molt increments resulting from improved feeding conditions lead to a gradual increase in mean size at maturity, and hence to increased *per capita* egg production. A feed-back response of this type would tend to regulate population egg production within limits generally defined by the carrying capacity of the environment.

In some lobster fisheries where minimum size limits may be too small, and where fishing effort is so high as to severely reduce egg and larval production, a declining trend in larval settlement (often masked by natural variability) is therefore likely to result in a gradual and progressive increase in mean size at maturity. This could lead to a situation where maturity size becomes appreciably larger than the (fixed) minimum size limit, so that more and more juveniles are taken by the fishery, and the reproductive potential of the stock is diminished even further.

Such a process could possibly account for the phenomenon noted by Chittleborough (1976), namely that the mean carapace length at first breeding of female *P. cygnus* in the southern and central parts of Western Australia appeared to increase from approximately 77 mm in the early 1950's to about 96 mm during the 1970's.

Recruitment overfishing will ultimately lead to stock extinction and clearly should be countered. However, experience has shown that socioeconomic constraints often make it extremely difficult to increase size limits after intensive fisheries have become established. A more effective policy might be to severely reduce fishing effort before attempting to increase size limits.

ACKNOWLEDGEMENTS

The author is grateful for the support and pleasant working atmosphere provided by the Directorate of the Sea Fisheries Research Institute, as well as for the encouragement given by colleagues in South Africa. The stimulus for this paper emanated from contacts made with fellow lobster researchers at the International Workshop on Lobster Recruitment held at St. Andrews, New Brunswick in July 1985. T. van Dalsen and D. Sewell are thanked for contributing the artwork.

REFERENCES

Acheson, J.M. & R. Reidman 1982. Biological and economic effects of increasing the minimum legal size of American lobster in Maine. *Trans. Amer. Fish. Soc.* 111: 1-12.

Aiken, D.E. & S.L. Waddy 1980. Reproductive biology. In J.S. Cobb & B.F. Phillips (eds), *The Biology and Management of Lobsters*. Vol. 1: pp. 215-276. New York: Academic Press.

Annala, J.H. 1980. Mortality estimates for the rock lobster, *Jasus edwardsii*, near Gisborne, New Zealand. *N.Z. J. Mar. Freshwater Res.* 14: 357-371.

Annala, J.H. 1983. New Zealand rock lobsters: biology and fishery. *Fish. Res. Div. Occ. Publ.* 42: 36 pp.

Annala, J.H. & B.L. Bycroft 1985. Growth rate of juvenile rock lobsters (*Jasus edwardsii*) at Stewart Island, New Zealand. *N.Z. J. Mar. Freshwater Res.* 19: 445-455.

Annala, J.H., J.L. McKoy, J.D. Booth & R.B. Pike 1980. Size at the onset of sexual maturity in female *Jasus edwardsii* (Decapoda : Palinuridae) in New Zealand. *N.Z. J. Mar. Freshwater Res.* 14: 217-227.

Arana, P.E. (ed.) 1985. *Investigaciones Marinas en el Archipiélago de Juan Fernández*. Valparaiso: Universidad Catolica de Valparaiso.

Arana, P.E. & M.F.C. Pizarro 1971. La langosta de Juan Fernández. 1. Caracteristicas morfometricas y distribucion de tallas y sexos de *Jasus frontalis* de la isla Robinson Crusoe. *Inv. Mar.* 2: 93-123.

Baez, P.R. 1973. Larvas phyllosoma del pacifico sur oriental (Crustacea, Macrura, Scyllaridea). *Rev. Biol. Mar. Valparaiso* 15: 115-130.

Baez, P.R. 1979. El puerulus de *Jasus frontalis* (H. Milne Edwards, 1837) (Crustacea: Decapoda: Palinuridae). *Rev. Biol. Mar. Dep. Oceanol. Univ. Chile*, 16: 225-228.

Berry, P.F. 1971. The biology of the spiny lobster *Panulirus homarus* (Linnaeus) off the east coast of southern Africa. *Invest. Rep. Oceanogr. Res. Inst.* (Durban) 28: 75 pp.

Beyers, C.J., De B. & P.C. Goosen 1987. Variations in fecundity and size at sexual maturity of female rock lobster *Jasus lalandii* in the Benguela ecosystem. *S. Afr. J. Mar. Sci.* 5: 513-521.

Booth, J.D. 1979. North Cape – a 'nursery area' for the packhorse rock lobster, *Jasus verreauxi* (Decapod: Palinuridae). *N.Z.J. Mar. Freshwater Res.* 13: 521-528.

Botsford, L.W. 1985. Models of growth. In A.M. Wenner (ed.) *Factors in Adult Growth: Crustacean Issues* 3: pp. 171-188. Rotterdam: Balkema.

Bradstock, C.A. 1950. A study of the marine spiny crayfish *Jasus lalandii* (Milne Edwards), including accounts of autotomy and autospasy. *Zool. Publ. Victoria Univ. Coll.* 7: 1-38.

Campbell, A. & D.G. Robinson 1982. Reproductive potential of three American lobster (*Homarus americanus*) stocks in the Canadian Maritimes. *Can. J. Fish. Aquat. Sci.* 40: 1958-1967.

Chapman, C.J. 1980. Ecology of juvenile and adult *Nephrops*. In J.S. Cobb & B.F. Phillips (eds), *The Biology and Management of Lobsters* Vol. II. pp. 143-178. New York: Academic Press.

Chapman, C.J. & A.L. Rice 1971. Some direct observations on the ecology and behaviour of the Norway lobster *Nephrops norvegicus*. *Mar. Biol.* 10: 321-329.

Chittleborough, R.G. 1976. Breeding of *Panulirus longipes cygnus* George under natural and controlled conditions. *Aust. J. Mar. Freshwater Res.* 27: 499-516.

Chittleborough, R.G. 1979. Natural regulation of the population of *Panulirus longipes cygnus* George and responses to fishing pressure. *Rapp. P-V Reun. Cons. int. Explor. Mer.* 175: 217-221.

Cobb, J.S. & D. Wang 1985. Fisheries biology of lobsters and crayfishes. In A.J. Provenzano (ed.), *The Biology of Crustacea* Vol. 10 *Economic Aspects: Fisheries and Culture:* pp. 168-248. New York: Academic Press.

Davis, G.E. 1975. Minimum size of mature spiny lobsters, *Panulirus argus*, at Dry Tortugas, Florida. *Trans. Amer. Fish. Soc.* 4: 675-676.

Davis, G.E. 1980. Juvenile spiny lobster management. *Fisheries* 5(4): 57-59.

Farmer, A.S.D. 1973. Age and growth in *Nephrops norvegicus* (Decapoda: Nephropidae). *Mar. Biol.* 23: 315-325.

Farmer, A.S.D. 1975. Synopsis of biological data on the Norway lobster *Nephrops norvegicus* (Linnaeus, 1758). *FAO Fish. Synop.* 122: 97 pp.

Fielder, D.R. & A.M. Olsen 1967. Synopsis of biological data on the southern spiny lobster *Jasus lalandei* (H. Milne Edwards) 1837. *Aust/N.Z. Meeting on Decapod Crustacea, Sydney Oct. 24-28, 1967.* 3(5): 47 pp.

Figueiredo, M.J. & H.J. Thomas 1967. On the biology of the Norway lobster, *Nephrops norvegicus* (L.) *J. cons. perm. int. Explor. Mer.* 31: 89-101.

France, R.L. 1983. Relationship of crayfish (*Orconectis virilis*) growth to population abundance and system productivity in small oligotrophic lakes in the experimental lakes area, northwestern Ontario. *Can. J. Fish. Aquat. Sci.* 42: 1096-1102.

Gilchrist, J.D.F. 1913. The Cape crawfish and the crawfish industry. *Mar. Biol. Rep. Cape Town* 1: 1-45.

Gore, R.H. 1985. Molting and growth in decapod larvae. In A.M. Wenner (ed.), *Larval Growth: Crustacean Issues* 2: pp. 1-66. Rotterdam: Balkema.

Grua, P. 1960. Les langoustes australes (*Jasus lalandii*). Biologie – milieu exploitation commerciale. Étude préliminaire. *Terr. Austra. Antar. France.* 10: 15-40.

Grua, P. 1963. Maturite-cycle sexual – soies ovigeres des langoustes australes femelles *Jasus paulensis* Heller, 1863 – étude statistique. *C.N.F.R.A. Inst. Geogr. Nat. Paris* 4: 1-35.

Gruvel, A. 1911. Contribution a l'étude systématique des Palinuridae. *C.R. Acad. Sci. Paris.* 152: 1350-1352.

Hartnoll, R.G. 1982. Growth. In L.G. Abele (ed.), *The Biology of Crustacea,* Vol. 2: pp. 111-196. New York: Academic Press.

Hartnoll, R.G. 1985. Growth, sexual maturity and reproductive output. In A.M. Wenner (ed.), *Factors in Adult Growth: Crustacean Issues* 3: pp. 101-128. Rotterdam: Balkema.

Heydorn, A.E.F. 1969. The rock lobster of the South Affrican west coast *Jasus lalandii* (H. Milne Edwards). 2. Population studies, behaviour, reproduction, moulting, growth and migration. *Invest. Rep. Div. Sea. Fish. S. Afr.* 71: 52 pp.

Hickman, U.V. 1946. Notes on the Tasmanian marine crawfish *Jasus lalandii* (Milne Edwards). *Pap. Roy. soc. Tasm.* 1945: 27-38.

Hillis, J.P. 1968. Larval distribution of *Nephrops norvegicus* (L.) in the Irish Sea and North Channel. *ICES C.M. 1968/K* 6: 5 pp. (unpubl.)

Hines, A.H. 1982, Allometric constraints and variables of reproductive effort in Brachyuran crabs. *Mar. Biol.* 69: 309-320.

Holthuis, L.B. & E. Sivertsen 1967. The Crustacea Decapoda, Mysidacea and Cirripedia of the Tristan da Cunha

Archipelago, with a revision of the 'frontalis' subgroup of the genus *Jasus*. *Results Norw. Scient. Exped. Tristan da Cunha* 52: 50 pp.

Jones, R. 1979. An analysis of a *Nephrops* stock using length composition data. *Rapp. P.-V. Cons. Perm. Int. Explor. Mer.* 175: 259-269.

Kensler, C.B. 1967. Fecundity in the marine spiny lobster *Jasus verreauxi* (H. Milne Edwards) (Crustacea: Decapoda; Palinuridae). *N.Z.J. Mar. Freshwater Res.* 1: 143-155.

Kensler, C.B. 1968. Notes on fecundity in the marine spiny lobster *Jasus edwardsii* (Hutton) (Crustacea: Decapoda: Palinuridae). *N.Z.J. Mar. Freshwater Res.* 2: 81-89.

Lesser, J.H.R. 1978. Phyllosoma larvae of *Jasus edwardsii* (Hutton) (Crustacea: Decapoda: Palinuridae) and their distribution off the east coast of the North Island, New Zealand. *N.Z.J. Mar. Freshwater Res.* 12: 357-370.

Lewis, R.K. 1977. Rock lobster puerulus settlement in the South East. *SAFIC* 13: 9-11.

Lewis, R.K. 1981. Southern rock lobster *Jasus novaehollandiae* – A review of the Zone N Fishery. *SAFIC* 5: 3-14.

Mauchline, J. 1980. The biology of mysids and euphausiids. *Adv. Mar. Biol.* 18: 1-677.

McKoy, J.L. 1985. Growth of tagged rock lobsters (*Jasus edwardsii*) near Stewart Island, New Zealand. *N.Z.J. Mar. Freshwater Res.* 19: 457-466.

McKoy, J.L. & D.B. Esterman 1981. Growth of rock lobsters (*Jasus edwardsii*) in the Gisborne regions, New Zealand. *N.Z.J. Mar. Freshwater Res.* 15: 121-136.

McWilliam, P.S. & B.F. Phillips 1987. Distinguishing the phyllosoma larvae of rock lobster species of the genus *Jasus* (Decapoda: Palinuridae) in the waters of Australia and New Zealand. *Crustaceana* 52: 1-24.

Miller, D.G.M. 1985. A late phyllosoma larva of *Jasus tristani* Holthuis (Decapoda: Palinuridae). *Crustaceana* 50: 1-6.

Momot, W.T. 1967. Population dynamics and productivity of the crayfish, *Orconectes virilis*, in a marl lake. *Amer. Midl. Nat.* 78: 55-81.

Momot, W.T. 1978. Annual production and production/biomass ratios of the crayfish, *Orconectes virilis*, in two Northern Ontario lakes. *Trans. Amer. Fish. Soc.* 107: 776-784.

Momot, W.T. 1986. Production and exploitation of the crayfish *Orconectes virilis*, in northern climates. *Can. Spec. Publ. Fish. Aquat. Sci.* 92: 154-167.

Momot, W.T. & H. Gowing 1977a. Results of an experimental fishery on the crayfish *Orconectes virilis*. *J. Fish. Res. Bd. Can.* 34: 2056-2066.

Momot, W.T. & H. Gowing 1977b. Response of the crayfish *Orconectes virilis* to exploitation. *J. Fish. Res. Bd. Can.* 34: 1212-1219.

Momot, W.T. & H. Gowing 1977c. Production and population dynamics of the crayfish *Orconectes virilis* in three Michigan lakes. *J. Fish. Res. Bd. Can.* 34: 2041-2055.

Paulian, P. 1957. La pêche autour des iles Saint-Paul et Amsterdam et son avenir. *La Terre et la Vie* 4: 267-282.

Phillips, B.F. & A.N. Sastry 1980. Larval ecology. In J.S. Cobb & B.F. Phillips (eds), *The Biology and Management of Lobsters*. Vol. II. pp. 11-57. New York: Academic Press.

Pollock, D.E. 1973. Growth of juvenile rock lobster *Jasus lalandii* (H. Milne Edwards). *Invest. Rep. Sea Fish. Brch. S. Afr.* 106: 16 pp.

Pollock, D.E. 1981. Population dynamics of rock lobster *Jasus tristani* at the Tristan da Cunha group of Islands. *Fish. Bull. S. Afr.* 15: 49-66.

Pollock, D.E. 1986. Review of the fishery for and biology of the Cape rock lobster *Jasus lalandii* with notes on larval recruitment. *Can. J. Fish. Aquat. Sci.* 43(11): 2107-2117.

Pollock, D.E. 1987. Simulation models of rock-lobster populations from areas of widely divergent yields on the Cape west coast. *S. Afr. J. Mar. Sci.* 5: 531-545.

Pollock, D.E. 1990. Palaeoceanography and speciation in the spiny lobster genus *Jasus*. *Bull. Mar. Sci.* 46: 387-405.

Pollock, D.E. & C.J. De B. Beyers 1981. Environment, distribution and growth rates of West Coast rock-lobster *Jasus lalandii* (H. Milne Edwards). *Trans. R. Soc. S. Afr.* 44(3): 379-400.

Pollock, D.E. & M.J. Roscoe 1977. The growth at moulting of crawfish *Jasus tristani* at Tristan da Cunha, South Atlantic. *J. Cons. int. Explor. Mer.* 37: 144-146.

Rae, B.B. 1967. The food of cod in the North Sea and on the west of Scotland grounds. *Mar. Res.* 1: 1-68.

Rae, B.B. 1968. The food of cod in Icelandic waters. *Mar. Res.* 6: 1-19.

Roscoe, M.J. 1979. Biology and exploitation of the rock lobster *Jasus tristani* at the Tristan da Cunha islands, South Atlantic, 1949-1976. *Invest. Rep. Sea Fish. Brch S. Afr.* 118: 47 pp.

Royce, W.F. 1984. *Introduction to the Practice of Fishery Science*. New York: Academic Press.

Sastry, A.N. 1983. Ecological aspects of reproduction. In F.J. Vernberg & W.B. Vernberg (eds), *The Biology of Crustacea* Vol. 8: 179-270. New York: Academic Press.

Silas, E.G. 1967. On the taxonomy, biology and fishery of the spiny lobster *Jasus lalandii frontalis* (H. Milne Edwards) from Saint-Paul and New Amsterdam islands in the southern Indian Ocean with an annotated bibliography on species of the genus *Jasus* Parker. *Proc. Sympos. Crustacea, Ernakulam, Jan. 12-15 1965* 4: 1466-1520.

Silberbauer, B.I. 1971. The biology of the South African rock lobster *Jasus lalandii* (H. Milne Edwards). 1. Development. *Invest. Rep. Div. Sea Fish. S. Afr.* 92: 70 pp.

Street, R.J. 1969. The New Zealand crayfish *Jasus edwardsii* (Hutton). *Fish. Tech. Rep. N.Z.* 30: 53 pp.

Sutcliffe, W.H. 1957. Observations on the growth rate of the immature Bermuda spiny lobster, *Panulirus argus*. *Ecology* 38: 526-529.

Winstanley, R.H., A.E. Caton, J.G.K. Harris & R.K. Lewis 1983. Southern rock lobster fishery: Fishery situation report. *SAFIC* 7: 7-13.

Egg production in the American lobster, *Homarus americanus*

S.L. Waddy & D.E. Aiken
Department of Fisheries and Oceans, Aquaculture and Invertebrate Fisheries Division, Biological Station, St. Andrews, New Brunswick, Canada

ABSTRACT

Egg production in most feral populations of the American lobster is synchronized and predictable. Smaller females spawn on an alternate year cycle, but those larger than 120 mm carapace length spawn more frequently, often three times in five years. Mating usually occurs immediately after the female molts, but males can recognize and copulate with uninseminated females prior to spawning. One copulation can provide sufficient sperm to fertilize two successive broods, which reduces the impact of reduced molting frequency as the females age. Temperature plays a major role in egg production, influencing size at maturity, time of spawning, coordination of the molting and spawning cycles, time of hatch, and egg attachment and incubation success. Higher temperatures have an adverse effect on reproduction; cultured females spawn infrequently and rarely carry their eggs to hatch. However, when they are returned to a temperature cycle that includes low winter and moderate summer water temperature, their egg production approximates that obtained with feral females.

1 INTRODUCTION

In the last hundred years, there have been three major periods of study on the reproductive biology of the American lobster, *Homarus americanus*. In the late 1800's and early 1900's there was considerable interest in artificial propagation to enhance wild stocks, and Herrick (1909) and his contemporaries recognized that culture would not succeed without a thorough understanding of lobster biology. Herrick described lobster reproductive biology in great detail and most of his observations have withstood the test of time. He was one of the first to report multiple extrusion and although it was later discounted, it is now well documented and accepted.

In later years, Wilfred Templeman studied lobster populations throughout Atlantic Canada. He was the first to recognize that temperature influenced size at maturity and time of spawning and hatching. He also provided the first information on maturity indices in American lobsters.

The third major effort occurred in the 1970's under the stimulus of fisheries management and aquaculture. Most of the earlier work had been observational and descriptive (Aiken & Waddy 1980, for review), but after 1970 the emphasis shifted to biological, environmental and physiological studies on lobster reproduction. This, in turn, made it possible to interpret, evaluate and extend earlier observations. The result was a number of conflicting reports on the mechanisms that control spawning in lobsters. Most of the contentious issues have since been resolved and are reviewed in this chapter.

2 EGG PRODUCTION

2.1 *Size at maturity*

The size or age at which American lobsters become sexually mature appears to be influenced primarily by temperature. Those areas with the highest summer water temperatures appear to produce the smallest sizes at first spawning, so the smallest ovigerous females are found in Northumberland Strait and western Long Island Sound. Size at onset of maturity varies from 59 mm carapace length (CL) in Northumberland Strait to 128 mm for some females in the Bay of Fundy (Fig. 1) (Campbell & Robinson 1983, Templeman 1936b), and it has been estimated that

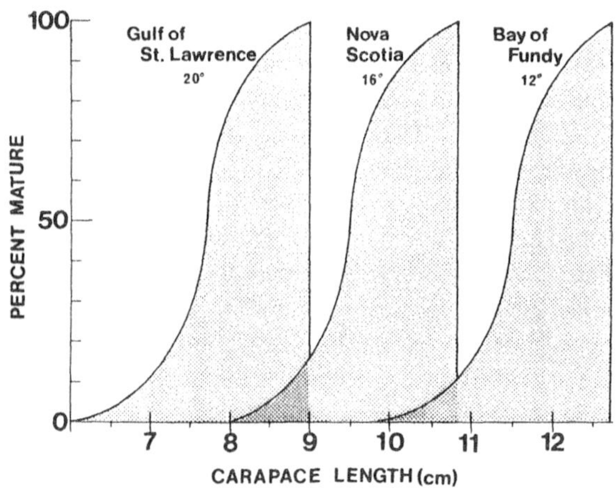

Figure 1. Size at maturity of female American lobsters from areas with different summer seawater temperatures.

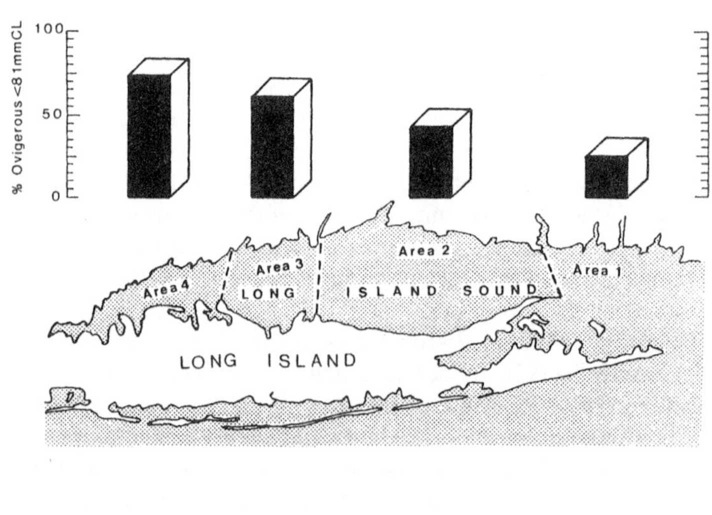

Figure 2. Relationship between geographic location and size of ovigerous females in Long Island Sound. Vertical bars show the percentage of small ovigerous females (less than 81 mm CL) in each of four sampling areas. (Data from Smith 1977).

half of the egg production in Northumberland Strait comes from lobsters that are smaller than the size at which Bay of Fundy lobsters start to mature (Campbell & Robinson 1983).

The notion that high summer water temperature induces maturation at a young age and small size was first articulated by Templeman, who further suggested that low summer water temperature delays the onset of maturity. This phenomenon occurs even in localized areas; lobsters found at the warmer western end of Northumberland Strait tend to be 1-3 cm shorter at onset of maturity than those at the cooler eastern end (Templeman 1936b). A similar variation is found from the western to the eastern end of Long Island Sound (Fig. 2, and Smith 1977), and between the Sound and the south shore of Long Island. Most of the females in Long Island Sound mature and produce eggs before their contemporaries from the south shore of Long Island start to mature (Briggs & Mushacke 1980). This apparently occurs in *H. gammarus* also (Simpson 1961).

High summer water temperature, while favorable for growth, is even more favorable for early maturation. Conversely, low summer water temperature delays onset of maturation, allowing the animal to grow (albeit more slowly) to a larger size before becoming sexually mature (as in the cool waters of the southern Bay of Fundy). When held for extended periods at constant 20°C, female *H. americanus* have matured and spawned at 52 mm CL, probably the physiological minimum for this species (Waddy & Aiken unpubl.).

Nelson (1986) suggested the duration of the intermolt period might determine size at maturity; e.g. once a female is large enough so that the intermolt period is longer than the egg incubation period, she could spawn without danger of losing her eggs at molt. In our view this is unlikely since at 20°C the molts are far enough apart even in lobsters of 45-50 mm CL to allow completion of egg incubation.

However, the relationship between temperature and maturity is not uncontested. It also happens that in warm water areas such as the Gulf of St. Lawrence and western Long Island sound, lobsters are found in relatively high density, suggesting that population density may influence size at maturity. However, this hypothesis is weakened by the existence of areas (such as the southern end of Grand Manan) where maturity occurs at a large size in spite of high population density. Other explanations have been proposed (e.g. intensive fishing pressure may exert genetic pressure for maturation at a smaller size), but the only factor that seems to be consistently associated with small size at maturity is high summer water temperature.

2.2 *Reproductive cycles*

The reproductive cycle of female American lobsters smaller than 120 mm CL typically takes

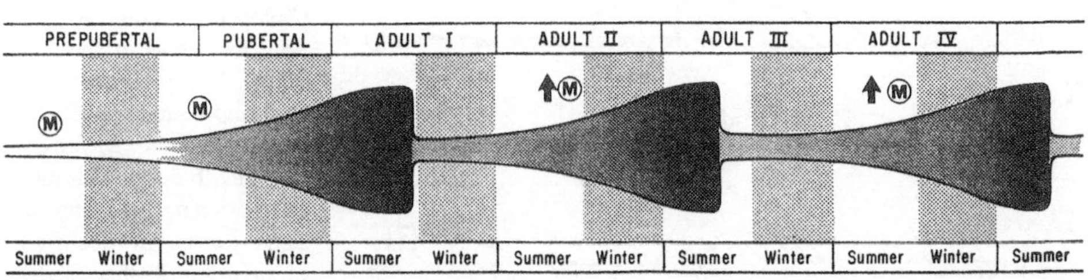

Figure 3. Reproductive cycle for typical inshore *Homarus americanus* over a 7-yr period. Ovary growth, indicated diagrammatically, culminates in spawning. Hatching (arrow) and molt (M) occur in the same summer and spawning in alternate summers.

two years, during which the female molts and mates one summer, spawns the following summer and carries the eggs on her pleopods until the third summer, when they hatch. Two to three months after her eggs hatch the female again molts and mates (to replace stored sperm lost at the molt), and the cycle is repeated (Fig. 3). At temperatures characteristic of inshore areas, molting and spawning are synchronized within the population. In smaller sizes a molt always occurs between successive spawnings, and successive molts are always separated by a spawning.

2.2.1 *Types of cycles*

2.2.1.1 *Adult Ia and Ib cycles.* The year in which spawning occurs for the first time is the Adult-I year. Although mature females normally spawn one summer and molt the next (Adult Ia, Fig. 4A), approximately 20% of the maturing Adult-I females from areas with high summer water temperature will molt and then spawn within two months (Adult Ib, Fig. 4B, C). This is rare in subsequent year-classes. Since it is associated with high summer water temperature, the Adult-Ib cycle is more common in the warm southern Gulf of St. Lawrence than along the cool Atlantic coast of Nova Scotia, and it is rare in the relatively cold Bay of Fundy. In Newfoundland waters the incidence appears to be influenced by the minimum size limit and the rate of exploitation (Ennis 1984b). The percent of lobsters that molt and spawn in the same year varies with both area and year and ranges from 0-38% of the ovigerous females (Ennis 1980). The pronounced yearly variations in summer water temperature recorded in Newfoundland may contribute to this variation.

It is not known whether single-season molting and spawning occurs because high summer temperature advances spawning by a full year or because the interaction between molting and reproductive cycles is modified. However, the mechanism allows small female lobsters that are still molting each summer to spawn as well. After this first spawning, all females adopt the normal alternate-year molting and spawning cycle. In the laboratory, we have been able to induce this same-year molt and spawn phenomenon in a similar percentage of older females (Adult III-IV) by holding them in abnormally high winter temperature regimes (Waddy & Aiken unpubl.).

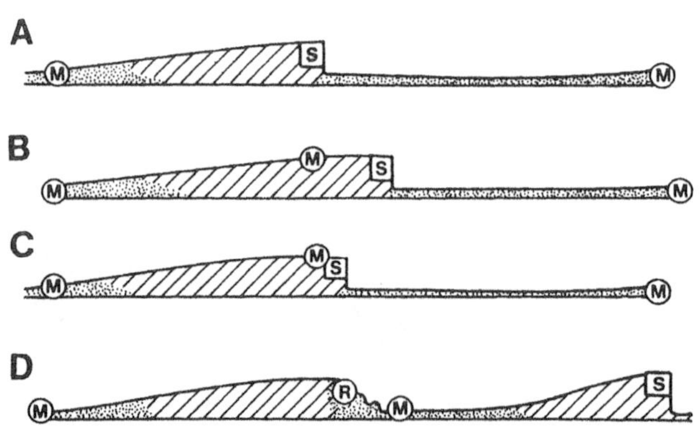

Figure 4. Cycles of molt (M), spawning (S) and oöcyte resorption (R) observed in females maintained in temperature and photoperiod conditions typical of nearshore areas of the Bay of Fundy. A. Typical alternate year molt and reproductive cycle of female lobsters < 120 mm CL. B, C. Adult Ib cycle, with molting and spawning occurring in the same season. D. Sequence of events in females that resorb their ovaries due to molt interference. Spawning is delayed until the following summer.

2.2.1.2 *Alternate-year spawning.* In 1976 we constructed a 'typical' 7-year molt-spawning cycle that started two molts (years) before the first spawning, encompassed three ovarian cycles, and ended with the Adult-V summer (Aiken & Waddy 1976). From the limited data available, we concluded that each maturity class had its own pattern of molt and reproductive events. We have since held individual lobsters for as long as 16 years (to the Adult-XIV year) and have found that year-class variation in reproductive cycles occurs at only two points: in the Adult-I year (as described above) and in 'jumbo' lobsters 120 mm CL or larger. Although we previously felt that single season molting and spawning could also occur in the Adult-III or subsequent age classes, we now know this is seldom the case in inshore areas. Exceptions occur in very warm areas or when the lobsters are exposed to abnormally high temperatures in the laboratory. In Long Island Sound, which has some of the highest summer temperatures in the lobster's range, the single season molt-spawn phenomenon apparently does occur beyond the Adult-I year (Briggs & McGroaty 1985), but the data are too limited to suggest the extent of this phenomenon. This could allow females smaller than 120 mm CL to spawn in consecutive years, a possibility that receives some support from the high incidence of ovigerous females (68-80%) in the larger size classes.

2.2.1.3 *Consecutive-year spawning.* Molting continues throughout the life of a lobster, but becomes less frequent with increasing size. Unlike small females, those over 120 mm CL often spawn twice without an intervening molt. This is termed consecutive spawning and it occurs in two forms: successive-year (spawning in two successive summers without an intervening molt), and alternate-year (spawning in alternate summers, but without an intervening molt) (Fig. 5). In both, females are usually able to fertilize the two broods with sperm from a single

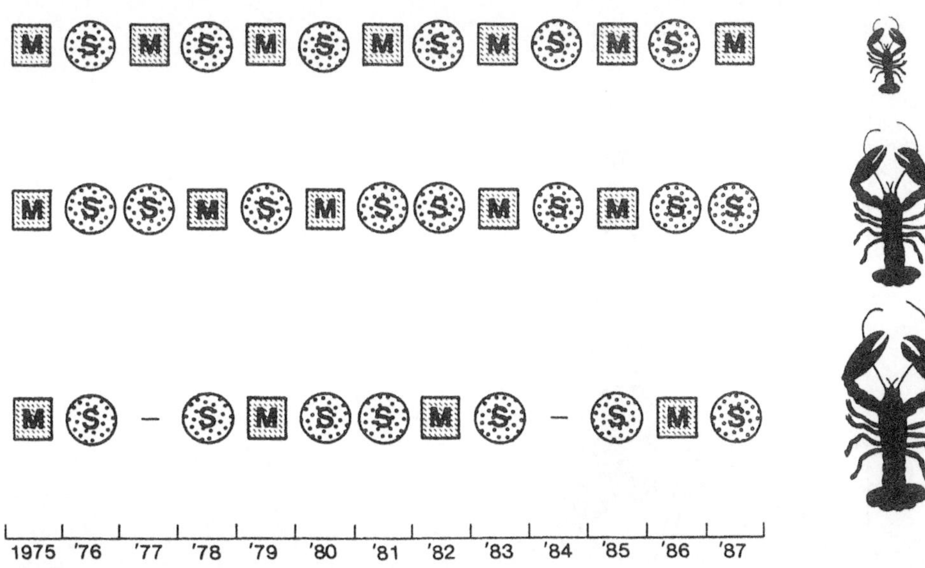

Figure 5. Molt and spawning cycles of female American lobsters. Molt is indicated by squares and spawning by circles. Upper row: molt and spawning cycles characteristic of small female lobsters (< 120 mm CL); Middle row: successive-year spawning cycles of large lobsters (≥120 mm CL); Lower row: alternate-year spawning cycles of lobsters molting on a 4yr cycle (≥160 mm CL).

insemination (i.e. multiple fertilization). If necessary, additional sperm can be obtained from an intermolt mating. All 24 females, 120 mm CL or larger, that we have held in the laboratory for sufficient time displayed one of these two types of consecutive spawning, and the type that occurred depended on the size of the female and the frequency of molting. Those on a 3-year molt cycle spawn in successive years, while those on a 4-year molt cycle spawn in alternate years. All large females spawned at least as frequently as smaller females (biennially), and females of 120-150 mm CL spawned more frequently. In these animals, the 3-year intermolt cycle alternates with a two year cycle, so that spawning occurs three times in five years. Consecutive spawning therefore enables 'jumbo' lobsters to spawn more frequently over the long term than their smaller counterparts, making their relative fecundity much greater than generally assumed (Waddy & Aiken 1986).

2.2.1.4 *Synchronous and asynchronous cycles.* Temperature appears to be the major synchronizer of spawning in feral populations. Lobsters from the southern Gulf of St. Lawrence held in our facility spawn between late June and early August, with 80% of females spawning in the first three weeks of July. But, as Templeman (1940) noted, spawning in the Gulf of St. Lawrence occurs in two pulses associated with different year-classes. Most females spawn in the first peak. The second peak consists of Adult-Ia females that molt just before spawning and therefore spawn later in the year (Fig. 4A, B). Except for the Adult-Ia cycle in warm water areas, spawning in a given area is synchronized and predictable. In the laboratory, 95% of females will spawn not only in the correct year but also at the expected time (Waddy unpubl.). The percentage should be even higher in feral populations because capture, confinement and handling cause stress that can reduce spawning incidence. Ennis (1984a) implied that a significant number of mature females in Newfoundland waters abort ovarian development by resorbing the yolk from their mature oöcytes before spawning. This may be caused by the marginal

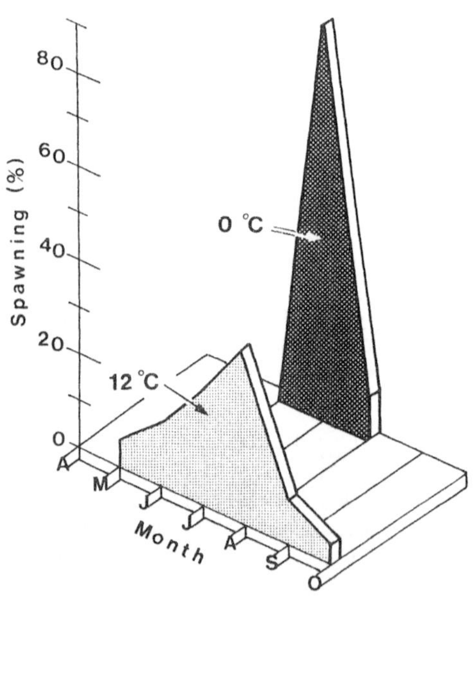

Figure 6. Effect of 0°C and 12°C winter seawater temperatures on the synchrony and success of spawning the following summer.

summer temperatures that occur in Newfoundland waters. We have observed an increased incidence of resorption (30-50%) when summer water temperature remains below normal and spawning is delayed.

Elevated temperature also affects the synchronization of spawning. When winter water temperature remains above 8°C, spawning occurs throughout a greater portion of the year (Fig. 6) and there is less synchrony and predictability. This is because high temperature stimulates both premolt and ovarian development, and since the two are antagonistic, only one can proceed normally. Thus some females spawn, some molt and then spawn, and some molt twice without spawning.

Spawning occurs in different months in different geographical areas depending on local temperature. It can start as early as May and end as late as October. Most spawning activity occurs from June through September, but in cold water areas like the Bay of Fundy it can extend into late October (Aiken & Waddy 1980).

2.2.2 *Interaction between molt and reproductive cycles*

We know a fair amount about the individual processes of molting and reproduction, but we know very little about how the two interact. It is important that the female does not molt while the eggs are attached to the pleopods, and that she is ready to molt soon after they hatch so that cuticle mineralization can be completed before the water temperature declines. Despite the potential for conflict, roughly 95% of female lobsters spawn on schedule when maintained in our local temperature and photoperiod regimes and it is extremely rare for them to molt while carrying eggs. It appears that the molt and reproductive cycles are regulated by temperature so as to minimize conflict, and that an extended period of low winter temperature is an essential part of this regulation.

Winter water temperature must remain below 9°C and perhaps even below 5°C to ensure proper synchronization of molting and reproduction. Above this temperature, molting often predominates (Fig. 7). At elevated temperatures, spawning occurs less frequently and when it

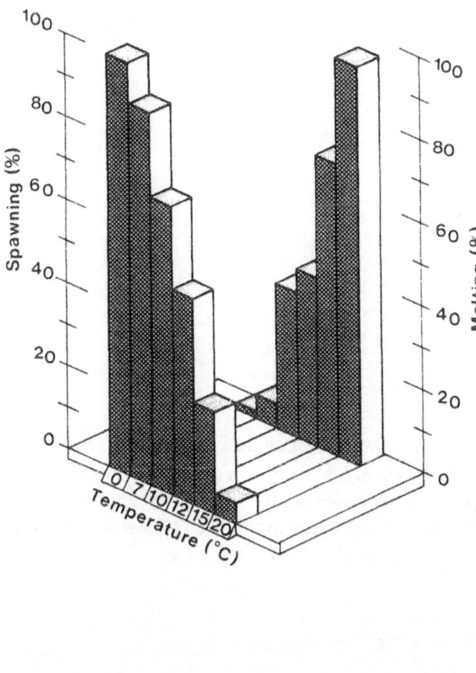

Figure 7. Effect of various winter seawater temperatures on the relative incidence of molting or spawning the following summer. Spawning incidence is inversely correlated with increasing winter temperature.

does occur it is distributed over a greater part of the year and is asynchronous (Waddy & Aiken 1984a, b). In the summer following a winter of elevated water temperature, the incidence of spawning will decrease and there will be a proportional increase in the number of lobsters resorbing their oöcytes and molting (Fig. 4D). This phenomenon also occurs in the European lobster, *H. gammarus*, which spawns throughout the year in the Channel Islands due to the high year-round temperatures (S.F. Bossy pers. comm.).

In contrast to feral stock, mature cultured females held at constant 20°C rarely spawn because of conflict between the molt and reproductive cycles. Indications are that vitellogenesis only occurs at certain molt stages and that if the molt cycle progresses into premolt (D_1-D_3), vitellogenesis stops and the ovarian yolk may be resorbed from the oöcytes. Although high temperature leads to both premolt induction and ovary maturation, molting usually takes precedence over spawning.

A major impediment to spawning in lobsters held at 20°C is the short intermolt period. Although the time between molts is long enough to allow spawning and egg incubation, the precise phasing required between somatic and reproductive events seldom occurs. Low winter temperature is the normal synchronizer of molting and spawning and if cultured females are transferred to conditions with a long period of low winter temperature, spawning incidence will increase (Fig. 8).

Contrary to common belief, neither spawning nor insemination is dependent on molting (see Bertran & Lorec 1986). However, lobsters are almost always in molt stage C_4 at the time of spawning (Aiken & Waddy 1980), which suggests an active relationship between the molt and reproductive cycles. Furthermore, this relationship appears to be associated with the eyestalk organs; females without eyestalks often spawn in early to mid-premolt (Aiken et al. 1981).

At normal photoperiod and temperature, ovigerous females will often be in mid-premolt when their eggs hatch. Females do occasionally molt before their eggs hatch, but it is an uncommon event. Of the thousands of females that have spawned in our laboratory under normal environmental conditions, only one molted while incubating eggs (Pl. 1). However, this

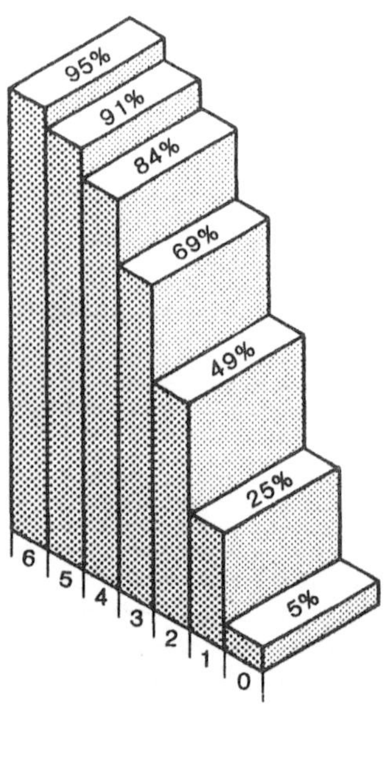

Figure 8. Incidence of spawning in cultured females held at 20°C compared to cultured females transferred to local seawater temperatures. Incidence of spawning increases with number of years spent in local temperatures; after 6 years spawning in cultured females occurs with the same frequency as in feral stock (95%).

has happened many times under abnormal photoperiod regimes. In these instances, molt occurred in the late winter or early spring, about 4-6 months early. Similar reports have come from studies in California where altered photoperiod and temperature regimes were used (Nelson et al. 1983).

Females are almost always in molt stage C_4 when they spawn, which suggests a formal relationship between the two. It may be that the transition from C_4 to D_0 is delayed when the ovary is in the final maturation stages. There are also indications that final ovary maturation will be terminated if it coincides with premolt (Fig. 4D), thereby conserving energy committed to the ovary and allowing the molt and reproductive cycles to be resynchronized so that spawning can occur the following summer.

2.2.2.1 *Oöcyte degeneration and resorption.* Massive oöcyte resorption causes ovarian regression and recrudescence similar to that in a spawned ovary. Resorption of oöcyte vitellin is common and readily apparent, because the dark green carotenoprotein in the hemolymph darkens the articular membranes and ventral abdomen and results in what is commonly called a 'black lobster'. This massive resorption of yolk proteins occurs during secondary vitellogenesis and causes elevated vitellogenin levels for 6-7 months (Byard 1985, Waddy unpubl.). Intact vitellin appears to be transferred from the resorbing oöcytes to the hemolymph, since the serum vitellogenin mobilities are similar in vitellogenic and resorbing lobsters (Dehn et al. 1983).

Resorption is often associated with unfavorable holding conditions or handling stress in the weeks prior to spawning (Aiken & Waddy 1985a, Templeman 1940). It is also common when lobsters are exposed to an abnormal environment. For example, lobsters at constant 20°C undergo successive cycles of ovarian development and resorption and have a reduced spawning rate (Hedgecock et al. 1978, Waddy & Aiken 1984a). However, resorption also occurs in the natural environment. Ennis (1984a, 1984d) concluded from cement gland development that mature females often fail to spawn, probably because of oöcyte resorption. He suggested (1984a) that lack of insemination was the cause, but we have not seen any relationship between insemination and either gonad development or spawning. Hundreds of uninseminated females

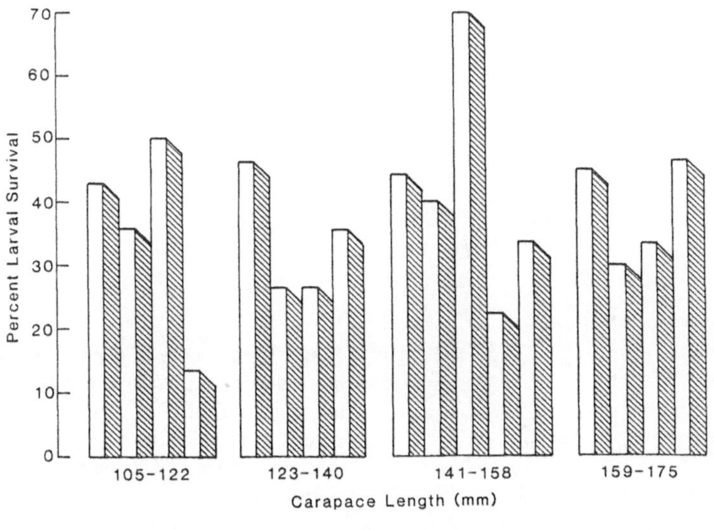

Figure 9. Survival of larvae from first to fourth stage. Each vertical bar indicates larvae from one female lobster from the Bay of Fundy (105-175 mm CL).

have spawned in our laboratory, and the only observed effect of infertility is the subsequent loss of the eggs.

The hemolymph of mature females is often pale to light green due to accumulation of vitellogenin. Whether this is due to excessive production, lack of uptake by the ovary or to minor resorption from the ovary is difficult to determine. Resorption on a small scale occurs immediately after spawning and reclaims residual oöcytes not extruded from the ovary and oviducts. As in other types of resorption, the oöcytes yield their green pigment and remain for at least two years as yellow or orange flecks in the developing ovary and the terminal portion of the oviduct, indicating a previous ovarian cycle (Herrick 1909, Aiken & Waddy 1980).

2.2.2.2 *Influence of size and age.* As noted previously, size influences the phasing of molt and spawning and large lobsters (120-150 mm CL) spawn more frequently than their smaller couterparts. Even the largest lobsters (200 mm CL) continue to spawn at least biennially, and the viability of eggs and larvae produced by these very large females is similar to those produced by younger lobsters (Fig. 9) (Aiken & Waddy 1986a). Moreover, large lobsters produce eggs with a higher energy content than do smaller lobsters (Attard & Hudon 1987).

2.3 *Spawning*

2.3.1 *Geographic variation*

Templeman (1940) was the first to relate variation in time of spawning to differing temperature regimes. Nearshore areas have pronounced seasonal temperature extremes and the temperature cycles vary from one area to another. As a result the molting and spawning cycles also vary from area to area. Spawning in the warm water of the southern Gulf of St. Lawrence occurs nearly a month before it does in the cool water around Grand Manan in the Bay of Fundy (Fig. 10). Offshore, where seasonal temperature changes are more moderate, molting and perhaps spawning (although reproductive cycles of offshore lobsters have not been described) are spread over a greater portion of the year (Cooper & Uzmann 1980). However, offshore lobsters are able to modify their cycles to some extent through seasonally directed movements (Campbell & Stasko 1986).

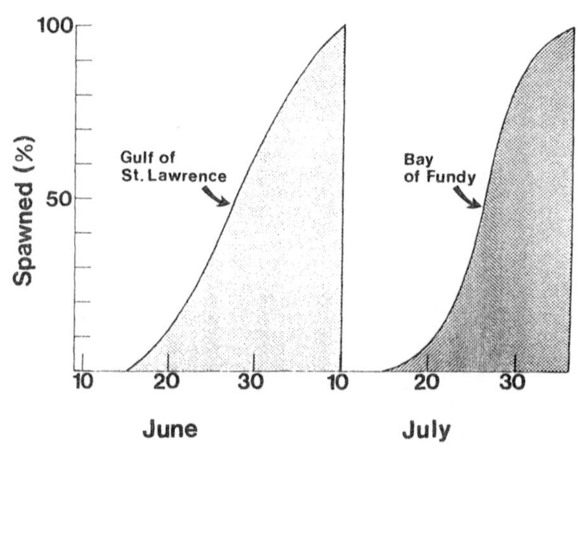

Figure 10. Influence of temperature on time of spawning. Water temperature in the southern Gulf of St. Lawrence may exceed 20°C. In the Bay of Fundy it seldom rises above 12°C (data from Templeman 1940).

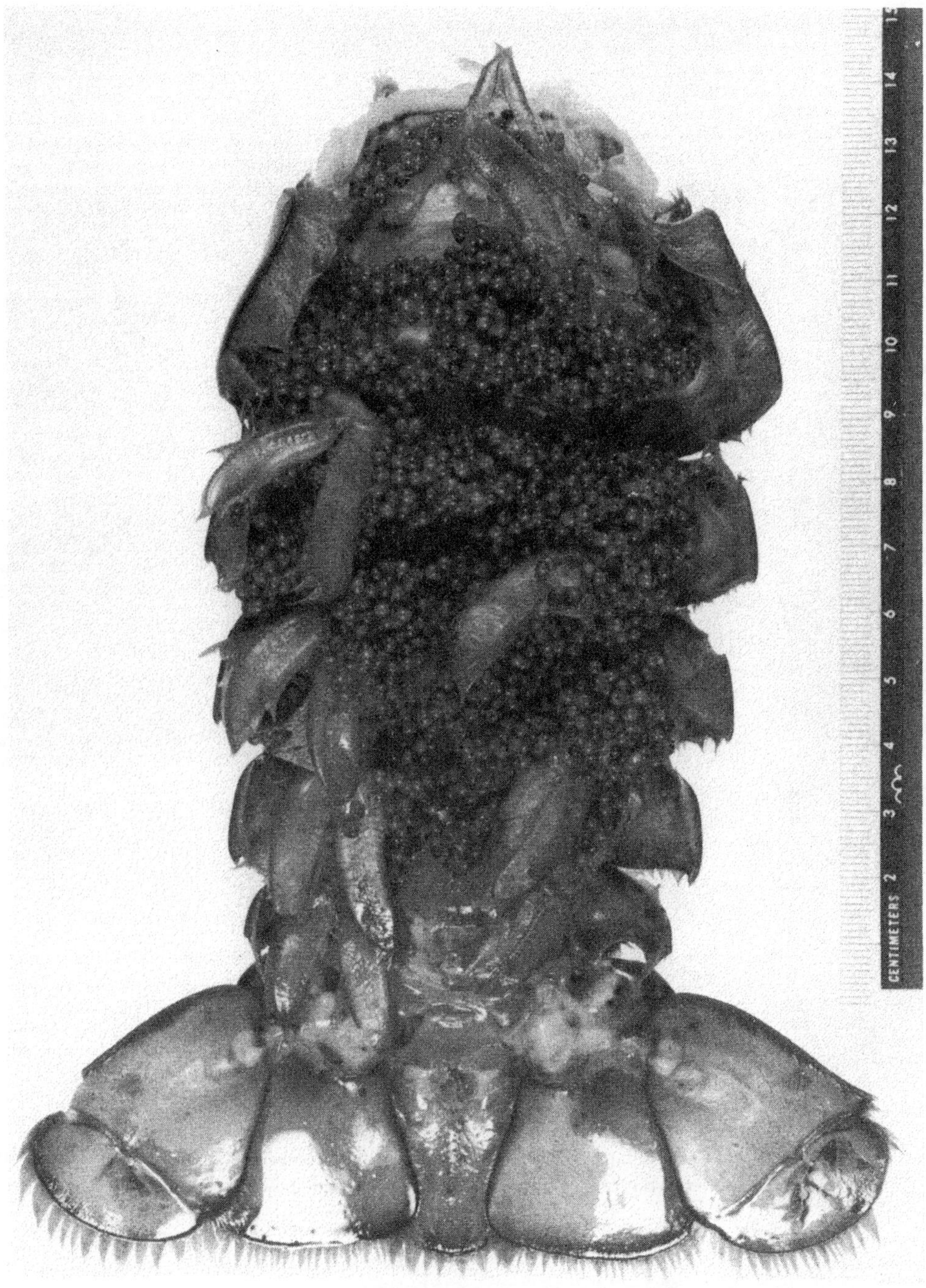

Plate 1. Exuvia of egg-bearing lobster that molted within two days of capture in the Bay of Fundy. Hatching would have occurred within two weeks.

Plate 2. Effect of temperature conditioning on egg attachment in cultured females reared at 20°C. Right – female conditioned for 6 years (3 spawning cycles) in seawater temperatures characteristic of the Bay of Fundy. Left – female recently transferred to Bay of Fundy temperature conditions and spawning eggs for the first time.

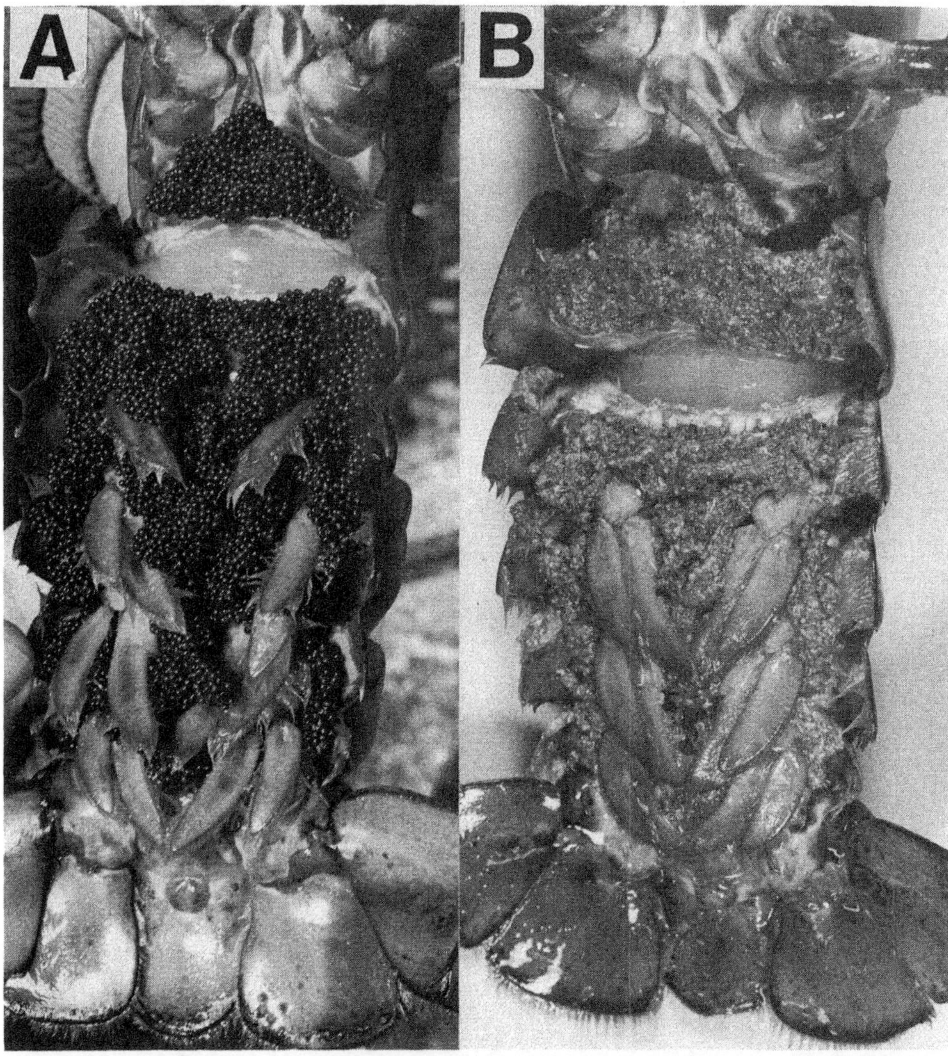

Plate 3. Massive infestation of a lobster egg mass by the nemertean *P. homari*. A. normal egg mass, free of nemerteans; B. egg mass heavily infested with nemerteans.

Plate 3 (continued). C. close view of B, showing various stages of nemertean development and numerous empty lobster egg membranes.

2.3.2 *Seasonality*

Somatic and reproductive growth is suspended when water temperature drops below 5°C, so in most inshore areas molting and spawning occur primarily between May and October. Spawning is more synchronized than molting and occurs before the peak of the molting season. In most areas, the egg hatching and spawning seasons overlap so that females carrying eggs spawned the previous year can be found at the same time as females carrying newly spawned eggs.

2.4 *Strategies for optimizing egg production*

2.4.1 *Movements of mature females*

Many benthic crustaceans undergo seasonal migrations involving inshore-offshore movement, and American lobsters are no exception. Molting and reproduction in lobsters have been linked to these migrations, but little is known about the relationship between seasonal movements and biological events. Offshore lobsters winter in the canyons of the continental slope, but in summer they move to shallow (and warmer) portions of the shelf and coastal areas where they molt and reproduce. During spring, lobsters tend to be concentrated in deep water in the Gulf of Maine and along the edge of the Continental Shelf, whereas in the fall, they are scattered on the shoal waters of Georges Bank and the Scotian Shelf, with lesser concentrations along the shelf edge (Burns & Clark 1980, Campbell et al. 1984, Cooper & Uzmann 1980, Fogarty et al. 1980, Uzmann et al. 1977). Deep waters in the Gulf of Maine tend to remain at 6-8°C throughout the year, perhaps sufficient to allow some degree of ovarian and egg development during the winter (Campbell 1986). Briggs & Mushacke (1980) felt that lobsters found along the south coast of Long Island were offshore animals that had moved inshore. These seasonal movements may allow the animals to remain within the optimal temperature range for synchronization of molting and spawning (Campbell & Stasko 1986, Cooper & Uzmann 1980). Seasonal movements might also congregate the lobsters and facilitate breeding either after the females molt or prior to spawning.

Even lobsters in the nearshore zone, often considered to be nonmigratory (Cooper & Uzmann 1980, for review), move to deeper water as the temperature declines in winter and return to shallow water as the temperature increases in spring. Mature and ovigerous females tend to move sooner, further, and deeper than immature females and mature males (Campbell 1984, 1986, Ennis 1983b, 1984a, Herrick 1909, Morrissey 1971), and mature females may migrate seasonally more than 20 km to shallow (20 m) water in summer and deep (200 m) water in winter.

These seasonal movements appear to be unnecessary in the Gulf of St. Lawrence since water temperatures are suitable for completion of reproduction (Campbell & Stasko 1986). Limited seasonal movements do occur off the Magdalen Islands and Newfoundland, but this may be to avoid winter storms and ice scouring (Ennis 1984c, Munro & Therriault 1983). In many tagging studies in the Gulf, lobsters have been recaptured close to their point of release. This is often cited as evidence that lobsters move very little. However, recent work by Pezzack & Duggan (1986) showed that lobsters can travel more than 120 km and return to within 9 km of their release site.

2.4.2 *Mechanisms to insure fertilization*

Mating usually occurs within hours of the female molt, and the spermatozoa are transferred to the seminal receptacle of the female where they are stored until spawning, which could be one to three years later. Sufficient sperm are transferred to fertilize at least two spawnings should they occur during one intermolt period (Waddy & Aiken 1986).

Conventional wisdom held that females which were not inseminated when they molted would remain uninseminated until they molted again two years later, and that such females would spawn infertile eggs (Atema 1986, Templeman 1936a, Waddy & Aiken 1984b). However, we now know that mating can also occur during intermolt (stage C) (Dunham & Skinner-Jacobs 1978), and that uninseminated preovigerous lobsters routinely mate before spawning. In one study in which we held 84 uninseminated females in a communal tank with males, all 84 were successfully inseminated in the two months prior to spawning (Waddy & Aiken 1987, in press). This is clearly a mechanism for ensuring the fertilization of spawned eggs. It can occur in females that for some reason did not mate at the previous molt, and in large females that are about to spawn for the second time during the same intermolt period and which have an insufficient number of stored spermatozoa for the second spawning.

Mature females without sperm in their seminal receptacles are reasonably common in some stocks (Krouse 1973). Ennis (1980) found that 11% of mature female lobsters from Bonavista Bay, Newfoundland were uninseminated. Our observations indicate that a high proportion of these uninseminated females would have mated prior to spawning.

Intermolt mating ensures insemination of unmated females and eliminates the problem of unsuccessful matings both in nature and in the laboratory. Our observations indicate that males continue to mate with females as long as they are not carrying a spermatophore (Waddy & Aiken in press) and that unsuccessful matings are therefore of little significance (see Saila & Flowers 1965).

Some males have an impressive capability to mate. In one study a male mated successfully with 11 newly molted females in 16 days and continued to mate frequently throughout the remainder of the summer (Waddy & Aiken in press). This capability combined with the male's ability to recognize unmated females reduces the probability that females will remain uninseminated and spawn infertile eggs (cf. Dubé & Grondin 1985).

2.4.3 *Multiple spawning*

Large females (≥120 mm CL) are capable of spawning more frequently than their smaller counterparts (Waddy & Aiken 1986), greatly increasing their relative contribution in areas such as the Bay of Fundy and the continental shelf, where they are most abundant. Campbell & Pezzack (1986) estimated that up to 77% of the eggs produced in the Bay of Fundy, southwest Nova Scotia, and the offshore areas of Georges and Browns Banks are from females that come from offshore areas where many are larger than the size at which multiple spawning begins. Because of the extra broods and the logarithmic relationship between size and fecundity, their contribution to the population is much greater than previously estimated. In the Bay of Fundy, Campbell & Robinson (1983) estimated that more than 70% of eggs are produced by females 100-140 mm CL. The impact of multiple spawning would increase this figure considerably.

2.5 *Fecundity*

2.5.1 *Relationship with animal size*

Most estimates of the number of eggs carried by American lobsters show a logarithmic relationship between carapace length and fecundity (reviewed in Aiken & Waddy 1980), with a range extending from as few as 5000 eggs to more than 100 000. Thus one jumbo lobster can produce as many eggs as 10-20 small ones. Discrepancies in published fecundity curves are probably due to differences in methodology, time of year the samples were taken and possibly even geographic variation.

2.5.2 *Geographic variation*

Some data suggest substantial geographic variation in size-fecundity relationships for lobsters, but as Ennis (1981) pointed out, valid data comparing different areas is difficult to obtain because of the requirement that all egg samples be taken at a comparable stage in development. In addition, the considerable individual variation in fecundity (Squires 1970, Herrick 1909) makes it necessary to sample many females to obtain statistically valid data. Data reported by Perkins (1971) show the variation in fecundity that can be obtained when broods at different stages of development are used. Complicating sampling even further is the considerable geographical variation in size at maturity and the difficulty in obtaining equivalent sizes of lobsters from different areas (see Campbell & Robinson 1983).

2.5.3 *Absolute versus relative and global fecundity*

Because of the size-fecundity relationship and the multiple spawning phenomenon, the contribution of 'jumbo' females is best expressed in terms of relative fecundity – the contribution that a size class of lobster makes to the total reproductive output of a population. The absolute fecundity of large lobsters at each spawning is impressive, 10-20 times that of a small female, but if they are few in number, spawn infrequently or produce infertile eggs or weak larvae, their relative fecundity would be small. On this basis the contribution of large lobsters has often been dismissed as insignificant (see Saila & Flowers 1965).

Intermolt mating, consecutive spawning and normal larval viability alter the equation. Large lobsters are capable of spawning and fertilizing eggs more frequently than smaller ones (Fig. 11), and their eggs and larvae are just as viable (Aiken & Waddy 1986a, Waddy & Aiken

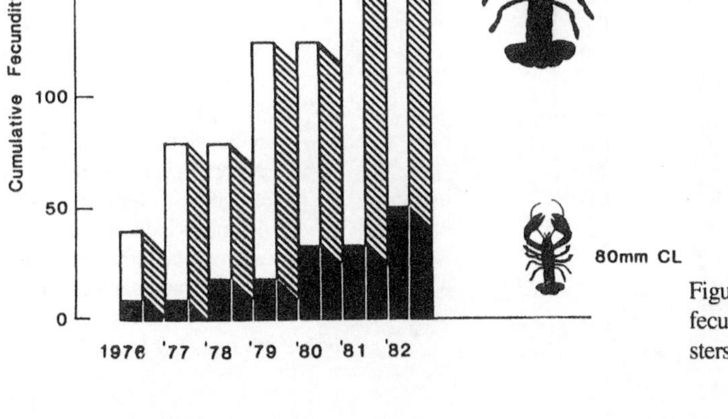

Figure 11. Relative cumulative fecundity of two sizes of female lobsters.

1986). Their relative fecundity is therefore a function of their absolute fecundity and the number of large females in the total population.

The number of large females in the total population is difficult to determine, but they do concentrate in certain areas. For example, 52% of the lobsters landed in 1979-80 at North Head, Grand Manan, were 127 mm CL or larger (Campbell & Duggan 1980). Comparable situations exist on Georges and Browns Banks. Mean CL of female lobsters in Corsair Canyon (Georges) was 127 mm and on SE Browns Bank 120 mm (Pezzack & Duggan 1983). Not only are the lobsters in these areas large, but there is a preponderance of ovigerous females (Fig. 12). Their relative contribution to a particular downstream area may therefore be much greater (and possibly more important) than their contribution to the population as a whole. Such a mechanism has been implied for the Newfoundland, Grand Manan, and Bay of Fundy fisheries (Campbell 1984, Campbell & Duggan 1980, Campbell & Robinson 1983, Ennis 1982a, Pring-

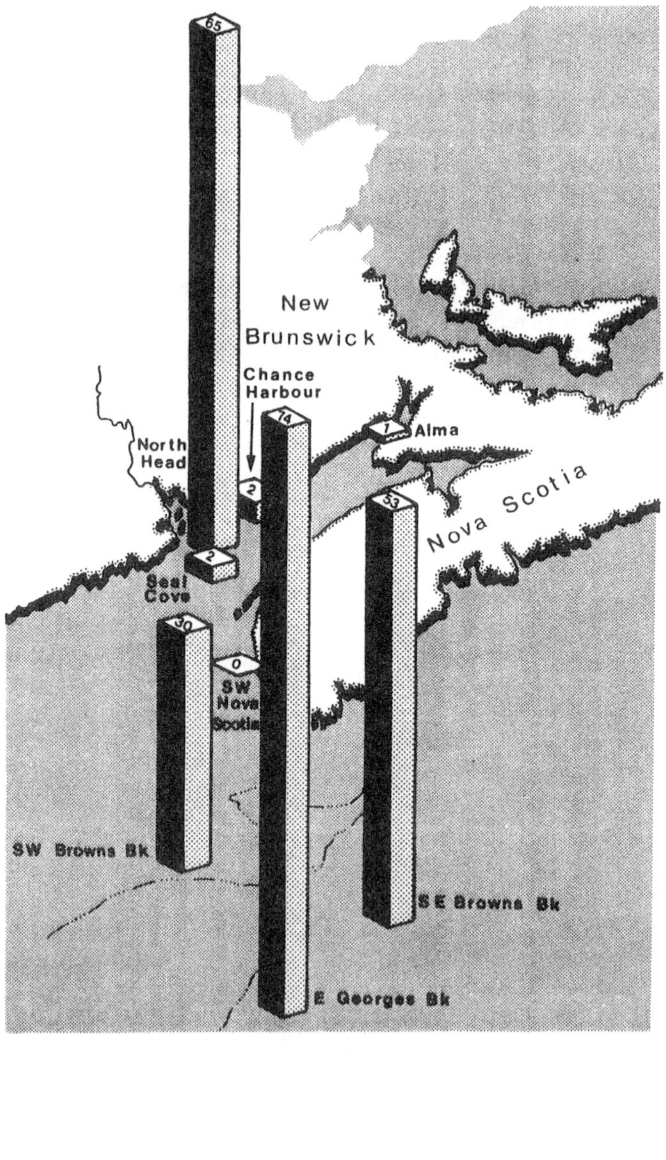

Figure 12. Relative abundance of ovigerous females in the Bay of Fundy and off southwestern Nova Scotia. In waters off southwestern Nova Scotia, ovigerous females are rare, but in areas such as Georges Bank, Browns Bank and North Head, Grand Manan concentrations of ovigerous females occur. Numbers indicate the percentage of total females caught in the commercial catch that were ovigerous (data from Campbell & Pezzack 1986).

le et al. 1983). Campbell & Pezzack (1986) estimated the median lobster size for egg production (50% of total egg production) was between 126 and 132 mm CL for North Head, Grand Manan. In view of the recent intensification of fishing effort directed at these large size classes in many areas (Campbell & Duggan 1980, Elner 1980, Ennis 1982a, Pezack & Duggan 1983), it would seem prudent to reassess the contribution that these large lobsters make to specific stocks and to the lobster population in general.

relative contribution to a particular downstream area may therefore be much greater (and possibly more important) than their contribution to the population as a whole. Such a mechanism has been implied for the Newfoundland, Grand Manan, and Bay of Fundy fisheries (Campbell 1984, Campbell & Duggan 1980, Campbell & Robinson 1983, Ennis 1982a, Pring-
le et al. 1983). Campbell & Pezzack (1986) estimated the median lobster size for egg production (50% of total egg production) was between 126 and 132 mm CL for North Head, Grand Manan. In view of the recent intensification of fishing effort directed at these large size classes in many areas (Campbell & Duggan 1980, Elner 1980, Ennis 1982a, Pezack & Duggan 1983), it would seem prudent to reassess the contribution that these large lobsters make to specific stocks and to the lobster population in general.

3 ENVIRONMENTAL REGULATION OF EGG PRODUCTION

Temperature is the major factor controlling egg production in American lobsters. Size at maturity, oöcyte maturation, spawning incidence, timing and synchronization, success of egg attachment and incubation, and time of hatching are all directly controlled by temperature. Other environmental factors, such as length of day, season, nutrition, social conditions, stress, pollutants and parasites, can also influence reproduction. However, we still know relatively little about the environmental control of reproduction and the effect than environmental factors can have on the various phases of the two year reproductive cycle. However, if conditions are not adequate, resorption of the oöcyte vitellin occurs and spawning is delayed for at least a year. Reproduction occurs within a narrower range of environmental conditions than those that are required for growth and survival, and environmental factors can interact to produce effects that are quite different. An example is the relationship between temperature and photoperiod in the control of spawning (Aiken & Waddy 1989).

3.1 *Temperature*

Ovarian maturation and spawning in the nearshore lobster stocks that have been studied requires an extended period below 9°C (and perhaps below 5°C) followed by an increase to 10°C or higher. If the spring increase in water temperature is delayed, final ovarian maturation and spawning will likewise be delayed. Under controlled conditions, females that normally spawn in July can be induced to spawn as late as December simply be delaying the normal increase in spring and summer water temperature (Fig. 13). If the water temperature remains at 2-3°C throughout the year, spawning will not occur at all.

At the northern limits of the lobster's range, summer water temperature occasionally remains sufficiently low to interfere with spawning. At Comfort Cove, Newfoundland, water temperature in mid-July can be anywhere from 4 to 18°C depending on year (Ennis 1982b). Year to year variations of this magnitude should have an important effect on egg production. Spawning would occur late in cold years and early in warm years, and in extreme cases perhaps

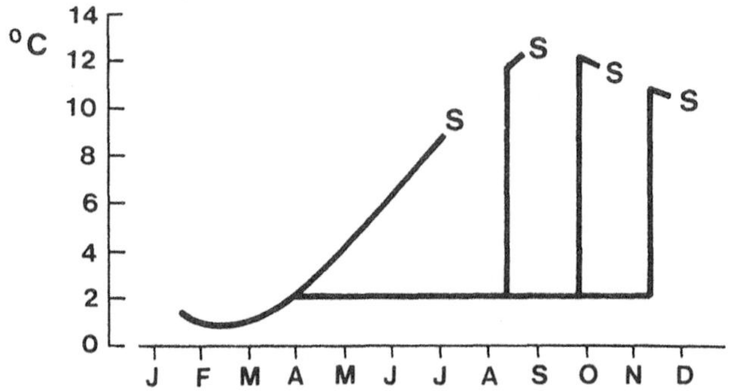

Figure 13. Effect of holding pre-ovigerous females at winter temperature (2°C) throughout the spring, summer and autumn and then transferring them to local sea-water temperature at various times from August through November. In each case, spawning occurred about 3 weeks after transfer to warmer temperatures.

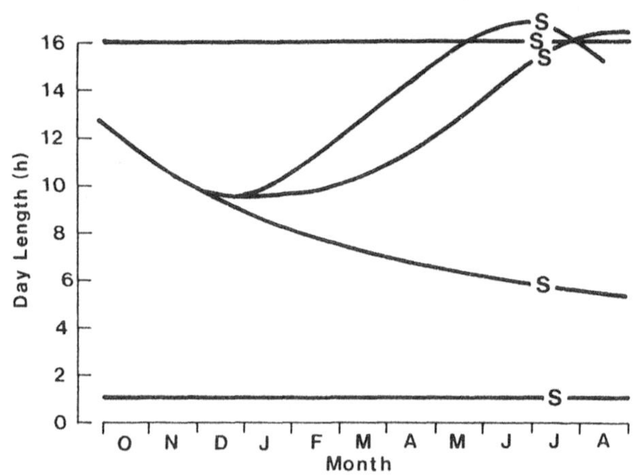

Figure 14. Effect of normal winter temperature (0-5°C) on the spawning response to altered photoperiod regimes. Regardless of daylength all lobsters spawned at the normal time (July).

not at all. This would produce comparable variations in the proportion of mature females that molt each year (22 to 73% in the 81-90 mm size range according to Ennis 1983a). Such differences in molting frequency undoubtedly affect the timing and incidence of spawning. Low summer temperature could inhibit or delay spawning by increasing the time required for vitellogenesis. Alternately, it could act indirectly by delaying the molt cycle. If temperature is sufficiently low to delay molt by a full year, the subsequent spawning might also be affected. This might account for some of the 2-30% variation Ennis (1980) reported in the percent of ovigerous females. An example of this may have been noted by Squires et al. (1971) in two arms of a Newfoundland bay that had different temperature regimes. The resident lobsters in the two bays had different reproductive cycles. Lobsters in the warmer of the two spawned biennially (the normal situation), but in the colder arm, they spawned only every third year.

Similarly, Ennis (1971) postulated that lobsters near the northern limit of their range, in the cold water of the northeast coast of Newfoundland, spawned every four years. This may well be the case, as summer temperatures in those areas are marginal for egg production. For this same reason it seems unlikely that the lobsters transplanted to Labrador in 1982 (Hoyles & Aggett 1984) would have maintained a normal spawning frequency. Bottom temperatures at the release sites were as low as 4°C in early July. Even if the temperature increased rapidly thereafter, it is doubtful that molting and spawning would occur on a biennial cycle.

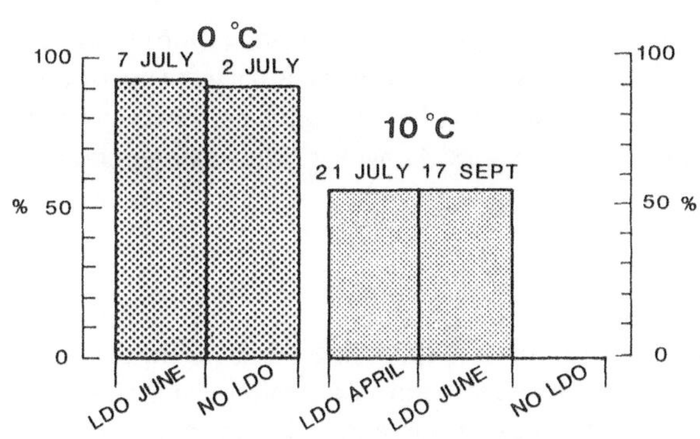

Figure 15. Effect of minimum winter seawater temperature on spawning in various photoperiods. Following a cold winter, females spawn irrespective of time of onset of long days (16 hr) (long day onset, LDO). Following a warm winter, females spawn at a set time after LDO and fail to spawn in the absence of LDO. Spawning incidence is greater in females exposed to cold winter temperature.

Low winter temperatures are equally essential for spawning success. If temperatures remain high year-round, spawning incidence is reduced (Waddy & Aiken 1984a, b, Hedgecock et al. 1978) in direct proportion to the degree of temperature elevation (Fig. 7).

3.2 *Photoperiod*

Photoperiod has no detectable effect on ovary development and spawning under temperature conditions that are typical of inshore areas (Fig. 14) (Aiken & Waddy 1985b, 1986b, in press). Low winter water temperature (5°C or lower) clearly dominates any influence of spring photoperiod in the control of spawning. However, when winter water temperature remains as high as 9-15°C, photoperiod becomes important in the control of spawning. At these elevated temperatures, lobsters will not spawn when held in short daylength (LD 8:16), and spawning can be delayed for at least six months in this way. However, if the short daylengths are extended to LD 16:8, spawning will occur 100-120 days later, depending on the stock (Fig. 15) (Aiken & Waddy 1989, Nelson et al. 1983, 1988a, b, Nelson 1986). A simple increase in daylength is not adequate to trigger this response. An increase to 10 hr of light per day is insufficient, 12 hr is marginal (12% of females spawn), and 14 or 16 hr are equally effective (50-80% of females spawn). Although Nelson (1986) found that photoperiodically-induced delay in spawning was followed by a comparable delay of the succeeding molt, we have found that a significant proportion of female lobsters exposed to 16 hr daylengths will molt prematurely with their eggs still attached to the cast shell (Aiken & Waddy unpubl.).

It may be significant that elevated winter temperature and long spring photoperiod cause a higher incidence of reproductive failure and a lower incidence of spawning (Fig. 15) (Aiken & Waddy 1989). These conditions are certainly less favorable for maturation and spawning than low winter temperature.

Results also indicate that the lobster's response to a combination of elevated temperature and short daylengths varies with the time of year. As spring progresses lobsters are more likely to react to elevated temperature than to short daylengths (Waddy & Aiken unpubl.).

Nelson (1986) felt that the stimulus of long day onset (LDO) removed the photoperiod inhibition of spawning only if it occurred near the time of molt, but under our conditions, LDO as late as 10-11 months after the molt was successful in inducing spawning (Aiken & Waddy 1989).

It appears that while low winter temperature is the normal regulatory parameter for vitellogenesis and spawning in nearshore populations, photoperiod can assume a regulatory role when winter water temperature remains inordinately high. Significantly, Nelson (1986) found that photoperiod synchronized spawning in offshore lobsters. These animals have the opportunity to experience relatively uniform year-round temperatures (8-10°C) and therefore may be unable to synchronize spawning without utilizing photoperiod. However, little is known about the reproductive cycles of offshore lobsters.

3.3 *Other factors*

Trapping, handling, and confinement in the month prior to spawning can cause the degeneration and resorption of oöcytes. In localities where lobsters mature at a small size and experience heavy fishing pressure during spring and summer, those that mature below the minimum legal size would be susceptible to oöcyte resorption as a result of repeated capture, confinement and release.

Reproductive success can also be influenced by lack of food or by deficiencies in dietary lipid or protein which can cause oöcyte resorption or negatively affect ovarian development, fecundity, or hatchability of eggs (Castell & Budson 1974, Castell & Kean 1986). During embryogenesis, yolk reserves are the sole source of nutrients for the developing embryo. Essential fatty acids and total lipid are very important to egg and larval development (Sasaki et al. 1986). These lipid reserves can serve as an important supplemental energy source during the immediate posthatch period, directly affecting both growth and survival of larvae. There are also indications that geographic location can influence fecundity but this is probably an indirect effect.

4 EMBRYOGENESIS AND HATCHING

4.1 *Environmental control*

Laboratory studies have shown that environmental conditions influence the security of egg attachment, the success of incubation, the rate of embryo development and the survival of the embryos. As Templeman (1936b) noted, hatching occurs earlier in areas of high summer water temperature and earlier in warmer years than colder ones.

4.1.1 *Temperature effects*

Lobster egg development rate is directly dependant on temperature (Templeman 1940, Perkins 1972). Embryogenesis takes 9-12 months in most native populations, but it can be shortened to as little as 3-4 months by holding ovigerous lobsters at 20-24°C. The effect of temperature is also modified by the degree of egg development; older or more advanced embryos develop at a slower rate than younger embryos (Perkins 1972).

During winter, when water temperatures can drop below 0°C, lobster eggs become dormant and the duration of the 'dormant' period depends upon the stage of the eggs when the temperature declines. It appears that egg development (like the molt cycle) is suspended at a specific point by low temperature, and the oldest eggs reach that point sooner and are thus dormant for a longer period (see Wear 1974). Perkins (1972) suggested that even if spawning in a given population were staggered over several weeks, this differential rate of development would compress the hatching period.

In general, the relationship between water temperature and the rate of embryonic development is linear at temperatures from 5-25°C, but below 6°C embryonic development proceeds slowly. Hatching can be delayed for as long as 6 months if temperature remains at 2-3°C throughout spring, summer and autumn, but the eggs will eventually hatch even at that temperature.

The time required for an entire brood to hatch varies from a few days to 3 weeks or more depending on temperature (Ennis 1975). Hatching occurs earlier where summer water temperature is high. Even in a relatively small area such as the southern Gulf of St. Lawrence, hatching time may vary by up to two months (Templeman 1936b).

Low temperature tends to synchronize embryo development. Broods held at low temperatures for many months will hatch over a brief period when the water is warmed, whereas those held at elevated temperature (20°C) will have a prolonged hatching period. Temperature can also affect lipid conversion efficiency in embryos, and as a result, the quantity of yolk available when the lobster hatches (Sasaki et al. 1986).

Campbell & Stasko (1986) estimated that lobster eggs require (on average) 1832 degree days (>3.4°C?) from spawning to hatching. They concluded that in areas such as Grand Manan there were insufficient degree-days (1594) to allow completion of egg development, and that ovigerous females therefore moved to warmer areas in winter to promote egg development.

4.2 *Causes of egg loss*

To survive the 9-12 months of normal incubation, an embryo must have sufficient organic reserves, be securely attached to the pleopods, and remain free from disease, predation, and exposure to water of poor quality.

Extruded eggs are attached to the non-plumose setae of the pleopods and ventral abdominal sterna. Attachment is usually secure in wild females and environmentally-conditioned cultured females. However, entire egg masses can be lost during incubation. Perkins (1971) estimated natural attrition in an offshore population at 36%, and in some females it is substantially greater (Ennis 1984a, Smith 1977). However, embryo loss from relatively undisturbed lobsters in the laboratory is lower (Waddy unpubl.).

Severe loss of embryos can occur when ovigerous females are captured in traps or trawls, taken onboard ship, and then thrown back. The female compounds this loss by flapping her tail, dislodging even more eggs. This effect is more pronounced immediately after spawning or late in development when the eggs are less securely attached. Egg loss is also higher in areas where fishing is intense.

Egg loss is significant in some localities. Smith (1977) and Ennis & Collins (1983) both found that 14-15% of ovigerous females had fewer than 150 embryos. It's worth noting that these reports from Newfoundland and Long Island Sound represent environmental conditions at the extremes of the lobster's range.

Ennis (1984a) felt that some of the egg loss might be due to lack of fertilization as he found that 11% of mature females at Arnold's Cove lacked spermatophores. However, if there are mature males in the population, those lobsters would normally be mated prior to spawning (Waddy & Aiken 1987). There is some evidence that environmental factors, particularly temperature, are involved in egg retention. There is a high incidence of egg loss among both wild and cultured lobsters held in California (Hedgecock 1983, Talbot et al. 1984) and France (Bertran & Lorec 1986), and our cultured lobsters (prior to temperature conditioning) retain far fewer eggs than feral ones (Pl. 2). For whatever reason, elevated temperature in the years prior to first spawning seem to affect egg retention. In a California study, only 2 of 23 lobsters held at elevated temperature retained and hatched a significant number of eggs (Talbot et al. 1984).

According to Talbot & Harper (1984), the morphology of the stalk (funiculus) that attaches the egg to the non-plumose setae (or to other eggs) is different in cultured lobsters than in feral lobsters. They concluded that egg stalks in cultured lobsters were morphologically abnormal and that this was a major cause of egg loss. The egg stalk is formed from the chorion, a membrane produced by the follicle cells. The factors that control chorion synthesis are not known but Talbot & Harper (1984), on the basis of differences in egg retention rates, suggested that environmental conditions such as temperature may be a factor. Our data implicate temperature history as an important factor in egg attachment and hatching success; cultured females maintained for three or more consecutive spawnings in our relatively low local winter water temperature hatch orders of magnitude more eggs than cultured females that are spawning eggs for the first or second time at this temperature (Pl. 2).

Nelson et al. (1983) postulated that clutch size and egg retention were directly correlated with the length of time between molting and the subsequent egg extrusion, but in areas such as the southern Gulf of St. Lawrence it is very common for females to molt and then spawn in the next few weeks. These females retain their eggs as efficiently as females that spawn a full year after molting (Waddy unpubl.).

4.2.1 *Disease and parasitism*

Fouling by both filamentous and non-filamentous microorganisms on the external surfaces of egg membranes causes anoxia. Microbial fouling is a result of nutrient load in the water rather than a disease caused by a specific pathogen (Fisher et al. 1978). The most common epibiont is the filamentous bacterium *Leucothrix mucor*, which can cause mass mortalities of lobster eggs under culture conditions. *Leucothrix* is ubiquitous in the marine environment and is occasionally found on embryos from wild-caught lobsters. However, its effect on natural populations has not been determined. It is particularly troublesome in a culture situation as its development is enhanced by high temperature. *Leucothrix* can prolong embryo development by 6 weeks or more and severe infestations can cause death of the entire brood. In less severe cases, the brood may hatch but the infection will transfer to the larvae. *Leucothrix* is seldom a problem when water quality is good.

A phycomycetous fungus, *Lagenidium*, can also infect lobster eggs and prevent hatching. This fungus occurs primarily on the eggs on the surface of the lobster egg mass (Fisher et al. 1978).

A decade ago, a previously undescribed nemertean (*Pseudocarcinonemertes homari*) threatened the eggs of Bay of Fundy lobsters. This nemertean feeds on lobster egg yolk (Waddy & Aiken 1985) and in the Grand Manan fishery as many as 74% of the ovigerous lobsters were infested (Campbell & Brattey 1986). Laboratory studies demonstrated that *P. homari* had the potential to cause significant problems for *Homarus* (Pl. 3) (Aiken et al. 1985), as has a similar nemertean on the Dungeness crab of the central California coast (Wickham 1986). Lobsters infected with this egg predator rarely retain any eggs to hatching. The nemertean can also survive in the lobster branchial chamber until another brood is spawned. However, unlike the *Carcinonemertes errans* of Dungeness crabs, *P. homari* has disappeared as mysteriously as it appeared. In the past six years there have been no reports of its presence on the eggs of wild lobsters.

5 SUMMARY

Temperature is the dominant regulator of egg production in American lobsters, influencing size at maturity, time of spawning, coordination of the molting and spawning cycles, time of hatch, and success of egg attachment and incubation. High summer temperature induces maturation at a young age and small size and spawning and hatching occur earlier in warm areas.

The reproductive cycle of female lobsters smaller than 120 mm CL requires two years, and molting and spawning occur in alternate years. Females from warm areas such as the southern Gulf of St. Lawrence and Long Island Sound that are spawning for the first time may molt and spawn in the same season. Large females (≥ 120 mm CL) molt less frequently than every second year, but are able to spawn twice without an intervening molt. In most cases, they obtain enough sperm from a single insemination to fertilize both broods. If not, they will be inseminated prior to the second spawning. Because of the extra broods produced by these large females and their concentration in certain areas, their contribution to the population may be much greater than has been assumed.

Ovarian maturation and spawning in nearshore stocks requires as extended winter period below 9°C (and perhaps below 5°C), followed by an increase in summer to 10°C or higher. Both low winter temperature and warm summer temperature are essential to the proper synchronization of the molt and reproductive cycles. If winter temperatures are too high, fewer females spawn and the incidence of molting increases. If summer temperatures are too low, spawning is delayed or may not occur at all.

Many environmental conditions influence the security of egg attachment, the success of incubation, the rate of embryo development and survival of the embryos. Temperature history is particularly critical; females reared at 20°C have impaired egg attachment and hatch few eggs. Several years of conditioning in appropriate temperatures (cold winter, warm summer) results in normal attachment and incubation success.

REFERENCES

Aiken, D.E. & S.L. Waddy 1976. Controlling growth and reproduction in the American lobster. *Proc. World Maricul. Soc.* 7: 415-430.

Aiken, D.E. & S.L. Waddy 1980. Reproductive biology. In J.S. Cobb & B.F. Phillips (eds), *The Biology and Management of Lobsters:* Vol 1, pp. 215-276. New York: Academic Press.

Aiken, D.E. & S.L. Waddy 1985a. Production of seed stock for lobster culture. *Aquaculture* 44: 103-114.

Aiken, D.E. & S.L. Waddy 1985b. The uncertain influence of spring photoperiod on spawning in the American lobster, *Homarus americanus. Can. J. Fish. Aquat. Sci.* 42: 194-197.

Aiken, D.E. & S.L. Waddy 1986a. Environmental influence on recruitment of the American lobster, *Homarus americanus:* a perspective. *Can. J. Fish. Aquat. Sci.* 43: 2258-2270.

Aiken, D.E. & S.L. Waddy 1986b. Oöcyte maturation and spawning in wild American lobsters (*Homarus americanus*): lack of evidence for significant regulation by photoperiod. *Can. J. Fish. Aquat. Sci.* 43: 1451-1453.

Aiken, D.E. & S.L. Waddy 1989. Interaction of temperature and photoperiod in the regulation of spawning by American lobster, *Homarus americanus. Can. J. Fish. Aquat. Sci.* 46: 145-148.

Aiken, D.E., S.L. Waddy & J.D. Castell 1981. Effect of temperature and eyestalk ablation on the reproductive organs of the lobster *Homarus americanus.* In W.H. Clark Jr & T.S. Adams (eds), *Developments in Endocrinology* Vol. 11, *Advances in Invertebrate Reproduction:* p. 330. New York: Elsevier/North Holland.

Aiken, D.E., S.L. Waddy & L.S. Uhazy 1985. Aspects of the biology of *Pseudocarcinonemertes homari* and its association with the American lobster, *Homarus americanus. Can. J. Fish. Aquat. Sci.* 42: 351-356.

Atema, J. 1986. Review of sexual selection and chemical communication in the lobster, *Homarus americanus*. *Can. J. Fish. Aquat. Sci.* 43: 2383-2390.

Attard, J. & C. Hudon 1987. Embryonic development and energetic investment in egg production in relation to size of female lobster (*Homarus americanus*). *Can. J. Fish. Aquat. Sci.* 44: 1157-1164.

Bertran, R. & J. Lorec 1986. Observations sur le cycle de reproduction du homard Européen (*Homarus gammarus*) en captivité. Int. Counc. Explor. Mer. C.M. 1986/K:12.

Briggs, P.T. & W.M. McGroaty 1985. Departures from the normal molting and reproductive cycle of female lobsters from Long Island Sound, New York. *NY Fish Game J.* 32: 211-212.

Briggs, P.T. & F.M. Mushacke 1979. The American lobster in western Long Island Sound. *NY Fish Game J.* 26: 59-86.

Briggs, P.T. & F.M. Mushacke 1980. The American lobster and the pot fishery in the inshore waters off the south shore of Long Island, New York. *NY Fish Game J.* 27: 156-178.

Burns, T. & S.H. Clark 1980. Application of US bottom trawl survey and commerical data to offshore lobster assessment. *Can. Tech. Rep. Fish. Aquat. Sci.* 932: 125-133.

Byard, E.H. 1985. The female specific protein and reproduction in the lobster, *Homarus americanus*. Ph.D. Thesis. London: University of Western Ontario, 106 pp.

Campbell, A. 1984. Aspects of lobster biology and fishery in the upper reaches of the Bay of Fundy. *Can. Tech. Rep. Fish. Aquat. Sci.* 1256: 469-488.

Campbell, A. 1986. Migratory movements of ovigerous lobsters, *Homarus americanus*, tagged off Grand Manan, Eastern Canada. *Can. J. Fish. Aquat. Sci.* 43: 2197-2205.

Campbell, A. & J. Brattey 1986. Egg loss from the American lobster, *Homarus americanus*, in relation to nemertean, *Pseudocarcinonemertes homari*, infestation. *Can. J. Fish. Aquat. Sci.* 43: 772-780.

Campbell, A. & D.R. Duggan 1980. Review of the Grand Manan lobster fishery with an analysis of recent catch and effort trends. *Can. Tech. Rep. Fish. Aquat. Sci.* 997: 1-20.

Campbell, A., D.E. Graham, H.J. MacNichol & A.M. Williamson 1984. Movements of tagged lobsters released on the continental shelf from Georges Bank to Baccaro Bank, 1971-1973. *Can. Tech. Rep. Fish. Aquat. Sci.* 1288: 16 pp.

Campbell, A. & D.S. Pezzack 1986. Relative egg production and abundance of berried lobsters, *Homarus americanus*, in the Bay of Fundy and off southwestern Nova Scotia. *Can. J. Fish. Aquat. Sci.* 43: 2190-2196.

Campbell, A. & D.G. Robinson 1983. Reproductive potential of three American lobster (*Homarus americanus*) stocks in the Canadian Maritimes. *Can. J. Fish. Aquat. Sci.* 40: 1958-1967.

Campbell, A. & A.B. Stasko 1986. Movements of lobsters (*Homarus americanus*) tagged in the Bay of Fundy. *Mar. Biol.* 92: 393-404.

Castell, J.D. & S.D. Budson 1974. Lobster nutrition: the effect on *Homarus americanus* of dietary protein levels. *J. Fish. Res. Board Can.* 31: 1363-1370.

Castell, J.D. & J.C. Kean 1986. Evaluation of the role of nutrition in lobster recruitment. *Can. J. Fish. Aquat. Sci.* 43: 2320-2327.

Cooper, R.A. & J.R. Uzmann 1980. Ecology of juvenile and adult *Homarus*. In J.S. Cobb & B.F. Phillips (eds), *The Biology and Management of Lobsters*, Vol. 2: pp. 97-142. New York: Academic Press.

Dehn, P.F., D.E. Aiken & S.L. Waddy 1983. Aspects of vitellogenesis in the lobster *Homarus americanus*. *Can. Tech. Fish. Aquat. Sci.* 1161: 24 pp.

Dubé, P. & P. Grondin 1985. Maturité sexuelle du homard (*Homarus americanus*) femelle aux Iles-de-la-Madeleine. *Can. Atl. Fish. Sci. Advis. Comm. Res. Doc.* 85/85, 37 pp.

Dunham, P.J. & D. Skinner-Jacobs 1978. Intermolt mating in the lobster (*Homarus americanus*). *Mar. Behav. Physio.* 5: 209-214.

Elner, R.W. 1980. Lobster gear selectivity – a Canadian overview. *Can. Rep. Fish. Aquat. Sci.* 932: 77-84.

Ennis, G.P. 1971. Lobster (*Homarus americanus*) fishery and biology in Bonavista Bay, Newfoundland 1966-70. *Fish. Res. Bd. Can. Tech. Rep.* 289: 1-46.

Ennis, G.P. 1975. Observations on hatching and larval release in the lobsters (*Homarus americanus*). *J. Fish. Res. Board Can.* 32: 2210-2213.

Ennis, G.P. 1980. Size maturity relationships and related observations in Newfoundland populations of the lobster (*Homarus americanus*). *Can. J. Fish. Aquat. Sci.* 37: 945-956.

Ennis, G.P. 1981. Fecundity of the American lobster, *Homarus americanus*, in Newfoundland waters. *US Dept. Comm. Fish. Bull.* 79: 796-800.

Ennis, G.P. 1982a. The Newfoundland lobster fishery: a review of statistics, stock structure and current management considerations. *Can. MS Rep. Fish. Aquat. Sci.* 1666: 11 pp.

Ennis, G.P. 1982b. Fisheries and population biology of lobsters (*Homarus americanus*) at Comfort Cove, Newfoundland. *Can. Tech. Rep. Fish. Aquat. Sci.* 1116: 45 pp.

Ennis, G.P. 1983a. Variation in annual growth in two Newfoundland lobster (*Homarus americanus*) populations in relation to temperature conditions. *Int. Counc. Explor. Sea* C.M. 1983/K:24.

Ennis, G.P. 1983b. Observations on the behavior and activity of lobster, *Homarus americanus*, in nature. *Can. Tech. Fish. Aquat. Sci.* 1165: 26 pp.

Ennis, G.P. 1984a. Comparison of physiological and functional size-maturity relationships in two Newfoundland populations of lobsters, *Homarus americanus*. *US Dept. Comm. Fish. Bull.* 82: 244-249.

Ennis, G.P. 1984b. Incidence of molting and spawning in the same season in female lobsters, *Homarus americanus*. *US Dept. Comm. Fish. Bull.* 82: 529-530.

Ennis, G.P. 1984c. Small-scale seasonal movements of the American lobster *Homarus americanus*. *Trans. Amer. Fish. Soc.* 113: 336-338.

Ennis, G.P. 1984d. Tag-recapture validation of molt and egg extrusion predictions based upon pleopod examination in the American lobster, *Homarus americanus*. *US Dept. Comm. Fish. Bull.* 82: 242-244.

Ennis, G.P. & P.W. Collins 1983. A computer program to assess egg production per recruit in a lobster (*Homarus americanus*) population. *Can. Tech. Rep. Fish. Aquat. Sci.* 1198: 15 pp.

Fisher, W.S., E.H. Nilson, J.F. Steenbergen & D.V. Lightner 1978. Microbial diseases of cultured lobsters: a review. *Aquaculture* 14: 115-140.

Fogarty, M.J., D.V.D. Borden & H.J. Russell 1980. Movements of tagged American lobster, *Homarus americanus*, off Rhode Island. *US Dept. Comm. Fish. Bull.* 78: 771-780.

Hedgecock, D. 1983. Maturation and spawning of the American lobster, *Homarus americanus*. In J.P. McVey (ed.), *CRC Handbook of Mariculture:* Vol. 1, pp. 261-290. Boca Raton: CRC Press.

Hedgecock, D., W.L. Moffett, W. Borgeson & K. Nelson 1978. Progress and problems in lobster broodstock development. *Proc. World Maricul. Soc.* 9: 497-506.

Herrick, F.H. 1909. Natural history of the American lobster. *Bull. US Bur. Comm. Fish.* 29: 149-408.

Hoyles, B. & D. Aggett 1984. An introduction of lobsters to St. Michael's Bay, Labrador during 1982. *Govt. Nfld. Lab. Dept. Fish., Dev. Br., Data Ser., Rep.* 17: 20 pp.

Krouse, J.S. 1973. Maturity, sex ratio and size composition of a natural population of American lobster, *Homarus americanus*, along the Maine coast. *US Dept. Comm. Fish. Bull.* 71: 165-173.

Morrissey, T.D. 1971. Movements of tagged American lobsters, *Homarus americanus*, liberated off Cape Cod, Massachusetts. *Trans. Am. Fish. Soc.* 100: 117-120.

Munro, J. & J.-C. Therriault 1983. Migrations saisonnières du homard (*Homarus americanus*) entre la côte et les lagunes des Iles-de-la-Madeleine. *Can. J. Fish. Aquat. Sci.* 40: 905-918.

Nelson, K. 1986. Photoperiod and reproduction in lobsters (*Homarus*). *Amer. Zool.* 26: 447-457.

Nelson, K., D. Hedgecock & W. Borgeson 1983. Photoperiodic and ecdysial control of vitellogenesis in lobsters (*Homarus*) (Decapoda, Nephropidae). *Can. J. Fish. Aquat. Sci.* 40: 940-947.

Nelson, K., D. Hedgecock & W. Borgeson 1988a. Factors influencing egg extrusion in the American lobster (*Homarus americanus*). *Can. J. Fish. Aquat. Sci.* 45: 797-804.

Nelson, K., D. Hedgecock & W. Borgeson 1988b. Effects of reproduction upon molting and growth in female American lobsters (*Homarus americanus*). *Can. J. Fish. Aquat. Sci.* 45: 805-821.

Perkins, H.C. 1971. Egg loss during incubation from offshore northern lobsters (Decapoda: Nephropidae). *US Dept. Comm. Fish. Bull.* 69: 451-453.

Perkins, H.C. 1972. Development rates at various temperatures of embryos of the northern lobster (*Homarus americanus* Milne Edwards). *US Dept. Comm. Fish. Bull.* 70: 95-99.

Pezzack, D.S. & D.R. Duggan 1983. The Canadian offshore lobster (*Homarus americanus*) fishery – 1971 to 1982. *Int. Counc. Explor. Sea* C.M. 1983/K:34.

Pezzack, D.S. & D.R. Duggan 1986. Evidence of migration and homing of lobsters (*Homarus americanus*) on the Scotian shelf. *Can. J. Fish. Aquat. Sci.* 43: 2206-2211.

Pringle, J.D., D.G. Robinson, G.P. Ennis & P. Dubé 1983. An overview of the management of the lobster fishery in Atlantic Canada. *Can. MS Rep. Fish. Aquat. Sci.* 1701: 103 pp.

Saila, S.B. & J.M. Flowers 1965. A simulation of sex ratios and regulations effects with the lobster, *Homarus americanus*. *Gulf Caribb. Fish. Inst., Univ. Miami, Proc.* 18: 66-78.

Sasaki, G.C., J.M. Capuzzo & P. Biesiot 1986. Nutritional and bioenergetic considerations in the development of the American lobster *Homarus americanus*. *Can. J. Fish. Aquat. Sci.* 42: 2311-2319.

Simpson, A.C. 1961. A contribution to the bionomics of the lobster (*Homarus vulgaris* Edw.) on the coast of north Wales. *Fish. Invest., (London), Ser. 2* 23: 1-28.

Smith, E.M. 1977. Some aspects of catch/effort, biology, and the economics of the Long Island Sound lobster fishery during 1976. *Final Report, USDC, NMFS, Connecticut*, Project No. 3-253-R-1: 97 pp.

Squires, H.J. 1970. Lobster (*Homarus americanus*) fishery and ecology in Port au Port Bay, Newfoundland, 1960-65. *Proc. Natl. Shellfish. Assoc.* 60: 22-39.

Squires, H.J., G.E. Tucker & G.P. Ennis 1971. Lobsters (*Homarus americanus*) in Bay of Islands, Newfoundland, 1963-65. *Fish. Res. Board Can. Tech. Rep.* 1151, 58 pp.

Talbot, P. & R. Harper 1984. Abnormal eyestalk morphology is correlated with clutch attrition in laboratory-maintained lobsters. *Biol. Bull.* 166: 349-356.

Talbot, P., C. Thaler & P. Wilson 1984. Spawning, egg attachment and egg retention in captive lobsters (*Homarus americanus*). *Aquaculture* 37: 239-249.

Templeman, W. 1936a. Further contributions to mating in the American lobster. *J. Biol. Bd. Can.* 2: 223-226.

Templeman, W. 1936b. Local differences in the life history of the lobster (*Homarus americanus*) on the coast of the Maritime provinces of Canada. *J. Biol. Bd. Can.* 2: 41-87.

Templeman, W. 1940. Embryonic developmental rates and egg-laying of Canadian lobsters. *J. Fish. Res. Bd. Can.* 5: 71-83.

Uzmann, J.R., R.A. Cooper & K.J. Pecci 1977. Migration and dispersion of tagged American lobsters, *Homarus americanus*, on the southern New England continental shelf. *NOAA Tech. Rep. NMFS* SSRF-705, 92 pp.

Waddy, S.L. & D.E. Aiken 1984a. Seed stock for lobster culture: the role of temperature in synchronizing the molt and reproductive cycle of cultured American lobsters. *J. World Maricul. Soc.* 15: 132-137.

Waddy, S.L. & D.E. Aiken 1984b. Broodstock management for year-round production of larvae for culture of the American lobster. *Can. Tech. Rep. Fish. Aquat. Sci.* 1272: 14 pp.

Waddy, S.L. & D.E. Aiken 1985. Immunofluorescent localization of American lobster egg yolk protein in the alimentary tract of the nemertean *Pseudocarcinonemertes homari*. *Can. J. Fish. Aquat. Sci.* 42: 357-359.

Waddy, S.L. & D.E. Aiken 1986. Multiple fertilization and consecutive spawning in large American lobsters, *Homarus americanus*. *Can. J. Fish. Aquat. Sci.* 43: 2291-2294.

Waddy, S.L. & D.E. Aiken 1987. Potential of intermolt mating in broodstock management for lobster culture. *Proc. Ann. Meet. Aqua. Assoc. Canada* 1: 26-27.

Waddy, S.L. & D.E. Aiken (in press). Intermolt insemination, an alternate mating strategy for the American lobster (*Homarus americanus*). *Can. J. Fish. Aquat. Sci.*

Wear, R.G. 1974. Incubation in British decapod Crustacea, and the effects of temperature on the rate and success of embryonic development. *J. Mar. Biol. Assoc. UK* 54: 745-762.

Wickham, D.E. 1986. Epizootic infestations by nemertean brood parasites on commercially important crustaceans. *Can. J. Fish. Aquat. Sci.* 43: 2295-2302.

Annual variation in egg production in a Newfoundland population of the American lobster, *Homarus americanus*

G.P. Ennis
Science Branch, Department of Fisheries and Oceans, St. John's, Newfoundland, Canada

ABSTRACT

Estimates of the number of eggs produced annually in a Newfoundland population of the American lobster for the 11-year period 1976-86 ranged from 7.4×10^6 to 86.8×10^6. Year to year fluctuations in egg production coincided with fluctuations in the estimated number of ovigerous females in the population, which was linked to annual variation in numbers entering the prerecruit and recruit portions of the population. Prerecruit females (< 81 mm carapace length) contributed from 26 to 58% of the total annual egg production. Among prerecruit females 71 to 80 mm CL, the percentage ovigerous ranged from 6.6 to 41.2 ($\bar{x} = 24.1$); among females ≥ 81 mm it ranged from 4.1 to 22.8 ($\bar{x} = 8.6$) in different years. There was a significant correlation between egg production and the percentage of ovigerous females. For both the prerecruit and recruit size groups there were significant negative correlations between the percentage molting during the summer and the percentage of ovigerous females in autumn. There was a significant negative correlation between standing stock estimates for the population and percentage molting, and a significant positive correlation resulted between standing stock and percentage of ovigerous females. Percentage molting appears to be influenced by population density and a reduced percentage molting leads to an increased percentage laying eggs. Fluctuations in the abundance of females and in the percentage of them that laid eggs accounted for most of the annual variation in egg production in this lobster population.

1 INTRODUCTION

Egg production in American lobster (*Homarus americanus*) populations is generally very low. Minimum legal sizes are small in relation to sizes at which females begin to lay eggs. In many areas most females do not spawn prior to being taken in the fishery, because exploitation rates are very high. Assessments have shown that increases in size limits or reductions in exploitation rates would result in substantial increases in egg production (Campbell & Robinson 1983, Campbell 1985, Ennis 1985). Low levels of egg production are such a common feature of lobster fisheries that ICES *Homarus* Working Groups (Anon. 1977, 1979) raised the possibility of widespread recruitment failure under current management regimes. Robinson (1979) concluded that recruitment failure due to inadequate egg production was the cause of stock collapse in two Atlantic Canada fishing areas.

Fogarty & Idoine (1986) found an asymptotic relationship between abundance of the final (fourth) larval stage and stock size 5-7 years later, indicating the possibility of density-dependent regulation between larval settlement and subsequent recruitment to the fishery. Such a capacity for compensatory population regulation would explain the apparent resilience of lobster populations at high levels of exploitation (Fogarty & Idoine 1986). This implies that relatively high levels of recruitment can be maintained despite low levels of egg production, although recruitment failure is inevitable at some critically low level.

Substantial year to year variation in recruitment was found in two heavily fished Newfoundland lobster populations (Ennis 1983, Ennis et al. 1986). Whether variation in egg production in earlier years contributed to this variation is unknown, but it is likely that the recruitment variation resulted in substantial year to year variation in egg production. The data used to estimate recruitment also provided a basis for estimating annual egg production. Estimates presented for one population for the period 1976 to 1986 document the magnitude of variation in egg production in a lobster population subjected to heavy fishing. Some of the factors which contribute to this variability are also examined.

2 MATERIALS AND METHODS

A study of various aspects of the biology and population dynamics of lobsters in the area of Arnold's Cove, Placentia Bay on the southeast coast of Newfoundland has been ongoing since 1969 (Ennis et al. 1986). In autumn, following the annual summer molting/spawning/hatching period, special fishing is carried out which involves tagging commercially legal lobsters (i.e. carapace length (CL) 81 mm or longer and non-ovigerous) as a basis for estimating standing stock (i.e. commercially legal portion of the population), recruitment (to the standing stock) and exploitation rate during the spring fishing season (April 20-June 30) the following year. This tagging has been done annually since 1975 and has included careful examination of all lobsters caught for sex, presence of eggs, carapace length, and shell condition.

Details of the method of estimating standing stock and recruitment and a consideration of the assumptions of the Petersen Model, which was used to generate the estimates, as they apply to the data, are provided in Ennis (1983) and Ennis et al. (1982, 1986). The numbers of males and females in the standing stock were determined separately. The proportion of ovigerous and non-ovigerous females by size ≥ 71 mm CL in the autumn trap-caught sample and the estimated number of non-ovigerous females ≥ 81 mm at the beginning of the fishing season the following spring were used to derive an estimate of the total number of non-ovigerous and ovigerous females at each one mm CL interval (from 71 mm) in the population. A size-fecundity relationship for Arnold's Cove lobsters (Ennis 1981) was used to estimate the number of eggs produced by the ovigerous females at each one mm size interval.

3 RESULTS

3.1 *Egg production estimates*

Estimates of the number of eggs produced in a year ranged from 7.4×10^6 in 1979 to 86.8×10^6 in 1976. Annual egg production fluctuated markedly from 1976 to 1980 but was much more stable, although at a lower level, from 1981 to 1986 (Fig. 1A). The estimated total number of

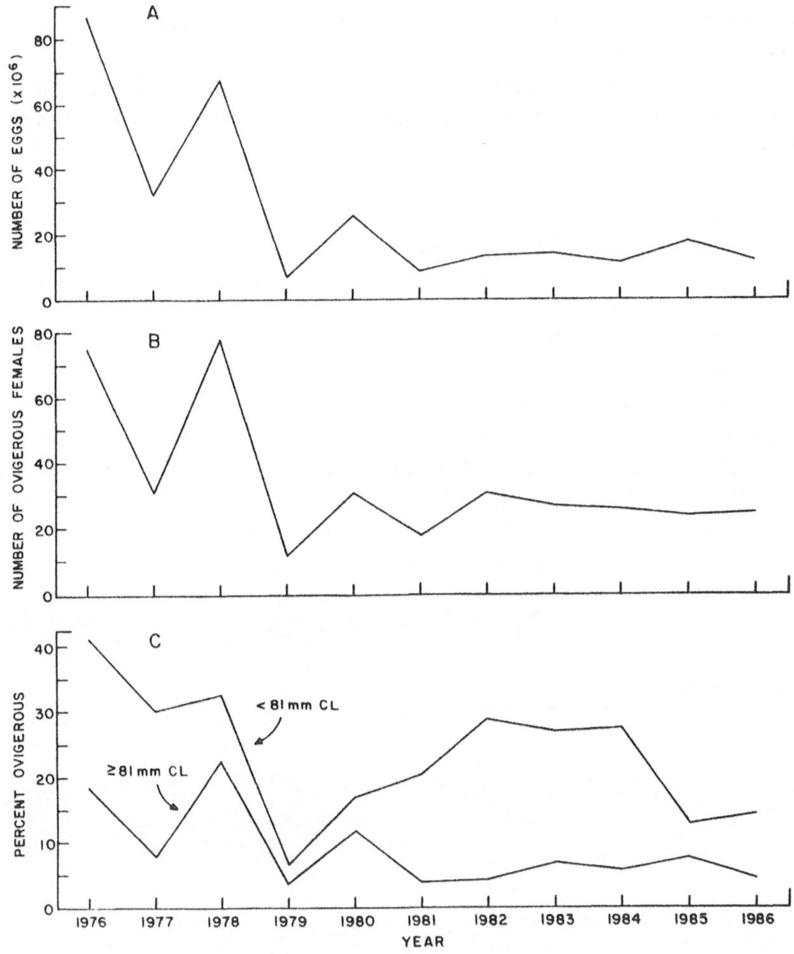

Figure 1. The American lobster population at Arnold's Cove, Newfoundland from 1976 to 1986. A. Estimates for the number of eggs produced annually. B. The observed number of ovigerous females. C. The percentage of ovigerous females.

ovigerous females ranged from 622 in 1979 to 7595 in 1976 and prerecruit females (< 81 mm CL) comprised from 33 to 68% of the total ovigerous. Fluctuations in the numbers of ovigerous females in the autumn samples (Fig. 1B) coincided with fluctuations in annual egg production ($r = 0.96$, $N = 11$, $P < 0.01$). Females < 81 mm contributed from 25.9 to 58.4% of the total number of eggs produced.

3.2 *Variation in percent ovigerous*

In the annual autumn sampling, percentage of females that were ovigerous ranged from 6.6 to 41.2 ($\bar{x} = 24.1$) for females <81 mm CL and from 4.1 to 22.8 ($\bar{x} = 8.6$) for females ≥81 mm (Fig. 1C). There were significant positive correlations between percentages of ovigerous females in the autumn samples and egg production estimates for the following year ($r = 0.72$, $N = 11$, P

<0.02 for females <81 mm; r=0.93, N=11, P <0.01 for females ≥81 mm; r=0.97, N=11, P <0.01 for all females).

A number of factors contribute to variation in percentage of ovigerous females in a lobster population. Annual variation in numbers of ovigerous females is linked to variation in recruitment. At Arnold's Cove, 50% of the non-ovigerous females were functionally mature (i.e. will extrude eggs during the upcoming spawning season) at 81 mm (Ennis 1984). The lower percentage ovigerous among females ≥ 81 mm (despite the sharply increasing proportions functionally mature at sizes beyond 81 mm) reflects the impact of the annual spring fishery, which reduces the relative abundance of mature non-ovigerous females just before the spawning season. Typically, the reproductive cycle in the American lobster is a two year one. Many of the non-ovigerous females present in the population in spring will spawn during the summer. These are the ovigerous females observed in samples taken in the autumn. Most of those that do not spawn molt and grow to larger sizes.

There was a significant negative correlation (r = -0.91, N = 12, P < 0.01) between the percentage molting among recruit females [81-90 mm, i.e. one molt increment above the minimum legal size (Ennis 1980, Ennis et al. 1986)] in summer and the percentage ovigerous in this size range in autumn sampling (Fig. 2). Females in the prerecruit (71-80 mm) size range in spring which molted during the summer were present in autumn as new-shelled (Ennis

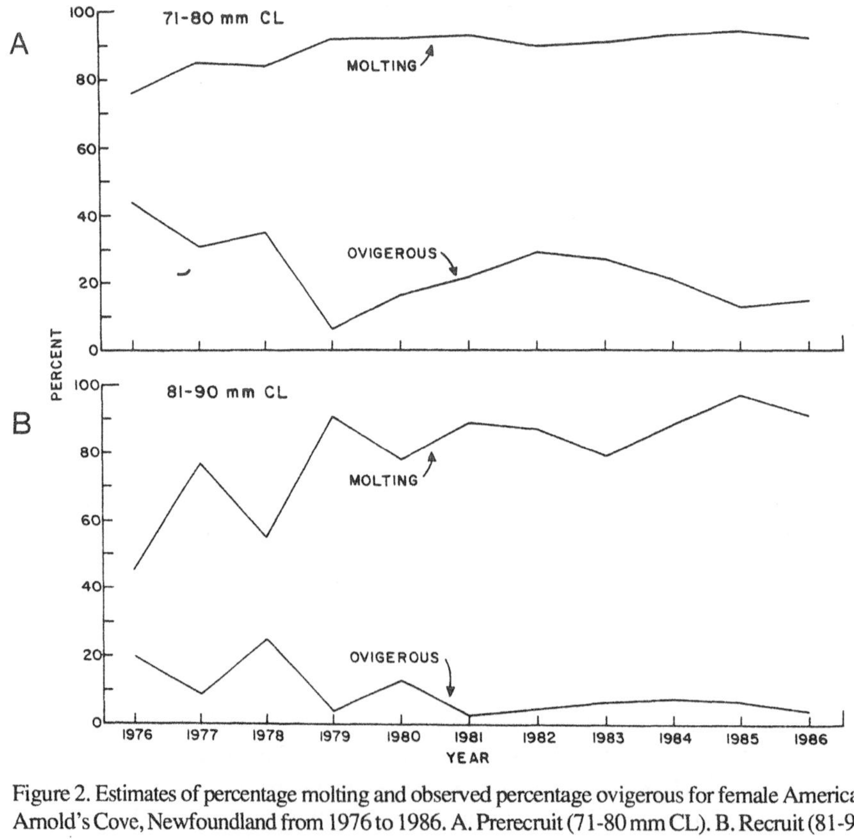

Figure 2. Estimates of percentage molting and observed percentage ovigerous for female American lobsters at Arnold's Cove, Newfoundland from 1976 to 1986. A. Prerecruit (71-80 mm CL). B. Recruit (81-90 mm CL).

1977), non-ovigerous females in the recruit size range. There was a significant negative correlation ($r = -0.76, N = 12, P < 0.01$) between percentage molting in the prerecruit size range during summer and the percentage of ovigerous females in the recruit size range in autumn. There was also a significant negative correlation ($r = -0.88, N = 12, P < 0.01$) between percentage molting among prerecruit females during the summer and percentage ovigerous in this size range in autumn (Fig. 2). There may be some year to year variation in the size-functional maturity relationship; however, variation in numbers growing from smaller and to larger sizes appears to be responsible for most of the variability in percentage ovigerous among prerecruit and recruit females.

3.3 *Effect of temperature and standing stock on percentage of ovigerous females*

Temperature conditions prior to the molting/spawning period varied from year to year. Accumulated degree-days above 0°C from April 20 to July 15 (from thermographs maintained on the bottom at a depth of 9 m) ranged from 320 in 1975 to 587 in 1981. Correlation coefficients between accumulated degree-days and percentage molting among females and between degree-days and percentage of ovigerous females were done for each 5 mm group over the 71 to 95 mm CL range (Table 1). Although none of these was significant (all P values < 0.2), they indicate a weak trend towards higher percentages molting and lower percentages ovigerous in warm water years.

The standing stock is composed mostly of lobsters that molted and grew from prerecruit sizes the preceding summer but also includes commercial lobsters that survived the previous year's spring fishery (less any females that laid eggs during the summer), as well as commercial size females that were ovigerous the previous year. The standing stock in a particular year is therefore a good indicator of abundance during the period preceding the previous year's molting/spawning season. Standing stock estimates for the Arnold's Cove lobster population ranged from 26,557 animals in 1976 to 15,003 in 1982 (Fig. 3). Correlation coefficients between

Table 1. Correlation coefficients between accumulated degree-days and percentage molting among females and between degree-days and percentage of ovigerous females in the American lobster population at Arnold's Cove, Newfoundland from 1975 to 1986. N: 12 in each correlation.

	Carapace length (mm) groups				
	71-75	76-80	81-85	86-90	91-95
% molting	.04	.32	.32	.36	.38
% ovigerous	–.10	–.33	–.26	–.36	–.37

Table 2. Correlation coefficients between estimates of standing stock and percentage molting among females and between standing stock and percentage of ovigerous females in the American lobster population at Arnold's Cove, Newfoundland from 1976 to 1986. N: 11 in each correlation.

	Carapace length (mm) groups				
	71-75	76-80	81-85	86-90	91-95
% molting	–.43	–.81	–.78	–.63	–.72
% ovigerous	.26	.54	.65	.65	.54

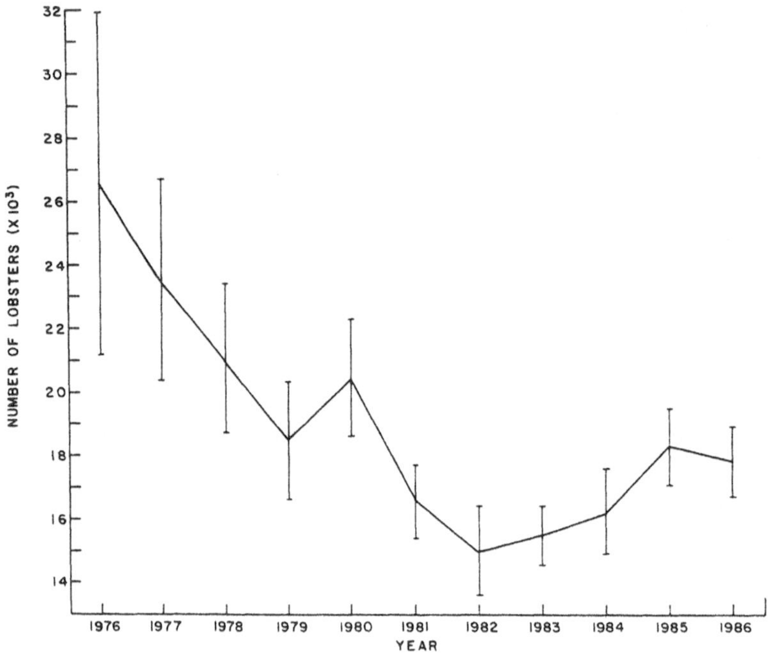

Figure 3. Standing stock estimates for the American lobster population at Arnold's Cove, Newfoundland from 1976 to 1986. Vertical lines indicate 95% confidence limits, which ranged from ±6.0 to ±20.4%.

standing stock and percentage molting among females during the previous summer and between standing stock and percentage of ovigerous females in sampling the previous autumn were calculated for 5 mm CL groups (Table 2).

Neither correlation was significant for the 71-75 mm group, but in the others between percentage molting and standing stock P values were < 0.01 for two and < 0.05 for the other two; while between percentage ovigerous and standing stock P values were < 0.05 for two and < 0.01 for the other two. These results indicate that percentage molting was lower and percentage of ovigerous females was higher at the higher levels of abundance.

In addition to variation in numbers of females growing out of smaller and into larger size groups, the foregoing analyses indicate that year to year variation in the percentage of non-ovigerous females that lay eggs in summer also contributes to variation in the percentage of ovigerous females in autumn samples as well as to variation in numbers of ovigerous females in the population. In generating the estimates of percentage molting, old-shelled [i.e. did not molt during preceding summer (Ennis 1977)] ovigerous females are included with non-molters at their particular size. If their numbers are included with the molters at that size, the data indicate that in most years 100% of the females over the size range considered here either molted or laid eggs (Table 3).

4 DISCUSSION

The general level of egg production in an American lobster (*Homarus americanus*) population

Table 3. Estimates of the percentage of females ≥ 71 mm CL in the American lobster population at Arnold's Cove, Newfoundland that molted or laid eggs annually from 1975-1986.

Year	No. females examined	No. molted	No. laid eggs	% that molted or laid eggs
1975	225	153	67	97.8
1976	177	145	29	98.3
1977	228	175	48	97.8
1978	222	202	10	95.5
1979	189	164	23	98.9
1980	232	211	18	98.7
1981	263	234	29	100.0
1982	208	181	26	99.5
1983	219	202	17	100.0
1984	183	173	10	100.0
1985	260	236	22	99.2
1986	240	228	9	98.0

depends primarily on the minimum legal size and the exploitation rate achieved by the fishery. In most fisheries, minimum legal sizes are quite small in relation to sizes at which females begin to lay eggs, and exploitation rates are high. Egg production is then a small fraction of the potential level for the population. Assessments have indicated that relatively small increases in minimum legal size and decreases in exploitation rate would result in substantial increases in egg production (Campbell & Robinson 1983, Campbell 1985, Ennis 1985).

In the Arnold's Cove population, 50% of the non-ovigerous females are functionally mature (i.e. will extrude eggs during the upcoming spawning season) at 81 mm CL (Ennis 1984), which is also the minimum legal size in the fishery. Exploitation rates have ranged from 78 to 96% over the 1976 to 1986 period (Ennis et al. 1986 and unpubl. estimates). Undersize females therefore account for a substantial proportion (up to 58%) of the annual egg production, although the current level of egg production is undoubtedly quite low in relation to what is potentially possible for this population (Ennis 1985). However, at the prevailing low level of egg production there is still considerable year to year variation in the numbers of eggs produced. This results from variation in the number of ovigerous females in the population, which is linked to recruitment as well as to variation in the percentage of mature females that lay eggs. The latter is inversely related to the percentage molting which appears to be influenced by population density and reduced at high levels of abundance.

The fishery appears to have relatively little impact on the annual variation in egg production in the Arnold's Cove population. The correlation between egg production and exploitation rate the preceding year was not significant ($r = -0.36$, $N = 10$, $P > 0.3$). There was, however, a significant correlation between the estimated number of commercially legal females left in the population at the end of the fishing season, which is a function of the level of recruitment to the standing stock in a particular year (as well as exploitation rate) and egg production the following year ($r = 0.69$, $N = 10$, $P < 0.05$).

Whether the annual variation in egg production contributes to subsequent variation in recruitment to the standing stock is unknown. Ennis (1985) suggested that, because current levels of landings are well below historical levels in most fisheries, parent stock sizes are probably at the low end of the range, i.e. on the ascending limb of a stock-recruitment curve. In that case it seems reasonable to assume that, within the limits of habitat carrying capacity, an increased level of egg production within a population would result in increased recruitment. In Newfoundland populations, recruitment to the standing stock may take as long as 8-9 years (Ennis 1980). Over such a lengthy period it is likely that variation in individual growth rates would result in recruitment of animals made up of a mixture of adjacent year-classes (Caddy 1977). Also, Attard & Hudon (1987) found that large female American lobsters both extrude and hatch their eggs earlier than smaller females and also that energy content per egg increased with female size. Larvae hatched early have a survival advantage associated with higher temperature (Hudon et al. 1986, Caddy 1979). The higher energy content of eggs from large females would likely enhance larval survival as well. If so, large females contribute relatively more to recruitment than their greater fecundity (Attard & Hudon 1987). This suggests that the annual variation in the proportion of total eggs produced by large females in the Arnold's Cove population would also contribute to recruitment variability.

It seems unlikely that the year to year variation in egg production from the Arnold's Cove population would be reflected in standing stock estimates for subsequent years. However, egg production in the Arnold's Cove population was much higher in 1976 to 1978 than in subsequent years. This may have contributed to increased recruitment, as reflected in standing stock estimates for Arnold's Cove, subsequent to a low point in 1982 (Fig. 3). If so, the low level of egg production beyond 1978 should also be reflected in standing stock estimates beyond 1986. A longer time series of egg production and standing stock estimates would be required before any pattern would be evident.

ACKNOWLEDGEMENTS

I gratefully acknowledge assistance with field work and data processing and analysis from G. Dawe and P.W. Collins. I am also indebted to the fishermen of Arnold's Cove whose cooperation and support over the years have made this study possible.

REFERENCES

Anonymous 1977. Report of the Working Group on *Homarus* stocks. *International Council for the Exploration of the Sea, Council Meeting* 1977/K:11, Copenhagen, Denmark.

Anonymous 1979. Report of the *Homarus* Working Group. *International Council for the Exploration of the Sea, Council Meeting* 1979/K:8, Copenhagen, Denmark.

Attard, J. & C. Hudon 1987. Embryonic development and energetic investment in egg production in relation to size of female lobster (*Homarus americanus*). *Can. J. Fish. Aquat. Sci.* 44: 1157-1164.

Caddy, J.F. 1977. Approaches to a simplified yield-per-recruit model for Crustacea, with particular reference to the American lobster, *Homarus americanus*. *Can. Fish. Mar. Serv. Man. Rep.* 1445, St. Andrews, Canada.

Caddy, J.F. 1979. The influence of variations in the seasonal temperature regime on survival of larval stages of the American lobster (*Homarus americanus*) in the southern Gulf of St. Lawrence. *Rapp. P.-V. Réun. Cons. Int. Explor. Mer.* 175: 204-216.

Campbell, A. 1985. Application of a yield and egg-per-recruit model to the lobster fishery in the Bay of Fundy. *N. Amer. J. Fish. Manage.* 5: 91-104.

Campbell, A. & D.G. Robinson 1983. Reproductive potential of three American lobster (*Homarus americanus*) stocks in the Canadian Maritimes. *Can. J. Fish. Aquat. Sci.* 40: 1958-1967.

Ennis, G.P. 1977. Determination of shell condition in lobsters (*Homarus americanus*) by means of external macroscopic examination. *Proc. Nat. Shellfish. Assoc.* 67: 67-70.

Ennis, G.P. 1980. Recent and current Canadian research on growth of lobsters in the wild. In V.C. Anthony & J.F.

Caddy (eds), *Proceedings of the Canada-US Workshop on Status of Assessment Science for NW Atlantic Lobster (Homarus americanus) Stocks, St. Andrews, N.B. Oct. 24-26, 1978:* pp. 9-15. *Can. Tech. Rep. Fish. Aquat. Sci.* 932, St. Andrews, Canada.

Ennis, G.P. 1981. Fecundity of the American lobster, *Homarus americanus,* in Newfoundland waters. *Fish. Bull.* 79: 796-800.

Ennis, G.P. 1983. Annual variations in standing stock in a Newfoundland population of lobsters. *N. Amer. J. Fish. Manage.* 3: 26-33.

Ennis, G.P. 1984. Comparison of physiological and functional size-maturity relationships in two Newfoundland populations of lobsters *Homarus americanus. Fish. Bull.* 82: 244-249.

Ennis, G.P. 1985. An assessment of the impact of size limit and exploitation rate changes on egg production in a Newfoundland lobster population. *N. Amer. J. Fish. Manage.* 5: 86-90.

Ennis, G.P., P.W. Collins & G. Dawe 1982. Fisheries and population biology of lobsters (*Homarus americanus*) at Comfort Cove, Newfoundland. *Can. Tech. Rep. Fish. Aquat. Sci.* 1116: iv + 45 p.

Ennis, G.P., P.W. Collins, G. Dawe & W.R. Squires 1986. Fisheries and population biology of lobsters (*Homarus americanus*) at Arnold's Cove, Newfoundland. *Can. Tech. Rep. Fish. Aquat. Sci.* 1438: iv + 34 p.

Fogarty, M.J. & J.S. Idoine 1986. Recruitment dynamics in an American lobster (*Homarus americanus*) population. *Can. J. Fish. Aquat. Sci.* 43: 2368-2376.

Hudon, C., P. Fradette & P. Legendre 1986. La répartition horizontale et verticale des larves de homard (*Homarus americanus*) autour des îles de la Madeleine, golfe du Saint-Laurent. *Can. J. Fish. Aquat. Sci.* 43: 2164-2176.

Robinson, D.G. 1979. Consideration of the lobster (*Homarus americanus*) recruitment overfishing hypothesis; with special reference to the Canso Causeway. In F.D. McCracken (ed.), *Canso Marine Environment Workshop. Part 3 of 4 parts, Fishery impacts:* pp. 77-79. *Can. Fish. Mar. Serv. Tech. Rep.* 834, St. Andrews, Canada.

Factors influencing fecundity and population egg production of *Jasus* species

John H. Annala
Fisheries Research Center, Wellington, New Zealand

ABSTRACT

The influences of body size, gonad size, egg size, size at onset of maturity, growth rate, and water temperature on fecundity and population egg production for seven species in the spiny or rock lobster genus, *Jasus*, were examined. The relationship between fecundity and body size was curvilinear for most species, with the slope of the curve generally decreasing at larger body sizes. For *J. edwardsii* total egg weight was almost directly proportional to gonad weight. Size at onset of maturity varied widely between species, from 56 mm carapace length for *J. lalandii* to 155-159 mm carapace length for *J. verreauxi*. For *J. edwardsii* there was an inverse relationship between size at onset of maturity and water temperature, while for *J. lalandii* differences in size at onset of maturity were unrelated to water temperature. There was a negative relationship between fecundity and size at onset of maturity for *J. edwardsii*, perhaps caused by smaller gonad sizes of large individuals in areas where size at maturity is large. Fecundity of *J. lalandii* was lower in areas where growth rate was slower. Population egg production of *J. edwardsii* was positively related to fecundity, but negatively related to size at onset of maturity.

1 INTRODUCTION

Seven species of the spiny or rock lobster genus, *Jasus*, have a circumpolar distribution in the temperate waters of the southern oceans (Fig. 1). The genus consists of two groups. One group comprises a single distinct species, *J. verreauxi*, which is found around New Zealand and along the east coast of Australia from New South Wales to South Australia. The other group, the '*lalandii*' group, contains the remaining six species. These six species were all regarded as *J. lalandii* until recognized as separate species by Holthuis (1963). This group was later separated into two subgroups (Holthuis & Sivertsen, 1967; George & Kensler, 1970). The '*lalandii*' subgroup includes *J. lalandii* (South Africa and Namibia), *J. novaehollandiae* (southern Australia), and *J. edwardsii* (New Zealand). The '*frontalis*' subgroup consists of *J. frontalis* (Juan Fernandez, San Felix and San Ambrosio islands), *J. paulensis* (St. Paul and Amsterdam islands) and *J. tristani* (Tristan da Cunha, Gough Island and Vema Seamount).

Kensler (1967, 1968) reviewed the data then available on the fecundity of *Jasus* species. This study reviews additional data on fecundity, population egg production, and associated reproductive characters of *Jasus*. Specifically, the influence of biological factors (body size,

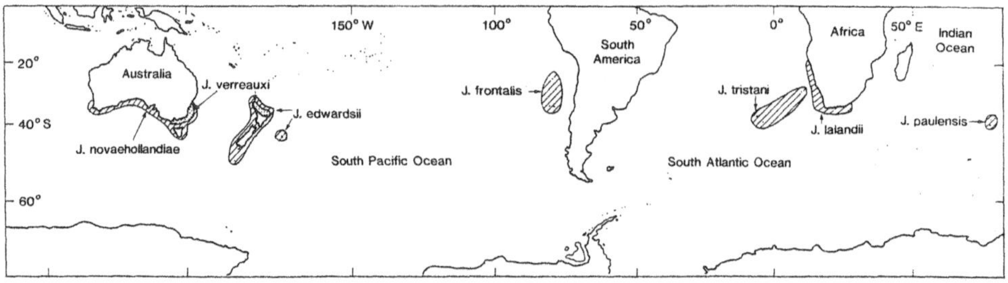

Figure 1. Distribution of *Jasus* species in the southern oceans.

gonad size, egg size, size at onset of maturity, and growth rates) and an environmental factor (water temperature) on fecundity and population egg prodcution are examined.

2 INFLUENCES ON FECUNDITY

2.1 *Body size*

The relationship between the total number of eggs carried externally under the tail of a species (its fecundity) and body size is determined by two interacting factors: 1) the total weight of eggs carried and 2) the number of eggs per gram of egg weight.

Data on the relationship between the total wet weight of eggs carried on the tail and carapace length (CL) are available for two species in the *J. lalandii* group (*J. lalandii* and *J. edwardsii*) and for *J. verreauxi* (Table 1). The relationship between the two variables appeared curvilinear for most species and areas based upon examination of the data points and residuals. A power function of the form, egg weight = aCL^b, was fit to the data using $\log_e$ transformed values. The data points for *J. verreauxi* from Spirits Bay (Fig. 3 of Kensler 1967), and of *J. edwardsii* from the Chatham Islands (Fig. 1 of Kensler 1968) and from the eight areas combined around mainland New Zealand (Annala & Bycroft unpubl. data) suggest that the slope of the curve decreases at larger CLs.

A two part continuous linear spline was fit to the transformed data for *J. edwardsii* from the eight areas combined. A process of trial and error was used to find the 'elbow' in the spline, i.e. the position which minimized the mean square error. Since this results in maximum likelihood estimates for the spline parameters, the standard likelihood ratio test (Kendall & Stuart 1967: 231) was used to test whether the spline gave a significantly better fit than the linear regression. The fitted spline had an elbow at 141 mm CL and gave a significantly better fit ($p < 0.05$) to the data than the linear regression.

J. lalandii and *J. edwardsii* carried a similar weight of eggs for a given CL, with both of these species carrying a greater weight of eggs for a given CL than similar sized females of *J. verreauxi*.

Data on the relationship between total wet weight of eggs carried and body wet weight are available only for *J. edwardsii* (Table 1). The CL, body weight and egg weight of each individual were measured, so the two data sets came from the same rock lobsters. The relationship for most areas appeared curvilinear. The data points for the eight areas combined suggest that the slope of the curve decreases at larger body weights, and a two part spline with an elbow at

1140 grams body weight provided a significantly better fit ($p < 0.05$) than the linear regression.

Data are available on the relationship between the number of eggs per gram of egg wet weight and CL for *J. edwardsii* and *J. verreauxi* (Table 2). For *J. edwardsii* from the Chatham Islands, there was no significant relationship between the two variables. However, there was a significant negative curvilinear relationship for one of the eight areas around mainland New Zealand (Stewart Island) and for all eight areas combined, with the slope of the curves decreasing at larger CLs. This negative relationship was due to the small number of eggs per gram carried by the few large females in the samples. A two part spline with an elbow at 126 mm CL provided a significantly better fit ($p < 0.01$) to the data for all eight areas combined than the linear regression. These results suggest that egg size increases as CL increases.

For *J. verreauxi* there was a significant positive curvilinear relationship between the number of eggs per gram egg weight and CL, which suggests that egg size decreases as CL increases.

Data on the relationship between number of eggs per gram egg wet weight and body wet weight are available only for *J. edwardsii* (Table 2). There was a significant negative curvilinear relationship between these two variables for two areas (Tauroa Point and Stewart Island) and for all eight areas combined, with the slope of the curves decreasing at larger body weights. A two part spline with an elbow at 970 grams body weight provided a significantly better fit ($p < 0.01$) to the data for all eight areas combined than the linear regression. Again, these results suggest that egg size increases as body weight increases.

Data are available on the relationship between fecundity and CL for four species in the *J. lalandii* group and for *J. verreauxi* (Table 3). The relationship appeared curvilinear for most species.

The most detailed data are for *J. edwardsii*, and the fecundity-size relationships of females from eight widely separated areas around New Zealand have been compared (Annala & Bycroft 1987). Significant differences were only found between a few areas. The slope of the curve fit to the data for all eight areas combined decreased at larger CLs, and a two part spline with an elbow at 128 mm CL provided a significantly better fit ($p < 0.05$) than the linear regression.

Data are also available on the fecundity-size relationships of *J. edwardsii* from Wellington and the Chatham Islands. The number of eggs carried by small females from these two areas was similar to the number carried by females of similar size from the eight other areas around New Zealand. However, for females larger than about 130 mm CL, the number of eggs carried by specimens from Wellington (up to 549 000) and the Chatham Islands (up to 538 000) was considerably greater than for the eight other areas. The reasons for this are unknown.

There was a great deal of overlap in the fecundity-size relationships between the four species in the *J. lalandii* group; however, there were some interesting differences. The fecundity-size relationships are similar for *J. lalandii* from South Africa, *J. frontalis* from San Felix and San Ambrosio, and *J. edwardsii* from New Zealand. However, *J. lalandii* from Namibia, *J. novae-hollandiae* from Tasmania, and *J. frontalis* from Juan Fernandez, have greater size-specific fecundity than the preceeding three groups.

The maximum fecundity of the four species in the *J. lalandii* group ranged between about 400 000-550 000 eggs. These maximum egg numbers were reached at different sizes in the various species, ranging from a low value of 111 mm CL for *J. frontalis* from Juan Fernandez (501 109 eggs) to a high value of 170 mm CL for *J. edwardsii* from the Chatham Islands (538 242 eggs). Assuming that the larger sizes found in the samples were close to the maximum sizes attained by the species, these results suggest that similar maximum fecundities were reached at widely different maximum sizes.

Table 1. Relationships between egg wet weight (g) and size for *Jasus* species. CL: Carapace length (mm), BW: body wet weight (g), a and b are parameters of the power function, r^2: coefficient of determination, –: no data. All equations significant at the 0.001 level.

Species (area)	Egg weight = a CL^b			Egg weight = a BW^b			n	Source
	a	b	r^2	a	b	r^2		
J. lalandii (Africa)								
Namibia	1.78×10^{-3}	2.25	0.967	–	–	–	13	Matthews (1962)
South Africa	4.53×10^{-4}	2.57	0.746	–	–	–	10	Heydorn (1965)
J. edwardsii (New Zealand)								
Chatham Is.	2.12×10^{-4}	2.62	0.954	–	–	–	14	Kensler (1968)
Wellington	8.87×10^{-5}	2.87	0.759	1.24×10^{-1}	0.950	0.794	16	Lesser (unpubl.)
Tauroa Pt.	4.13×10^{-7}	4.00	0.915	6.74×10^{-3}	1.40	0.946	38	Annala & Bycroft (unpubl.)
Ngawi	8.32×10^{-7}	3.87	0.771	1.26×10^{-2}	1.31	0.866	25	Annala & Bycroft (unpubl.)
Otago	5.56×10^{-6}	3.45	0.845	1.91×10^{-2}	1.24	0.879	16	Annala & Bycroft (unpubl.)
Milford Sd.	1.51×10^{-5}	3.29	0.488	2.09×10^{-2}	1.25	0.732	19	Annala & Bycroft (unpubl.)
Kaikoura	2.26×10^{-6}	3.70	0.608	1.11×10^{-2}	1.36	0.780	21	Annala & Bycroft (unpubl.)
Gisborne	1.37×10^{-5}	3.35	0.770	3.76×10^{-2}	1.19	0.836	14	Annala & Bycroft (unpubl.)
Whitianga	2.03×10^{-3}	2.19	0.567	2.00×10^{-1}	0.878	0.676	27	Annala & Bycroft (unpubl.)
Stewart I.	3.19×10^{-4}	2.62	0.767	8.29×10^{-2}	1.03	0.827	22	Annala & Bycroft (unpubl.)
Last 8 areas combined	2.75×10^{-5}	3.13	0.813	4.28×10^{-2}	1.12	0.871	182	Annala & Bycroft (unpubl.)
J. verreauxi (New Zealand)								
Spirits Bay	1.12×10^{-5}	3.13	0.896	–	–	–	21	Kensler (1967)

Comparisons of the fecundities of the species in the *J. lalandii* group with *J. verreauxi* are difficult because the latter species reaches a much larger size. The number of eggs carried by small *J. verreauxi* was similar to the number caried by large *J. edwardsii* from the Chatham Islands, while the maximum number of eggs carried by *J. verreauxi* (2,000,000) was much greater than the maximum carried by the other species.

The only data available on the relationship between fecundity and body weight are for *J. edwardsii* from eight areas around New Zealand (Table 3). The relationship appeared curvilinear, and the slope of the curve decreased at larger body weights. A two part spline with an elbow at 990 grams body weight provided a significantly better fit ($p < 0.001$) than the linear regression.

2.2 *Gonad size*

Data on the influence of gonad size on egg weight are available only for *J. edwardsii* from Wellington (Lesser unpubl. data). Unfortunately, direct measurements of gonad size and subsequent egg production by the same individual are not possible. Indirect estimates were made using predictive regressions of gonad wet weight of mature females on CL and egg wet weight on CL, which were combined to estimate the relationship between egg weight and gonad weight.

The relationship between gonad weight and CL appeared linear, and the equation gonad weight = -104 + 1.28 CL, $r^2 = 0.759$, $n = 45$, was fit to the data. A linear function of the form egg weight = -92.5 + 1.45 CL, $r^2 = 0.745$, $n = 16$, was fit to the egg weight-CL data so that the two equations could be combined to estimate the parameters of a linear function relating egg weight to gonad weight. The resulting equation was: egg weight = 25.0 + 1.13 gonad weight. This equation suggests that the increase in egg weight was almost directly proportional to the increase in gonad weight.

2.3 *Egg size*

The observed egg diameters for *J. lalandii* (von Bonde 1936, Paterson 1969, Silberbauer 1971), *J. frontalis* (Arana et al. 1985), and *J. edwardsii* (Kensler 1968) were similar, ranging between 0.75 and 0.8 mm. The egg diameter of *J. verreauxi* was considerably smaller, averaging 0.6 mm (Kensler 1968). Kensler (1968) noted that the larger egg sizes in *J. edwardsii* compared to *J. verreauxi* could be accounted for by differences in egg stage development, as the *J. edwardsii* eggs were well developed (with eye spots), while the *J. verreauxi* eggs were newly laid (without eye spots). Most of the published studies on the egg sizes of *Jasus* species lack information on the developmental stages of the eggs sampled, and conclusions cannot be made concerning the effects of egg size on fecundity and egg production.

2.4 *Size at onset of maturity*

The size at onset of maturity (SOM) is defined as the size at which at least 50% of the females bear ovigerous setae on the endopods of the pleopods. The SOM of the six species in the *J. lalandii* group (Table 4) covered a wide range, from 56 mm CL for *J. lalandii* from Namibia to 120-124 mm CL for *J. edwardsii* from the Otago area of New Zealand. The SOM values of *J. lalandii, J. tristani,* and *J. paulensis* were relatively small, generally < 70 mm CL. The SOM values for *J. edwardsii* covered a wide range, from 60-64 mm CL to 120-124 mm CL.

Table 2. Relationships between number of eggs per gram total egg wet weight and size for *Jasus* species. CL: carapace length (mm), BW: body wet weight (g), a and b are parameters of the power function, r^2: coefficient of determination, –: no data.

Species (area)	No. eggs per g = aCL^b			Signific.	No. eggs per g = aBW^b			Signific.	n	Source
	a	b	r^2	level	a	b	r^2	level		
J. edwardsii (New Zealand)										
Chatham Is.	3262	0.060	0.018	n.s.	–	–	–	–	14	Kensler (1968)
Tauroa Pt.	12210	-0.253	0.100	n.s.	6974	-0.096	0.122	$p < 0.05$	38	Annala & Bycroft (unpubl.)
Ngawi	9045	-0.200	0.048	n.s.	5884	-0.080	0.076	n.s.	25	Annala & Bycroft (unpubl.)
Otago	3328	-0.071	0.029	n.s.	2864	-0.028	0.036	n.s.	16	Annala & Bycroft (unpubl.)
Milford Sd.	3905	-0.111	0.005	n.s.	4146	-0.090	0.033	n.s.	19	Annala & Bycroft (unpubl.)
Kaikoura	40135	-0.584	0.108	n.s.	12582	-0.249	0.186	n.s.	21	Annala & Bycroft (unpubl.)
Gisborne	15835	-0.393	0.100	n.s.	6836	-0.156	0.136	n.s.	14	Annala & Bycroft (unpubl.)
Whitianga	508	0.421	0.133	n.s.	1669	0.118	0.078	n.s.	27	Annala & Bycroft (unpubl.)
Stewart I.	29733	-0.512	0.328	$p < 0.01$	9136	-0.189	0.316	$p < 0.01$	22	Annala & Bycroft (unpubl.)
Last 8 areas combined	18215	-0.388	0.086	$p < 0.001$	8103	-0.155	0.114	$p < 0.001$	182	Annala & Bycroft (unpubl.)
J. verreauxi (New Zealand)										
Spirits Bay	935	0.393	0.285	$p<0.05$	–	–	–	–	21	Kensler (1967)

Table 3. Relationships between fecundity and size for *Jasus* species. CL: carapace length (mm), BW: body wet weight (g), a and b are parameters of the power function, r^2: coefficient of determination, –: no data. All equations significant at the 0.001 level.

Species (area)	Fecundity = a CL^b			Fecundity = a BW^b			CL range	Range in fecundity	n	Source
	a	b	r^2	a	b	r^2	(mm)			
J. lalandii (Africa)										
Namibia	8.0849	2.28	0.974	–	–	–	44-127	50000-490000	13	Matthews (1962)
South Africa	0.0136	3.58	0.930	–	–	–	59-101	32800-180252	8	Heydorn (1965)
J. novaehollandiae (Australia)										
Tasmania	0.1023	3.17	0.952	–	–	–	74-124	65170-413220	148	Hickman (1946)
J. frontalis (Chile)										
Juan Fernandez	0.0005	4.40	0.997	–	–	–	75-111	61359-501109	91	Arana et al. (1985)
San Felix and San Ambrosio	0.0002	2.75	0.961	–	–	–	112-140	190000-524000	204	Báez et al. (1985)
J. edwardsii (New Zealand)										
Chatham Is.	0.6921	2.69	0.938	–	–	–	95-170	124278-538242	14	Kensler (1968)
Wellington	0.0048	3.80	0.858	–	–	–	83-135	86000-549000	28	Bradstock (1948)
Tauroa Pt.	0.0050	3.75	0.935	46.99	1.30	0.955	82-132	75761-407032	38	Annala & Bycroft (in press)
Ngawi	0.0072	3.67	0.798	75.19	1.22	0.879	78-115	37499-257331	25	Annala & Bycroft (in press)
Otago	0.0189	3.38	0.887	54.60	1.21	0.919	79-145	42523-353479	16	Annala & Bycroft (in press)
Milford Sd.	0.0567	3.18	0.592	86.49	1.16	0.820	90-118	80535-215926	19	Annala & Bycroft (in press)
Kaikoura	0.0889	3.11	0.663	141.17	1.11	0.802	75-99	67722-200372	21	Annala & Bycroft (in press)
Gisborne	0.2101	2.95	0.769	257.24	1.03	0.810	74-102	64375-184764	14	Annala & Bycroft (in press)
Whitianga	1.0000	2.61	0.790	333.62	0.996	0.854	79-120	95338-278494	27	Annala & Bycroft (in press)
Stewart I.	9.2999	2.11	0.728	757.48	0.837	0.806	102-157	143151-402917	22	Annala & Bycroft (in press)
Last 8 areas combined	0.4887	2.74	0.807	347.23	0.969	0.838	74-157	37499-407032	182	Annala & Bycroft (in press)
J. verreauxi (New Zealand)										
Spirits Bay	0.0102	3.52	0.933	–	–	–	155-235	386611-2040125	21	Kensler (1967)

Table 4. Size at onset of maturity of *Jasus* species. CL: carapace length, *smallest size at which at least 50% of the females bore eggs during the egg-bearing season (= size at first breeding).

Species (area)	Size at 50% maturity (mm CL)	Source
J. lalandii (South Africa and Namibia)		
South Africa	65-69	Heydorn (1965)
	65-69 to 70-74	Newman & Pollock (1974)
	66-69	Pollock & Beyers (1981)
	67	Pollock (1982)
	65	Pollock et al. (1982)
	59	Pollock (1986)
	66	Pollock (1986)
Namibia	56-62.5	Matthews (1962)
	56	Matthews & Smit (1979)
	57	Beyers (1979)
	57	Pollock & Beyers (1981)
	57	Pollock (1986)
J. tristani (South Atlantic Ocean)		
Tristan da Cunha Island	< 60	Roscoe (1979)
J. paulensis (Indian Ocean)		
St. Paul Island	64	Grua (1963)
New Amsterdam Island	62	Grua (1963)
J. frontalis (Chile)		
Juan Fernandez	77	Arana et al. (1985)
J. novaehollandiae (Australia)		
South Australia	100-104	Fielder (1964)
J. edwardsii (New Zealand)		
North and South Islands	< 60-64 to 120-124	Annala et al. (1980)
Chatham Islands	85-89	Booth (1983)
J. verreauxi (New Zealand)		
Spirits Bay	155-159	Kensler (1967)
New Zealand wide	160-164 to 180-184*	Booth (1984)

The SOM of *J. verreauxi* from Spirits Bay was much larger than the SOM values of the species in the *J. lalandii* group, averaging 155-159 mm CL. However, Booth (1984) found that the smallest size at which at least 50% of female *J. verreauxi* bore eggs during the egg bearing season (= size at onset of breeding) ranged between 160-164 mm and 180-184 mm CL from different areas around New Zealand. This size was usually at least 10 mm CL greater than the size at which setae first developed on the endopods of the pleopods.

Annala et al. (1980) found an inverse relationship between the SOM of *J. edwardsii* and the average of the winter and summer bottom water temperatures near the 100 m isobath. They suggested that SOM is probably a function not only of temperature, but of an interaction between temperature and one or more other factors, such as growth rate, age, metabolic rate, population density, food availability, and other environmental factors. Pollock (1986) noted that the SOM of *J. lalandii* was smaller towards the northern limit of its distribution (in Namibia and the northern coast of South Africa); this trend was unrelated to water temperature, because the entire coastline was influenced by the cool Benguela upwelling system.

Annala & Bycroft (1987) suggested that the fecundity of *J. edwardsii* from eight areas around New Zealand did not appear correlated with any broad-scale biological or environmen-

Table 5. Fecundity at 100 mm carapace length (CL), size at onset of maturity, growth constant 'b' in the equation postmolt CL = a + b premolt CL, annual growth increment at 100 mm CL, and average of the summer and winter surface water temperatures for female *J. edwardsii* from 10 areas around New Zealand. *from equations in Table 3. **data from Annala & Breen (1988), +data from Annala et al. (1980).

Area	Fecundity at 100 mm CL*	Size at onset of maturity (mm CL)+	Growth constant**	Growth increment**	Surface water temperature (°C)+
Tauroa Point	158 114	92.5	0.878	2.8	17.75
Ngawi	157 518	82.5	–	–	15.0
Otago	108 758	122.5	0.929	2.8	12.5
Milford Sound	129 892	87.5	0.966	2.7	13.0
Kaikoura	147 537	72.5	0.943	2.3	13.0
Gisborne	166 888	72.5	1.01	2.9	17.25
Whitianga	165 959	77.5	0.951	3.0	17.75
Steward Island	154 340	107.5	0.953	2.8	12.75
Wellington	191 091	82.5	–	–	15.0
Chatham Islands	166 023	87.5	–	–	13.0

tal variable, such as SOM or water temperature. Their data have been re-analysed with the addition of data from Wellington and the Chatham Islands. The fecundity at 100 mm CL for these 10 areas was calculated from the equations in Table 3, and the results are shown in Table 5. (This size was chosen because at least some females from all areas reached maturity at 100 mm CL). There was a negative correlation (r = -0.622) between fecundity at 100 mm CL and SOM for these areas. This correlation was not significant ($r_{0.05,8} = 0.632$) but nearly so, suggestive of a negative relationship between fecundity and SOM.

Is there any biological basis for a negative relationship between fecundity and SOM? One possible explanation is that gonad size does not begin to increase until sexual maturity is reached. Gonad weight data from Wellington support his suggestion; the gonad weights of immature females of a given size were much less than the gonad weights of equivalent sized mature females (Lesser unpubl. data). Therefore, gonad size of mature females would be smaller at a given CL in areas where SOM is large than in areas where SOM is small. If fecundity is directly correlated with gonad size, as is suggested by the significant positive correlation between egg weight and gonad weight for Wellington, then fecundity will be smaller at a given size in areas where SOM is large because gonad size is smaller. Unfortunately, no data are available on gonad sizes from areas with different SOM values, so this hypothesis cannot be tested.

2.5 *Growth rate*

Growth rate may also affect fecundity. For example, in areas where growth rates are fast, less energy may be put into reproduction and fecundity may be lower than in areas where growth rates are slower. Conversely, in areas where growth rates are fast, more energy may also be available for reproduction and fecundity may be greater than in areas where growth rates are slower.

Pollock (1986), citing studies by Beyers & Goosen (in prep.), suggested that the fecundity of *J. lalandii* varied regionally in the Cape area of South Africa, with fewer eggs produced by

Table 6. Number of eggs-per-rectuit (EPR) of an unfished stock, EPR at existing mortality rates (instantaneous fishing mortality rate = 1.0, instantaneous natural mortality rate = 0.1), percentage of total population egg production less than the current minimum legal size (equivalent to 79.0 mm carapace length (CL) for Otago, 93.0 mm elsewhere), fecundity at 100 mm CL, size at onset of maturity, the growth constant 'b' in the equation postmolt CL = a + b premolt CL, annual growth increment at 100 mm CL, and average of the summer and winter surface water temperatures for female *J. edwardsii* from eight areas around New Zealand. *Data from Annala & Breen (1988), **From equations in Table 3, +Data from Annala et al. (1980).

Area	EPR unfished*	EPR at existing mortality rates*		% egg production <legal size*	Fecundity 100 mm CL**	Size onset maturity (mm CL)+	Growth constant*	Growth increment*	Water temperature (°C)+
		Total	% of unfished						
Tauroa Point	1.357×10^9	0.232×10^9	17.1	52	158 114	92.5	0.878	2.8	17.75
Western Bay of Plenty	1.155×10^9	0.247×10^9	21.4	73	165 959	77.5	0.951	3.0	17.75
Mahia East	0.910×10^9	0.265×10^9	29.1	86	166 888	72.5	0.962	2.8	15.75
Mahia West	0.758×10^9	0.262×10^9	34.6	90	166 888	72.5	0.994	2.9	15.75
Kaikoura	0.775×10^9	0.285×10^9	36.8	87	147 537	72.5	0.943	2.3	13.0
Otago	0.056×10^9	0.001×10^9	1.2	0	108 758	122.5	0.929	2.8	12.5
Stewart Island	0.329×10^9	0.006×10^9	1.7	30	154 340	107.5	0.953	2.8	12.75
Fiordland	0.580×10^9	0.108×10^9	18.7	73	129 892	87.5	0.966	2.7	13.0

females in areas where growth rates were slower. However, he did not suggest a causal mechanism for this relationship.

Growth rate does not appear to influence the differences in fecundity observed between areas for *J. edwardsii*. The relationship between fecundity at 100 mm CL and growth was investigated for the six areas for which growth data on mature females were available (Table 5). Two different measures of growth were used; the constant 'b' in the growth equation postmolt CL = a + b premolt CL and the annual growth increment at 100 mm CL. There was no significant relationship between fecundity and either the growth constant ($r = 0.192$, $n = 7$) or the growth increment ($r = 0.259$, $n = 7$). This is not surprising, as the growth rate of mature female *J. edwardsii* is slow, generally ranging between 2-3 mm CL per year, and there was little difference in growth rate between areas.

2.6 *Water temperature*

There was a positive correlation ($r = 0.511$) between fecundity at 100 mm CL and surface water temperature for *J. edwardsii* from 10 areas around New Zealand (Table 5). This correlation was not significant ($r_{0.05,8} = 0.632$) but is suggestive of a positive relationship between fecundity and temperature. This is not surprising, given the significant negative correlation between SOM and water temperature for this species and the nearly significant negative relationship between fecundity and SOM observed earlier.

3 INFLUENCES ON POPULATION EGG PRODUCTION

A number of biological and environmental factors may influence the egg production of a population, including fecundity, SOM, growth rate, natural mortality rate, water temperature, and food availability. In a fished population, factors such as exploitation rate, the minimum legal size (MLS), mortality of discards, etc. will also affect egg production.

Pollock (1986) calculated the relative egg production per size-class of female *J. lalandii* at the Robben Island sanctuary in South Africa. Because of the relatively low SOM of 60-65 mm CL, compared to the existing minimum legal size of 89 mm CL, egg production is unlikely to be less than about 90% of the maximum egg production of an unfished stock even in fished areas.

Table 6 contains data for *J. edwardsii* on the number of eggs per recruit (EPR) for an unfished stock, the EPR at existing mortality rates, EPR at existing mortality rates as a percentage of the EPR of an unfished stock, fecundity at 100 mm CL, growth rate, and surface water temperature for eight areas. The EPR model (Annala & Breen 1988) was used to estimate the potential number of eggs produced by 1000 individual female recruits entering the fishery. The model was used to assess the effects of changes in size at first capture and exploitation rate on egg production. The results can also be used to qualitatively assess if current egg production is sufficient to sustain the population by comparing current EPR to the EPR of an unfished population.

The EPR of an unfished stock at an instantaneous natural mortality rate (M) = 0.1 varied considerably between areas (EPR for Otago was 4.1% and for Stewart Island was 24.2% of that for Tauroa Point). There was a significant positive relationship between unfished EPR and both fecundity at 100 mm CL and surface water temperature (Table 7). Moreover, there was a negative relationship between unfished EPR and SOM which was not significant, but nearly so

Table 7. Correlation coefficients between the number of eggs-per-recruit (EPR) of an unfished stock, EPR at existing mortality rates (instantaneous fishing mortality rate = 1.0, instantaneous natural mortality rate = 0.1), and total population egg production less than the current minimum legal size (equivalent to 79.0 mm carapace length (CL) for Otago, 93.0 mm CL elsewhere) v.s. size at onset of maturity, fecundity at 100 mm CL, growth constant 'b' in the equation postmolt CL = a + b premolt CL, annual growth increment at 100 mm CL, and average of summer and winter surface water temperatures for female *J. edwardsii* from eight areas around New Zealand. n.s.: not significant, *p < 0.05, **p < 0.01.

Factor	EPR of unfished stock	EPR at existing mortality rates	% of egg production < current minimum legal size
Size at onset of maturity	- 0.666 n.s.	- 0.916**	- 0.988**
Fecundity at 100 mm CL	0.723*	0.699 n.s.	0.667 n.s.
Growth constant 'b'	- 0.310 n.s.	0.078 n.s.	0.455 n.s.
Growth increment at 100 mm CL	0.137 n.s.	- 0.145 n.s.	- 0.190 n.s.
Surface water temperature	0.866**	0.651 n.s.	0.389 n.s.

($r_{0.05,6}$ = 0.707). There was no significant relationship between unfished EPR and either measure of growth rate. Given the negative relationships between SOM and water temperature and between fecundity at 100 mm CL and SOM discussed earlier, these results suggest that the unfished EPR is positively related to fecundity but negatively to SOM.

The exploited EPR at existing mortality rates (instantaneous fishing mortality rate, F = 1.0, and M = 0.1) also varied considerably between areas (exploited EPR for Otago was 0.4% and for Stewart Island was 2.1% of that for Tauroa Point) (Table 6). The exploited EPR as a percentage of the unfished EPR ranged from 1.2% to 36.8%. There was a significant negative relationship between exploited EPR and SOM (Table 7). Moreover, there was a positive relationship between exploited EPR and fecundity at 100 mm CL which was not significant, but nearly so ($r_{0.05,6}$ = 0.707). There was no significant relationship between exploited EPR and either measure of growth rate or water temperature.

The available data for *J. edwardsii* suggest that F and M do not vary much between areas (Annala & Breen 1988), so the differences in EPR probably are not the result of any differences in mortality rates. Moreover, the MLS is uniform throughout New Zealand and equals 152 mm tail length (= 93 mm CL for females), except for Otago where the MLS = 127 mm tail length (= 79 mm CL for females). Otago is also the area with the lowest unfished EPR, lowest exploited EPR, lowest fecundity at 100 mm CL, and largest SOM. Removal of the Otago data from the analysis did not change the significance of the negative relationship between exploited EPR and SOM ($r = -0.883, p < 0.01$), which indicates that this relationship is not influenced by the smaller MLS in Otago.

The percentage of the exploited population egg production produced by females smaller than the existing MLS ranged from 0 to 90% (Table 6). For Otago, where the MLS (79.0 mm CL) is much smaller than the SOM (122.5 mm CL), and no females mature at sizes smaller than the MLS, no egg production comes from females smaller than the MLS. For Stewart Island, where the MLS (93.0 mm CL) is smaller than the SOM (107.5 mm CL), but where some females mature at sizes smaller than the MLS, about 30% of egg production comes from females smaller than the MLS. For areas where the MLS is larger than the SOM, at least 73% of egg production comes from females smaller than the MLS.

There was a significant negative relationship between egg production below the MLS and

SOM (Table 7). Moreover, there was a positive relationship between egg production below the MLS and fecundity at 100 mm CL which was not significant, but nearly so ($r_{0.05,6}=0.707$). There was no significant relationship between egg production below the MLS and either measure of growth rate or water temperature. Removal of the Otago data did not change the significance of the negative relationship between egg production below the MLS and SOM ($r = -0.973$, $p < 0.01$).

4 DISCUSSION AND CONCLUSIONS

The relationship between total egg weight/fecundity and body size was curvilinear for most species/area combinations, and was fit by a power function.

For *J. edwardsii* from mainland New Zealand, there was a significant negative relationship between the number of eggs per gram egg weight and body size, which suggests that egg size increases as body size increases. An apparent result of this relative decline in egg number was a decrease in the slope of the relationship between total egg weight/fecundity and body size at larger body sizes.

Significant differences were found in the fecundity-CL relationship for *J. edwardsii* between a few areas, but no significant differences were observed between most. The data suggest that there is a negative relationship between fecundity and SOM, perhaps caused by smaller gonad sizes of large individuals in areas where SOM is large. Moreover, there is a positive relationship between fecundity and water temperature, which is not surprising given the significant negative correlation between SOM and water temperature for this species. Growth rate does not appear to affect the fecundity of *J. edwardsii*.

For *J. lalandii* from South Africa, SOM appears unrelated to water temperature. However, fecundity is lower in areas where growth rates are slower.

As stated previously, most of the published studies on the fecundity of *Jasus* species lack information on the development stages of the eggs sampled. Egg size and weight increase during development, which will affect the number of eggs carried per gram of egg weight. Egg loss occurs during incubation, which reduces the weight and number of eggs carried. Annala & Bycroft (1987) observed an egg loss of 20% between the early and middle stages of egg development for *J. edwardsii* at Kaikoura, and they attempted to sample females in the early stages to minimize errors caused by egg loss during incubation. Theirs is the only published study in which egg stages are specified, and in which the sampling procedure was designed to minimize the effects of differences in development stage. Therefore, the lack of data on egg development stages makes it difficult to conclusively compare the relationships between total egg weight, the number of eggs per gram of egg weight, and hence fecundity, and body size, between areas for the same species and between species.

Both SOM and fecundity appear to influence the population egg production of *J. edwardsii*. The EPR of an unfished stock was low in areas where SOM was large because of the small number of individuals that reach maturity and contribute to the egg-bearing population. Furthermore, the negative relationship between fecundity and SOM results in a smaller number of eggs at a given size in areas with a large SOM. Exploitation further reduces egg production in large SOM areas to a small percentage of that of an unfished stock. Most egg production from these areas comes from individuals larger than the minimum legal size (e.g. from Otago and Stewart Island).

Fecundity and population egg production may also be affected by a number of other biological and environmental factors. Characteristics of the tail, such as tail area, tail shape, pleopod size and shape, etc., may change relative to body size as body size increases. Egg mortality, through the action of egg parasites and predators, disease, etc., may vary between areas and between years for the same area. Food availability, which is a potentially important limiting factor, may also vary temporally and spatially. Unfortunately, no data are available on the effects of these factors for any *Jasus* species.

REFERENCES

Annala, J.H. & P.A. Breen 1988. Yield-and egg-per-recruit data inputs and model results for the rock lobster *Jasus edwardsii* from 10 areas around New Zealand. *Fisheries Research Centre Internal Rep.* No. 87: 46.

Annala, J.H. & B.L. Bycroft 1987. Fecundity of the New Zealand red rock lobster, *Jasus edwardsii*. *NZ J. Mar. & Freshwat. Res.* 21: 591-597.

Annala, J.H., J.L. McKoy, J.D. Booth & R.B. Pike 1980. Size at the onset of sexual maturity in female *Jasus edwardsii* (Decapoda: Palinuridae) in New Zealand. *NZ J. Mar. & Freshwat. Res.* 14: 217-227.

Arana, P., E. Dupré & V. Gaete 1985. Ciclo reproductivo, talla de primera madurez sexual y fecundidad de la langosta de Juan Fernandez (*Jasus frontalis*). In P. Arana (ed.), *Investigaciones marinas en el Archipielago de Juan Fernandez:* pp. 187-211. Valparaiso: Universidad Católica de Valparaiso.

Bez, P., A. Weinborn & C. Martínez 1985. Contribución al conocimiento del recurso langosta de las islas San Félix y San Ambrosio (Crustacea, Decapoda, Palinuridae). In P. Arana (ed.), *Investigaciones marinas en el Archipielago de Juan Fernandez:* pp. 279-290. Valparaiso: Universidad Católica de Valparaiso.

Beyers, C.J. de B. 1979. Stock assessment and some morphometric and biological characteristics of the rock lobster *Jasus lalandii* on Marshall Rocks, its main commercial fishing area off South West Africa, 1971-1974. *Investl Rep. Sea Fish. Brch. S. Africa.* 117: 1-26.

Booth, J.D. 1983. Chatham Islands rock lobster study. *Catch '83.* 10(1): 13-16.

Booth, J.D. 1984. Size at onset of breeding in female *Jasus verreauxi* (Decapoda: Palinuridae) in New Zealand. *NZ J. Mar. & Freshwat. Res.* 18: 159-169.

Bradstock, C.A. 1948. A study of the marine spiny crayfish *Jasus lalandii* (Milne Edwards). M.Sc. Thesis, Victoria University of Wellington, Wellington, New Zealand.

Fielder, D.R. 1964. The spiny lobster, *Jasus lalandei* (H. Milne Edwards), in South Australia. II. Reproduction. *Aust. J. Mar. & Freshw. Res.* 15: 133-144.

George, R.W. & C.B. Kensler 1970. Recognition of marine spiny lobsters of the *Jasus lalandii* group (Crustacea: Decapoda: Palinuridae). *NZ J. Mar. & Freshwat. Res.* 4: 292-311.

Grua, P. 1963. Maturité-cycle sexuel- soies ovigères des langoustes australes femelles *Jasus paulensis* (Heller 1862) - étude statistique. *CNFRA Instit. Géogr. Nat., Paris.* 4: 1-35.

Heydorn, A.E.F. 1965. The rock lobster of the South African west coast *Jasus lalandii* (H. Milne Edwards). 1. Notes on the reproductive biology and the determination of minimum size limits for commercial catches. *Invest Rep. Div. Sea Fish. S. Afr.* 53: 1-32.

Hickman, V.V. 1946. Notes on the Tasmanian marine crayfish, *Jasus lalandii* (Milne Edwards). *Pap. Proc. Roy. Soc. Tasmania, 1945:* 27-38.

Holthuis, L.B. 1963. Preliminary descriptions of some new species of Palinuridae (Crustacea, Decapoda, Macrura, Reptantia). *Proc. K. Ned. Akad. Wet. Ser. C Biol. Med. Sci. (Sect. C)* 66: 54-60.

Holthuis, L.B. & E. Sivertsen 1967. The Crustacea, Decapoda, Mysidacea and Cirripedia of the Tristan da Cunha Archipelago with a revision of the '*frontalis*' subgroup of the genus *Jasus*. *Results Norw. Sci. Exped. Tristan da Cunha 1937-38.* 52: 1-55.

Kendall, M.G. & A. Stuart 1967. *The advanced theory of statistics. Vol. 2: Inference and relationship.* London: Griffin.

Kensler, C.B. 1967. Fecundity in the marine spiny lobster *Jasus verreauxi* (H. Milne Edwards) (Crustacea: Decapoda: Palinuridae). *NZ J. Mar. & Freshwat. Res.* 1: 143-155.

Kensler, C.B. 1968. Notes on fecundity in the marine spiny lobster *Jasus edwardsii* (Hutton) (Crustacea: Decapoda: Palinuridae). *NZ J. Mar. & Freshwat. Res.* 2: 81-89.

Mathews, J.P. 1962. The rock lobster of South West Africa (*Jasus lalandii*) (Milne Edwards). Size frequency, reproduction, distribution and availability. *Investl. Rep. Mar. Res. Lab. S.W. Afr.* 7: 1-61.

Mathews, J.P. & N.L. Smit 1979. Trends in the size composition, availability, egg-bearing and sex ratio of the rock lobster *Jasus lalandii* in its main fishing area off South West Africa, 1958-1969. *Investl. Rep. Sea Fish. Brch. S. Afr.* 103: 1-38.

Newman, G.G. & D.E. Pollock 1974. Biological cycles, maturity and availability of rock lobster *Jasus lalandii* on two South African fishing grounds. *Investl. Rep. Sea Fish. Brch. S. Afr.* 107: 1-16.

Paterson, N.F. 1969. The behaviour of captive Cape rock lobsters, *Jasus lalandii* (H. Milne Edwards). *Ann. S. Afr. Mus.* 52: 225-264.

Pollock, D.E. 1982. The fishery for and population dynamics of west coast rock lobster related to the environment in the Lamberts Bay and Port Nolloth areas. *Investl. Rep. Sea Fish. Inst. S. Afr.* 124: 1-57.

Pollock, D.E. 1986. Review of the fishery for the biology of the Cape rock lobster *Jasus lalandii* with notes on larval recruitment. *Can. J. Fish. Aquat. Sci.* 43: 2107-2117.

Pollock, D.E., C.J. Augustyn & P.C. Goosen 1982. The rock lobster *Jasus lalandii* and its environmental biology on the Saldanha-Columbine fishing ground off the Cape West Coast, 1978-1981. *Investl. Rep. Sea Fish. Inst. S. Afr.* 125: 1-30.

Pollock, D.E. & C.J. de B. Beyers 1981. Environment, distribution and growth rates of West Coast rock-lobster *Jasus lalandii* (H. Milne Edwards). *Trans. Roy. Soc. S. Afr.* 44: 379-400.

Roscoe, M.J. 1979. Biology and exploitation of the rock lobster *Jasus tristani* at the Tristan da Cunha Islands, South Atlantic, 1949-1976. *Investl. Rep. Sea Fish. Brch. S. Afr.* 118: 1-47.

Silberbauer, B.I. 1971. The biology of the South African rock lobster *Jasus lalandii* (H. Milne Edwards). 1. Development. *Investl. Rep. Div. Sea Fish. S. Afr.* 92: 1-70.

Von Bonde, C. 1936. The reproduction, embryology and metamorphosis of the Cape crawfish *Jasus lalandii* (Milne Edwards) Ortmann. *Investl. Rep. Fish. Mar. Biol. Surv. Div. Un. S. Afr.* 6: 5-25.

Reproductive dynamics of three species of Penaeidae in tropical Australia, and the role of reproductive studies in fisheries management

Peter J. Crocos
CSIRO Division of Fisheries Research, Cleveland, Queensland, Australia

ABSTRACT

The reproductive dynamics of three species of commercially important penaeids (*Penaeus merguiensis*, *P. semisulcatus* and *P. esculentus*) are presented as three case histories. Reproductive output depends on the interactions of population abundance, size structure and reproductive condition. Reproductive output varies geographically and from year to year, both within and between species. In the Gulf of Carpentaria, spawning seasonality in *P. semisulcatus* is bimodal, with the major peak (in spring) restricted to a particular area. By contrast, *P. esculentus* tends to have more protracted and variable egg production. *P. merguiensis* has a bimodal spawning pattern, but the dominant peak occurs at a different time of year according to geographic location.

Comparison of seasonal patterns of egg production with the abundance of later life history phases indicates that the periods of greatest reproductive output are not necessarily most important for stock renewal. Only periods of spawning leading to recruitment of sub-adults to the fishery should be considered when formulating management measures. In the Gulf of Carpentaria, fishing has been banned in the period prior to the spawning season in order to maximize spawning stock.

1 INTRODUCTION

Central to studies of spawning stock/recruitment relationships in penaeid prawns is the need to understand the reproductive dynamics of each species and to be able to accurately define the spawning stock. Annual stock sizes fluctuate widely in penaeid species, and to date few spawning stock/recruitment relationships have been demonstrated for the Penaeidae (Garcia 1983, 1985). There is, however, increasing concern over significant declines in some penaeid stocks (Penn & Caputi 1985, 1986, Somers 1988), which has encouraged closer examination of the biology and exploitation of these stocks with a view to long-term management.

The life history of penaeids involves offshore and coastal phases: the highly fecund adults spawn at sea; eggs hatch as a primitive first larval stage which undergoes several changes in morphology and feeding habits before entering estuarine or coastal nursery grounds as post-larvae. Juveniles spend several months in these grounds before emigrating to the open sea as sub-adults. Estimation of reproductive output in a population is difficult because the eggs are not brooded but released free into the open sea. The apparent reproductive condition of individuals, based on measures of ovary condition, does not necessarily describe the dynamics of population reproductive output because it does not take into account important factors such as the change in the numbers of eggs released with changes in population abundance and size composition.

The present paper summarizes the development of an index of population egg production for ascertaining the reproductive output of a population. Calculated on the temporal and spatial basis this index provides a description of spawning seasonality and spawning areas for a population. Case histories of three commercially important penaeids from the Gulf of Carpentaria (*Penaeus semisulcatus, P. esculentus* and *P. merguiensis*) are presented for comparison of their reproductive patterns. The relationships between egg production and subsequent recruitment to the later life history phases, and the effective reproductive output in terms of population renewal, are then examined. The role of such information in life history studies in support of management strategies is then assessed.

2 ESTIMATION OF REPRODUCTIVE OUTPUT

2.1 *Egg production index*

Most previous studies of reproduction in penaeid prawn populations have relied on ovary condition of females (King 1948, Tuma 1967) as an indication of reproductive activity (Cummings 1961, Rao 1967, Teng 1971, Munro 1975, Chong 1979, Buckworth 1985, Robertson et al. 1985). These studies have used size, color and turgidity of the ovary to describe reproductive activity of individual prawns. Similarly, a gonado-somatic index, incorporating ovary weight, has been used in several studies to describe differences in ovarian maturity, and therefore presumed reproductive activity (e.g. Thomas 1974, O'Connor 1979).

These methods have relied on a single parameter, the proportion of females with developed ovaries, to define reproductive output of a population. However, this approach can give a biased picture, since a true measure of population reproductive output depends on the abundance of spawners, their fecundity and spawning frequency. Penaeid populations exhibit large seasonal fluctuations in abundance, the fecundity of females varies with size and potential spawning frequency is related to the length of the molt-cycle (Crocos & Kerr 1983).

In the present studies, an index of egg production for a population has been developed which takes into account most of these factors. The index has been calculated as the number of spawners (percentage of the females population actively spawning × abundance) multiplied by their individual fecundity (egg production as a function of size). Accordingly, for a population in a particular study region, at a particular point in time:

$$I = \frac{1}{N} \sum_{i=1}^{N} a_i \, p_i \, \frac{1}{n_i} \left[\sum_{j=1}^{n_i} f\,(CL_{ij}) \right]$$

where I = egg production index, N = number of sampling stations in the region, a_i = abundance of females at station i, p_i = proportion of actively spawning females in the sample from station i, n_i = number of spawners in the sample from station i, CL_{ij} = carapace length of the jth active spawner caught at station i, and $f(CL)$ = fecundity of an active spawner with carapace length CL, estimated from the fecundity/carapace length relationship (for *P. merguiensis*, Crocos & Kerr

1983; *P. semisulcatus*, Crocos 1987a; *P. esculentus*, Crocos 1987b). The index is relative, since generally only a measure of relative abundance of a wild population can be obtained. Calculation of this index on a temporal and spatial basis provides for the description of spawning seasons and areas.

2.2 *Maturation, spawning and the molt cycle*

The sequence of events in the reproductive cycle of closed-thelycum penaeids is linked to the molting cycle (*P. merguiensis*, Crocos & Kerr 1983; *P. esculentus*, Crocos 1985; *P. latisulcatus*, Penn 1980; *P. indicus*, Emmerson 1980). Figure 1 summarizes both a reproductive molting cycle, and a non-reproductive molting cycle for the growth of immature females and for mature females where maturation and spawning does not occur in successive molting cycles. In closed-thelycum penaeids, insemination occurs immediately after ecdysis (molt stage A), when females have undeveloped ovaries. In a non-reproductive molting cycle, insemination may occur but the ovaries do not mature and the cycle proceeds to the next ecdysis. In a reproductive molting cycle, ovarian maturation proceeds through intermolt (Stage C) and early premolt (Stage D), with spawning occurring during premolt. Ovaries are in the spent stage at the next

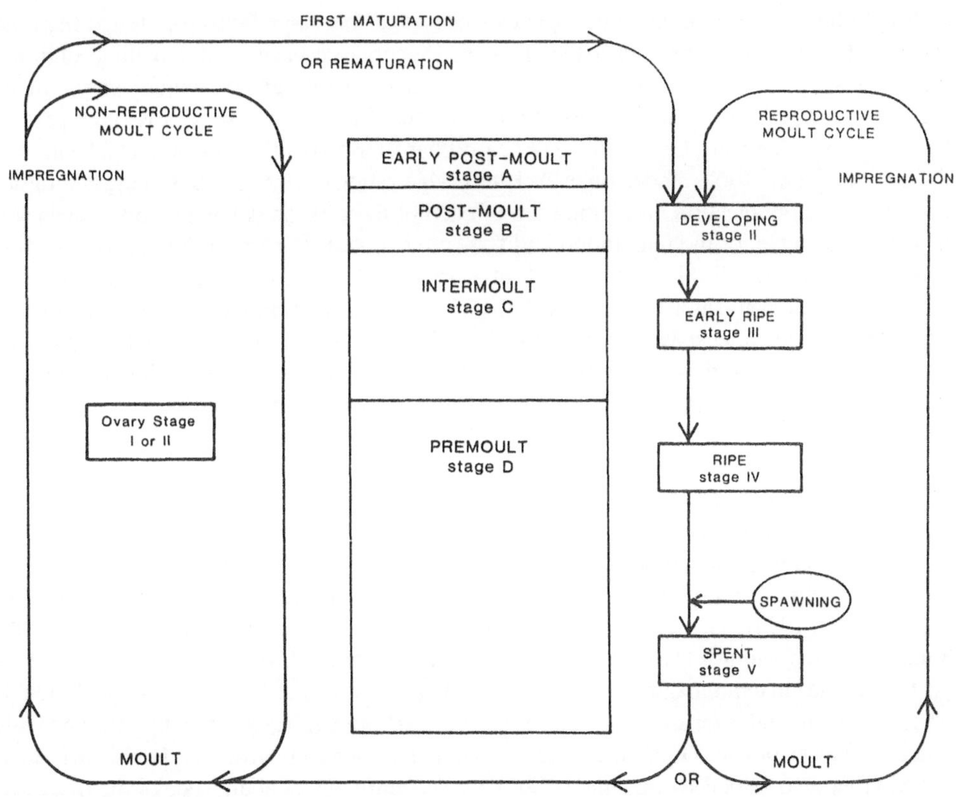

Figure 1. Sequence of events in the reproductive cycle in relation to the molting cycle for closed-thelycum penaeids. Ovary stages after Tuma (1967), molting cycle stages after Drach & Tchernigovtzeff (1967) (from Crocos & Kerr 1983).

ecdysis. The spent stage is transitional, since the small number of unspawned ova are resorbed and redevelopment from stage II may begin. The open-thelycum species, e.g. *P. stylirostris, P. vanamei,* differ mainly in that mating occurs immediately before spawning rather than immediately following ecdysis.

Multiple spawning during a spawning season, over successive molts, is presumed to occur in penaeids (Hudinaga 1942, Lindner & Anderson 1956, Arnstein & Beard 1975, Beard et al. 1977, Penn 1980, Emmerson 1980). Although difficult to test under natural conditions, the presence of high proportions of ripe females throughout the spawning season supports this presumption. Under natural conditions there is no evidence for more than one spawning per molt-cycle, although two or three spawnings per molt-cycle can be induced in penaeids in captivity (e.g. *P. monodon*, Beard & Wickins 1980, *P. indicus*, Emmerson 1980). Further, since ovary stages III and IV coincide with molting stages C and D respectively, it seems unlikely that within a particular molting cycle another ovarian maturation occurs after spawning (Crocos & Kerr 1983, Crocos unpubl.)

2.3 *Assessment of spawning activity*

2.3.1 *Characterization of spawners*

As maturation begins, the posterior lobes of the ovary become visible, and, in these species, yellow to light green in color. Vitellogenesis occurs at this stage (early-ripe), and then the post-vitellogenic oöcytes develop to the ripe stage. Accurate description of spawning condition requires histological examination to determine whether a visible, green-colored ovary is in the early-ripe or ripe stage, since neither the color nor the shape of the ovary provides a precise boundary. Histologically, the ripe stage is clearly characterized by the cortical specialization in the oöcytes (Tuma 1967, Anderson et al. 1985). Maturation of *P. esculentus* in the laboratory has provided an indication of the duration of each of the ovarian developmental stages and hence a time scale for the maturation and spawning process (Crocos 1985, Crocos & Kerr 1986). Cortical specialization in the oöcyte at ovary stage IV usually occurs 7 to 9 days prior to spawning (Anderson et al. 1985, Crocos 1985). The spent ovary stage is recognizable histologically for 1 to 3 days after spawning (Crocos 1985). Thus the combined number of females in the ripe and spent stages (i.e. within approximately 8 to 12 days of spawning) provides an estimate of the number of active spawners in a particular population at a particular time.

2.3.2 *Spawning periodicity*

The egg production index method assumes that there is no short-term periodicity in spawning in a population. Potential spawning frequency has been assumed to be one spawning per monthly sampling period, based on the observed molt-cycle duration (Crocos 1985). Synodic rhythms of reproduction have been documented for several marine invertebrates (Korringa 1947, Palmer 1974), but there is little available evidence of such periodicity in penaeids. In the Gulf of Mexico, Munro et al. (1968) found higher abundances of *P. duorarum* first zoeae in the last half of the lunar month. Price (1979) reported peaks of abundance of *Penaeus sp.* (probably *P. semisulcatus* and *P. latisulcatus*) eggs and nauplii at both lunar quarters in the Arabian Gulf, but abundances of later larval stages were too small to interpret. However, Rothlisberg et al. (1987) found no evidence of a lunar periodicity in occurrence or abundance of the first zoeal stages of four species of penaeids (*P. merguiensis, P. esculentus, P. semisulcatus, P. latisulcatus*) in the Gulf of Carpentaria. A study of *P. esculentus* in south-eastern Queensland, Australia

(Crocos 1985), demonstrated irregular short-term variations in the percentage of actively spawning females, but when pooled over common lunar quarters there was no evidence of lunar periodicity. There is, therefore, no strong evidence for strict lunar periodicity in spawning of penaeids.

3 REPRODUCTIVE DYNAMICS OF *P. SEMISULCATUS*, *P. ESCULENTUS* AND *P. MERGUIENSIS*: THREE CASE HISTORIES

3.1 *Penaeus semisulcatus*

The most intense spawning activity for *P. semisulcatus* occurred in the offshore region in the north-eastern sector of the study area (Crocos 1987a). The intensity of egg production is a consequence of the post-summer migration of young adults from nursery areas in Blue Mud Bay and the northern coastline of Groote Eylandt into a relatively small area offshore from Cape Grey (Fig. 2). This distribution of adults is also related to substrate type (Somers 1987).

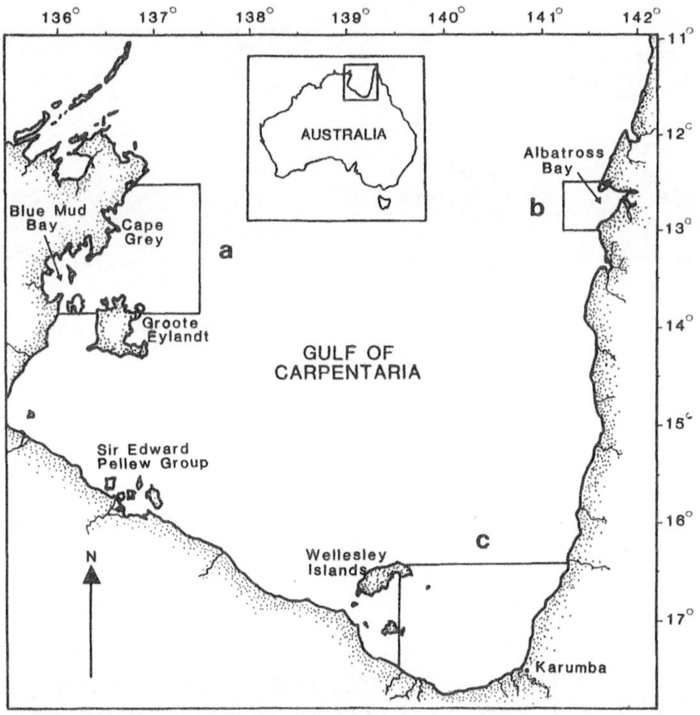

Figure 2. The Gulf of Carpentaria, Australia, sampling locations for the case history studies. (a) North-western Gulf of Carpentaria, study area for *P. semisulcatus* and *P. esculentus*, details of the studies and sub-regions within this area given in Crocos (1987a, b). (b) Albatross Bay, north-eastern Gulf of Carpentaria, study area for the northern population of *P. merguiensis*, details of the study and sampling transects given in Crocos & Kerr (1983). (c) South-eastern Gulf of Carpentaria, study area for the southern population of *P. merguiensis*, with sampling details as for (b).

The magnitude of egg production was found to vary seasonally as a result of the interaction of population abundance, size structure and reproductive condition of individuals, with a major peak of egg production occurring from about August to October in both years studied (Fig. 3d). In this case, even though the mean size of spawners varied little for most of the year (Fig. 3b), both the population abundance (Fig. 3a) and the proportion of spawners (Fig. 3c) increased markedly from August to October, resulting in a strong peak of egg production (Fig. 3d). A smaller peak of egg production in January-February was only evident in the second year of the study. The absence of this peak in 1984 was primarily due to the very low proportion of actively spawning females (3-5%) between January and March 1984 compared to 20-30% of females spawning at this time in 1985.

Closer inshore, north of Groote Eylandt ('North Groote' region, Crocos 1987a), annual egg production was strongly bimodal, with the August to October and February peaks in egg production being about the same size, although egg production was only around 25% that of the offshore Cape Grey region (Crocos 1987a). Tagging studies by Somers & Kirkwood (1984) showed that *P. semisulcatus* pass through the North Groote region during migration from

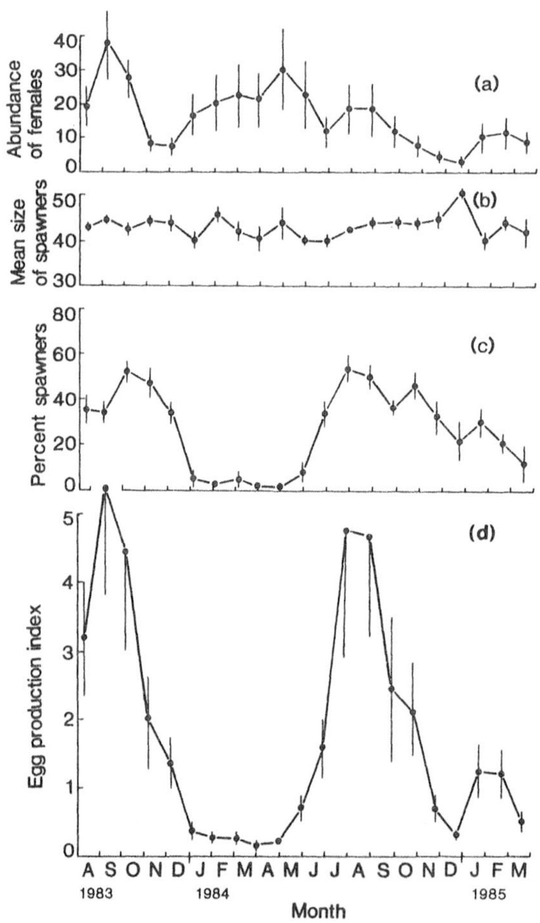

Figure 3. Reproductive dynamics of *P. semisulcatus*, Cape Grey regions, north-western Gulf of Carpentaria. (a) Relative abundance of females (mean of catch per standard trawl). (b) Mean size of spawning females (mm CL). (c) Percentage of females spawning. (d) Mean egg production index. (From Crocos 1987a).

inshore nursery areas to the offshore Cape Grey region. Young adult *P. semisulcatus* recruit to the inshore North Groote region in the warmer months of the year, beginning in October (Somers et al. 1987), and by January-February some of these females have begun to spawn, producing a small peak of egg production. By August, more fecund, although less numerous, large females begin to spawn, resulting in a peak of egg production in August, September and October.

The basic pattern of spawning for *P. semisulcatus* is therefore an initial inshore spawning, when the first individuals of a cohort breed at approximately 6 months of age, then a massive offshore spawning at about 10-12 months of age when both the proportion of spawners in the population and the fecundity of these large-sized adults is high.

3.2 *Penaeus esculentus*

Egg production for *P. esculentus* was highest in the inshore region to the north of Groote Eylandt, and very low in shallow Blue Mud Bay and the offshore regions of the study area (Crocos 1987b). The distribution of adults in the study area is dependent on substrate type, with *P. esculentus* showing preference for the less muddy substrates to the north of Groote Eylandt (Somers 1987, Somers et al. 1987). Consequently, most spawning occurred in this region.

Rather than the very distinct seasonal pattern of egg production observed in *P. semisulcatus*, egg production by *P. esculentus* showed considerable interannual variation over the period August 1983 to March 1985 (Fig. 4). In the first year of the study, egg production was relatively high throughout spring and summer (September to February), but in the following year it peaked earlier (July-August) and then declined during spring before another peak in late summer (February-March). Thus the spring-summer patterns of egg production were not the same in each year. Egg production is more protracted for *P. esculentus*, when compared to *P. semisulcatus* (Fig. 3d). Indications of protracted spawning for *P. esculentus*, based on the proportion of ripe females in the population, has also been observed by Buckworth (1985) in the Groote Eylandt area, and by Robertson et al. (1985) in the south-eastern Gulf of Carpentaria.

3.3 *Penaeus merguiensis*

3.3.1 *South-eastern Gulf of Carpentaria*

Annual egg production of *P. merguiensis* in the south-eastern Gulf was weakly bimodal (Fig. 5a). During spring (September-October), this population has a high proportion of spawning females, which were large in size and highly fecund; however, population abundance was low, so the peak in egg production was small. Population abundance increased sharply as young

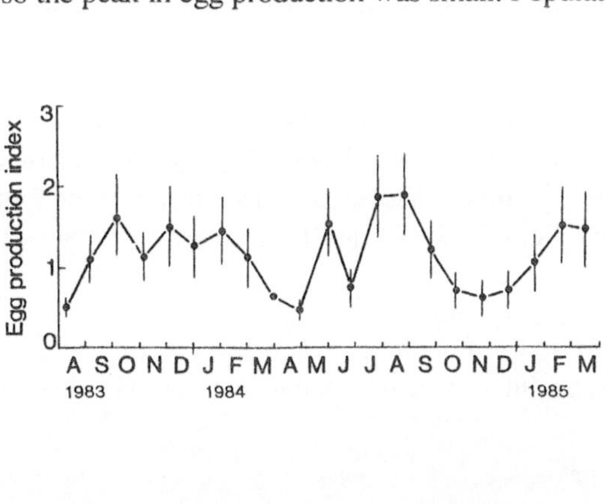

Figure 4. Monthly egg production index values for the tiger prawn *P. esculentus*, in the major spawning area.

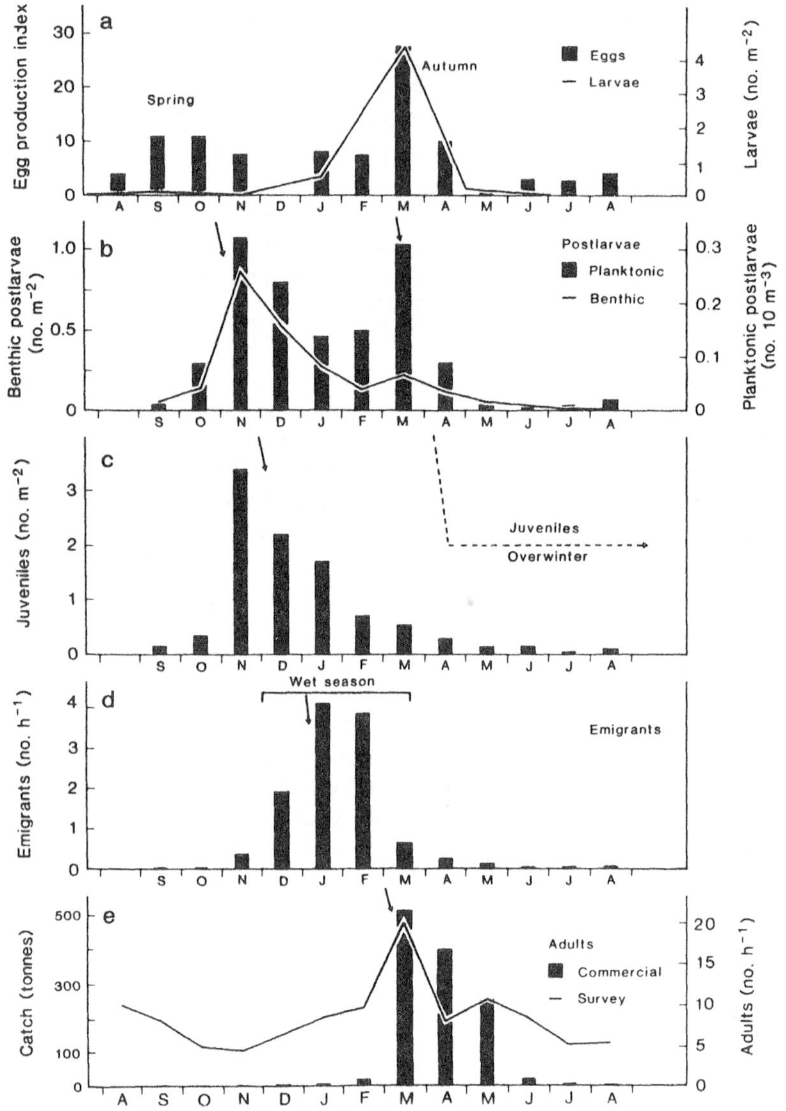

Figure 5. Seasonal abundance of *P. merguiensis* in the south-eastern Gulf of Carpentaria. (a) Eggs and larvae. (b) postlarvae. (c), (d) juveniles. (e) adults. Solid arrows indicate main cohorts. (From Rothlisberg et al. 1985).

adults recruited through the summer months, and reached a peak in March, followed by a sharp decline in April after the opening of the commercial fishery. The peak of egg production in March was of greater magnitude than the September-October peak as a result of the large population at this time, even though the proportion of spawners and the fecundity of individuals was not as high (Crocos & Kerr 1983).

3.3.2 *North-eastern Gulf of Carpentaria*

Annual egg production in the north-eastern Gulf was also bimodal (Fig. 6a). However, in contrast to the south-eastern Gulf, the principal period of egg release was during spring (September), with a lesser peak in autumn (March). The September peak was due to the high proportion of large, highly fecund females in this population. During March, although the population was larger, proportionately fewer, and smaller, females were spawning than in September.

3.4 *Spawning patterns: Summary*

The egg production patterns of the three penaeid species examined in the Gulf of Carpentaria vary both within and between species, with geographic location, and from year to year. The egg production at a particular time of year or location is determined by the dynamic interactions of the proportion of spawners, the size of the population, and the size structure of the population as it affects the fecundity of spawners. Any one, or any combination, of these parameters can be the major factor determining the magnitude of a particular peak of egg production. For example, the marked increase in the proportion of *P. semisulcatus* females spawning resulted in the peak of egg production in August-October. By contrast, the March peak of egg production by *P. merguiensis* in the south-eastern Gulf was due primarily to the very large population abundance at that time. The major spring peak of egg production by *P. merguiensis* in the north-eastern Gulf was a result of the relatively large size of the spawners, despite their being only a moderate proportion in a moderately sized population.

Seasonal and interannual changes in the parameters used to calculate the egg production index therefore determine the reproductive patterns observed. The strength of the egg production index as an assessment of reproductive output is its ability to take into account these changes.

4 ROLE OF REPRODUCTIVE INFORMATION IN MANAGEMENT-ORIENTED LIFE HISTORY STUDIES

4.1 *Significance of temporal and spatial spawning patterns in the life history*

4.1.1 *Penaeus merguiensis: South-eastern Gulf*

Detailed studies of the life history of *P. merguiensis* in the south-eastern Gulf have shown that the major spawning period in terms of highest egg production is not the effective spawning for stock renewal (for review see Rothlisberg et al. 1985). Two peaks of egg production were observed: one in September-October (spring) and another, much larger, peak in March (autumn) (Fig. 5a). Seasonal timing and relative abundance of larvae followed a similar pattern. Planktonic postlarval abundance was also bimodal, with a large peak in November (spring) and a slightly smaller one in March (autumn) (Fig. 5b). However, over the four years studied, the spring peak occurred each year, the autumn peak only once (Staples 1980a). Thus spring is the main period of postlarval immigration, when the abundance of females offshore is relatively low.

One of the reasons for the variable success of postlarval immigration in this region may be the effects of tidal- and wind-driven currents. Using an advection model based on vertical migration and the migratory behaviour of larvae in combination with observed current regimes, Rothlisberg (1982) and Rothlisberg et al. (1983) found evidence of offshore transport of larvae during the March spawning period, which may explain why postlarval recruitment to coastal rivers is low at this time in most year (Fig. 5b). Further, the mortality of the postlarvae of the February-March spawning that did reach the estuaries was high, and relatively few juveniles survived (Figs. 5b, 5c; Staples 1980b).

Juveniles from the spring spawning adults remained in the estuaries for 1 to 4 months and began emigrating during the wet season (December to March) (Fig. 5d), culminating in March and April (Fig. 5e) when most of these adults were 6 to 7 months old; these autumn-spawning females produced the large peak in egg production. During the September-October spawning season, the large size of the spawning females suggests that they were mainly adults that had escaped from the autumn fishery.

Therefore, due to seasonal differences in the mortality of the spring and autumn generations, imposed by environmental factors and fishing, the bimodal pattern of spawning results in a unimodal pattern of recruitment to the inshore juvenile and offshore adult populations. Consequently, a substantial amount of the total reproductive output is lost.

4.1.2 *Penaeus merguiensis: North-eastern Gulf*

Annual egg production in the north-eastern Gulf was also bimodal, but in contrast to the south-eastern Gulf, the spring peak was dominant (Fig. 6a). However, larvae were not found offshore during spring; only the autumn peak was apparent (Fig. 6a). Planktonic postlarvae arrived in the Embley estuary adjacent to Albatross Bay in three pulses over a prolonged period from November to July (Fig. 6b). Staples (1979) suggested the autumn influx of postlarvae to rivers in the north-eastern Gulf was consistent year to year, and larger than the spring influx. Thus in this region some juveniles overwintered in the estuary and emigrated in spring (Fig. 6c), while some adults recruited to the offshore population during winter (Fig. 6d), resulting in a more abundant population of larger-sized females compared to the south-eastern Gulf.

In *P. merguiensis*, seasonal differences in the survival rates of all life history stages determine the importance of a particular period of egg release to the life history. Although the general timing of peaks of egg production, and peaks of larval and juvenile abundance in both these regions agree in most respects (the apparent absence of larvae from the spring spawning in the north-eastern Gulf being the main unexplained discrepancy), there are large seasonal differences in the relative proportions of the different stages. The main population in the south-eastern Gulf is derived from a small spring spawning, whereas 300 nautical miles to the north the large autumn juvenile population arises from the smaller autumn spawning peak. Therefore, effective reproductive output can be assessed only in terms of the survival of later life history stages.

4.1.3 *Penaeus semisulcatus*

The relative success of the two main spawning periods per year of *P. semisulcatus* can be assessed from the subsequent life history stages. Egg production peaked in September with a smaller peak in February. Larval abundance was also bimodal but fewer larvae were observed from the September spawning (Rothlisberg et al. 1987). Juvenile *P. semisulcatus* were found in shallow-water intertidal nursery areas within the study area from September through to April, and were most abundant in December. There was little ingress of postlarvae during the winter months (Staples unpubl.). Sub-adults (20 mm CL) recruited to the offshore population from November to April, with a major peak in January and indications of a lesser peak in March-April (Somers et al. 1987). The high abundance of larvae observed in January suggests a possible underestimate of the egg production in January, or the discrepancy may reflect the effects of interannual variability as the Rothlisberg et al. (1987) study was not carried out in the same year

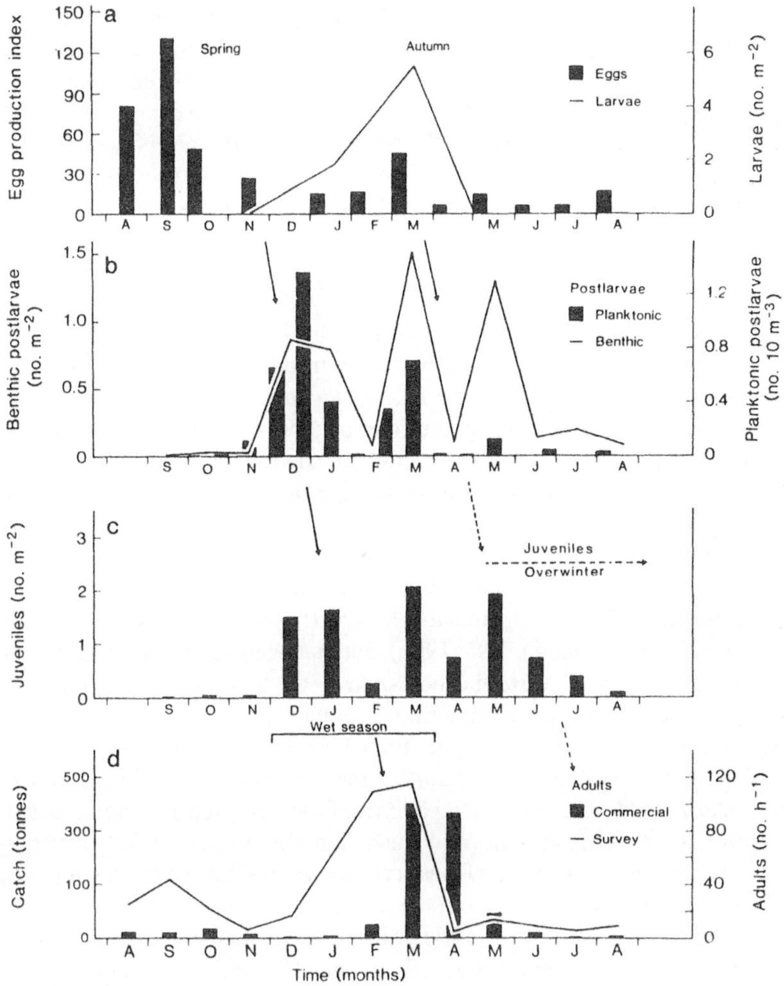

Figure 6. Seasonal abundance of *P. merguiensis* in Albatross Bay, north-eastern Gulf of Carpentaria. (a) (b) (c) as for Figure 6, (d) adults. (From Rothlisberg et al. 1985).

as the adult and juvenile studies. Possible environmental effects on egg hatching or larval advection must be considered, since the seasonal abundances of postlarvae and juveniles, and subsequent recruitment of young adults, do not reflect the January larval peak.

Thus for *P. semisulcatus*, the major spring spawning results in the highest postlarval and juvenile abundances in the coastal nursery areas, followed by the major period of recruitment back to the adult population. Spring thus appears to be the effective spawning period for stock renewal.

4.1.4 *Penaeus esculentus*

Egg production by *P. esculentus* does not show a strongly bimodal annual pattern. The annual pattern was not consistent from year to year, and egg production was more prolonged than in the other species examined. Egg production was high during the spring (September onwards) of the first year, and in late-winter and early-spring (July to September) of the second year. Although larvae were most abundant in January (Rothlisberg 1987), egg release in late-winter and early-spring was consistent with peaks of abundance of juveniles from October to February (Staples unpubl.) and with recruitment of sub-adults to the offshore population between November and March (Somers et al. 1987). Despite prolonged low-level spawning in this species, effective spawning which contributes to stock renewal occurs over a more restricted period.

4.2 *Fishery management measures and life history studies*

Commercial prawn fishing began in the Gulf of Carpentaria in the late 1960's, initially concentrating on the banana prawn, *P. merguiensis*, along the southern and eastern coasts of the Gulf. By the 1970's, the tiger prawns *P. esculentus* and *P. semisulcatus* were being fished in the western Gulf. Tiger prawns are now the most important component of the catch. Early management measures for the penaeid prawn fisheries in the Gulf, directed at *P. merguiensis*, reduced fishing effort by restricting licences, and controlled the size of prawns at first capture by seasonal and area closures to prevent the uneconomic capture of juvenile stock. A variable opening date for the fishing season has been used to maximize value per recruit (Bowen & Hancock 1985, Somers 1985). Essentially, these measures have an economic basis.

Until recently, spawning stock/recruitment relationships had not been identified in penaeids (Garcia 1983). However, Penn & Caputi (1985, 1986) demonstrated such a relationship for populations of *P. esculentus* in Western Australia. Indications of declining stocks of *P. esculentus* and *P. semisulcatus* in the Gulf of Carpentaria, and concern about possible recruitment-overfishing led to changes in the management strategy for these fisheries. Due to concern over low levels of the spawning stocks, biological management measures have been introduced to protect pre-spawning stocks. Although spawning patterns of the three species vary in detail, there is a common basic pattern. Penaeids in tropical Australian waters spawn chiefly in spring and autumn. The relative size and importance of these two spawnings varies between regions, species, and years. In a large region such as northern Australia, detailed surveys of the reproductive dynamics of all prawn stocks are not practicable. Information from the spatially restricted studies for *P. merguiensis, P. esculentus* and *P. semisulcatus* must therefore be extrapolated for the purposes of management.

In the Gulf of Carpentaria, pre-spawning stocks of *P. escultentus* and *P. semisulcatus* are protected by total bans on trawling from mid-June to the end of July. A closure during the peak spawning season in spring (August to October) is economically impractical, since this is also the peak of the fishing season. It would also be less effective at maximizing the number of spawners. The pre-spawning closure has the effect of maintaining stock levels before the spawning season to achieve a larger spawning population at the onset of the effective spawning period. As the fishery is closed from December until mid-April each year, the autumn spawning stocks are already protected. The autumn closure was introduced to maximize value per recruit of *P. merguiensis*, but it also has value as a biological measure.

The life cycle dynamics outlined for the three penaeid species examined in the Gulf of Carpentaria indicate that a number of biological and physical factors determine the extent to which a period of major egg production is reflected in stock renewal. A detailed understanding of spawning times and locations, larval and juvenile dynamics, and timing of recruitment to the adult population is necessary to identify the critical periods and locations of egg production, and to ascertain the factors that affect recruitment at each life history phase. In a fishery where there is concern that fishing pressure on the spawning stock is limiting stock renewal, reproductive information to enable protection of spawning stocks is critical.

ACKNOWLEDGEMENTS

Drs Burke Hill and Peter Rothlisberg of CSIRO made many constructive comments on the manuscript.

REFERENCES

Anderson, S.L., W.H. Clark & E.S. Chang 1985. Multiple spawning and molt synchrony in a free-spawning shrimp *Sicyonia ingentis:* (Penaeoidea). *Biol. Bull.* 168: 377-394.

Arnstein, D.R. & T.W. Beard 1975. Induced maturation of the prawn *Penaeus orientalis* Kishinouye in the laboratory by means of eyestalk removal. *Aquaculture* 5: 411-412.

Beard, T.W. & J.F. Wickins 1980. Breeding of *Penaeus monodon* Fabricius in laboratory recirculation systems. *Aquaculture* 20: 79-89.

Beard, T.W., J.J. Wickins & D.R. Arnstein 1977. The breeding and growth of *Penaeus merguiensis* de Man in Laboratory recirculation systems. *Aquaculture* 10: 275-289.

Bowen, B.K. & D.A. Hancock 1985. Review of penaeid prawn fishery management regimes in Australia. In P.C. Rothlisberg, B.J. Hill & D.J. Staples (eds), *Second Australian National Prawn Seminar, NPS2,* pp. 247-265. Cleveland, Australia.

Buckworth, R.C. 1985. Preliminary results of a study of commercial catches, spawning and recruitment of *Penaeus esculentus* and *Penaeus semisulcatus* in the western Gulf of Carpentaria. In P.C. Rothlisberg, B.J. Hill & D.J. Staples (eds), *Second Australian National Prawn Seminar, NPS2,* pp. 85-93. Cleveland, Australia.

Chong, V.C. 1979. The biology of the white prawn *Penaeus merguiensis* de Man (Crustacea: Penaeidae) in the Palau Angsa-Klang Strait waters (Straits of Malacca). M.Sc. Thesis, University of Malaya, 141 pp.

Crocos, P.J. 1985. Appraisal of some factors relevant to the development of penaeid prawn population reproductive models. In P.C. Rothlisberg, B.J. Hill & D.J. Staples (eds), *Second Australian National Prawn Seminar, NPS2,* pp. 159-164. Cleveland, Australia.

Crocos, P.J. 1987a. Reproductive dynamics of the grooved tiger prawn, *Penaeus semisulcatus*, in the north-western Gulf of Carpentaria, Australia. *Aust. J. Mar. Fresh. Res.* 38: 79-90.

Crocos, P.J. 1987b. Reproductive dynamics of the tiger prawn *Penaeus esculentus*, and a comparison with *P. semisulcatus* in the north western Gulf of Carpentaria, Australia. *Aust. J. Mar. Fresh. Res.* 38: 91-102.

Crocos, P.J. & J.D. Kerr 1983. Maturation and spawning of the banana prawn *Penaeus merguiensis* de Man (Crustacea: Penaeidae) in the Gulf of Carpentaria, Australia. *J. Exp. Mar. Biol. Ecol.* 69: 37-59.

Crocos, P.J. & J.D. Kerr 1986. Factors affecting induction of maturation and spawning of the tiger prawn *P. esculentus* (Haswell), under laboratory conditions. *Aquaculture* 58: 203-214.

Cummings, W.C. 1961. Maturation and spawning of the pink shrimp, *Penaeus duorarum. Trans. Amer. Fish. Soc.* 90: 462-468.

Drach, P. & C. Tchernigovtzeff 1967. Sur la méthode de determination des stades d'intermue et son application générale aux Crustaces. *Vie et Milieu* 18A: 595-610.

Emmerson, W.D. 1980. Induced maturation of the prawn *Penaeus indicus. Mar. Ecol. Prog. Ser.* 2: 121-131.

Garcia, S. 1983. The stock-recruitment relationship in penaeid prawns: reality of artifacts and misinterpretations. *Oceanogr. Trop.* 18: 25-48.

Garcia, S. 1985. Reproduction, stock assessment models and population parameters in exploited penaeid prawn popultions. In P.C. Rothlisberg, B.J. Hill & D.J. Staples (eds), *Second Australian National Prawn Seminar, NPS2,* pp. 139-158. Cleveland, Australia.

Hudinaga, M. 1942. Reproduction, development and rearing of *Penaeus japonicaus* Bate. *Jpn. J. Zool.* 10: 303-347.

King, J.E. 1948. A study of the reproductive organs of the common marine prawn, *Penaeus setiferus*. *Biol. Bull.* 94: 244-262.

Korringa, P. 1947. Relations between the moon and periodicity in the breeding of marine animals. *Ecol. Monog.* 17: 349-381.

Lindner, J.J. & W.W. Anderson 1956. Growth, migrations, spawning and size distribution of prawn *Penaeus setiferus*. *US Fish. Wildl. Serv. Fish. Bull.* 56: 553-645.

Munro, I.S.R. 1975. Biology of the banana prawn, *Penaeus merguiensis*, in the south east corner of the Gulf of Carpentaria. In P.C. Young (ed.), *First Australian National Prawn Seminar, Maroochydore, 1973*, pp. 60-78. Canberra: AGPS.

Munro, I.S.R., A.C. Jones & D. Dimitriou 1968. Abundance and distribution of the larvae of the pink prawn, *Penaeus duorarum*, on the Tortugas Shelf of Florida, August 1962-October 1964. *US Fish. Wildl. Serv. Fish. Bull.* 67: 165-181.

O'Connor, C.Y. 1979. Reproductive periodicity of a *Penaeus esculentus* population near Low Islets, Queensland, Australia. *Aquaculture* 16: 153-162.

Palmer, J.D. 1974. *Biological Clocks in Marine Organisms*. Wiley, New York.

Penn, J.W. 1980. Spawning and fecundity of the western king prawn, *Penaeus latisulcatus* Kishinouye, in Western Australian waters. *Aust. J. Mar. Fresh. Res.* 31: 21-35.

Penn, J.W. & N. Caputi 1985. Stock recruitment relationships for the tiger prawn, *Penaeus esculentus*, fishery in Exmouth Gulf, Western Australia, and their implications for management. In P.C. Rothlisberg, B.J. Hill & D.J. Staples (eds), *Second Australian National Prawn Seminar, NPS2*, pp. 165-173. Cleveland, Australia.

Penn, J.W. & N. Caputi 1986. Spawning stock recruitment relationships and environmental influences on the tiger prawn *Penaeus esculentus* fishery in Exmouth Gulf, Western Australia. *Aust. J. Mar. Fresh. Res.* 37: 491-505.

Price, A.R.G. 1979. Temporal variations in abundance of penaeid prawn larvae and oceanographic conditions of Ras Tamura, western Arabian Gulf. *Est. Coast. Mar. Sci.* 9: 451-465.

Rao, P.V. 1967. Maturation and spawning of the penaeid prawns of the south west coast of India. *FAO Fish. Report* 57: 285-302.

Robertson, J.W.A., R.G. Coles & G.B. Goeden 1985. Distribution patterns of commercial prawns and reproduction of *Penaeus esculentus* around the Wellesley Islands in the southeastern Gulf of Carpentaria. In P.C. Rothlisberg, B.J. Hill & D.J. Staples (eds), *Second National Prawn Seminar, NPS2*, pp. 71-75. Cleveland, Australia.

Rothlisberg, P.C. 1982. Vertical migration and its effects of dispersal of penaeid prawn larvae in the Gulf of Carpentaria, Australia. *Fish. Bull. US.* 80: 541-554.

Rothlisberg, P.C., J.A. Church & A.M.G. Forbes 1983. Modeling the advection of vertically migrating prawn larvae. *J. Mar. Res.* 41: 511-538.

Rothlisberg, P.C., C.J. Jackson & R.C. Pendrey 1987. Larval ecology of penaeids of the Gulf of Carpentaria, Australia. I. Assessing the reproductive activity of five species of *Penaeus* from the distribution and abundance of the zoeal stages. *Aust. J. Mar. Fresh. Res.* 38: 1-17.

Rothlisberg, P.C., D.J. Staples & P.J. Crocos 1985. A review of the life history of the banana prawn, *Penaeus merguiensis*, in the Gulf of Carpentaria. In P.C. Rothlisberg, B.J. Hill & D.J. Staples (eds), *Second National Prawn Seminar, NPS2*, pp. 125-136. Cleveland, Australia.

Somers, I.F. 1985. Maximising value per recruit in the fishery for banana prawns, *Penaeus merguiensis*, in the Gulf of Carpentaria. In P.C. Rothlisberg, B.J. Hill & D.J. Staples (eds), *Second National Prawn Seminar, NPS2*, pp. 185-191. Cleveland, Australia.

Somers, I.F. 1987. Sediments as a factor in the distribution of commercial prawn species of the western Gulf of Carpentaria, Australia. *Aust. J. Mar. Fresh. Res.* 38: 133-149.

Somers, I.F. 1988. The Northern Prawn Fishery., In *Proceedings of the workshop; Scientific advice for fisheries management*. July 1987. Department of Primary Industry, Bureau of Rural Science, Canberra, Australia, July 1987.

Somers, I.F., P.J. Crocos & B.J. Hill 1987. Distribution and abundance of tiger prawns, *Penaeus esculentus* and *P. semisulcatus*, in the north-western Gulf of Carpentaria, Australia. *Aust. J. Mar. Fresh. Res.* 38: 63-78.

Somers, I.F. & G.P. Kirkwood 1984. Movement of tagged tiger prawns, *Penaeus* spp., in the Western Gulf of Carpentaria. *Aust. J. Mar. Fresh. Res.* 35: 713-723.

Staples, D.J. 1979. Seasonal migration patterns of postlarval and juvenile banana prawns, *Penaeus merguiensis* de Man, in the major rivers of the Gulf of Carpentaria, Australia. *Aust. J. Mar. Fresh. Res.* 30: 143-157.

Staples, D.J. 1980a. Ecology of juvenile and adolescent banana prawns, *Penaeus merguiensis*, in a mangrove estuary and adjacent offshore area of the Gulf of Carpentaria. I. Immigration and settlement of postlarvae. *Aust. J. Mar. Fresh. Res.* 31: 635-653.

Staples, D.J. 1980b. Ecology of juvenile and adolescent banana prawns, *Penaeus merguiensis*, in a mangrove estuary and adjacent offshore area of the Gulf of Carpentaria. II. Emigration, population structure and growth of juveniles. *Aust. J. Mar. Fresh. Res.* 31: 653-665.

Teng, Seng Keh 1971. Observations on certain aspects of the biology of *Metapenaeus brevicornis* (H. Milne Edwards) and *Penaeus merguiensis* (de Man) in the Brunei estuary system. M.Sc. Thesis University of Singapore, pp. 125.

Thomas, M.M. 1974. Reproduction, fecundity and sex ration of the green tiger prawn, *Penaeus semisulcatus* de Haan. *Indian J. Fish.* 21: 152-163.

Tuma, D.J. 1967. A description of the development of primary and secondary sexual characteristics in the banana prawn *Penaeus merguiensis* de Man (Crustacea: Penaeidae). *Aust. J. Mar. Fresh. Res.* 18: 73-88.

Estimating egg production in multi-brooding populations

Adrian M. Wenner & Jenifer Dugan
Department of Biological Sciences, University of California, Santa Barbara, USA

Harrington Wells
Department of Biology, Tulsa University, Oklahoma, USA

ABSTRACT

Crustacean biologists have traditionally described egg production as a function of size without regard for the year classes present in the population sample. That is because age determination of crustaceans is very difficult, due in part to the molting process by which all hard parts are lost with each increment in growth. By the use of probability graph paper, one can often separate year classes from one another in an analysis of size-frequency distributions. If one knows the length of the reproductive season and egg development time, the number of broods per year can also be deduced for each year class. Egg counts can then be made and the egg number data for each year class can be plotted on log-normal graph paper as a function of crustacean size (expressed as a percentile) within each year class. The mean egg number per mean size of animals in a particular age class, coupled with the proportion of each year class in the population and the number of broods within each year class, then permits estimation of total egg number produced per representative 1000 female crustaceans for any given year and/or locality. The same technique can yield estimates of egg production for a cohort of crustaceans which produces only one brood per year but more than one brood per lifetime.

1 INTRODUCTION

Age determination of individuals in many crustacean species is difficult, especially in those cases where the correlation between age and size has not been established. Crustaceans grow by molting, a variable incremental growth process in which all hard parts are lost with each increase in size (see contributions in Wenner 1985 for details). The crustacean molting process thus results in the loss of anatomical structures by which age can be determined; also, tags placed on individuals may be lost with each molt. Consequently, even basic ecological questions dealing with age-fecundity relationships in crustaceans become more difficult to resolve.

That difficulty in resolution of egg production patterns is particularly prevalent in species where more than one brood is produced per year by some individuals, where egg number per given size or year class varies with season, and/or where crustaceans live more than one year (e.g. Paris & Pitelka 1962, Penn 1980). However, it is still possible to estimate the number of eggs produced by a representative number of individuals in a population by a variety of techniques. To illustrate one such technique we have herein estimated, by means of a graphical technique, egg production for several species of crustaceans which have different patterns of reproduction.

1.1 *Graphical analysis*

Size frequency distributions can be analyzed and separated into constituent modal units (e.g. molt and/or age classes) by the use of normal probability graph paper (e.g. Harding 1949, Cassie 1950, 1954, Wenner & Fusaro 1979, Siegel & Wenner 1985). In addition, various aspects of egg production as a function of size or age can be characterized quantitatively by probit analysis (e.g. Wenner et al. 1974, Siegel & Wenner 1985, Wenner et al. 1985, 1987). Probit analysis has already proven valuable in studies of the influence of environmental factors on crustacean population ecology (e.g. Wenner & Fusaro 1979, Fusaro 1978, Siegel 1985, Wenner 1988). By this technique one can quantitatively compare representative egg production between populations or within a population through time.

1.2 *Initial assumptions*

The probability graph paper approach to analysis of size frequency distributions requires only that the organism studied has a restricted period of recruitment and/or a growth rate which varies with season, as well as a normally distributed size within each modal size-age class. Seasonal recruitment, of course, is a common phenomenon, as are normally distributed quantitative traits. Normally distributed quantitative traits are the product of multiple genes and/or environmental factors acting in an additive fashion in determining each character (see Johannsen 1903, Yule, 1906, Nilsson-Ehle 1909, East 1916, Mather 1949, Stern 1953, Falconer 1960; see also the appendix in Wenner et al. 1974). Finally, statistics associated with each size or age group can be obtained even with greatly overlapping size-age class modes in populations by using the probability graph paper modal separation techniques (e.g. Harding, 1949, Cassie 1950, 1954).

1.3 *Extension of graphical probit analysis*

Our contribution proposes an extension of the graphical analysis approach initially proposed by Harding (1949) and Cassie (1950, 1954). They applied their technique to the modal analysis of size frequency distributions. Once modal size classes have been separated by that technique, one can also consider population characteristics within each mode (e.g. egg production as a function of size). That is, if individual sizes in a sub-population are normally distributed and if egg production increases logarithmically with a linear increase in size, then egg production is a log-normal function. Log-normal graph paper is available to assist in that extension of the Harding-Cassie technique of probit analysis and readily yields a geometric mean for egg number.

Herein, then, we consider the applicability of using the graphically derived geometric mean for estimating number of eggs in species which have indeterminate growth. For those forms, egg production as a function of size can be displayed as a straight line on log-normal probability graph paper. That is because animals with indeterminate growth may have an approximately exponential increase in 'clutch size' or ovary weight when compared with a linear measure of organism size (see Wenner et al. 1987 and Somers this volume). The exponential relationship is not surprising when one considers length to volume ratios (e.g. allometry of growth; Sneath & Sokal 1973) whenever egg size is relatively constant.

2 POSSIBLE PATTERNS AND PROBLEMS INHERENT IN EACH CASE

Consider the following levels of complexity:

2.1 *One year class, one brood per year, fixed number per brood*

Perhaps the most simple case exists when female organisms produce only one brood in their lifetime, live only one year, and produce only a fixed number of eggs during that time (as in many insects and in some crustaceans). To estimate the total number of eggs produced in a representative population, one then need only multiply the total number of females present by the mean number of eggs produced per female.

2.2 *One year class or modal size class, two or more broods per season, egg number a function of size*

A slight change in conditions introduces additional important complications, even with only a single year class; the passage of time provides for growth as well as mortality. While there is an increased number of eggs in later broods due to the larger size attained during that growth, there is also a smaller number of females available for egg production due to mortality during that passage of time. An estimate of the mean number of eggs produced in each of the season's broods can nevertheless be obtained by plotting the number of eggs per female as a function of percentile rank in the population with the use of lognormal graph paper (see example in Section 3.2).

2.3 *Two or more broods per season, two or more year classes, variable number of eggs per brood, egg number a function of size*

Under this set of circumstances, estimates of total egg production for a population can still be made but only if one thoroughly understands the biology of a species. Certain assumptions must be made for calculations of population fecundity; the more information available, the fewer and the less tenuous the assumptions (see Sections 3-5).

2.4 *Variation in egg size*

For most crustaceans, any differences in egg size between populations appears to be relatively unimportant compared to other factors which contribute to variation in egg numbers. Egg numbers can vary by as much as an order of magnitude for a given size of female at any one time, while egg size usually varies by only a fraction of its diameter.

When a female crustacean grows after sexual maturity, its relative growth often changes little (e.g. Hartnoll 1982), but many environmental factors may select for an optimal egg size (e.g. Smith & Fretwell 1974, Brockelman 1975, Lawlor 1976). Environmental factors include predation, parasitism, various aspects of the physical environment, and physiological dynamics of metabolism within the egg (e.g. Lack 1947, Svardson 1949, Gotto 1962, Kaplan & Salthe 1979).

The most important consideration for our estimates of population fecundity is that egg mass volume (one measure of egg number) can rise geometrically with any linear measure of adult growth. Our proposed technique to estimate fecundity thus rests on the assumption that variations in egg size are not appreciable when compared to the much broader variation in egg number found in different environmental situations. For illustration, we will use species which have a relatively constant egg size.

3 ESTIMATING EGG NUMBER WITHIN YEAR OR SIZE CLASSES

In any consideration of egg number as a function of size, one should ideally compare the estimated number of eggs to the volume of the female, since both involve a three-dimensional mass. However, that is not usually done; measurements made of females, although sometimes expressed in terms of weight, are more often linear and expressed as either carapace (or body) length or width. Occasionally, however, one finds a volume estimate, 'carapace length cubed,' on the abscissa of a graph (see Somers, this volume).

There is apparently no convention among crustacean biologists when it comes to displaying egg production data. The result has been a paucity of theory development, a circumstance encountered less often in other areas of fisheries biology than in crustacean studies (Somers this volume). This shortcoming can be corrected.

An exponential relationship may exist between egg mass and 'size' (linear width or length) of an individual. Consequently, describing this geometric process (an exponential increase in egg number with a linear increase in size) by a geometric mean, rather than an arithmetic mean, may quite effectively portray relationships in data sets (e.g. Croxton & Crowden 1944, Meyer 1975). That reasoning constitutes the theoretical argument.

We suggest here that the deliberate use of the geometric mean should be more widespread than it had been in studies of size or age related fecundity. For unimodal distributions, both the exponential relationship between fecundity and normally distributed size/age classes and the geometric mean and variance for egg number can be estimated in a single exercise by using log-normal probability graph paper or its equivalent display (i.e. conversion of size-frequency percentiles to probits). The ease of use and intuitive result provides a practical argument for employing the geometric mean.

3.1 *Percentile conversions and polymodality*

Size frequency distributions are often polymodal. Normal probability graph paper displays facilitate (and simplify) analyses of population structure by revealing the presence of polymodality in those distributions. Polymodal distributions can then be split into component size/age classes for meaningful calculations of egg number as a function of size and/or age (e.g. Wenner et al. 1987). The graphical analysis technique permits a splitting of year class or size class modes and provides a simple basis for estimating the mean and variance in both size and egg number for each mode.

3.2 *Log-normality and log-normal graph paper*

The analysis of size or age related fecundity in a population, as well as effects of environment on egg production and adult survival, can be further clarified by the use of log-normal probability graph paper or its equivalent (i.e. conversion of percentiles to probits). The technique is analogous to the use of normal probability paper for problems associated with the study of population ecology (e.g. in the analysis of size-frequency distributions). The technique advocated here also corresponds, in a graphical sense, to that same concept as proposed earlier by Sokal & Rohlf (1981: 421) 'Logarithmic transformations are frequently needed in the analysis of variables related to the growth of organisms...'

In the following sections we provide discrete examples of the analysis of fecundity with the aid of log-normal graphical displays. We start with the least complicated situation, that of an analysis of egg production in a population sample which exhibits a unimodal population structure and a logarithmic increase in egg number (ovary weight in this case), coincident with linear increase in size.

3.3 *Egg production in a population of penaeid shrimps*

Data gathered by Penn (1980) on prawn fecundity can be used to illustrate a simple case in which mean egg number can be estimated for a population sample with the use of log-normal graph paper.

3.3.1 *Normality of size frequency distribution in the sample*

Penn measured 38 western king prawns (*Penaeus latisulcatus*) and weighed the ovaries from each of them. When size frequency data (extracted from his Fig. 2) are plotted in the conventional histogram format (Fig. 1), it appears that his particular sample was normally distributed.

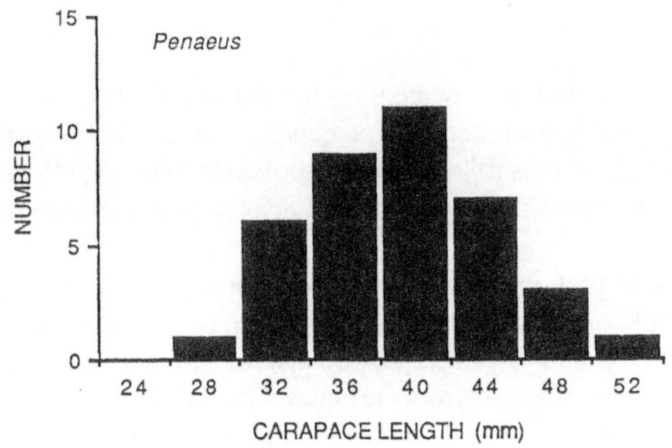

Figure 1. Size frequency distribution for data gathered on *Penaeus latisulcatus* along the Western Australian coast (Penn 1980). The data appear to be normal when grouped in this manner.

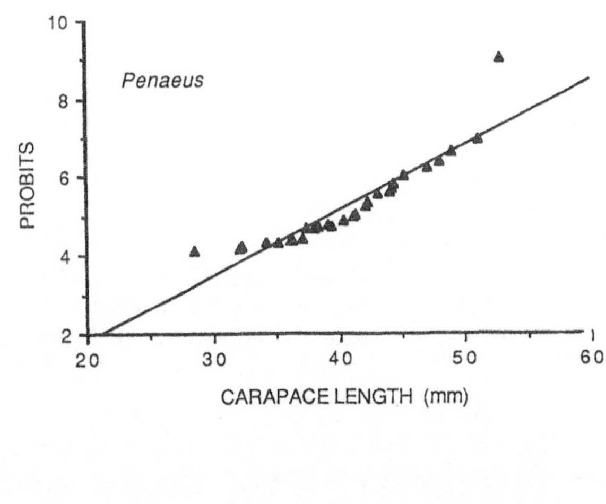

Figure 2. Percentile plot of data shown in Figure 1, when probits are expressed as a function of carapace length. Again, the data are close to normal.

When those same data are converted to percentiles and plotted on probability graph paper (Fig. 2), it is evident that his sample was in fact normal, since the percentiles (expressed here as probits) fall on a straight line.

The large apparent deviation for a few points at either end of the display matters little during line fitting; in probit displays, values at the extreme ends of the display represent only a very small percentage of the data. In the event of any doubt about closeness of fit, one can run either a Lillifor's test for normality (Conover 1980) or some equivalent test. (In the Penn example, the greatest deviation from a line representing normality, permitted at the 1% significance level by the Lillifor's test, is 14.3%. By contrast, the greatest deviation from the line in Figure 2 is only 6%).

3.3.2 *Egg production as a logarithmic function of size*

The next step in illustrating the process is a plot of the egg number data as a function of prawn length on semi-logarithmic graph paper (Fig. 3). For this comparison, we have plotted an estimate of egg numbers rather than ovary weight, using the conversion value provided by Penn (1980: 25). The display obtained indicates that a logarithmic function represents the data quite well. (See Somers this volume for a comparison of various displays, including egg number versus logarithm of body size).

In summarising the previous analysis, two facts are evident: the size frequency distribution is normal, and egg number increases with size in an exponential manner. It follows from those two facts that the Penn fecundity data form a log-normal distribution.

3.3.3 *Lognormality*

Once the size frequency data have been converted to percentiles (as was done in the test for normality), egg number can be plotted directly as a function of percentiles on log-normal graph paper (Fig. 4). The techniques proposed here thus follow those described by Harding (1949) and Cassie (1950, 1954) for use of normal probability paper in the display of size frequency distributions.

In the present circumstance, one can plot egg number as a function of size frequency percentile rank in the population sample or subsample (rather than directly as a function of linear size measurement) on log-normal probability graph paper. Then the geometric mean of the egg numbers $X_g = (\Pi X_i)^{1/N}$ can be obtained directly from the graph at the 50% abscissa level (or 5 on the probit scale). The standard deviation (S_g) corresponding to the log (X_g) is the

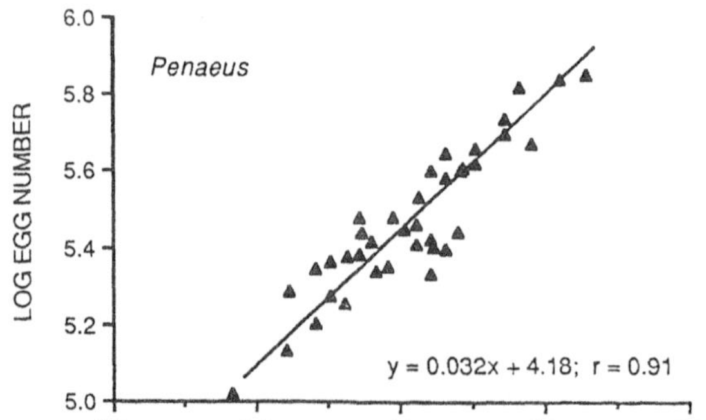

Figure 3. Logarithm of egg number as a function of size (CL) for *Penaeus latisulcatus* (data from Penn 1980). A semi-logarithmic plot accurately depicts the relationship.

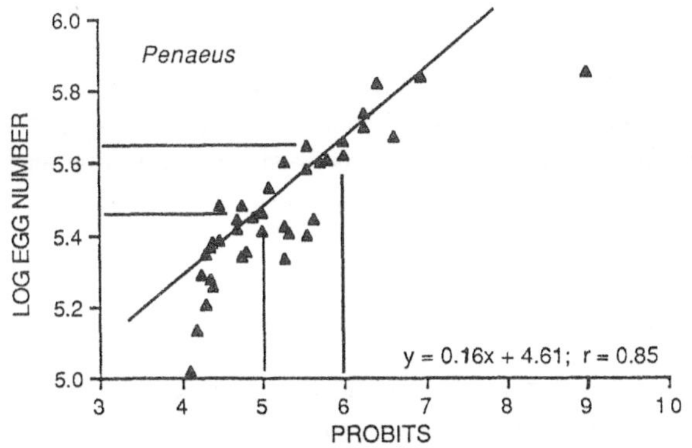

Figure 4. Log-normal plot of the relationship shown in Figure 3. Log of the egg number is shown as a function of percentile rank in the population. One set of lines delineates the mean value (probit = 5); the other set of lines is one standard deviation removed from the mean value for egg number.

difference between the intercepts at the 84% (or 16%) and 50% abscissal levels (6 or 4 on the probit scale, respectively).

In this particular display (Fig. 4), the percentiles were converted to probits by use of standard conversion tables. That conversion renders the percentile scale linear and thus permits calculation of linear regressions.

The mean logarithm of egg number for this example ($\text{Log } X_g = \sim 5.42$) can thus be read directly off the ordinate of the graph and represents the estimated egg number produced by the mean size of female (i.e. corresponding to the 50th percentile on the abscissa and expressed here as a value of 5 on the probit scale).

Standard deviations can also be calculated from values provided in the graphical display. That can be done for this example by finding the logarithm of the estimated number of eggs equivalent to the 84% level on log-normal graph paper (or a value corresponding here to 6 on the probit scale) and subtracting the mean logarithm of egg number from that estimate. The net result is that the logarithm of mean egg number can be expressed as follows:

$$\text{Log } X_g = \sim 5.42 + 0.13$$

4 ANECDYSIS EXAMPLE: PRIMIPAROUS VS MULTIPAROUS TANNER CRABS

Majid crabs differ from most other decapods in that they have a terminal molt at maturity (O'Brien & Van Wyk 1985: 199). Egg number thus does not increase with growth, since growth no longer occurs; nevertheless, egg number can vary with time in successive broods after crabs reach adulthood.

Majid crabs, with their terminal molt, thus provide good material for some types of comparative studies. Primiparous crabs (those bearing their first brood) have molted recently and can usually be distinguished from older crabs by the lack of epiphytes on their carapaces. Multiparous females with established epiphytes on their shells, by contrast, are likely brooding their second or later batch of eggs. The distinction between primiparous and multiparous females permits direct comparisons between the mean clutch size of first brood and the mean clutch size produced in subsequent broods.

Somerton & Meyers (1983) compared egg production in tanner crabs (*Chionoecetes bairdi*) which had different breeding histories; they provided us with their original data for the analysis which follows.

4.1 *Unimodality of the data*

Although a slight trimodality could be perceived in the Somerton & Meyers data, overall the tanner crab size frequency distribution data did not deviate significantly from a normal distribution ($p < 0.01$, Lillifor's test for data in Fig. 5). Since no difference in size distributions was evident in the two female categories, data for primiparous and multiparous females were combined in the next step.

4.2 *Egg number data as a logarithmic function*

A plot of egg number as a function of size for both primiparous and multiparous females (Fig. 6) indicates that a semi-logarithmic plot is also a good representation of those data. (Somerton & Meyers used a curvilinear plot on linear graph paper). This stage in the analysis process, however, can be omitted if one uses a log-normal display, as illustrated in Section 4.3.

4.3 *Logarithm of egg number as a function of percentile rank*

Figure 7 shows egg number as a function of percentile rank in the size frequency distribution of

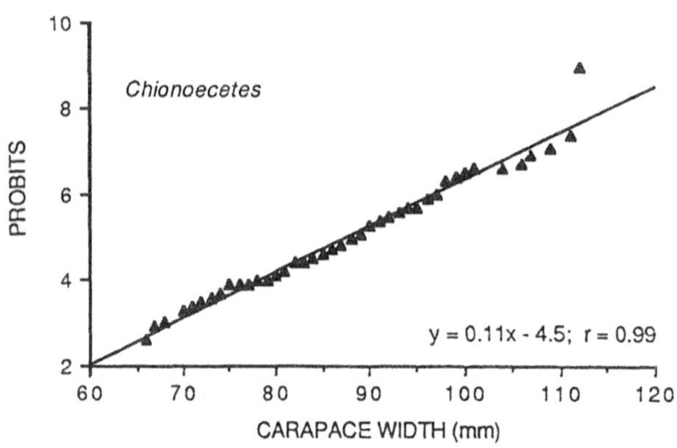

Figure 5. Percentile ranking in the population as a function of body size (CW) for a tanner crab (*Chionoecetes bairdi*) population (data from Somerton & Meyers 1983). The data are clearly normal.

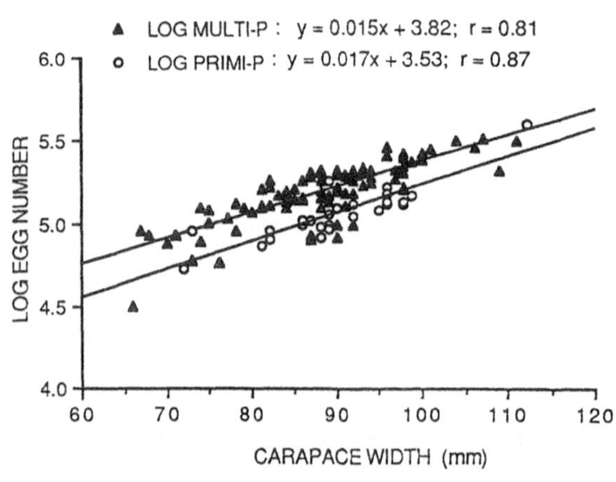

Figure 6. A semi-logarithmic plot for egg number as a function of crab size (CW) in tanner crabs. On average, the first brood (primiparous) contained fewer eggs than subsequent broods.

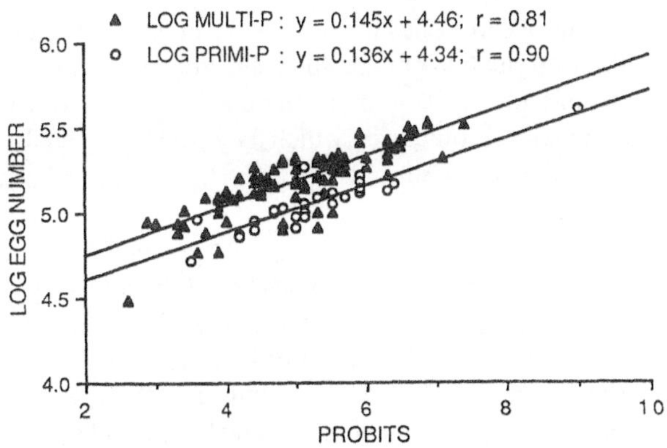

Figure 7. The same data shown in Figure 6 as a log-normal plot; the logarithm of egg number is expressed as a function of percentile rank in the population.

transformed log-normal graph paper. One can now extract means and standard deviations for the two variables directly off the axes.

If log-normal graph paper had been used rather than its equivalent transformations as shown in Figure 7 (logarithm of egg number as a function of probit rank), the mean and standard deviation for size of each type of egg bearing females would have been more evident, since those would be the size measurements corresponding to the percentile values found at the 50% and 84% (or 16%) levels. By running perpendicular lines to the ordinate, as shown in Figure 4, the geometric mean and standard deviations for egg number could be derived simultaneously from the same display. For the example shown, those values are as follows:

	Primiparous	Multiparous
Mean egg number X_g	1.01×10^5	1.64×10^5

On the basis of mean values, Somerton & Meyers calculated that primiparous females had produced 70% as many eggs as multiparous females. By a direct reading from the original graph on log-normal paper, we obtained the value of 62%.

5 POLYMODALITY AND FECUNDITY

The examples treated above involved only unimodal data sets. However, the graphical technique presented also allows geometrically-based polymodality to be recognized readily and treated quantitatively after modal separation.

Strictly arithmetic manipulations give only a single mean for a population sample. However, that arithmetic mean is misleading, particularly if it represents data from a skewed, polymodal, or otherwise non-normal size frequency distribution. The technique advocated here (a graphical representation of all data on log-normal graph paper) can permit a more accurate characterization of results. After modal separation, each mode can be treated as a normal distribution, permitting meaningful seasonal, geographic, and yearly population comparisons (Wenner 1988).

5.1 *The Paris and Pitelka study of pillbug fecundity*

Paris & Pitelka (1962) were the first investigators to estimate the total number of eggs which could be produced by a field population of crustaceans. They recognized the presence of more than one year class in their samples, the likely production of more than one brood per year, and the complications those factors introduced into the problem of estimating total egg number. Although they lacked the graphical probit analysis technique by which they could have analyzed size frequency distributions, they nevertheless published a paper which has not been excelled in substance since that time.

The pillbugs which Paris & Pitelka studied (*Armadillidium vulgare*) apparently live an absolute maximum of 4 years and have a high mortality rate, with only about 11% surviving to the end of their first year. That rate matches what has been found for a variety of other animals (Hoenig 1983). By measuring several factors, Paris & Pitelka recognized that two broods could be produced within a single season by mature females. Their data also indicated that a smaller percentage of mature females produced eggs during the time of the second brood, and that the number of eggs produced varied considerably between broods. Brood sizes also increased with age and with size of females, coincident with a decrease in the number of reproductive females with time as a consequence of mortality.

5.2 *Sand crab egg production*

In their pioneering study Paris & Pitelka were able to conclude that the second year age class contributed the greatest number of recruits to the population and that the large first year age class made the second greatest contribution. The next example extends their analysis and also illustrates the value of using the polymodal graphical analysis technique to estimate the mean number of eggs within each mode for a population sample. The crustacean of choice was the common sand crab (*Emerita analoga*). For simplicity, a single sample will be considered, one obtained at one date during the reproductive season (late July).

5.2.1 *Polymodal size-frequency histogram*

The set of data shown in Figure 8 is a fairly typical condition encountered in any crustacean population sample in which there is either a series of recruitment pulses during one season and/or in which there is a multiple number of year classes. For sand crabs, laboratory and field studies of growth rates (Siegel & Wenner 1985) indicate that sand crabs live a maximum number of 3 years on the beach. Data representing more than one year class thus appear in the histogram.

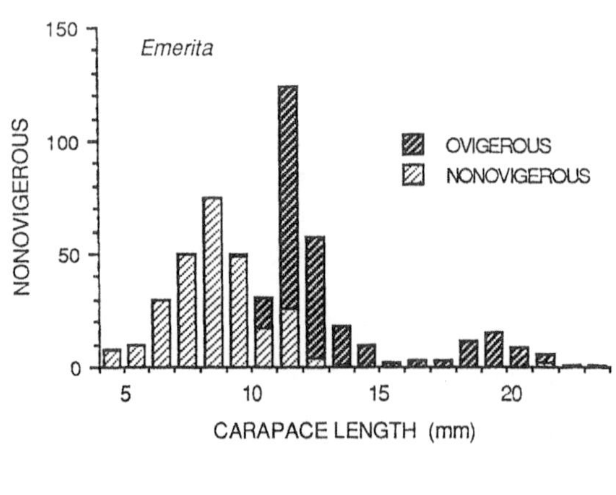

Figure 8. Size frequency distribution for a population sample of common sand crabs (*Emerita analoga*) collected in late July 1982 at Goleta Beach, California.

In the spring one can often readily distinguish between those modes which represent third year and second year female sand crabs. In addition, the large number of newly arriving recruits dominates the size frequency histogram obtained. For the data shown (late July), however, third year female crabs usually can no longer be distinguished from second year crabs in the upper mode of the polymodal distribution. By mid-summer first year females (5-15 mm CL) had separated into two minor overlapping modes, with most of the larger of those individuals bearing eggs. The double peak seen in the data for first year female crabs may represent two separate influxes onto the beach during spring recruitment from the plankton.

5.2.2 *Evidence of strong bimodality*

When percentile rank in the population (expressed as probits) is plotted as a function of female length (Fig. 9), two major modes emerge and correspond to the two year classes known to prevail on the beach at that time of year in southern California (Siegel & Wenner 1985). At localities north of Point Conception, all three year classes may remain distinct throughout the reproductive season.

The two constituent modes can be separated by the Cassie-Harding technique (not shown here). After that separation, one can calculate regressions for each set of points (Wenner et al. 1987), as well as testing each set for normality by application of the Lillifor's test (Conover 1980).

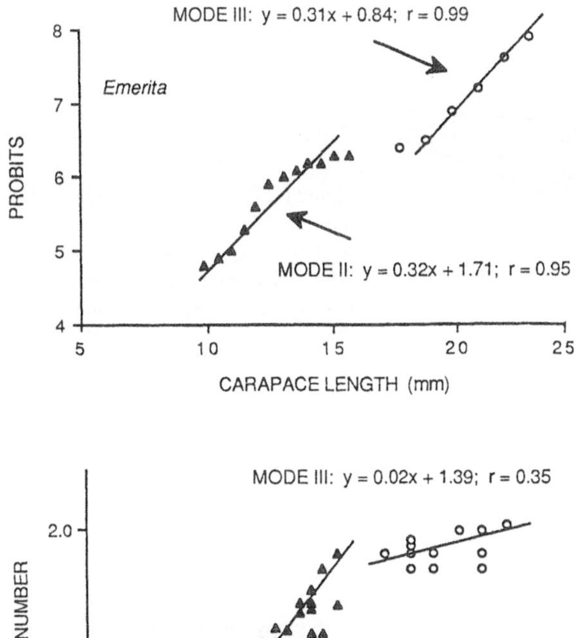

Figure 9. Percentile rank in the population plotted as a function of size (CL) for the data shown in Figure 8. This modal split permits examination of egg production within each year class (Wenner et al. 1987); see also Figure 10.

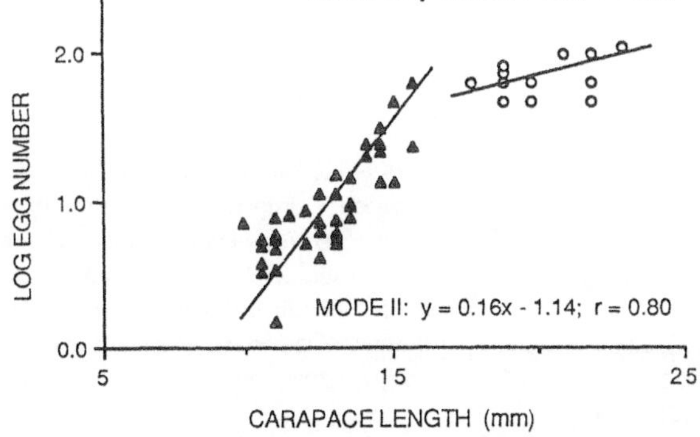

Figure 10. Semilogarithmic plot of egg number as a function of size for the egg bearing population shown in Figure 9.

5.2.3 *Logarithm of egg number as a function of size*

After distinguishing between the two modes (age classes in this case), one can plot egg number as a function of size (Fig. 10) on semi-log graph paper (or as a semi-log plot with the appropriate computer program, as shown). However, this step can be eliminated completely by moving directly to use of the log-normal display (next section).

5.2.4 *Log-normal representation*

By plotting egg number as a function of percentile rank in the sub-populations on log-normal graph paper, both mean number of eggs and mean size of female (together with standard deviations for both) can again be read directly from one display. If the plot is done instead by computer as shown in Figure 11 (with logarithm of egg number plotted as a function of size frequency percentiles expressed as probits), regression lines can be fitted to the data and provide the opportunity to make quantitative comparisons between populations.

5.3 *Additional considerations for estimating annual egg production*

5.3.1 *Egg development time*

As with the Paris & Pitelka study, we had to know how many broods could be produced within a reproductive season. To estimate number of broods possible during a season we measured the

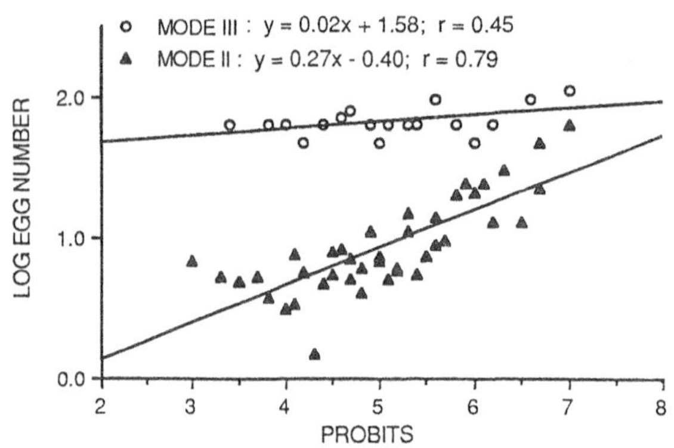

Figure 11. The same data as shown in Figure 10 but displayed as a log-normal plot, with log of egg number plotted as a function of percentile ranking in the population. Mean and standard deviation of each variable for each age class can now be read directly off the graph.

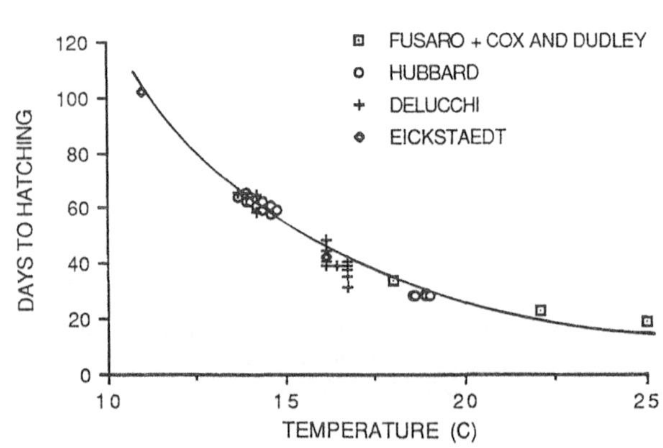

Figure 12. Egg development time as a function of temperature for common sand crabs (*Emerita analoga*). Data are from various sources (see text). Displays such as this permit estimation of maximum number of broods possible during the reproductive season.

time required for egg development at different temperatures and now have good estimates for sand crabs (Fig. 12). The unpublished data were obtained by Carla Delucchi, Craig Fusaro, and David Hubbard and were supplemented by a few readings obtained at more extreme temperatures by Cox & Dudley (1968) and Eickstaedt (1969).

Egg development time is, of course, highly dependent on temperature (Q-10 law), but water temperatures tend to remain fairly constant at any one site along the central California coastline during the reproductive season. In Santa Barbara, for example, temperatures rarely move outside the 15° to 20°C range. Brood development time in that locality would thus be approximately 40 days. Therefore, approximately 3 broods could be produced by mature female crabs during the reproductive season in Santa Barbara.

5.3.2 *Seasonal pattern of egg production by each generation*

Given all of the above information, supplemented by growth rate studies in laboratory and field (Siegel & Wenner 1985), we present a model depicting the reproductive pattern for mature female sand crabs (Fig. 13).

Most sand crab recruits arrive as megalopa during the winter and early spring. Some young-of-the-year female crabs become reproductively competent by the time older female crabs are producing their second brood of the season. By the end of the first summer, during favorable years in the Southern California Bight, a high percentage of those first-year female crabs can be bearing eggs.

During their second year on the beach, surviving female crabs can produce a full three broods (Fig. 13). However, egg production is not synchronous; hence, separations between broods are not evident in the percentage of females bearing eggs. those crabs which survive into their third year produce one brood and perhaps a second, but their numbers are considerably fewer due to mortality in the intervening months. The few third-year female crabs present along southern California beaches by the time of the second brood are often no longer distinguishable as a distinct mode, since second-year female crabs have grown faster than third-year crabs and mask the presence of the latter.

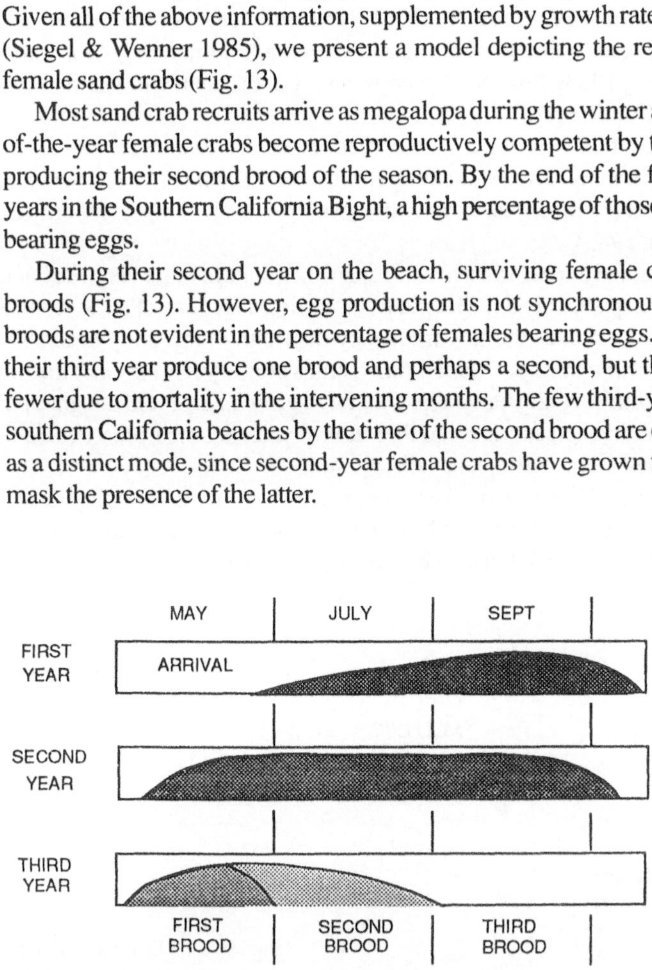

Figure 13. Egg production for long-lived sand crabs during their lifetimes. The height of any part of the shaded portion indicates relative percentage of populations brooding eggs at any one time. The two broods produced by first-year female crabs and the three broods produced by second-year female crabs overlap one another and permit no separation in this display. The separation shown for third-year females depicts the fact that an ever smaller percentage of females remains in the habitat.

5.3.3 *Variation in clutch size during the season*

Eickstaedt (1969) found striking differences in the number of eggs produced by a given size of female during the reproductive season. We found the same differences to prevail for every year in which we have counted eggs throughout the season (Wenner, Dugan & Hubbard in prep.). Considerably more eggs were produced in the second brood than in the first. Egg number dropped again during the third brood of the season but not to the spring level.

6 POLYMODALITY, LONGEVITY AND MORTALITY

Estimates of longevity and/or mortality must be included in any calculation of egg production by a population. Because of their growth by molting, any attempt to estimate mortality rates for crustaceans can be a most frustrating exercise. Studies by Hoenig (1983), on the other hand, provide the basis for generating an estimate of mortality when maximum longevity is know. Figure 14 (from Fig. 1 in Hoenig 1983) illustrates the consistency of the data he has gathered on mortality as a function of maximum age.

Paris & Pitelka (1962) found that only about 11% of recruits to the habitat were still alive a year later and that 4 years was the maximum longevity for *Armadillidium vulgare*. That value matches what one would expect from a comparison with mortality rate and longevity comparisons for other species (Hoenig 1983).

The sand crabs we have studied appear to have a life span and mortality rate very similar to *Armadillidium* and to the other species used in Hoenig's comparison. By June of each year, the first year class (new recruits) dominates the size frequency distribution on sandy beaches in southern California. After that time third year female sand crabs are very hard to find; the few left cannot be distinguished from second year crabs, who have grown faster and caught up in size. A very few sand crabs could even survive into their fourth year, but their presence could not be detected by modal analysis of size frequency distributions.

The above considerations thus allow us to make some assumptions about mortality rates in sand crab populations. A mortality rate of 90%-99% per year would match those rates reported by Hoenig (1983) for other animals which live a maximum of 3-4 years.

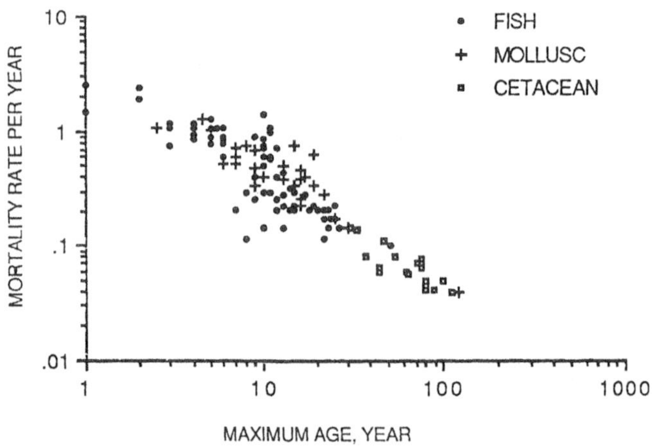

Figure 14. Derived mortality rates for a variety of animals plotted as a function of maximum age of each species (from Fig. 1 of Hoenig 1983).

7 ESTIMATING TOTAL EGG NUMBER/YEAR FOR A REPRESENTATIVE POPULATION

The foregoing treatment provides information on several of the components necessary for estimating total egg production in multi-brooding sand crab populations, as follows:

1) Populations with a polymodal size frequency distribution may prove amenable to modal separation.

2) Knowledge of egg development time may be useful for determining the number of broods which can be produced within a reproductive season. If the ovary of ovigerous females is ripe, another brood can be produced after egg release (Kuris 1971).

3) The number of eggs produced by a given size of female may differ within a reproductive season.

4) Mortality rate may be estimated if one knows the maximum age attained.

The next stage in the process involves determining the mean clutch size per modal size of female crab.

7.1 *Representative sampling*

Ideally, one should attempt to collect a representative quantitative sample of the population. In our sand crab studies, the sampling space is too large and crab distribution is too dynamic for accurate quantitative sampling. We could only estimate ratios in numerical abundance of different ages of crabs on the basis of maximum longevity (Hoenig 1983).

7.2 *Polymodality and percentiles*

Studies of growth rates in both laboratory and field may permit the assignment of modal classes found within the size frequency distributions of natural populations to actual age groups, as illustrated in Section 5.

7.3 *Mean number of eggs per age class*

Once modal size classes have been separated and new percentiles representative of the individual year classes have been obtained, the number of eggs found on each female can be plotted as a function of the percentile ranking (probit rank) of that female for each of the sub-populations.

Figures 15-17 illustrate the results of such an analysis for three 1985 Goleta Beach samples, representing in turn the mean number of eggs in each of the three broods of eggs which were produced by the mean size of female of each age class during that reproductive season. Estimates of mean egg number can now be read directly off the ordinate; standard deviations can be estimated on the basis of the difference between ordinate equivalents of the values at the 5 probit and the 4 to 6 probit levels.

Table 1 summarizes the values for mean egg numbers derived from those three figures in the example under consideration.

7.4 *Percentage of each year class producing eggs*

Knowing the mean number of eggs for the mean size of female crabs which produce eggs is not

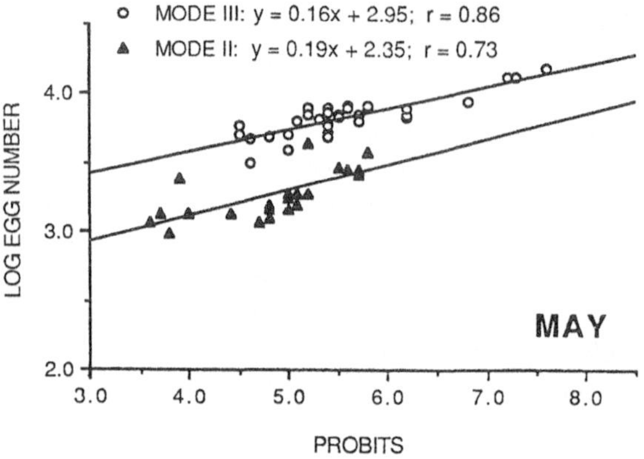

Figure 15. Log-normal plots for egg number as a function of percentile ranking in the population for sand crabs in May 1985 (see Figs. 16 and 17, Table 1).

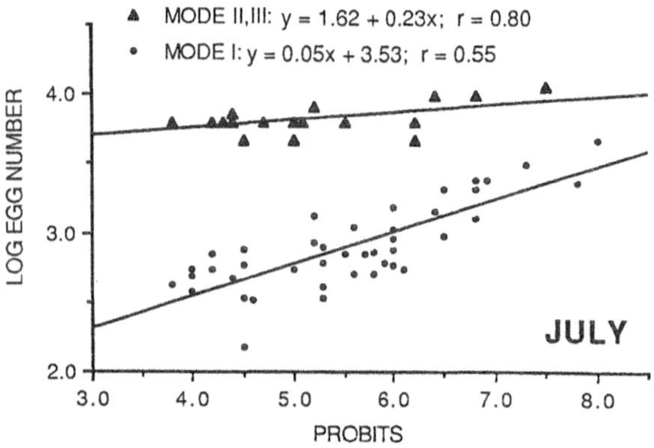

Figure 16. Log-normal plots for egg number as a function of percentile ranking in the population for sand crabs in July 1985. By July second and third year females become indistinguishable; at that time, however, first year females begin producing eggs. Logarithm of mean number of eggs for each age class can now be estimated directly for each brood (see Figs. 15 and 17, Table 1).

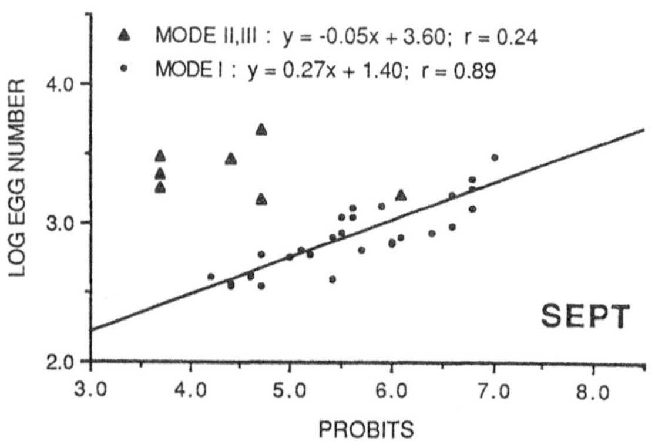

Figure 17. Log-normal plots for egg number as a function of percentile ranking in the population for sand crabs in September 1985 (see Figs. 15 and 16, Table 1).

sufficient for calculations of total egg production. One must also have an estimate of the percentage of females within each year class which produce eggs at each time of sampling (at the time of each of the three main broods in the present example).

Table 2 contains the percentages of females within each year class which carried eggs during

Table 1. Estimated mean number of eggs produced by each of the three year classes of sand crabs (*Emerita analoga*) in each of the three portions of the breeding season.

Year class	May (I)	July (II)	September (III)
1	–	6.3×10^2	5.5×10^2
2	1.9×10^3	6.5×10^3	2.2×10^3
3	5.3×10^3	[6.5×10^3]	[2.2×10^3]

Table 2. Estimated percentage of the three year classes of female sand crabs which brood eggs in each of the three portions of the breeding season.

Year class	May	July	September
1	–	42	81
2	88	96	92
3	93	?	–

the 1985 Goleta Beach sampling season. These values provided part of the rationale for construction of Figure 13.

7.5 *Basis for calculations, with examples*

We now have a composite set of information, as follows:

1) The number of broods which each year class can produce.

2) The percentage of females of each age class which produce eggs at the time of each brood.

3) The mean number of eggs the mean size of female can produce at each of the three times of year.

Altogether these numbers provide the basis for estimating total egg production in a multi-brooding population.

7.5.1 *Total egg production during the second year of life*

Table 3 contains estimates obtained for total egg production by an original hypothetical count of 1000 female sand crabs during the second year of their life. From Figure 13 we can see that a female in the second year of her life will produce approximately 3 full broods. However, at the 90% mortality rate per year which we postulated (on the basis of a steady, semi-logarithmic, decline in numbers), fewer than 100 female crabs remain at the beginning of the second year to produce those three broods, as indicated in the 2nd column of Table 3. The percentages of those crabs which survive and carry eggs are shown in Table 2 (middle row), and the mean numbers of eggs per mean size of female crab are shown in Table 1. Overall, the second brood of the season contributes more than one half of the total eggs produced in all three of the broods, despite the loss due to mortality of females between the time of the first brood and the time of the second brood.

7.5.2 *Total egg production in the second brood of the year*

With a 90% mortality rate per year, the number of females surviving within each year class decreases markedly with time (second column in Table 4). The higher representation by first year females is partially offset by the fact that more than one half of them have not yet begun to

Table 3. Comparison of the total number of eggs produced in each of the three broods by sand crab females in their second year on the beach.

Brood number	% alive	% ovigerous	Total eggs ($\times 10^5$)
I	90	88	1.5
II	62	96	3.9
III	42	92	1.0
Total			6.4×10^5

Table 4. Comparison of the total number of eggs produced by each of the three year classes of sand crab females during the second brood of the season.

Year class	N. alive	% ovigerous	Total eggs ($\times 10^4$)
1	620	42	2.6
2	62	96	38.7
3	6	(96)	3.9
Total			4.5×10^5

produce eggs (third column in Table 4). In addition, first year females are not as large as second and third year females and thus cannot produce nearly the number of eggs (Table 1).

The preceding calculations reveal that second year females dominated egg production in that example during the second brood of the year; they produced more than 5 times the number of eggs contributed by the first and third year females combined.

8 A MATHEMATICAL MODEL FOR EGG PRODUCTION IN CRUSTACEA

Assume that the crustacean species being modeled reproduced only at specific times during each year. Furthermore, the probability of reproducing, as well as fecundity of those individuals which do reproduce at a specified time, may be age related. Define a reproductive period as a specific time of the year at which eggs are released by females. Those females which do not survive to the time at which eggs are released do not effect the future of the species genetically or ecologically, and thus are not considered to have reproduced even though they may have mated.

Let E equal the total egg production of an individual over its life. However, total egg production is simply the sum of eggs produced at each reproductive period. Let T = time, and define $\{t_r | r = 1, \ldots, m\}$ the specific times at which reproduction can occur in the species. Let ε_T = the number of eggs produced by an individual at time T (note: $\varepsilon_T = 0$ when $T \neq \{t_r\}$. *A priori* prediction of E is dependent on the probability of the individual surviving to time T (denoted S_T), and the likelihood of reproducing at time T – given the individual is still alive at time T (denoted R_T). Total egg production by an individual over its life can thus be predicted through equation (1).

$$E = \sum_{T=\{t_r\}} S_T \, R_T \, \varepsilon_T \tag{1}$$

Two basic statistics are of interest in predicting population dynamics of Crustacea: 1) mean egg production per individual, and 2) the variance in individual total egg production. The expected egg production (mean = μ_E) is dependent on egg density per female. Let g_E be the probability that an individual produces E eggs in its life. Then:

$$\mu_E = \sum_{E=0}^{\infty} E g_E \tag{2}$$

Now, let f_{ε_T} be the probability that an individual produces ε_T eggs at time T, given that it is still alive and will reproduce. Total mean egg production can thus be expressed as (3) since the mean of a variable which is the sum of a set of variables is the sum of the mean of each variable of which it is composed, where μ_{ε_T} = mean egg production by a surviving reproducing individual during reproductive period T.

$$\mu_E = \sum_{T=\{t_r\}} S_T\ R_T \sum_{\varepsilon_T=0}^{\infty} \varepsilon_T\ f_{\varepsilon_T} = \sum_{T=\{t_r\}} S_T\ R_T\ \mu_{\varepsilon_T} \tag{3}$$

The variance in E, σ_E^2, can also be expressed as a function of the variance and covariance in egg production of individuals in each reproductive period. Denote the variance in egg production at time T of individuals surviving and reproducing as $\sigma_{\varepsilon_T}^2$ and the covariance in egg production between time T_1 and T_2 in these individuals as $\mathrm{cov}\left(\varepsilon_{T_1}\ \varepsilon_{T_2}\right)$. Then:

$$\sigma_E^2 = \sum_{E=0}^{\infty} g_E\ (E-\mu_E)^2 =$$

$$= \sum_{T=\{t_r\}} \left[S_T\ R_T\left(\sigma_{\varepsilon_T}^2 - \mu_{\varepsilon_T}^2\right) + \mu_{\varepsilon_T}^2\right] + \sum_{T_1,T_2=\{t_r\}} \sum_{T_1 \neq T_2} cov\left(\varepsilon_{T_1}\ \varepsilon_{T_2}\right) \tag{4}$$

Using the relationship that the correlation between X and Y, r_{XY}, is $r_{X,Y} = \mathrm{cov}\,(X,Y)/(\sigma_X\,\sigma_Y)$, the $\mathrm{cov}\left(\varepsilon_{T_1}\ \varepsilon_{T_2}\right)$ can be expressed as:

$$r_{\varepsilon_{T_1}\ \varepsilon_{T_2}}\left[S_{T_1}R_{T_1}\ (\ \sigma_{\varepsilon_{T_1}}^2 - \mu_{\varepsilon_{T_1}}^2) + \mu_{\varepsilon_{T_1}}^2\right]^{1/2}\left[S_{T_2}\ R_{T_2}\left(\sigma_{\varepsilon_{T_2}}^2 - \mu_{\varepsilon_{T_2}}^2\right) + \mu_{\varepsilon_{T_2}}^2\right]^{1/2} \tag{5}$$

Thus, mean (μ_E) and variance (σ^2_E) in total egg production can be calculated and can be expressed in terms of a set of easily estimated parameters.

Consider, as an example, the sand crab *Emerita analoga*. First, survival can be described as a density independent function of time (see prior discussion). Between 90 and 95% mortality occurs each year in adult sand crab populations. Values of S_T for the adult sand crabs at $\{t_r\}$ can be found in Table 5 or estimated, using the function $S_T = a^T$ where T = time in months and a = Darwinian fitness ($0.78 \leq a \leq 0.83$).

Second, sand crabs reproduce relatively synchronously and reproduce at most three times, each year, at predictable times, on average (end of May, end of July, end of September). Values of R_T for adult sand crabs at $\{t_r\}$ can be found in Table 5. Probability of reproducing appears to increase linearly during the first year a sand crab can reproduce: $R_t \cong 0.21.T$ where T = time in months, and T = 0 is the first May an adult can reproduce. Thereafter, the percent of sand crabs reproducing appears to fluctuate regularly about a mean of 94%. Estimation of the probability of third year adults reproducing in July and September when they are indistinguishable from

Table 5. Probability of egg production among female sand crabs as a function of age in months (i.e. time spent on beach).

Time (T)	Probability of survival to T (S_T)	Probability of repro-ducing at T (R_T)	Log-normal distri-bution, eggs $\mu_\in$	$\delta_\in$	Normal distribu-tion, length μ_λ	δ_λ	In (θ) = λ In (A) + In (B) relation of λ to ∈ In (B)	In (A)
0	1.000	0	0	0	17.0	4.0	–	–
2	.681	.42	5372	9862	31.0	3.2	-5.5508	.4305
4	.464	.81	7327	2246	37.0	3.1	4.9912	.1045
12	.100	.88	751	481	14.1	2.8	3.3526	.2271
14	.068	.96	6288	711	20.0	2.8	7.935	.0405
16	.046	.92	12537	159	37.0	3.2	7.935	.0405
24	.0100	.93	830	400	22.2	2.8	2.6849	.1773
26	.0068	(.96)	12207	1296	36.4	2.6	7.935	.0405
28	.0046	(.92)	12207	1296	36.4	2.6	7.935	.0405

second year adults thus can be obtained by use of the function $R_T \cong 0.94\text{-}0.02.\ \sin(\pi\,[T+1]/2)$: where the angle is in radians, T = time in months as before, and T > 4.5 (estimated $R_{26} = 0.96$, $R_{28} = 0.92$).

Third, eggs per surviving reproducing female is correlated with age and sand crab length (see prior discussion). The logarithm of number of eggs carried by a female is directly proportional to sand crab length within an age class (log-linear relation). Since length in an age class is normally distributed, then *ipso facto*, eggs per female is log-normally distributed in each age class. (In fact, both a true log-log and a true log-linear relation between length and eggs exists simultaneously). Thus, $f_{\varepsilon T}$ can be approximated using a log-normal density ϕ_{ε_T} so that $\mu_{\varepsilon_T} = \int \varepsilon_T\, \phi\varepsilon_T\, d\varepsilon_T$ and $\sigma^2_{\varepsilon_T} = \int \left(\varepsilon_T - \mu\varepsilon_T\right)^2 \phi_{\varepsilon_T}\, d\varepsilon_T$. Values of μ_{ε_T} and σ_{ε_T} given in Table 5 were calculated, based on the relation of eggs to sand crab length and age.

Assuming covariances are 0, using Equations (4) and (5), and values of S_T, R_T, μ_{ε_T}, σ_{ε_T} found in Table 5: mean egg production of adult sand crab females $F128\mu_E$ (sand crabs which survive at least to May of their first reproductive year) is 5385, with a standard deviation (σ_E) of 20 511. Therefore probability of surviving from hatching to an adult (May of 1st reproductive year) is approximately 0.0004 (assuming 1/2 of survivors are male) given assumption that migration does not occur and population size is stable.

Consider a theoretical harvest situation using the *Emerita* data with 50% of the individuals present harvested in May, in July and again in September. How would fecundity then be affected? If harvest consisted of taking individuals at random, then mean egg production per adult entering the breeding population for the first time would be 50% of normal (5385/2 = 2693). However, if larger individuals were selected in harvesting by sieve or net size, then by altering the distribution of reproducing individuals (lower half of normal curve for carapace length) mean egg production can be shown (Equations 4 and 5) to be only 30% of normal mean eggs per female (1756). In fact, if carapace length between age groups is correlated, fecundity would be lowered still further.

9 CONCLUDING REMARKS

Estimating the total number of eggs produced through time by an individual, by a cohort of a population, or by a given number of females can be a very difficult task. The problem is particularly difficult whenever an individual female produces more than one brood per year and/or whenever an individual produces eggs for more than one year. An estimation of total egg production, however, can be obtained if one divides the problem into constituent parts; estimates of mean values for each of the variables under consideration can then be obtained and fitted into a composite estimate of egg production.

The technique illustrated in this chapter provides a means by which populations living in different habitats can be compared with one another. Such estimates can prove especially valuable when comparing populations in environmentally impacted localities with those living in more undisturbed areas. Fisheries management can also be improved with application of the estimating technique described here.

ACKNOWLEDGEMENTS

We thank D. Hubbard, M. Page and A. Kuris for their comments on the manuscript. Studies which provided the original data included herein were funded in part by the marine Review Committee, Inc., the CS Fund, the Lerner Gray Fund for Marine Research (to J. Dugan), and both the Marine Science Institute and the Faculty Research Committee of the University of California at Santa Barbara. Special thanks go to the core of student volunteers and helpers from the UCSB Sand Crab Lab through the years who played such an important part in furthering our understanding of sand crab biology. Most recently this list included K. Albright, J. Coyne, C. Culver, B. Hanley, A. Harris, S. Mason, T. McCallister, Y. Ricard, J. Shoffner, D. Shumate, T. Stevens, K. Vines and A. Wagner.

REFERENCES

Brockelman, W.Y. 1975. Competition, the fitness of offspring, and optimal clutch size. *Amer. Nat.* 109: 667-699.

Cassie, R.M. 1950. The analysis of polymodal frequency distributions by the probability paper method. *New Zealand Science Review* 8: 89-91.

Cassie, R.M. 1954. Some uses of probability paper in the analysis of size frequency distributions. *Aust. J. Mar. Freshwater Res.* 5: 513-522.

Conover, W.J. 1980. *Practical Nonparametric Statistics*. John Wiley and Sons.

Cox, G.W. & G.H. Dudley 1968. Seasonal pattern of reproduction of the sand crab *Emerita analoga* in southern California. *Ecology* 49: 746-751.

Croxton, F.E. & D.J. Crowden 1944. *Applied General Statistics*. New York: Prentice-Hall, Inc.

East, E.M. 1916. Studies on size inheritance in *Nicotiana*. *Genetics* 1: 164-176.

Eickstaedt, L.L. 1969. The reproductive biology of the sand crab *Emerita analoga* (Stimpson). Ph.D. Thesis, Stanford University.

Falconer, D.S. 1960. *Introduction to Quantitative Genetics*. New York: The Ronald Press Co.

Fusaro, C. 1978. Growth rate of the sand crab, *Emerita analoga* (Hippidae) in two different environments. *US Fish. Bull.* 76: 369-375.

Gotto, R.V. 1962. Egg-number and ecology in commensal and parasitic copepods. *Ann. Mag. Nat. Hist.* (13)5: 97-107.

Harding, J.P. 1949. The use of probability paper for the graphical analysis of polymodal frequency distributions. *J. Mar. Biol. Assoc. UK.* 28: 141-153.

Hartnoll, R.G. 1965. The biology of spider crabs: a comparison of British and Jamaican species. *Crustaceana* 9: 1-16.

Hartnoll, R.G. 1982. Growth. In L.G. Abele (ed.), *The Biology of Crustacea*, Vol. 2. New York: Academic Press.

Hoenig, J.M. 1983. Empirical use of longevity data to estimate mortality rates. *US Fish. Bull.* 82: 898-902.

Johannsen, W. 1903. *Uber Erblichkeit in Populationen und in reinen Linien*. Jena: G. Fischer.

Kaplan, R.H. & S.N. Salthe 1979. The allometry of reproduction: an empirical view in salamanders. *Amer. Nat.* 113: 671-689.

Kuris, A.M. 1971. Population interactions between a shore crab and two symbionts. PhD dissertation, University of California, Berkeley.

Lack, D. 1947. The significance of clutch-size. *Ibis* 89: 302-352.

Lawlor, L.R. 1976. Parental investment and offspring fitness in the terrestrial isopod *Armadillidium vulgare* (Lats.) (Crustacea; Oniscoidea). *Evolution* 30: 775-785.

Mather, K. 1949. *Biometrical Genetics*. London: Methuen and Co.

Meyer, S.L. 1975. *Data Analysis for Scientists and Engineers*. New York: John Wiley & Sons.

Nilsson-Ehle, H. 1909. Kreuzungsuntersuchungen an Hafer und Weizen. *Lunds Univ. Aarskr. NF Afd.* 5(2): 1-122.

O'Brien, J.J. & P.M. Van Wyk 1985. Effects of crustacean parasitic castrators (epicaridean isopods and rhizocephalan barnacles) on growth of crustacean hosts. In A.M. Wenner (ed.), *Crustacean Growth: Crustacean Issues* 3: 191-218. Rotterdam: Balkema.

Paris, O.H. & F.A. Pitelka 1962. Population characteristics of the terrestrial isopod *Armadillidium vulgare* in California grassland. *Ecology* 43: 229-248.

Penn, J.W. 1980. Spawning and fecundity of the western king prawn, *Penaeus latisulcatus* Kisinouye, in Western Australian waters. *Aust. J. Mar. Freshwater Res.* 31: 21-35.

Siegel, P.R. 1985. Food-induced size-specific molt synchrony of the sand crab, *Emerita analoga* (Stimpson). *Biol. Bull.* 167: 579-589.

Siegel, P.R. & A.M. Wenner 1985. Field and laboratory studies of sand crab growth. In A.M. Wenner (ed.), *Crustacean Growth: Crustacean Issues* 3: 165-170. Rotterdam: Balkema.

Smith, C.C. & S.D. Fretwell 1974. The optimal balance between size and number of offspring. *Amer. Nat.* 108: 499-506.

Sneath, P.H.A. & R.R. Sokal 1973. *Numerical Taxonomy*. San Francisco: W.H. Freeman & Company.

Sokal, R.R. & F.J. Rohlf 1981. *Biometry: The Principles and Practice of Statistics in Biological Research*. 2nd Ed. San Francisco: W.H. Freeman & Company.

Somerton, D.A. & W.S. Meyers 1983. Fecundity differences between primiparous and multiparous female Alaskan tanner crabs (*Chionoecetes bairdi*). *J. Crust. Biol.* 3: 183-186.

Stern, C. 1953. Model estimates of the frequency of white and near-white segregants in the American Negro. *Acta Genet. Stat. Med.* 4: 281-298.

Svardson, G. 1949. Natural selection and egg number in fish. *Rept. Inst. Freshwater Res. Drottningholm* 29: 115-122.

Warner, R.R. 1975. The reproductive biology of the protogynous hermaphrodite, *Pimelometopon pulchrum* (Pisces: Labridae). *Fish. Bull.* 73: 262-283.

Wenner, A.M. (ed.) 1985. *Crustacean Growth*. In *Crustacean Issues* 3: 1-362. Rotterdam: Balkema.

Wenner, A.M. 1988. Crustaceans and other invertebrates as indicators of beach pollution. In D. Soule & G. Kleppel (eds), *Marine Organisms as Indicators*. pp. 199-229. Berlin: Springer-Verlag.

Wenner, A.M. & C. Fusaro 1979. An analysis of population structure in Pacific mole crabs (*Hippa pacifica* Dana). *Biol. Bull.* 157: 205-220.

Wenner, A.M., C. Fusaro & A. Oaten 1974. Size at onset of sexual maturity and growth rate in crustacean populations. *Can. J. Zool.* 52: 1095-1106.

Wenner, A.M., D.M. Hubbard, J. Dugan, J. Shoffner & K. Jellison 1987. Egg production by sand crabs (*Emerita analoga*) as a function of size and year class (Decapoda, Hippidae). *Biol. Bull.* 172: 225-235.

Wenner, A.M., H.M. Page & P.R. Siegel 1985. Variation in size at onset of egg production. In A.M. Wenner (ed.), *Crustacean Growth: Crustacean Issues* 3: 149-164. Rotterdam: Balkema.

Yule, G.U. 1906. On the theory of inheritance of quantitative compound characters on the basis of Mendel's laws. *Rept. 3rd Intern. Confer. Genet.* pp. 140-142.

Characterizing size-specific fecundity in crustaceans

Keith M. Somers
Department of Zoology, University of Toronto, Ontario, Canada

ABSTRACT

Studies of crustacean fecundity are compromised by the absence of a general model for estimating the relationship between fecundity and female size. Coefficients of determination (i.e. r^2) of five different regression models were compared in analyses of 20 published data sets to determine if any one model of size-specific fecundity was superior. No one model was best, possibly because variability in fecundity increases with female size such that each model accounted for a similar amount of the total variance. Of these five, the allometric model of log fecundity and log size that is commonly used in fisheries biology was further examined to establish whether the allometric egg number (volume) -female size relationship was characterized by a slope of 3.0. The observed slope was not significantly different from the theoretical slope of 3.0 in 60% of the data sets. No consistent deviation from the expected value was apparent within taxa; however, the frequent occurrence of clusters of points representing size classes with possibly different size-specific fecundity may have affected these results. Although studies of crustacean life history recognize the importance of size at maturity and fecundity, the absence of a general model of size-specific fecundity limits broad comparisons. The allometric model is recommended because 1) analytical simplicity, 2) relative scale independence and 3) it provides two life-history parameters: the intercept (or adjusted mean) describing absolute size-specific fecundity, and the slope estimating relative size-specific fecundity. The existence of differences in size-specific fecundity among discrete size classes within populations suggests new directions for future studies of crustacean life history.

1 INTRODUCTION

Early studies of crustacean fecundity established a positive relationship between fecundity and female size (e.g. Tack 1941, Goellner 1943, Hoglund 1943, Forster 1951, Hynes 1954, Jensen 1958). Presently, crustacean biologists estimate size-specific fecundity with a number of regression models ranging from simple linear regressions to rather complex asymptotic curves (e.g. Corey 1984, Emmerson 1985, Somerton & MacIntosh 1985). The absence of a general model of crustacean fecundity contrasts with studies of fish where size-specific fecundity is assumed *a priori* to constitute an allometric or log-log relation (e.g. Bagenal 1967, Mann & Mills 1979, Wootton 1979). This inconsistency among crustacean studies is surprising since similar biological processes are being modelled in both fish and crustaceans. In the following paper, I review current methods and problems of estimating size-specific fecundity. I then evaluate the efficiency of five common approaches to modelling size-specific fecundity using 20 published data sets. Lastly, I examine recent interpretations of size-specific fecundity with an emphasis on the consequences of choosing among the different regression models. I propose that because of analytical simplicity, relative scale independence and the estimation of two life-history parameters (i.e. relative and absolute size-specific fecundity), the allometric or log-log model of size-specific fecundity that is used in fisheries biology should also be used in studies of crustacean fecundity.

1.1 *Size-specific fecundity*

Eggs carried by females from the same population are generally similar in size (e.g. Mason 1978, Hines 1982), although egg size often differs among populations (e.g. Efford 1969, Mason 1978, Corey 1981). Consequently, fecundity, which is usually represented as the total number of eggs in the ovary, marsupium or externally brooded on the pleopods, also represents a volume measure. In this context, each egg is assumed to have a similar volume such that the number of eggs times this volume of an individual egg provides a volumetric estimate of the ovary or the egg mass. Since egg size is relatively constant within a population, fecundity expressed as the number of eggs per female should also be highly correlated with fecundity estimated as clutch volume (e.g. see Corey 1981, 1987a). As a result, the fecundity of large clutches of eggs is frequently estimated by volumetric or weight approximations (e.g. Ennis 1981, Somerton 1981, Diaz et al. 1983) and occasionally, fecundity is presented as a volume (e.g. ml or mm^3, Hartnoll 1985, Corey 1987a). Since fecundity is a volumetric measure tending to increase exponentially with female size (e.g. Goellner 1943), Forster (1951) proposed that fecundity is limited by both the volume of the female and egg size (i.e. egg diameter). Subsequent comparison of regressions of the number of eggs and female volume for a variety of crustaceans indicated that female volume affected the relative number of eggs, but environmental factors influenced the absolute number (Jensen 1958).

Crustacean fecundity is also characterized by greater variation in the number of eggs carried by large females (e.g. Goellner 1943, Hynes 1954, Jensen 1958). Such size-specific variation in fecundity confounds comparisons between populations. Hynes (1954) acknowledged this problem and used a square-root transformation of the number of eggs to eliminate the correlation between the variance in egg number and female size. More recently, the same result has been achieved with a logarithmic transformation (e.g. Saila et al. 1969, Ennis 1981, Du Preez & McLachlan 1984); however, there is some disagreement as to whether female size should also be transformed (e.g. see Wenner et al. 1987).

Somerton (1981) instead interpreted size-specific variation in fecundity as evidence of an asymptotic relationship between fecundity and female size (i.e. the number of eggs asymptotically approaches a maximum as female size increases; also see Somerton & MacIntosh 1985). Although a significant polynomial regression indicated that fecundity might decrease beyond some threshold female size, Somerton argued that decreased fecundity is unlikely and that an asymptotic number of eggs is more realistic (cf. Woodhead 1979). This asymptotic-fecundity proposal contrasts with results from the study of a number of amphipod species (i.e. *Gammarus* spp.) where fecundity declined among the oldest females (Kinne 1961). This reproductive senility or senescence interpretation was also advanced by Katre (1977) for a freshwater prawn.

Increased variation in fecundity with female size may also contribute to the recent observation that size-specific fecundity varies among year classes (e.g. Wenner et al. 1987). Differences in fecundity between populations are common (e.g. Hynes 1954, Jensen 1958, Saila et al. 1969), but evidence of fecundity differences between year classes is rare (e.g. Paris & Pitelka 1962, Lawlor 1976). This is probably because crustacean age classes are estimated by length-frequency analysis of often overlapping size classes (e.g. Momot 1967, Goldman & Rundquist 1977; but also see Wenner et al. 1987).

1.2 *Current methods of characterizing fecundity*

Numerous methods are available for estimating crustacean fecundity. Surprisingly, the dependence of fecundity on female size is occasionally ignored and only mean or median fecundity is reported (e.g. Momot & Gowing 1977, McGriff 1983, France 1985). Subsequent comparisons between populations involve t-tests or one-way analysis of variance with little regard for the female-size covariate (e.g. McGriff 1983, Momot 1986, Corey 1987b), although ratios have also been used to correct for size effects (e.g. see France 1987, Ivanova & Vassilenko 1987). Conclusions based on such comparisons are weakened by the fact that female sizes may differ between samples or that the relationships (i.e. slopes) between fecundity and female size may differ and yet go undetected.

Often a simple, least-squares regression is used to estimate the number of eggs based on female size (e.g. Tack 1941, Paris & Pitelka 1962, Corey 1981). On theoretical grounds this regression model is inappropriate since the number of eggs is predicted *a priori* to be a volume measure tending to increase exponentially with female size (e.g. see Goellner 1943, Hoglund 1943, Saila et al. 1969). As a result, simple linear regressions using raw data underestimate the number of eggs produced by the largest females (e.g. see Corey 1981, Hamr & Berrill 1985, Corey 1987b), or the smallest females (e.g. Brewis & Bowler 1985). This problem is rarely recognized because outliers (e.g. Smart 1962, Corey 1984) and increased scatter in the fecundity of large females (e.g. Donn & Croker 1986) often obscure the pattern. The magnitude of this underestimation is also partly dependent on the range of female sizes, but this effect should not be confused with problems associated with using Model I regressions when both variables are not perfectly correlated (see Jensen 1986 for details).

Hynes (1954) compared the size-specific fecundity of two amphipod populations using an analysis of covariance (ANCOVA). Hynes recognized that size-specific variation in fecundity confounded the analysis so the variance in fecundity was standardized with a square-root transformation. Currently, a logarithmic transformation is generally employed to eliminate the correlation between the variance in female fecundity and female size (see Bartlett 1947 or Green 1979: 12-15 for a review of transformations). This logarithmic transformation also changes fecundity to a scale-independent measure representing the proportional change in fecundity (see Bryant 1986 and below for details). Removing scale dependence with the log transformation enables comparisons across taxa since the absolute number of eggs is converted to a relative measure (e.g. see Hines 1982).

Many researchers employ the power function or allometric model suggested by early studies of crustaceans (e.g. Goellner 1943, Hoglund 1943, Forster 1951), and generally used by fisheries biologists (e.g. Bagenal 1957, 1966, 1967, Wootton 1979). The power function (i.e. $Y = aX^b$) becomes a simple least-squares regression when fecundity and female size are both logarithmically transformed (i.e. the allometric model of $\log Y = \log a + b [\log X]$, where a is the intercept and b is the slope; see Teissier 1960, Gould 1966, Pasternack & Gianutsos 1969, Schmidt-Nielsen 1984 for details). Both Teissier (1960) and Gould (1966) emphasized that the allometric model has an empirical basis with intuitive simplicity and interpretability.

In brief, the allometric model represents the proportional change in fecundity with respect to the proportional change in female size (e.g. Bagenal 1967). Because logarithms are employed, the slope is independent of the original scale of measurement. Moreover, since fecundity is a volume and female size is a linear measure, the slope based on the allometric model should approximate 3.0 (i.e. b = 3.0) and the intercept will represent the absolute number of eggs independent of female size (i.e. a; see White & Gould 1965, Bagenal 1967, Wootton 1979). Consequently, allometric models of size-specific fecundity should have slopes of 3.0, whereas slopes differing from 3.0 imply the absence of a simple volumetric relationship. The latter pattern may reflect changes in egg size with respect to female size (e.g. see Begon & Parker 1986, Attard & Hudon 1987), or a disproportionate change in the size of the ovary (i.e. female shape changes with respect to overall size). Although the allometric model is often used by crustacean biologists (e.g. Efford 1969, Saila et al. 1969, Ennis 1981, Ivanova & Vassilenko 1987), values of the slope are rarely examined to generate hypotheses regarding size-specific differences in egg or ovary sizes.

The exponential model (i.e. $Y = a^{bx}$) is also used to estimate size-specific fecundity (e.g. Daborn 1975, Wenner et al. 1987). This model becomes a least-squares regression when fecundity, but not females size, is logarithmically transformed (i.e. on semi-logarithmic axes where log Y = log a + b [X]; see Pasternack & Gianutsos 1969). Unfortunately, size-specific fecundity on semi-logarithmic axes is linear over only a small range of female sizes (e.g. single year classes), and as a result, the size-specific fecundity relationship is curvilinear when female size spans a large range (e.g. see Daborn 1975, Wenner et al. 1987). Unlike the allometric model, the slope of the exponential model is not independent of the scale of measurement because only one axis is log transformed. Consequently, interpretations of size-specific fecundity estimated from an exponential model are limited when compared with results from the allometric model (e.g. see Gould 1966, Hines 1982, Rayner 1985, Shea 1985).

Jensen (1958) extrapolated from general allometric concepts and regressed fecundity on female volume (expressed as female carapace length cubed) to examine whether female volume limited absolute fecundity (or only relative fecundity). This approach is based on the fact that the relationship between fecundity and female volume is linear (i.e. two volume measures), whereas the relation between fecundity and female size is curvilinear (e.g. see Jensen 1958). However, Jensen failed to log transform fecundity and female volume. Had the log transformation been employed, the slopes of the allometric regressions should approximate 1.0 and consequently, comparisons among taxa would have been simplified (e.g. see Hines 1982, Schmidt-Nielsen 1984). Regression estimates of size-specific fecundity based on raw data (e.g. Katre 1977, Corey 1981) have less general applicability than those based on log-transformed data because estimates of the slope are only scale independent when both variables are log transformed (e.g. Kinne 1961, Hartnoll 1985; also see Wootton 1979). Lastly, Somerton (1981) proposed that an asymptotic curve best approximates size-specific fecundity. The asymptotic estimate of maximum fecundity has heuristic value as a population parameter (e.g. Somerton 1981, Somerton & MacIntosh 1985), but contrasts with a log-log slope of 1.0 or 3.0 that is based on allometric principles (e.g. see Schmidt-Nielsen 1984).

1.3 *Common problems in estimating size-specific fecundity*

The analysis of crustacean size-specific fecundity is complicated by problems rarely encountered in fisheries biology. Aside from the confusion associated with choosing an appropriate regression model, the common practice of counting externally brooded eggs on ovigerous females (i.e. 'berried' females with eggs attached to their pleopods) is perhaps the foremost problem. Goellner (1943) noted that counts of yolked eggs in the ovary of the crayfish, *Orconectes immunis*, were much less variable than counts of external eggs. Ovigerous females caught in traps, nets or trawls occasionally display abnormally low numbers of external eggs, possibly due to dislodgement during capture (e.g. see Perkins 1971, Abello & Sarda 1982). Additionally, both disturbances during egg laying and a variety of natural factors may also result in reduced numbers of external eggs (e.g. Balasundaram & Pandian 1982, Corey this volume, Kuris this volume).

Unless females with incomplete clutches are identified and removed from subsequent analyses (e.g. Somerton 1981, Brewis & Bowler 1985, France 1987), parametric regression-based estimates of size-specific fecundity may be noticeably biased by outlying observations Unfortunately, removing outliers introduces subjectivity into the analysis and may also influence the results (e.g. see discussion of outliers by Beckman & Cook 1983).

As an alternative, resistant or robust regression models could be employed. Resistant regressions use statistics like the median that are less sensitive to the influence of outliers (e.g. see Ryan et al. 1981: 23, Hoaglin et al. 1983: 129-165). In contrast, robust regressions inversely weight data points according to their relative deviation from the general elliptical pattern, giving outlying observations less influence in the regression solution (e.g. see McNeil 1977: 149-179, Mosteller & Tukey 1977: 356-365, Rey 1983: 112). Nonparametric approximations are also possible (e.g. Hollander & Wolfe 1973: 200-218). Such methods are designed for 'messy' data and consequently, these techniques should be useful for confirming estimates of the size-specific fecundity relationship.

The presence of atypical data points is often complicated by small sample sizes. When size-specific fecundity is estimated from a few observations (e.g. Kensler 1968, Corey 1981, 1984, Du Preez & McLachlan 1984), a single observation can appreciably influence the regression (e.g. see Corey 1984). The fact that any number of regression models estimate the size-specific fecundity relationship equally well (e.g. see below) only accentuates this problem, particularly when comparisons between populations or taxa are attempted (e.g. Corey 1981, 1987b, Hines 1982).

In conclusion, a number of different regression models are currently used to estimate the relationship between fecundity and female size. The prevalence of many models may contribute to the relative absence of comparative studies of crustacean fecundity (but see Ivanova & Vassilenko 1987). Factors such as increased variation in fecundity among larger females, outliers, and small sample sizes complicate attempts to estimate and compare size-specific fecundity among populations. The obvious unanswered question is whether any one regression model works best. In the next section I compare results of five regression approaches using 20 data sets to determine whether any one method best summarizes patterns in size-specific fecundity across a variety of crustacean taxa.

2 CHOOSING AMONG REGRESSION MODELS

Six regression models that are used in studies of crustacean size-specific fecundity have been described above. Five of these models have been selected for further comparison: 1) a simple least-squares regression of fecundity versus female size with no transformations, 2) the exponential model or simple least-squares regression with fecundity log transformed (i.e. a semi-log model), 3) the square-root transformation model [i.e. like (2) but with fecundity square-root

Table 1. Taxa, sample sizes, and data sources for the 20 data sets used to compare regression models. Approximate female size range is included; TL: total length, CL: carapace length, CW: carapace width.

Taxon	N	Scientific name	Source of data	Size range
Isopods	65	*Armadillidium vulgare*	Paris & Pitelka	6.5-13 mm TL
Amphipods	63	*Haustorius canadensis*	Donn & Croker	6-9 mm TL
Mysids	96	*Mysidopsis bigelowi*	Allen	4.4-8.1 mm TL
Shrimp	18	*Crangon septemspinosa*	Corey (1987a)	7-12 mm CL
	26	*Crangon septemspinosa*	Corey (1987a)	6.5-13 mm CL
	31	*Palaemon pacificus*	Emmerson	41-59 mm TL
	50	*Macrobrachium heterochirus*	Ching & Velez	4-20 mm CL
	81	*Leander squilla*	Hoglund	29-51 mm TL
	117	*Macrobrachium lamarrei*	Katre	4.1-5.3 cm TL
Crabs	25	*Uca thayeri*	Salmon	1.4-2.4 cm CW
	28	*Ovalipes punctatus*	Du Preez & McLachlan	28-63 mm CW
	83	*Emerita analoga*	Wenner et al.	10-23 mm CL
	145	*Paralithodes platypus*	Somerton & MacIntosh	95-155 mm CL
Crayfish	59	*Austropotamobius pallipes*	Brewis & Bowler	27-42 mm CL
	102	*Orconectes immunis*	Goellner	25-41 mm CL
	121	*Orconectes immunis*	Goellner	24-40 mm CL
	135	*Procambarus clarkii*	Penn	31-52 mm CL
Lobsters	9	*Homarus americanus*	Taylor	8-12 cm CL
	14	*Jasus edwardsii*	Kensler	9.5-17 cm CL
	77	*Homarus americanus*	Saila et al.	76-192 mm CL

transformed], 4) a simple least-square regression of fecundity versus female volume (where female volume is approximated by the cube of female size) and 5) the allometric model of log-transformed fecundity versus log-transformed female size. The sixth model incorporating the curvilinear, asymptotic regression proposed by Somerton (1981) was not included because of its arithmetic complexity and limited use.

2.1 *Sources of data on size-specific fecundity*

Fortunately, many studies of crustacean fecundity include tables of raw data or figures from which the original data can be estimated. Wenner et al. (1987) and Ivanova & Vassilenko (1987) list a number of such studies for a variety of crustaceans. To compare the 'performance' of the five regression models, I compiled 20 data sets on fecundity and female size from a number of different sources (see Table 1). Sixteen genera are represented. These include isopods, amphipods, mysids, shrimp, crabs, crayfish and lobsters. Data were selected from published sources according to two criteria: 1) the total data set must represent a number of different taxa with only limited replication and 2) the data were readily available from tables or clear figures. The number of observations in each data set was not important, nor were the sampling or enumeration methods. I hoped that this selection would provide a diverse data set without focusing on any particular taxon or any one individual's research. Readers interested in comparisons with other taxa should consult Ivanova & Vassilenko (1987).

Two features of the data should be emphasized. First, because sample sizes range from 9 to 145, some data sets may produce results biased by outliers (i.e. particularly those sets with less than 30 observations). In some cases, sample sizes are less than the published sources because multiple points in figures could not be distinguished. Second, a variety of measures of female size are included (e.g. carapace width and length, and total length) and the range of female sizes also varies (e.g. see Table 1). The type of measure of female size is not important because these size variables are generally highly correlated (e.g. Hines 1982, Corey 1987b), but the range of female sizes may influence how well the various regression models summarize variation in fecundity (see below). For comparative purposes, the 'best' model should not be affected by the range and scale of female size.

2.2 *Is any one regression model 'best'?*

Choosing among regression models to estimate size-specific fecundity appears to depend on a mix of traditional, theoretical and statistical criteria. As detailed above, no one regression model is used by all crustacean biologists; and thus, 'traditional' simply refers to the repeated use of the same model (e.g. the allometric model is routinely used in studies of lobster fecundity; Saila et al. 1969, Perkins 1971, Ennis 1981).

To evaluate the efficiency of the five regression models I employed the coefficient of determination (or r^2) associated with each regression. This measure is often interpreted as an estimate of the percentage of the total variation in the dependent variable (i.e. Y) incorporated by the regression. No attempt was made to evaluate whether any particular regression model was significantly different from any other (i.e. statistically) because the regression results are not independent. Instead, I was more interested in 1) the number of times a given model provided the highest coefficient of determination relative to the other four models and 2) whether any one model was better for some taxa, but not others. I also determined whether the range in female sizes in each data set influenced the coefficient of determination in any predictable manner (e.g. the difference in r^2 values for the five models in each set of data).

Coefficients of determination were quite variable, ranging from 0.16 for one dataset to 0.96 for another (Table 2). Where two data sets were available for the various taxa, no one taxon showed more or less variability in size-specific fecundity than any other taxon. Several points should be emphasized. First, all regression models tended to account for a similar amount of the total variation in fecundity in each data set. The difference between the maximum and minimum r^2 values ranged from 1 to 17% with an average of 5-6%. Consequently, each of the five regression models tended to perform equally well. Second, examination of the r^2 values across data sets indicated that no one model provided the highest r^2 values consistently. Four of the five models: the raw-data regression, the square-root model, the volume regression and the log-log model, each contributed the highest r^2 value five times. In contrast, the semi-log model never provided the highest r^2 value, but provided the lowest r^2 value 7 times and the second-lowest value 9 times. The log-log method gave the lowest r^2 value twice, the raw-data model had the lowest r^2 value three times, and the volume model had the lowest r^2 value three times, and the volume model had the lowest r^2 value 8 times. The square-root regression fit the data with the highest or second-highest r^2 value in 15 of the 20 data sets and never contributed the lowest r^2 value. However, differences among the r^2 values were so subtle that the third decimal place was often required to rank the performance of the regressions (e.g. see Table 2).

No one regression model was consistently best for any particular taxon (Table 2). This may have occurred because female size range varied within taxa (Table 1). The regression using raw data produced the highest r^2 values (i.e. within a data set) when the range in female sizes was small. In contrast, the log-log allometric model provided high coefficients of determination whenever the range in female sizes was large. The semi-log model produced the lowest or

Table 2. Coefficients of determination associated with the five regression models for each of the 20 data sets.

Taxon	N	r^2 values Raw	Semi-log	Sq.root	Volume	Log-log	Difference (max.-min.)
Isopods	65	0.744	0.770	0.773	0.724	0.787	0.063
Amphipods	63	0.592	0.512	0.574	0.579	0.526	0.080
Mysids	96	0.765	0.744	0.766	0.760	0.757	0.022
Shrimp	18	0.866	0.844	0.911	0.891	0.869	0.067
	26	0.745	0.704	0.736	0.701	0.731	0.044
	31	0.719	0.708	0.728	0.768	0.696	0.072
	50	0.814	0.774	0.816	0.745	0.794	0.071
	81	0.918	0.931	0.936	0.934	0.937	0.019
	117	0.183	0.160	0.175	0.179	0.162	0.023
Crabs	25	0.570	0.567	0.576	0.565	0.570	0.011
	28	0.741	0.805	0.786	0.811	0.774	0.070
	83	0.928	0.896	0.946	0.929	0.934	0.050
	145	0.587	0.588	0.595	0.544	0.616	0.072
Crayfish	59	0.859	0.870	0.878	0.879	0.873	0.020
	102	0.784	0.686	0.752	0.796	0.688	0.110
	121	0.360	0.211	0.302	0.377	0.204	0.173
	135	0.621	0.631	0.636	0.620	0.640	0.020
Lobsters	9	0.878	0.847	0.874	0.851	0.869	0.031
	14	0.964	0.896	0.938	0.895	0.938	0.069
	77	0.858	0.850	0.881	0.833	0.887	0.054

second-lowest r^2 value irrespective of female size range, whereas the regression of fecundity on female volume provided lower r^2 values when the smallest individual was half or less than half the size of the largest female.

Thus, no one model has a clear advantage in providing the best fit to the data. However, the regression of fecundity versus female volume had the lowest r^2 value in 8 out of the 20 data sets, and the semi-log regression had the lowest r^2 value in 7 of the 20 data sets. The results show that the five regression models have relatively similar efficiency in summarizing the size-specific fecundity relation (i.e. on average choosing among the different models affected the r^2 value by only 5-6%). Consequently, irrefutable support for only one model of size-specific fecundity does not arise from a purely heuristic basis.

3 INTERPRETING PATTERNS IN SIZE-SPECIFIC FECUNDITY

Hynes (1954) compared the size-specific fecundity of two populations of amphipods using an ANCOVA. This procedure contrasts the slopes and intercepts of two or more regressions to determine whether the lines differ (e.g. see DeLury 1948, Cochran 1957, Overall & Woodward 1977, Huitema 1980, Cox & McCullagh 1982 for details). Unfortunately, this statistical approach is not routinely used by crustacean biologists (but see Saila et al. 1969, Hines 1982, Wenner et al. 1987). More commonly, a number of regression equations are presented and the reader is left to decide whether the relationships do or do not differ (e.g. see Efford 1969, Corey 1981, 1987b, Somerton & MacIntosh 1985, Ivanova & Vassilenko 1987). Often, only a single population is described; thus, size-specific fecundity is presented as a simple graph with little elaboration (i.e. here is the size-specific relation, period; e.g. see Brewis & Bowler 1985, Salmon 1987). No hypotheses are advanced or tested (cf. Begon & Parker 1986). As a result, the scope of studies of crustacean fecundity is often limited when compared with other taxa (e.g. Tilley 1968, Salthe 1969, Smith & Fretwell 1974, McPhail 1977, Kaplan & Salthe 1979, Reznick & Endler 1982).

3.1 *Consequences of choosing different models – Simulated data*

Since the five regression models discussed above seem to account for similar amounts of variation in female fecundity, the consequences of selecting one model in favor of the others are not immediately apparent. The first evaluation of these consequences incorporates a simple simulation of a small data set. Arbitrarily, female size was assigned the 17 integers ranging from 4 to 20 which span the size range reported for *Macrobrachium heterochirus* (see Table 1, Ching & Velez 1985). Because much of this discussion focuses on the observation that fecundity represents a volumetric function relative to female size (e.g. Hoglund 1943, Forster 1951, Jensen 1958), fecundity was simulated as the cube of female size (i.e. ranging from 64 to 8000 eggs). These 17 simulated observations were plotted with the corresponding least-squares regressions for five models of size-specific fecundity: 1) raw data with no transformations, 2) the semi-log model with fecundity log transformed, 3) the square-root approach with fecundity expressed as a square root, 4) the volume model with fecundity versus female size cubed and 5) the allometric log-log regression.

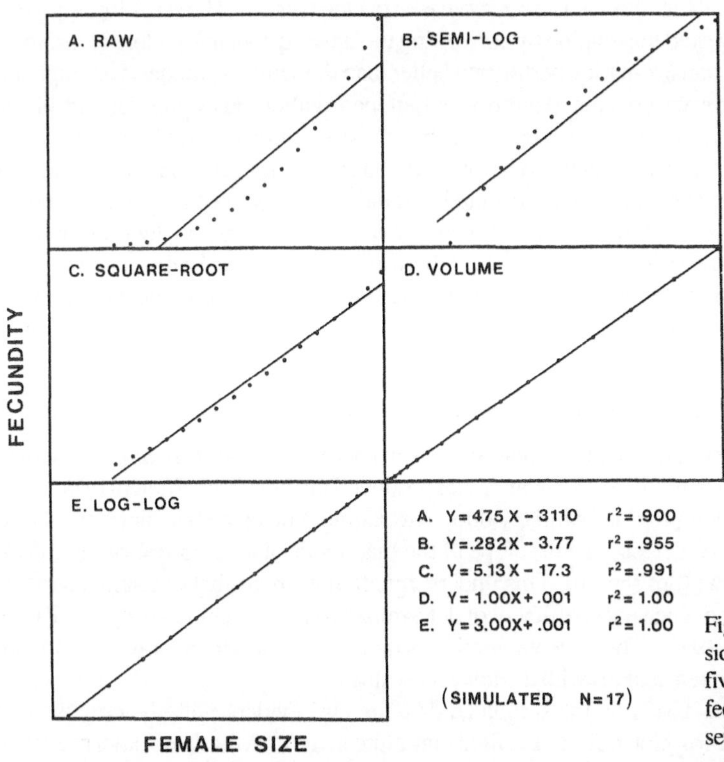

Figure 1. Scatterplots, regressions and equations based on five models of size-specific fecundity for a simulated data set with 17 observations.

The least-squares regression for the raw data failed to fit the simulated data well because fecundity increases at an exponential rate with respect to female size (Fig. 1). In contrast, the semi-log model provided a higher coefficient of determination ($r^2 = 0.955$), but the calculated regression overestimated the fecundity of small and large females. Among the large females, this overestimation may be erroneously interpreted as reproductive senescence. [Note that the two previously cited studies inferring reproductive senescence used forms of the volumetric model and not the semi-log approach (i.e. Kinne 1961, Katre 1977).]The regression based on the square-root transformation displayed a pattern similar to the raw data since the fecundity of the smallest and largest females was underestimated, but the coefficient of determination is 9% higher (i.e. $r^2 = 0.991$). Both the volume and the allometric models have r^2 values of 1.000 and slopes of 1.0 because simulated fecundity equaled female volume, by definition.

Surprisingly, the volumetric model gave the fecundity of large females more influence in the regression because these data are spread out. The cubic transformation of female size to approximate female volume resulted in greater spacing between data from large females than from small females. This observation is statistically troublesome, since variation in fecundity among large females is greater than variation among small females (e.g. see Hynes 1954). In comparison, the allometric or log-log model brought data points for the largest females closer together, thereby giving greater importance to observations based on the smaller females (Fig. 1). The semi-log model also tended to compress the spacing between the largest observations, whereas the square-root approach tended to maintain the original interpoint spacing.

The simulated data also provided an opportunity to evaluate the relative success of the various models in straightening the theoretical size-specific fecundity relation (i.e. assuming that fecundity does increase volumetrically with respect to female size). The raw data illustrated the ineffectiveness of a linear model when a curve is expected (Fig. 1). The semi-log approach dramatically over-corrected the relation by converting the upward concave pattern to a convex form. The square-root transformation performed better than the semi-log model (i.e. a higher r^2 value), because the upward concave pattern was flattened without inverting the curvilinear trend. [Given that the simulation incorporated a cubic transformation of female size to achieve the hypothetical fecundity data, a cube-root transformation of fecundity would have straightened the relationship and also would have resulted in equal spacing between consecutive points (i.e. $r^2 = 1.000$, slope = 1.00, not shown). Since my objectives do not include introducing additional models of size-specific fecundity the cube-root model will not be discussed further.] Both the volumetric and allometric models straightened the simulated curvilinear relationship between fecundity and female size.

3.2 *Consequences of choosing different models – Actual data*

The first example consists of counts of ovarian eggs from 135 crayfish, *Procambarus clarkii* (Penn 1943, Fig. 2). The allometric log-log model produced the largest coefficient of determination, but all five models provided similar values. An examination of the five plots reveals that the regression lines all underestimate the slope of the major axis of an ellipse surrounding the general scatter of points (not shown). This underestimation is an artifact of using a Model I least-squares regression when the coefficient of determination is less than 1.00 (e.g. see Ricker 1973, 1984, Jolicoeur 1975). The underestimation is most apparent with the raw data and the volume model, the two regressions with the lowest r^2 values.

Several observations based on the simulated data are also evident with the crayfish data (Fig. 2). First, the semi-log plot indicated a slight curvature in the data that is not apparent in the

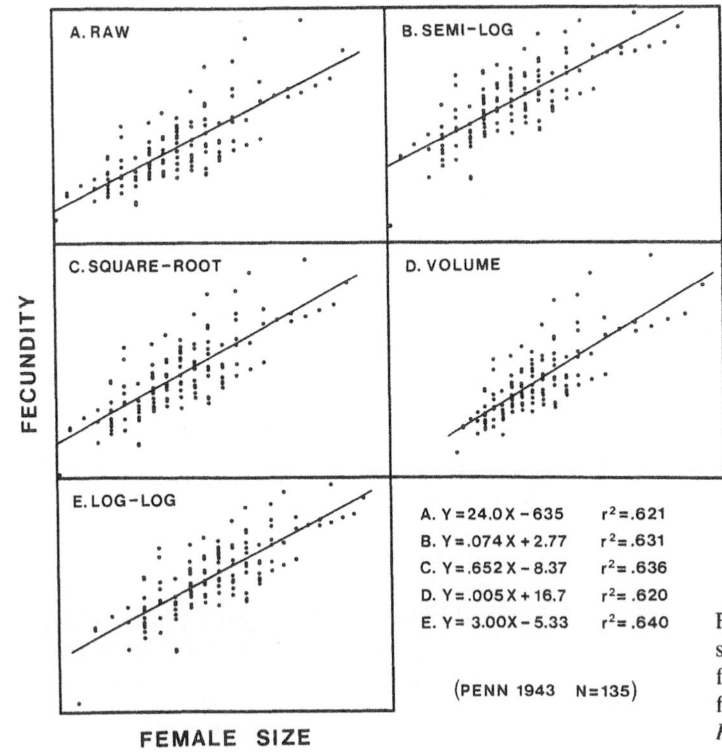

Figure 2. Scatterplots, regressions and equations based on five models of size-specific fecundity for 135 crayfish, *Procambarus clarkii*.

raw data. Second, the semi-log, square-root and log-log models stabilized the variance in fecundity across all female sizes. In contrast, the variance in fecundity increased with female size in the raw data and volume plots. Third, the inter-point spacing between large females was expanded in the volume model, whereas the log-log approach contracted this spacing. Additionally, the slopes of the first four regressions are difficult to compare because the various transformations changed the scale of the X and Y axes affecting the values of the slopes. The scale independence associated with the log-log transformations in the allometric model resulted in a slope equivalent to the theoretical value of 3.0, implying a volumetric relationship between fecundity and female size.

A cursory examination of the shape of the scatter of points gives some support for a decline in fecundity with female size that might be interpreted as evidence of reproductive senescence (Fig. 2). However, there is no evidence of asymptotic fecundity or discrete clusters of values that might represent different age classes. The absence of these latter two patterns may simply reflect the problem of distinguishing year classes of crayfish because the size ranges of older individuals overlap substantially (e.g. see Momot 1967, Goldman & Rundquist 1977).

The crayfish data contrast with data on the shrimp, *Leander squilla* (Hoglund 1943, Fig. 3). The shrimp data can be separated into two discrete size groups that may represent distinct reproductive age groups. Patterns in this data also reflect problems revealed in the analysis of the simulated data. The regression based on the raw data had the lowest coefficient of determination, whereas the log-log model had the largest r^2 value. But as before, all five regressions tended to fit the data equally well (i.e. a 2% difference in the r^2 values). Since there is little

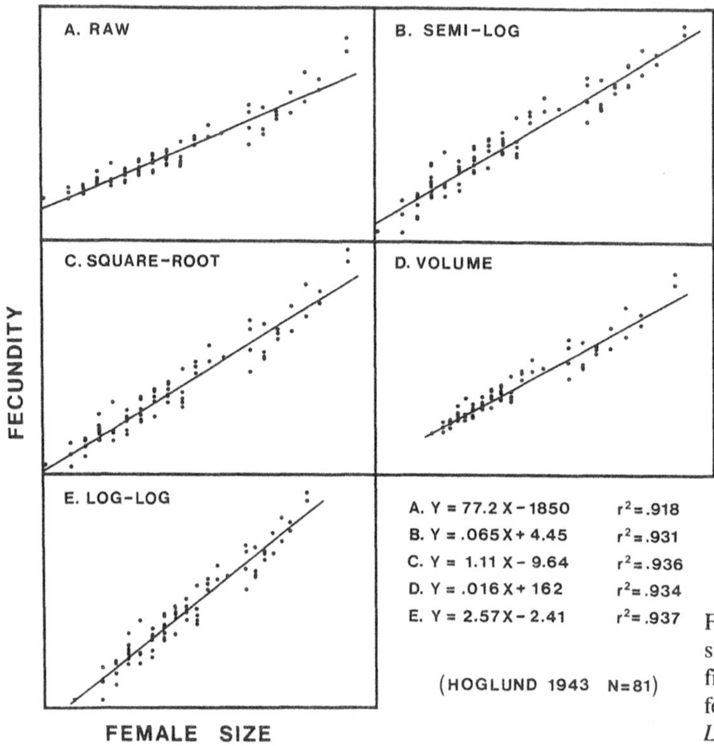

Figure 3. Scatterplots, regressions and equations based on five models of size-specific fecundity for 81 shrimp, *Leander squilla*.

evidence of a correlation between variation in fecundity and female size, the various transformations have little effect on stabilizing this variance.

Interestingly, the allometric slope for the shrimp data is less than 3.0. This observation may reflect the existence of two discrete size groups, since separately the two groups display steeper slopes than the pooled slope for the combined data. The two size groups are apparent in all five plots. Moreover, it is tempting to propose that another group consisting of the two largest females also exists.

The third data set consists of abdominal egg counts from 83 crabs, *Emerita analoga* (see Wenner et al. 1987 for details). Plots of the raw and volume-transformed data illustrate that the variance in fecundity increases with female size (Fig. 4). As with the simulated data, the semi-log plot over-corrected the curvature in the raw data, and as a result, tended to emphasize the existence of three groups in the data. Although these three groups are relatively contiguous, separation is most apparent in the raw, the semi-log and the volume plots. However, irrespective of the transformation, three size groups are apparent, and size-specific fecundity differs among these groups (see Wenner et al. 1987 for details). As a result, the regression for the pooled or combined data represents a compromised approximation of three different patterns, suggesting that the three size classes experience different selective forces (e.g. size-specific predation or competition). This conclusion is supported by an allometric slope greater than 3.0, implying that the relative sizes of the eggs or ovaries may change with increasing female size (i.e. the allometric slopes for the three groups are 5.3, 4.1 and 2.1, respectively, where only the third slope does not differ significantly from 3.0; but since this slope is less than 3.0, reproductive

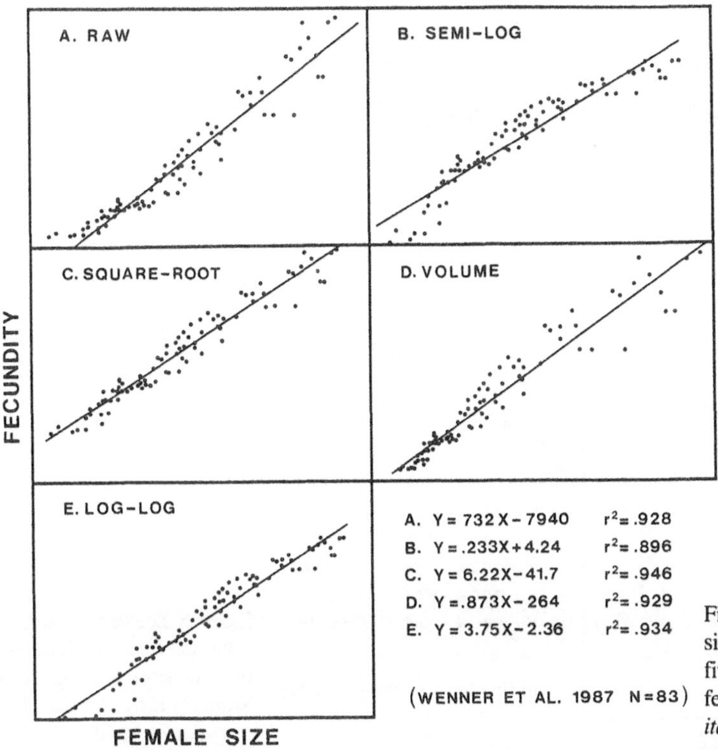

Figure 4. Scatterplots, regressions and equations based on five models of size-specific fecundity for 83 crabs, *Emerita analoga*.

senescence may be indicated). Such observations and subsequent interpretations are not unrealistic, but additional studies are required to reveal the factors contributing to different patterns of size-specific fecundity among size classes (e.g. Wenner 1988).

In the absence of distinct group structure within the *Emerita* data, the semi-log plots might be interpreted as providing evidence of either reproductive senescence or asymptotic fecundity (Fig. 4). From results of the simulation, we can argue that the transformation associated with the semi-log plot can produce a curvilinear or asymptotic pattern. However, a similar point distribution in the log-log plot was not evident in the simulation (Fig. 1). As a result, an asymptotic interpretation is plausible, but from a theoretical basis (i.e. an expected allometric slope of 3.0) three discrete size classes with differing size-specific fecundities seems more tenable.

A similar pattern is evident in data from 77 lobsters, *Homarus americanus* (Saila et al. 1969, Fig. 5). Variation in fecundity increases with female size, but the semi-log, square-root and log-log plots display stabilized variances due to the respective transformations. Again, the semi-log scatter is some-what curved, implying reduced fecundity with increasing female size which might be interpreted as evidence of reproductive senescence or asymptotic fecundity. The log-log model provided the best fit to the data, but the differences between the five approaches were small (i.e. 5-6%). Both volume and log-log plots illustrate changes in inter-point distances that give influence to data from the larger or the smaller females, respectively.

The allometric model of size-specific fecundity provided a slope of 2.86 that is not significantly different from a theoretical slope of 3.0 (Fig. 5, $P > 0.05$ and see below). However, in contrast with the *Emerita* data (Fig. 4) at least four groups or clusters of points can be identified

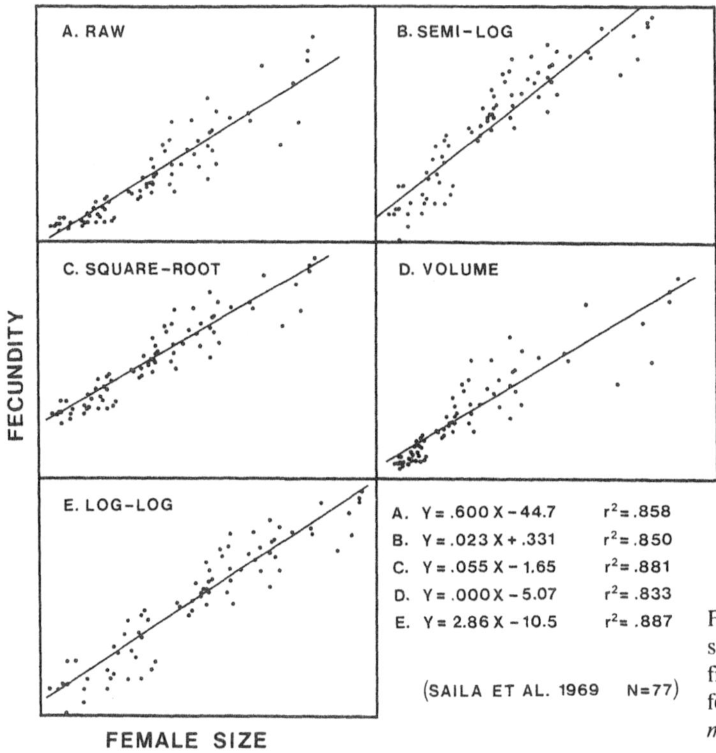

Figure 5. Scatterplots, regressions and equations based on five models of size-specific fecundity for 77 lobsters, *Homarus americanus*.

in the lobster data, yet the pooled allometric slope equals the expected value. Unfortunately, there are too few points in the separate size classes of the lobster data to evaluate whether individual allometric slopes differ from 3.0. However, it is interesting to note that the majority of the data points for the largest females fall below the log-log regression line, thereby suggesting reproductive senescence. Although differences in size-specific fecundity may exist among size classes within populations, it is premature to assume that patterns in different size classes will always differ.

The final data set consists of counts of externally brooded eggs from 145 king crabs, *Paralithodes platypus* (Somerton & MacIntosh 1985, Fig. 6). Increasing variation in fecundity with increasing female size is evident from the plot of the raw data. The semi-log, square-root and log-log approaches reduced, but did not eliminate the correlation between the variance in fecundity and female size. The semi-log plot illustrates a curvilinear pattern that was first observed in the analysis of the simulated data (Fig. 1). As before, the spacing among the largest females was expanded in the volume model, and in this instance likely contributed to the lowest r^2 value. In contrast, the log-log regression produced the largest coefficient of determination, although the allometric slope is significantly less than 3.0 (i.e. $P < 0.05$, see below for details).

An examination of the five plots suggests that the data might consist of two separate groups. As a result, the observed allometric slope that differs from 3.0 may represent a compromise estimate of the slopes of two subgroups (i.e. compare Fig. 6 with the patterns in Figs. 4 and 5). The biological significance of these two groups remains to be established, but a number of interpretations are possible (e.g. the size classes experience different mortalities; also see Wen-

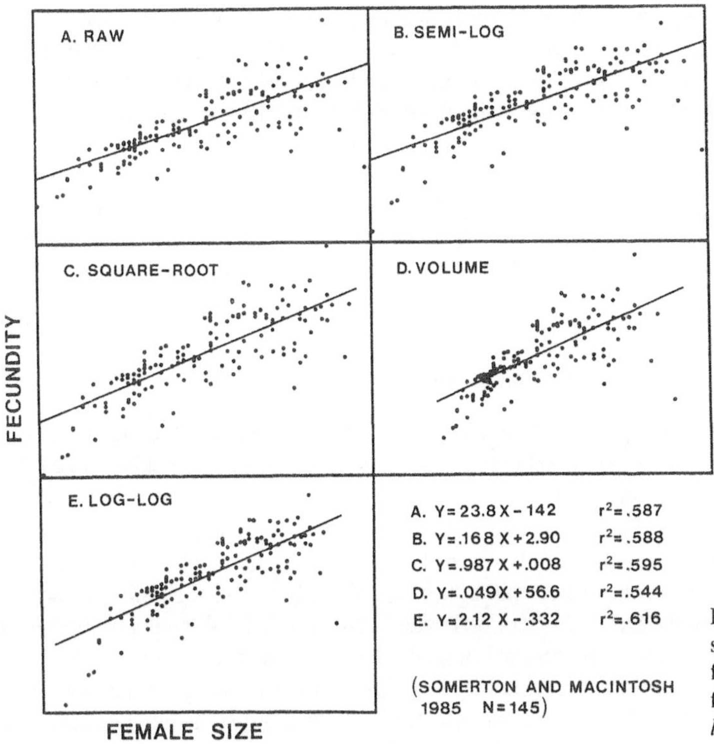

Figure 6. Scatterplots, regressions and equations based on five models of size-specific fecundity for 145 crabs, *Paralithodes platypus*.

ner et al. 1987). In any case, the asymptotic model may fit this data because of the existence of two size classes with different size-specific fecundities (e.g. see Somerton & MacIntosh 1985).

To summarize, the five regression models incorporate variation in size-specific fecundity with similar efficiency, but interpretations based on the five models may differ. The semi-log model tended to overcorrect the curvature in the raw data and may contribute to the identification of curvilinear patterns of size-specific fecundity mimicking reproductive senescence or asymptotic fecundity. The model of fecundity versus female volume fails to remove size-specific variation in fecundity and gives greater influence to the fecundity of large females because inter-point spacing is enhanced. The square-root transformation eliminates the correlation between the variance in fecundity and female size, and does not overcorrect the curvature in the raw data. The square-root model, however, is rarely used (e.g. see Hynes 1954). The greatest single detriment of these models is their scale dependence (i.e. changes in the scale or units of the X or Y axes affect the slope and intercept). The log-log model performed as well as the other approaches, but does not exhibit scale dependence. As a result, allometric regressions can be compared irrespective of the original measurement units or scales. In addition, the allometric log-log model can be contrasted with a theoretical slope of 3.0 to test for a volumetric relationship between fecundity and female size (cf. Jensen 1958).

Plots based on the five data sets indicate the existence of a number of different patterns of size-specific fecundity. Results from the five case studies indicate that varying numbers of clusters of points may be present. Such clusters may represent separate year classes or age classes (e.g. see Wenner et al. 1987), but this interpretation remains to be verified with known-age animals. As a result, regressions based on data that are pooled or combined irrespective of the existence of size classes may produce atypical results (e.g. see Figs. 4 and 6). Such patterns may be recognized when the allometric slope differs from the theoretical value of 3.0 (but see Fig. 5); however, the overall slope depends on the slope of each size class as well as the intercept of each group. Differences in individual slopes, intercepts or both will affect the pooled slope (e.g. see Rahn et al. 1975, Rayner 1985, Shea 1985). Alternate interpretations of reproductive senescence and asymptotic fecundity may simply explain differences in the size-specific fecundity of separate size classes, and indeed, four of the five datasets presented here (i.e. Figs. 2, 4, 5 and 6), provide evidence that the fecundity of the largest females falls below the expected values on average.

3.3 *Interpreting results from the allometric model*

In the previous section, I have emphasized that the allometric model of log fecundity and log female size has several advantages over other regression models. Among these advantages are: 1) the logarithmic transformation eliminates the correlation between the variation in fecundity and female size, 2) the two log transformations make the slope of the regression scale independent thereby permitting comparisons between populations or taxa and 3) the allometric model of size-specific fecundity has a theoretical slope of 3.0 implying that fecundity is a simple volumetric function of female size. In this section I examine results of the allometric regressions of size-specific fecundity for the 20 data sets described above (see Table 1). In particular, I evaluate the hypothesis that crustacean fecundity is a volumetric function of female size (e.g. see Forster 1951, Jensen 1958). This hypothesis is tested by determining whether the allometric slope differs significantly from the theoretical value of 3.0.

Slopes from log-log regressions for the 20 data sets ranged from 2.09 to 5.23 (Table 3). The associated 95% confidence limits permit an evaluation of the hypothesis that the observed slope equals 3.0, since an observed slope is not significantly different from 3.0 when the upper and lower 95% confidence limits enclose the expected value (i.e. a t-test). Twelve of the data sets provided slopes that do not differ from 3.0, implying that size-specific fecundity approximates a volumetric relationship. These results are not associated with the coefficients of determination or the location of the eggs (i.e. ovarian eggs and externally brooded eggs tend to show similar relationships with female size). Similarly, an allometric slope of 3.0 is not restricted to any particular taxon.

The reason why 40% of the regressions differ from an expected allometric slope of 3.0 is not clear; however, observations based on the five case studies suggest that the presence of groups or clusters of points within the data sets may affect the overall regression slope. Rahn et al. (1975) identified this sort of problem when comparing allometric regressions for a number of orders of birds, and Wenner et al. (1987) described discrete size classes associated with separate age groups in the crab, *Emerita analoga*. A re-examination of bivariate scatterplots of fecundity versus female size confirms that clusters of points are present in almost every data set (not shown). Unfortunately, the number of points within each cluster is usually small, so statistical evaluations were not attempted.

In any case, the allometric pattern of size-specific fecundity should approximate a volumetric relationship with a slope of 3.0 (e.g. see Jensen 1958, Bagenal 1967, Wootton 1979). Deviations from this pattern may reflect the presence of discrete reproductive groups displaying separate size-specific fecundities. Such an interpretation is not unrealistic since size classes within a population can experience different mortalities associated with size-selective preda-

Table 3. Regression statistics associated with the allometric model to evaluate whether the observed slope equals 3.0.

Taxon	N	Slope	95% CL on slope	Ho: b=3.0 (t-test)	r^2 value	Egg location
Isopods	65	2.78	2.41-3.14	Yes	0.787	Marsupium
Amphipods	63	3.15	2.39-3.92	Yes	0.526	Brood pouch
Mysids	96	3.06	2.71-3.42	Yes	0.757	Marsupium
Shrimp	18	5.23	4.16-6.31	No	0.869	On pleopods
	26	3.42	2.55-4.30	Yes	0.731	On pleopods
	31	2.98	2.23-3.72	Yes	0.696	On pleopods
	50	2.47	2.11-2.84	No	0.794	On pleopods
	81	2.57	2.42-2.72	No	0.937	On pleopods
	117	2.23	1.30-3.17	Yes	0.162	On pleopods
Crabs	25	2.09	1.31-2.87	No	0.570	On pleopods
	28	3.09	2.41-3.76	Yes	0.774	On pleopods
	83	3.75	3.53-3.98	No	0.934	On pleopods
	145	2.12	1.84-2.39	No	0.616	On pleopods
Crayfish	59	4.23	3.80-4.66	No	0.873	On pleopods
	102	2.58	2.23-2.92	No	0.688	Ovary
	121	2.49	1.60-3.99	Yes	0.204	On pleopods
	135	3.00	2.61-3.38	Yes	0.640	Ovary
Lobsters	9	2.82	1.83-3.80	Yes	0.867	On pleopods
	14	2.69	2.25-3.12	Yes	0.938	On pleopods
	77	2.86	2.62-3.09	Yes	0.887	On pleopods

tors (e.g. Paine 1976, Stein 1977), and predation can affect fecundity and life history in general (e.g. see Stearns 1976, Mann & Mills 1979, Reznick & Endler 1982).

Alternatively, unusual relationships may reveal atypical life-history strategies (see Stearns 1976 for a review). Differences in intercepts indicate varying absolute fecundity (i.e. more or less eggs at a given size), whereas differences in slopes reflect modification of relative fecundity and possible changes in the proportionate sizes of the eggs or ovaries (i.e. changes in female shape relative to female size; e.g. see Begon & Parker 1986, Wenner et al. 1987). This latter scenario implies that egg size and ovary size may vary like other life-history traits, such as the number of eggs and size at maturity (e.g. McPhail 1977, Mann & Mills 1979, Reznick & Endler 1982, Wenner et al. 1985).

4 SUMMARY AND CONCLUSIONS

Although the theoretical basis for modelling size-specific fecundity is established in fisheries biology (e.g. Bagenal 1967, Wootton 1979), crustacean biologists use a number of regression models to summarize the relationship between fecundity and female size. In this study, I analyzed 20 published data sets with five common regression approaches to evaluate whether any one approach provides higher coefficients of determination and greater interpretive insight. All five models provided similar coefficients of determination suggesting that statistically each method summarized patterns within the data equally well, although this conclusion is partly based on the observation that variation in crustacean fecundity tends to increase with female size.

Detailed examination of the results of the five different regression models using simulated data and five case studies indicated that some approaches transform size-specific fecundity to produce unusual patterns that prompt conflicting interpretations. Analyses based on the raw data or fecundity versus female volume were compromised by increased variation in fecundity with increasing female size. In contrast, approaches employing logarithmic or square-root transformations of fecundity tended to stabilize the variance in fecundity across all female sizes. The semi-log model over-corrected the curvilinear pattern evident from plots of the raw data and produced an asymptotic scatter of points that might be erroneously interpreted as evidence of reproductive senescence or asymptotic fecundity. The volume model failed to stabilize size-specific variation in fecundity and also expanded inter-point spacing between data from the largest females. This inter-point dilation gave greater influence to the fecundity of the largest females that tend to be most variable. As a result, estimates of size-specific fecundity based on the volume model may be unduly influenced by the fecundity of a few of the largest females.

Unfortunately, the raw data, semi-log, square-root and volume models are all scale dependent, such that changes in the units or scales used to measure fecundity and female size will affect regression slopes and intercepts. Consequently, results from these four approaches are not easily compared across populations or taxa. The log-log model contrasts with the other methods because the logarithmic transformations produce regression slopes that are scale independent. In addition, allometric regressions with slopes equal to 3.0 imply that fecundity is a volumetric function of female size. The log-log model also stabilizes size-specific variation in fecundity and reduces the influence of the fecundity of the largest females by contracting inter-point distances between these data.

Twelve of the allometric slopes from log-log regressions based on the 20 published data sets equalled the theoretical value of 3.0. Examination of scatterplots from the 20 data sets indicated that allometric slopes differing from 3.0 were often based on data displaying discrete groups or clusters of points. Such clusters may represent separate reproductive groups with distinct allometric slopes and intercepts reflecting different selective pressures (i.e. mortality) on each group. Some data sets with slopes equal to 3.0 also displayed clusters of points suggesting the widespread existence of discrete size classes with varying patterns of size-specific fecundity (e.g. Wenner et al. 1987).

The log-log model of size-specific fecundity is preferred because the logarithmic transformations stabilize the variance in fecundity and also produce a scale independent slope that should approximate 3.0. The log-log model also contracts inter-point distances among data for the largest females that tend to be most variable and most influential in estimating the size-specific fecundity regression. Clusters of data representing distinct size classes may compromise estimates of size-specific fecundity for the entire population since size-specific fecundity may differ among size classes (e.g. reproductive senescence among the largest females). Differences in size-specific fecundity among discrete size classes within populations suggest new directions for future studies of crustacean life history.

ACKNOWLEDGEMENTS

The ideas presented in this paper evolved through discussions with R.H. Green, D.A. Jackson, D.P.M. Stechey and H.H. Harvey. I thank M. Berrill, S. Corey, R.L. France, J.V. Huner, D.A. Jackson, K.A. Keenleyside, A.M. Kuris, W.T. Momot, G.E. Morgan, K. Nelson, A.M. Wenner and two anonymous reviewers for comments and helpful suggestions on various drafts of this manuscript. Portions of this research were funded by NSERC Operating Grants to R.H. Green and H.H. Harvey, and an NSERC PDF to the author.

REFERENCES

Abello, P. & F. Sarda 1982. The fecundity of the Norway lobster [*Nephrops norvegicus* (L.)] off the Catalan and Portuguese coasts. *Crustaceana* 43: 13-20.

Allen, D.M. 1984. Population dynamics of the mysid shrimp *Mysidopsis bigelowi* W.M. Tattersall in a temperate estuary. *J. Crust. Biol.* 4: 25-34.

Attard, J. & C. Hudon 1987. Embryonic development and energetic investment in egg production in relation to size of female lobster (*Homarus americanus*). *Can. J. Fish. Aquat. Sci.* 44: 1157-1164.

Bagenal, T.B. 1957. Annual variations in fish fecundity. *J. Mar. Biol. Assoc. UK* 36: 377-382.

Bagenal, T.B. 1966. The ecological and geographical aspects of the fecundity of the plaice. *J. Mar. Biol. Assoc. UK* 46: 161-186.

Bagenal, T.B. 1967. A short review of fish fecundity. In S.D. Gerking (ed.), *The Biological Basis of Freshwater Fish Production*. pp. 89-111. Edinburgh: Blackwell.

Balasundaram, C. & T.J. Pandian 1982. Egg loss during incubation in *Macrobrachium nobilii* (Henderson & Mathai). *J. Exp. Mar. Biol. Ecol.* 59: 289-299.

Bartlett, M.S. 1947. The use of transformations. *Biometrics* 3: 39-52.

Beckman, R.J. & R.D. Cook 1983. Outlier.......s. *Technometrics* 25: 119-149.

Begon, M. & G.A. Parker 1986. Should egg size and clutch size decrease with age? *Oikos* 47: 293-302.

Brewis, J.M. & K. Bowler 1985. A study of reproductive females of the freshwater crayfish *Austropotamobius pallipes*. *Hydrobiologia* 121: 145-149.

Bryant, E.H. 1986. On use of logarithms to accommodate scale. *Syst. Zool.* 35: 552-559.

Ching, C.A. & M.J. Velez Jr 1985. Mating, incubation and embryo number in the freshwater prawn *Macrobrachium heterochirus* (Wiegmann, 1936) (Decapoda, Palaemonidae) under laboratory conditions. *Crustaceana* 49: 42-48.

Cochran, W.G. 1957. Analysis of covariance: its nature and uses. *Biometrics* 13: 261-281.

Corey, S. 1981. Comparative fecundity and reproductive strategies in seventeen species of the Cumacea (Crustacea: Peracarida). *Mar. Biol.* 62: 65-72.

Corey, S. 1984. The comparative fecundity of two species of Cumacea from the shallow waters of Florida. *Can. J. Zool.* 62: 514-515.

Corey, S. 1987a. Reproductive strategies and comparative fecundity of *Crangon septemspinosa* Say (Decapoda, Caridea). *Crustaceana* 52: 25-28.

Corey, S. 1987b. Comparative fecundity of four species of crayfish in southwestern Ontario, Canada (Decapoda, Astacidea). *Crustaceana* 52: 276-286.

Cox, D.R. & P. McCullagh 1982. Some aspects of analysis of covariance. *Biometrics* 38: 541-561.

Daborn, G.R. 1975. Life history and energy relations of the giant fairy shrimp *Branchinecta gigas* Lynch 1937 (Crustacea: Anostraca). *Ecology* 56: 1025-1039.

DeLury, D.B. 1948. The analysis of covariance. *Biometrics* 4: 153-170.

Diaz, H., J.E. Conde & M. Bevilacqua 1983. A volumetric method for estimating fecundity in Decapoda. *Mar. Ecol. Prog. Ser.* 10: 203-206.

Donn, Jr T.E. & R.A. Croker 1986. Life-history patterns of *Haustorius canadensis* (Crustacea: Amphipoda) in northern New England. *Can. J. Zool.* 64: 99-104.

Du Preez, H.H. & A. McLachlan 1984. Biology of the three spot swimming crab, *Ovalipes punctatus* (De Hann) III. Reproduction, fecundity and egg development. *Crustaceana* 47: 285-297.

Efford, I.E. 1969. Egg size in the sand crab, *Emerita analoga* (Decapoda, Hippidae). *Crustaceana* 16: 15-26.

Emmerson, W.D. 1985. Fecundity, larval rearing and laboratory growth of *Palaemon pacificus* (Stimpson) (Decapoda, Palaemonidae). *Crustaceana* 49: 277-289.

Ennis, G.P. 1981. Fecundity of the American lobster, *Homarus americanus*, in Newfoundland waters. *Fish. Bull.* 79: 796-800.

Forster, G.R. 1951. The biology of the common prawn, *Leander serratus* Pennant. *J. Mar. Biol. Assoc. UK* 30: 333-360.

France, R.L. 1985. Relationship of crayfish (*Orconectes virilis*) growth to population abundance and system productivity in small oligotrophic lakes in the Experimental Lakes Area, northwestern Ontario. *Can. J. Fish. Aquat. Sci.* 42: 1096-1102.

France, R.L. 1987. Reproductive impairment of the crayfish *Orconectes virilis* in response to acidification of Lake 223. *Can. J. Fish. Aquat. Sci.* 44(1): 97-106.

Goellner, K.E. 1943. The life cycle and productivity of the crayfish *Cambarus immunis* Hagen. Ph.D Dissertation, University of Michigan, Ann Arbor.

Goldman, C.R. & J.C. Rundquist 1977. A comparative ecological study of the California crayfish, *Pacifastacus leniusculus* (Dana), from two subalpine lakes (Lake Tahoe and Lake Donner). *Freshwat. Crayfish* 3: 51-80.

Gould, S.J. 1966. Allometry and size in ontogeny and phylogeny. *Biol. Rev.* 41: 587-640.

Green, R.H. 1979. *Sampling Design and Statistical Methods for Environmental Biologists*. New York: Wiley.

Hamr, P. & M. Berrill 1985. The life histories of north-temperate populations of the crayfish *Cambarus robustus* and *Cambarus bartoni*. *Can. J. Zool.* 63: 2313-2322.

Hartnoll, R.G. 1985. Growth, sexual maturity and reproductive output. In A.M. Wenner (ed.), *Factors in Adult Growth: Crustacean Issues* 3: 101-128. Rotterdam: Balkema.

Hines, A.H. 1982. Allometric constraints and variables of reproductive effort in brachyuran crabs. *Mar. Biol.* 69: 309-320.

Hoaglin, D.C., F. Mosteller & J.W. Tukey 1983. *Understanding Robust and Exploratory Data Analysis*. New York: Wiley.

Hoglund, H. 1943. On the biology and larval development of *Leander squilla* (L.) *forma typica* de Man. *Svenska Hydrograph. Biol. Komm. Skr. NY Serie. Biol. Bd.* 2: 1-43.

Hollander, M. & D.A. Wolfe 1973. *Nonparametric Statistical Methods*. New York: Wiley.

Huitema, B.E. 1980. *The Analysis of Covariance and Alternatives*. New York: Wiley.

Hynes, H.B.N. 1954. The ecology of *Gammarus duebeni* Lilljeborg and its occurrence in fresh water in western Britain. *J. Animal Ecol.* 23: 38-84.

Ivanova, M.B. & S.V. Vassilenko 1987. Relationship between number of eggs, brood weight, and female body weight in Crustacea. *Int. Revue ges. Hydrobiol.* 72: 147-169.

Jensen, A.L. 1986. Functional regression and correlation analysis. *Can. J. Fish. Aquat. Sci.* 43: 1742-1745.

Jensen, J.P. 1958. The relation between body size and number of eggs in marine malacostrakes. *Meddel. Danmarks Fisk-og Havundersog. NY Serie*, 2: 1-25.

Jolicoeur, P. 1975. Linear regression in fishery research: Some comments. *J. Fish. Res. Bd. Can.* 32: 1491-1494.

Kaplan, R.H. & S.N. Salthe 1979. The allometry of reproduction: An empirical view in salamanders. *Amer. Nat.* 113: 671-689.

Katre, S. 1977. The relation between body size and number of eggs in the freshwater prawn, *Macrobrachium lamarrei* (H. Milne Edwards) (Decapoda, Caridea). *Crustaceana* 33: 17-22.

Kensler, C.B. 1968. Notes on fecundity in the marine spiny lobster *Jasus edwardsii* (Hutton) (Crustacea: Decapoda: Palinuridae). *NZ J. Mar. Freshwat. Res.* 2: 81-89.

Kinne, O. 1961. Growth, molting frequency, heart beat, number of eggs, and incubation time in *Gammarus zaddachi* exposed to different environments. *Crustaceana* 2: 26-36.

Lawlor, L.R. 1976. Parental investment and offspring fitness in the terrestrial isopod *Armadillidium vulgare* (Latr.) (Crustacea: Oniscoidea). *Evolution* 30: 775-785.

Mann, R.H.K. & C.A. Mills 1979. Demographic aspects of fish fecundity. In P.J. Miller (ed.), *Fish Phenology: Anabolic Adaptiveness in Teleosts* pp. 161-177. London: Academic.

Mason, J.C. 1978. Significance of egg size in the freshwater crayfish, *Pacifastacus leniusculus* (Dana). *Freshwat. Crayfish* 4: 83-92.

McGriff, D. 1983. Growth, maturity, and fecundity of the crayfish, *Pacifastacus leniusculus*, from the Sacramento-San Joaquin delta. *Calif. Fish and Game* 69: 227-242.

McNeil, D.R. 1977. *Interactive Data Analysis. A Practical Primer*. New York: Wiley.

McPhail, J.D. 1977. Inherited interpopulation differences in size at first reproduction in threespine stickleback, *Gasterosteus aculeatus* L. *Heredity* 38: 53-60.

Momot, W.T. 1967. Population dynamics and productivity of the crayfish *Orconectes virilis* in a marl lake. *Amer. Midl. Nat.* 78: 55-81.

Momot, W.T. 1986. Production and exploitation of the crayfish, *Orconectes virilis*, in northern climates. In G. Jamieson & N. Bourne (eds), *North Pacific Workshop on Stock Assessment and Management of Invertebrates. Can. Spec. Publ. Fish. Aquat. Sci.* 92: 154-167.

Momot, W.T. & H. Gowing 1977. Response of the crayfish *Orconectes virilis* to exploitation. *J. Fish. Res. Bd. Can.* 34: 1212-1219.

Mosteller, F. & J.W. Tukey 1977. *Data Analysis and Regression. A Second Course in Statistics.* Reading MA: Addison-Wesley.

Overall, J.E. & J.A. Woodward 1977. Common misconceptions concerning the analysis of covariance. *Multivariate Behav. Res.* 12: 171-185.

Paine, R.T. 1976. Size-limited predation: An observational and experimental approach with the *Mytilus-Pisaster* interaction. *Ecology* 57: 858-873.

Paris, O.H. & F.A. Pitelka 1962. Population characteristics of the terrestrial isopod *Armadillidium vulgare* in California grassland. *Ecology* 43: 229-248.

Pasternack, B.S. & R.R. Gianutsos 1969. Application of the exponential and power functions to the study of allometric growth, with particular reference to doryline ants. *Amer. Nat.* 103: 225-234.

Penn, Jr G.H. 1943. A study of the life history of the Louisiana red-crawfish, *Cambarus clarkii* Girard. *Ecology* 24: 1-18.

Perkins, H.C. 1971. Egg loss during incubation from offshore northern lobsters (Decapoda: Homaridae). *Fish. Bull.* 69: 451-453.

Rahn, H., C.V. Paganelli & A. Ar 1975. Relation of avian egg weight to body weight. *Auk* 92: 750-765.

Rayner, J.M.V. 1985. Linear relations in biomechanics: the statistics of scaling functions. *J. Zool., Lond.* (A) 206: 415-439.

Rey, W.J.J. 1983. *Introduction to Robust and Quasi-Robust Statistical Methods*. Berlin: Springer-Verlag.

Reznick, D. & J.A. Endler 1982. The impact of predation on life history evolution in Trinidadian guppies (*Poecilia reticulata*). *Evolution* 36: 160-177.

Ricker, W.E. 1973. Linear regression in fisheries research. *J. Fish. Res. Bd. Can.* 30: 409-434.

Ricker, W.E. 1984. Computation and uses of central trend lines. *Can. J. Zool.* 62: 1897-1905.

Ryan, Jr T.A., B.L. Joiner & B.F. Ryan 1981. *Minitab Reference Manual.* Boston: Duxbury.

Saila, S.B., J.M. Flowers & J.T. Hughes 1969. Fecundity of the American lobster, *Homarus americanus. Trans. Amer. Fish. Soc.* 98: 537-539.

Salmon, M. 1987. On the reproductive behavior of the fiddler crab *Uca thayeri*, with comparisons to *U. pugilator* and *U. vocans*: evidence for behavioral convergence. *J. Crust. Biol.* 7: 25-44.

Salthe, S.N. 1969. Reproductive modes and the number and sizes of ova in the urodeles. *Amer. Midl. Nat.* 81: 467-490.

Schmidt-Nielsen, K. 1975. Scaling in biology: the consequences of size. *J. Exp. Zool.* 194: 287-308.

Schmidt-Nielsen, K. 1984. *Scaling: Why is Animal Size so Important?* Cambridge: Cambridge Univ. Press.

Shea, B.T. 1985. Bivariate and multivariate growth allometry: statistical and biological considerations. *J. Zool., Lond.* (A) 206: 367-390.

Smart, Jr G.C. 1962. The life history of the crayfish *Cambarus longulus longulus*. *Amer. Midl. Nat.* 68: 83-94.

Smith, C.C. & S.D. Fretwell 1974. The optimal balance between size and number of offspring. *Amer. Nat.* 108: 499-506.

Somerton, D.A. 1981. Contributions to the life history of the deep-sea king crab, *Lithodes couesi*, in the Gulf of Alaska. *Fish. Bull.* 79: 259-269.

Somerton, D.A. & R.A. MacIntosh 1985. Reproductive biology of the female blue king crab *Paralithodes platypus* near the Pribilof Islands, Alaska. *J. Crust. Biol.* 5: 365-376.

Stearns, S.C. 1976. Life-history tactics: A review of the ideas. *Quart. Rev. Biol.* 51: 3-47.

Stein, R.A. 1977. Selective predation, optimal foraging, and the predator-prey interaction between fish and crayfish. *Ecology* 58: 1237-1253.

Tack, P.I. 1941. The life history and ecology of the crayfish *Cambarus immunis* Hagen. *Amer. Midl. Nat.* 25: 420-446.

Taylor, C.C. 1947. Some observations on the weight and numbers of eggs of female lobsters. *Dept. Sea and Shore Fish., Fish. Circ.* 1: 1-3, Fish. Lab. Boothbay Harbor, ME.

Teissier, G. 1960. Relative growth. In T.H. Waterman (ed.), *The Physiology of Crustacea. I. Metabolism and Growth:* pp. 537-560. New York: Academic.

Tilley, S.G. 1968. Size-fecundity relationships and their evolutionary implications in five desmognathine salamanders. *Evolution* 22: 806-816.

Wenner, A.M. 1988. Crustaceans and other invertebrates as indicators of beach pollution. In D.F. Soule & G.S. Kleppel (eds), *Marine Organisms as Indicators:* pp. 199-229. New York: Springer-Verlag.

Wenner, A.M., D.M. Hubbard, J. Dugan, J. Shoffner & K. Jellison 1987. Egg production by sand crabs (*Emerita analoga*) as a function of size and year class (Decapoda, Hippidae). *Biol. Bull. 172: 225-235.*

Wenner, A.M., H.M. Page & P.R. Siegel 1985. Variation in size at onset of egg production. In A.M. Wenner (ed.), *Factors in Adult Growth: Crustacean Issues* 3: 149-164. Rotterdam: Balkema.

White, J.F. & S.J. Gould 1965. Interpretation of the coefficient in the allometric equation. *Amer. Nat.* 99: 5-18.

Woodhead, A.D. 1979. Senescence in fishes. In P.J. Miller (ed.), *Fish Phenology: Anabolic Adaptiveness in Teleosts:* pp. 179-205. London: Academic.

Wootton, R.J. 1979. Energy costs of egg production and environmental determinants of fecundity in teleost fishes. In P.J. Miller (ed.), *Fish Phenology: Anabolic Adaptiveness in Teleosts:* pp. 133-159. London: Academic.

Crustacean egg production and fisheries management

Louis W. Botsford
Department of Wildlife and Fisheries Biology, University of California, Davis, USA

ABSTRACT

Egg production and its impact on recruitment are critical aspects of the management of crustacean harvest, but our understanding of them involves considerable uncertainty. Management has therefore adopted a conservative strategy of simply protecting reproductive capacity using one or more of several *ad hoc* tactics. The effect of crustacean egg production on recruitment has been described by stock-recruitment relationships in a few cases, but there is not a general, consistent form; each case involves substantial environmental variability. In addition to the obvious dangers of management allowing egg production to become too low (lower yield and higher probability of collapse), there are also undesirable consequences of egg production being too high (lower yield and greater variability). Future management of crustacean egg production could be refined by considering the age or size dependence of fecundity as well as life history characteristics that are unique to crustaceans.

1 INTRODUCTION

While the other chapters in this volume present various physiological and ecological aspects of our current understanding of crustacean egg production, the purpose of this chapter is to explore the implications of some of that knowledge for management of crustacean harvest. This is particularly appropriate, because much of the available information on crustacean egg production has been gathered for management purposes.

The approach to management of crustacean populations is similar to the approach to management of fish populations, and many of the techniques used are the same. However, some of the unique characteristics of crustaceans lead to different management techniques. Differences between fish and crustaceans that are relevant to management include the discrete nature of growth in crustaceans, the relationship of egg production to that process, ease of differentiating between sexes, the ability to easily recognize ovigerous females, the possibility of a terminal molt at a range of sizes, and possible male/female size relationships required for mating.

Egg production is a primary consideration in the management of crustacean fisheries. In these populations, as in fish populations, the harvest rate depends on the rate of recruitment into the population, which in turn depends on the total rate of egg production by the population (Fig. 1). The shape of the latter dependence varies, as described below. Total egg production

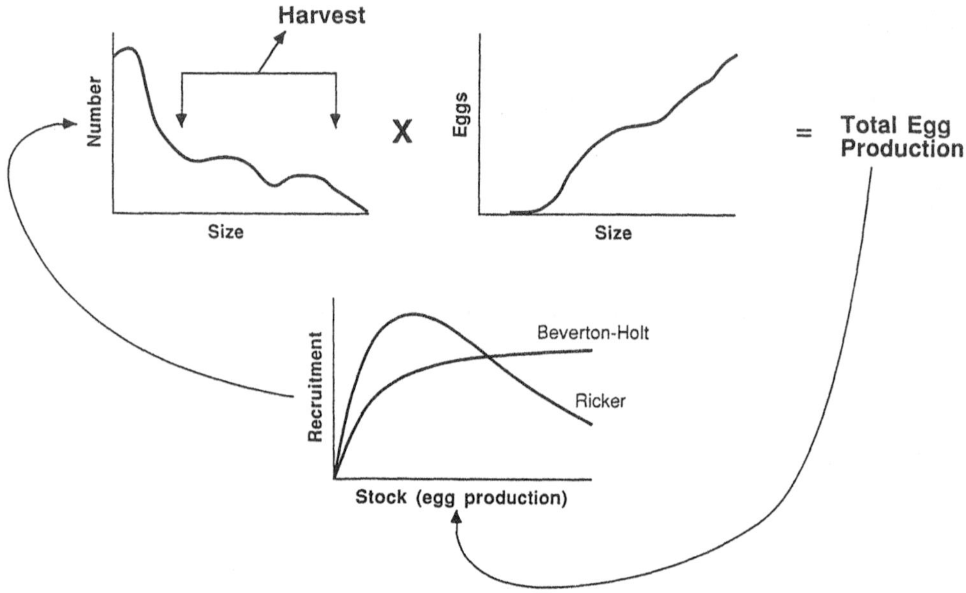

Figure 1. A schematic view of how egg production affects management. Harvest influences the size structure. Total egg production, which depends on size structure and fecundity at each size, influences recruitment through an egg-recruitment relationship. Recruitment in turn affects the size structure and harvest.

depends on the number of individuals at each size or age, which depends on harvest policy in the recent past (Fig. 1).

2 MANAGEMENT OF CRUSTACEAN FISHERIES

The methods that have been used by fishery managers to account for egg production in crustacean fisheries are basically a conservative response to uncertainty, designed to avoid recruitment overfishing (i.e. low egg production due to excessive harvest) and to minimize the risk of recruitment failure. The uncertainty involved in crustacean fisheries stems from both an incomplete understanding of population dynamics (e.g. the dependence of recruitment on egg production, and the effects of harvest on size structure) and unknown future variability due to a fluctuating environment (e.g. inter-annual variability in recruitment). The basic management strategy for dealing with this uncertainty has been to protect egg production as much as possible, using a variety of specific management tactics.

A common tactic is single-sex harvesting, in which only males are allowed to be landed. This policy is possible for management of some crustaceans because of the relative ease of differentiating the sexes. In some fisheries, this means of protection is facilitated by the low marketability of females. A second tactic, a limited season, is used where reproduction is concentrated in one part of the year (see Conan 1985 for a review of seasonal periodicity of molt and reproduction). The fishing season begins after the reproductive season and does not continue into the next reproductive season; thus egg production is greater than it would be with effort distributed over all months (cf. Crocos 1987). Another tactic that sometimes discriminates against females is a lower size limit, which reduces the number of females in the catch in species in which females grow more slowly or achieve a lower final size. Use of a lower size limit in a single sex fishery requires demonstration that a sufficient number of males escape the fishery to mate with females. In at least one case, the red king crab (*Paralithodes camtschatica*) in Alaska, a single sex fishery on males was deemed responsible for a large number of mature females not bearing eggs (McMullen & Yoshihara 1969); however, in most cases single sex fisheries have been found not to affect egg production (e.g. Ennis et al. 1988). An upper size limit has been used in some fisheries based on the questionable argument that reproductive value of large individuals is greater than their market value (see Botsford & Hobbs 1986 for a discussion of the suitability of this argument). Another tactic is protection of females that are obviously about to reproduce, by not allowing 'berried' (i.e. ovigerous) females to be landed. Lastly, a rarely used tactic is the practice of allowing fishermen to permanently mark the carapace in a specific way (e.g. v-notching the tail of lobsters), then not allowing these individuals to be landed. In most cases the individuals marked by fishermen are berried females. The distribution of these tactics over various crustacean species is shown in Table 1. In addition to these management tactics, overall fishing effort is controlled in some crustacean fisheries.

Current management of the lobster (*Homarus americanus*) fishery in Maine provides an example of the effects of most of these. In that fishery the lower size limit is currently 81 mm, the upper size limit is 127 mm, and neither berried nor v-notched lobsters can be landed (the lower size limit is in the process of being changed). As fishing effort has increased over the years, so has concern that egg production will fall below a critical level and recruitment will decline.

The effects of various changes in management are usually evaluated using population models that describe the processes of growth and egg production at varying levels of detail. Initially, continuous models developed for fish harvest were borrowed for crustaceans, but accurate evaluation of crustacean harvest requires a realistic description of the traits specific to crustaceans, such as growth by molting and the dependence of reproduction on that process [see Caddy (1979) and Ennis (1985) for two examples of realistic crustacean models]. The effects

Table 1. Examples of the use of various management options designed to protect egg production.

Species	Location	Single sex	Season	Lower limit	Upper limit	Berried protection	V-notch
Paralithodes camtschatica	Bering Sea	x	x	x			
Chionoecetes bairdi	Eastern Pacific	x	x	x			
Chionoecetes opilio	Eastern Pacific	x	x	x			
Cancer magister	Western US	x	x	x			
Cancer magister	Western Canada		x	x			
Cancer pagurus	Europe		x	x			
Callinectes sapidus	Eastern US			x			
Homarus americanus	Eastern N.A.			x	Maine	x	x
Homarus gammarus	Europe			x		In some locations	
Panulirus argus	Florida		x	x			
Panulirus cygnus	Western Australia		x	x		x	
Jasus lalandii	South Africa			x			
Jasus edwardsii	New Zealand			x		x	

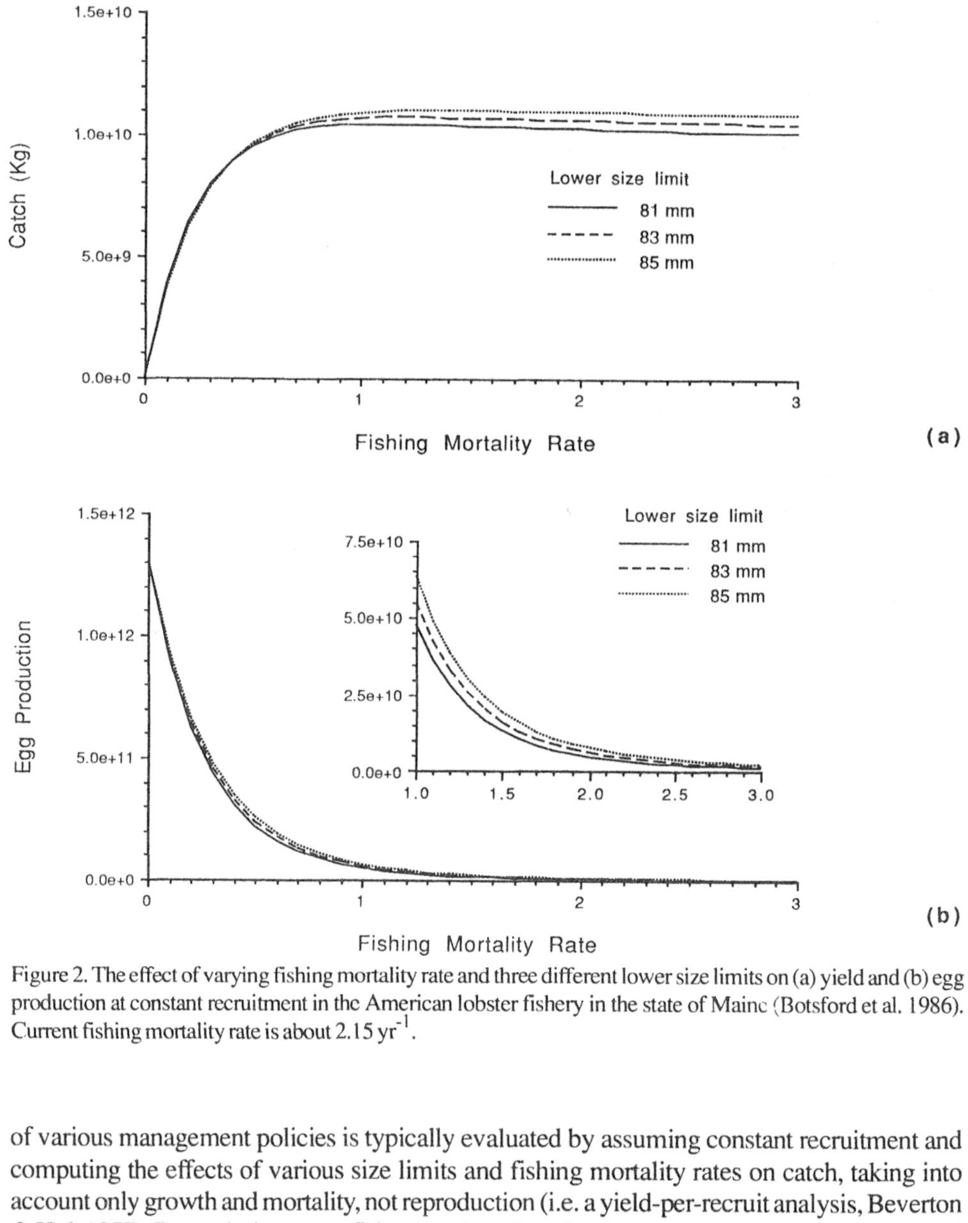

Figure 2. The effect of varying fishing mortality rate and three different lower size limits on (a) yield and (b) egg production at constant recruitment in the American lobster fishery in the state of Maine (Botsford et al. 1986). Current fishing mortality rate is about 2.15 yr^{-1}.

of various management policies is typically evaluated by assuming constant recruitment and computing the effects of various size limits and fishing mortality rates on catch, taking into account only growth and mortality, not reproduction (i.e. a yield-per-recruit analysis, Beverton & Holt 1957). Recently, however, fishery analysts have begun to evaluate the potential effect of various harvest policies on reproduction as well by computing egg production (i.e. eggs-per-recruit) (Saila & Flowers 1965, Bennet & Edwards 1981, Campbell 1985, 1986, Ennis 1985, Botsford et al. 1986).

For many heavily exploited species the projected results of changes in various management tactics are similar to those obtained in a recent analysis of the Maine lobster fishery (Botsford et al. 1986). Egg production per recruit is currently far below that of an unharvested population (1%), and the current yield could be taken with far less effort (cf. Campbell 1985, 1986, Ennis 1986) (Fig. 2). Increasing the lower size limit would lead to long term increases in both yield

Table 2. The effect of v-notching American lobsters in the Maine fishery (from Botsford et al. 1986).

Fraction of berried females caught that are returned	Catch (millions of pounds)	Total egg production (billions)
0.00	22.46	5.53
0.05	22.45	7.35
0.10	22.44	9.04
0.15	22.42	10.61
0.20	22.40	12.06
0.25	22.39	13.41
0.50	22.33	18.80
1.00	22.26	25.20

-per-recruit and eggs-per-recruit at the expense of a short term decrease in yield. Removal of the upper size limit in this case increased catch by an insignificant amount, and decreased egg production by 30%. Thus, having an upper size limit, which increased egg production but reduced catch, did not compare favorably to increasing the lower size limit, which increased both long term catch and egg production. The effects of v-notching on this fishery were evaluated for the case in which only a certain fraction of berried females caught by fishermen were v-notched. The results were a dramatic increase in egg production with only a small decline in catch as the fraction v-notched increased (Table 2).

3 EFFECT OF EGG PRODUCTION ON RECRUITMENT

Comprehensive evaluation of the implications of egg production on management and the effectiveness of existing management tactics require an understanding of how egg production affects recruitment. Recruitment, as used here, means abundance of a cohort at the age at which relative cohort abundance has been set. This age is difficult to define for crustaceans in general, but typically will be shortly after settlement. Recruitment to a crustacean population depends on egg production through a more or less deterministic stock-recruitment relationship but is also affected by egg predation, the random pre-recruitment environment, and migration of planktonic larvae.

A discussion of stock and recruitment must be prefaced by some mention of the general difficulties involved in determining the form of such a relationship, or even establishing that one exists. In general, considerable uncertainty in parameter values of stock-recruitment relationships results from errors incurred in measuring stock and recruitment (Walters & Ludwig 1981). Walters (1985) and Kope (1988) have shown that there is substantial bias in stock-recruitment parameter values obtained from short series, although the magnitude of the bias is less in the presence of substantial environmental variability (Caputi 1988). Even for fish populations there are few good examples of clearly defined stock-recruitment relationships, and in some cases analysts have begun to employ probabilistic descriptions of this relationship (e.g. Getz & Swartzmann 1981, Evans & Rice 1988). For crustaceans specifically, the fact that little was known regarding crustacean stock-recruitment relationships by 1970 (Hancock 1973) and that little progress was made in the ensuing 15 years (Caddy 1986) are in large part due to these difficulties [also see Garcia (1983)].

A model of the stock-recruitment relationship should describe the dependence of recruitment on stocks as their level and composition change due to harvest. For our purposes we will represent stock in terms of total egg production by the stock. The stock-recruitment relationship describes the intuitively reasonable proposition that at low levels of egg production, as the number of eggs produced increases, the number of recruits will increase in direct proportion, but will not continue to do so at higher levels of egg production. Although stock recruitment curves can have a variety of shapes, for the purposes of evaluating dynamic behavior these are usually separated into two general types: one in which recruitment increases with increasing egg production up to a level at which it remains constant for any further increase in egg production, and a second in which recruitment first increases then declines with increasing egg production. The former is known as a Beverton-Holt type (Beverton & Holt 1957), and the latter is known as a Ricker type (Ricker 1954) (the two examples in Fig. 1). There is typically not enough data, neither in general nor in specific cases, to clearly determine which is the appropriate form, neither for fish nor for curstaceans (Caddy 1986).

Expressing stock in terms of total egg production is more realistic than using numbers or biomass, because it more accurately reflects the influence of size structure on the potential contribution to recruitment. For most crustaceans, egg production is an allometric function of size for sizes greater than the size of maturity (see Hartnoll 1985, Lipcius 1985 and the reviews by Ivanova & Vassilenko 1987 and Somers, this volume). The age of first reproduction varies within genera (e.g. Orensanz & Gallucci 1988) and even within the same species (e.g. the American lobster along the east coast of North America, Campbell & Robinson 1983; *Emerita analoga* along the coast of southern California, Wenner et al. 1985). Total egg production can be expressed as the sum, over all sizes, of the number of individuals at each size times egg production at that size. In at least one species, spiny lobster in the Hawaiian Islands, the size of first maturity has changed with exploitation (Polovina 1989), hence the way in which eggs are summed over age must vary with density.

Effective total egg production in some species is dramatically influenced by the presence of egg predators. In particular, several species of nemerteans have been found on various harvested crustaceans [see Wickham (1986) for a review].

Although egg production affects recruitment in crustaceans, its effect is often overshadowed by abiotic and biotic factors in the subsequent pre-recruitment environment. This may occur after settlement through mechanisms such as limited resources (Chittleborough & Phillips 1975, Fogarty & Idoine 1986), cannibalism, either by adults or other juveniles (Winnor 1966), or abiotic factors such as rainfall or freshwater runoff (Garcia 1983, Penn & Caputi 1985, Sutcliffe 1973). During the larval stages, the environment can affect later recruitment through variation in food (Incze et al. 1987, Paul & Paul 1980, Dawirs 1984), predation (Botsford et al. 1982), and transport (Jamieson et al. 1988, Johnson et al. 1986, Johnson 1985, Nichols et al. 1982, Rothlisberg 1982, Rothlisberg & Miller 1983, Phillips & McWilliam 1986).

In addition to adding random variability to the stock-recruitment relationship, the possibility of transport over long distances can cast doubt on the identity of the stock. For example, the potential for transport by the counter-clockwise currents in the Gulf of Maine make it difficult to determine the source of recruitment to the juvenile lobster populations in both Canadian and US waters (Harding et al. 1983, 1987, Harding & Trites 1988). As a result the effects of adult harvest of the various stocks is difficult to determine.

Relatively few stock-recruitment curves have been established for crustaceans, and most of them are obscured by measurement errors and environmental variability. The relationship between puerulus settlement and spawning stock in the western rock lobster (*Panulirus cygnus*) is one of the better ones (Morgan et al. 1982) (Fig. 3a). In this plot, there is clear evidence for the

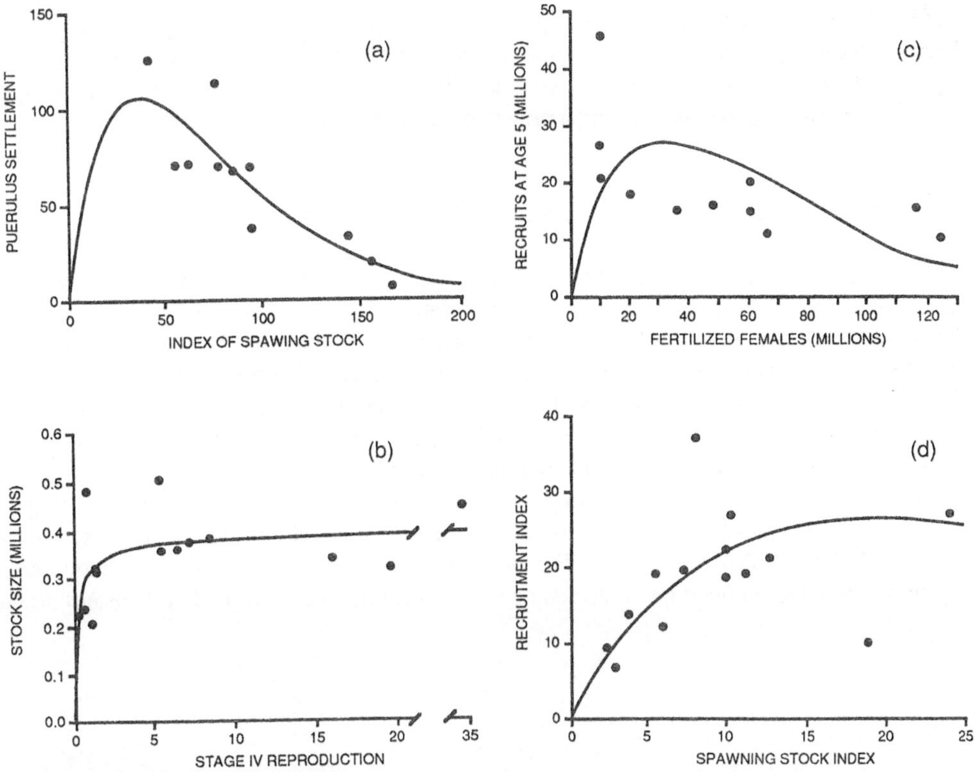

Figure 3. Examples of stock-recruitment relationships that have been formulated for various harvested crustacean species: (a) the western rock lobster in Western Australia, *Panulirus cygnus* (redrawn from Morgan et al. 1982), (b) the American lobster (*Homarus americanus*) in eastern Canada (redrawn from Fogarty & Idoine 1986), (c) the red king crab (*Paralithodes camtschatica*) in the Bering Sea (Redrawn from Reeves 1985), and (d) the tiger prawn (*Penaeus esculentus*) in Western Australia (Redrawn from Penn & Caputi 1986).

descending limb necessary for a Ricker type relationship. [Australian researchers currently (1990) have little confidence in these data (N. Caputi pers. comm.)]. However, recruitment to fishery may approach an upper limit (i.e. its dependence on puerulus settlement may be of the Beverton-Holt type) because of limited resources in the post-settlement environment (Chittleborough & Phillips 1975, Phillips 1986). Scarratt's (1973) larval data provide the basis for a stock-recruitment relationship in the American lobster (*Homarus americanus*) in the Gulf of St. Lawrence (Fig. 3b). That this relationship is of the Beverton-Holt type has been attributed to limited resources in the post-settlement environment (Fogarty & Idoine 1986). For crabs, the only published relationship is that for red king crab (*Paralithodes camtschatica*), which is not a tight fit to the data, but appears to be of the Ricker type (Reeves 1985, Jamieson 1986) (Fig. 3c). A Ricker type stock-recruitment relationship is implied in several analyses of cyclic fluctuations of Dungeness crab in the western United States, but has never been demonstrated (Botsford 1986a). Most attempts to establish a stock-recruitment relationship for shrimp have resulted in a linear dependence of recruitment on female abundance (Garcia 1983). Garcia

(1983) concluded that because of year-round spawning and intraseries correlation in most data, none of the previous attempts adequately demonstrate a stock-recruitment relationship for shrimp. However, Penn & Caputi (1985) developed a stock-recruitment relationship for a penaeid (*Penaeus esculentus*), the tiger prawn in Western Australia (Fig. 3d).

4 EFFECT OF EGG PRODUCTION ON POPULATION DYNAMICS

With some understanding of the nature of the egg-recruitment relationship, we can evaluate the effect of various levels of egg production on population dynamics. From the previous section, the shape of the egg-recruitment relationship is uncertain (i.e. they may be of either the Beverton-Holt or the Ricker type), and it is also substantially influenced by environmental variability. Management's efforts to cope with this uncertainty have tended to guard egg production, based on the assumption that having too few eggs is a serious problem, without much concern for having too many eggs. However, both are potential problems.

It is obvious that having too few eggs can lead to recruitment overfishing and collapse (having no eggs leads to no recruits, except in areas where larval immigration occurs), but its occurrence is difficult to demonstrate. For the American lobster, collapses to two fisheries have been attributed to reduced adult stocks, central Northumberland Straight and eastern Nova

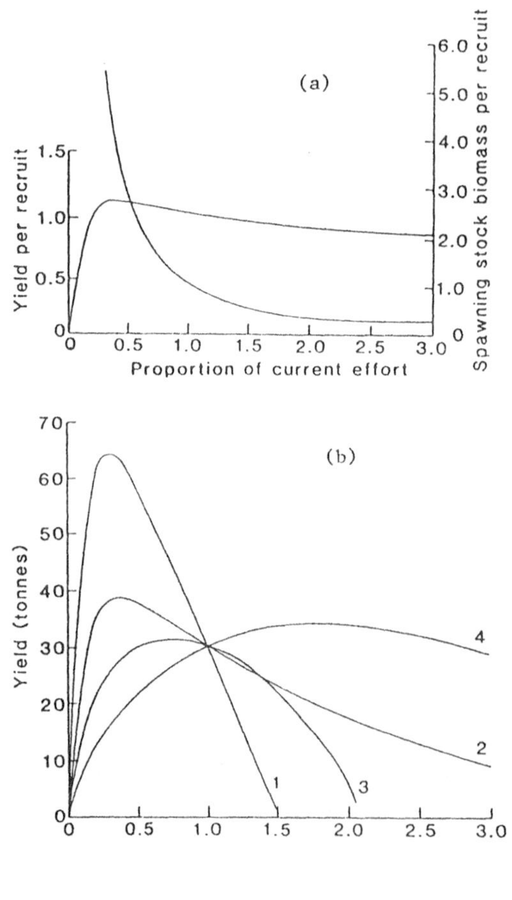

Figure 4. (a) Yield per recruit and spawning stock biomass per recruit plotted versus fishing mortality rate (normalized to current rate) at the current size limit, for female lobsters, *Homarus gammarus*, in Yorkshire, 1972-80, (b) yield from a combination of the yield-per-recruit model that generated part (a) and several different stock--recruitment models. Curve number 2 uses a Beverton-Holt type and curve number 4 uses a Ricker type (the others are intermediate forms). (Redrawn from Bannister & Addison 1986).

Scotia (Robinson 1979, Stasko 1980, Ennis 1986). Shrimp populations have been characterized as very resilient, hence relatively invulnerable to overharvest, with the exception of the Louisiana population of *Penaeus setiferous* (Neal 1975, Neal & Maris 1985). The decline in catches of the rock lobster off South Africa in the mid 1960's was initially attributed to recruitment overfishing, but other possibilities have recently been suggested (Pollock 1987, Pollock & Shannon 1987). The central California Dungeness crab fishery collapsed in the late 1950's, but Methot & Botsford (1982) showed that harvest rates were the same as those for adjacent populations, and other possible causes have been suggested (physical environment; an egg predator, Wickham 1979; density-dependent growth, Botsford 1981).

The potential negative effects of total egg production being too high are twofold: a decrease in yield and greater sensitivity to environmental variability. Both of these depend on the shape of the stock-recruitment curve. In the presence of a descending limb on the stock-recruitment curve (i.e. it being of the Ricker type rather than the Beverton-Holt type) higher egg production can actually result in lower yield. As an example, Bannister & Addison (1986) combined several hypothetical stock-recruitment curves for the lobster (*H. gammarus*) with a model of growth and mortality whose yield-per-recruit and eggs-per-recruit characteristics were as shown in Figure 4a. They then examined the effects of various levels of fishing effort (both sexes harvested) on yield from the combined models.

The addition of recruitment to the model changed the yield characteristics in different ways depending on the shape of the stock-recruitment curve (Fig. 4a, b). Curve 2 in Figure 4b is the yield that would result from the yield-per-recruit and egg-per-recruit relationships in Figure 4a if recruitment were computed from a stock-recruitment curve of the Beverton-Holt type. The shape of the yield declines more rapidly with increasing effort. Similarly, curve 4 in Figure 4b is the yield that would result if recruitment were computed from a stock-recruitment curve of the Ricker type. The region of higher yield is shifted to the right, resulting in low yield at high egg production (low fishing effort). Thus when egg production and its effect on recruitment are accounted for, the yield versus effort relationship is considerably different, especially when the egg-recruit relationship is of the Ricker type. Having too many eggs leads to recruitment, hence yield, being lower than it would be at higher effort levels. The behavior of the combined model (Fig. 4b) is an example of typical behavior, but details in any specific case will depend on growth, mortality and fecundity schedules. In this case both sexes were fished. When egg production is protected by other means, such as single sex harvest, the effect on yield would depend on whether the density-dependence in the stock recruitment relationship were due to female density or both male and female density.

Having egg production too high also leaves a population more sensitive to random environmental influences on recruitment, especially when the stock-recruitment relationship is of the Ricker type. Botsford (1986b) used a model in which recruitment each year was proportional to the product of total egg production, a recruitment survival function that depended on adult density (Fig. 5a), and multiplicative environmental noise, to evaluate the sensitivity of recruitment to environmental variability. He showed that this sensitivity defined as the coefficient of variation in recruitment divided by the coefficient of variation of the environmental variable that influences recruitment survival, increased as the slope of the recruitment survival function at equilibrium (i.e. K in Fig. 5a) increased. When only males are harvested, the equilibrium point does not change and this slope remains steep. However, when both sexes are harvested the equilibrium moves in the direction shown by the arrow in Figure 5a, to points with lower slope, thus reducing sensitivity to the random pre-recruitment environment. Single sex harvesting does not reduce population density, but rather allows it to remain at the high levels at

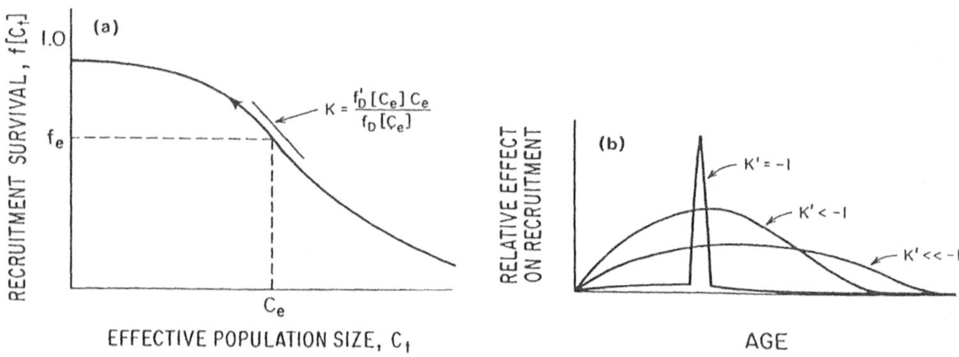

Figure 5. (a) A hypothetical recruitment survival function, the ratio of recruits to total population egg production, and (b) the age structure as it might affect density-dependent recruitment. Sensitivity of the population to environmental perturbations increases with increasingly negative slope, K, and increasingly narrow, peaked age structure. (From Botsford 1986b).

which density-dependent effects are strongest. He also showed that populations with narrower, more peaked age structure were more sensitive to environmental fluctuations than populations with broader, flatter age structure (i.e. more positive values of K' in Fig. 5b are more sensitive). Thus, size-selective harvest can make a population more sensitive to environmental fluctuations.

The implications of these results are that for a population with a Ricker type stock-recruitment function, size-selective harvest tends to increase the sensitivity of recruitment to environmentally induced variability in recruitment survival. However, harvesting both males and females tends to reduce this sensitivity. If we protect egg production by not harvesting females, recruitment (and therefore abundance and harvest) will be more variable. For example, in a model of the northern California Dungeness crab (*Cancer magister*) fishery, increasing female harvest results in reduced variability in recruitment (Fig. 6).

5 ACCOUNTING FOR THE DEPENDENCE OF EGG PRODUCTION ON SIZE

Information reviewed thus far indicates that, in spite of the fact that the relationship between recruitment and egg production is extremely variable and poorly understood, the problems associated with egg production being too low or too high imply that reducing egg production from the high levels of an unharvested population, to a moderate level, will often yield higher catch. However, there are many ways to reduce egg production. Our choice of management tactic should take into account as much information as possible. Management approaches formulated in the past have used *ad hoc* tactics for protecting egg production, coupled with size limits and fishing effort, which are based on the size and age dependence of individual growth and mortality. These methods of protecting egg production have not accounted for the size- or age-dependent fecundity schedules.

Conventional means of formulating optimal fishery policy can be modified to account for the size or age dependence of egg production. Instead of determining the size limits and effort that maximize yield, based on the age dependence of size and mortality (i.e. a standard yield-per-recruit analysis), one can determine the size limits and fishing effort that maximize yield

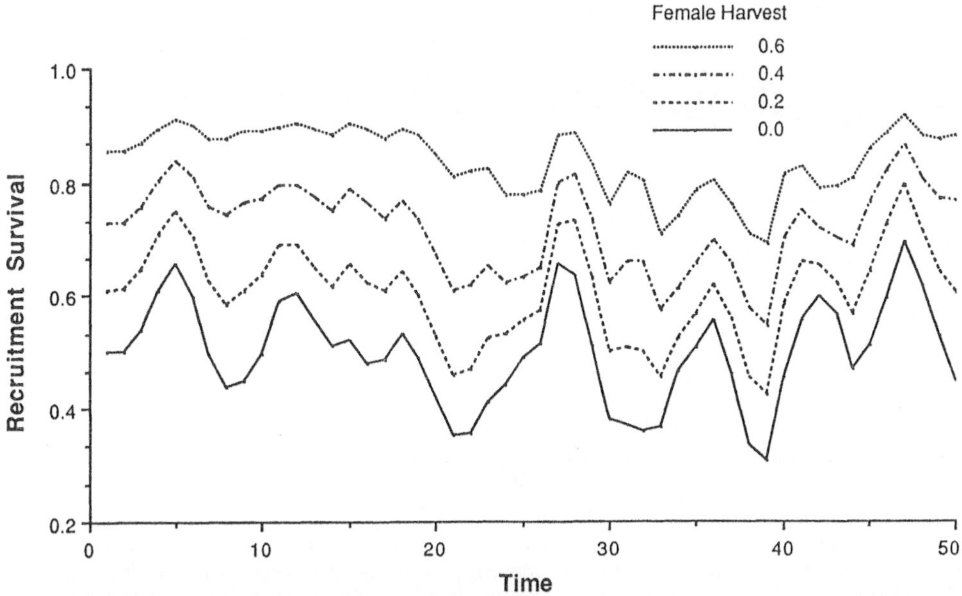

Figure 6. Recruitment survival from an age structured model of the Dungeness crab with recruitment dependent on density and environment. The harvest rate for males is 0.8.

under the constraint that total population egg production, based on the size or age specific fecundity schedule, is held constant. This constant level of egg production can be chosen to be a moderate level based on cursory examination of the available recruitment versus egg production data.

The resulting policy can be displayed graphically on a plot of the combinations of size limit and fishing effort that lead to specific constant values of egg production (Fig. 7, Botsford & Hobbs 1986). The optimal policy follows a curve roughly perpendicular to these iso-egg production lines (see the example in Fig. 7). When a more conservative policy (in terms of total egg production) is required, one moves to the lower right on this line, and vice versa. The shape of this line depends on parameters such as the cost of fishing, market price and recruitment (see Botsford & Hobbs 1986 for details).

6 DISCUSSION

Egg production plays a critical role in crustacean population dynamics, hence must be accounted for in the management of crustacean fisheries. However, our understanding of crustacean egg production is far from complete. Because of this, and the substantial impact of an unpredictable pre-recruitment environment, a conservative approach has been taken to management of egg production in crustacean fisheries. As we have described here the strategy of protecting egg production through various *ad hoc* tactics is not always the best.

In the future one would hope that more information will be available on various aspects of crustacean egg production (e.g. through volumes such as this one), and that this information will

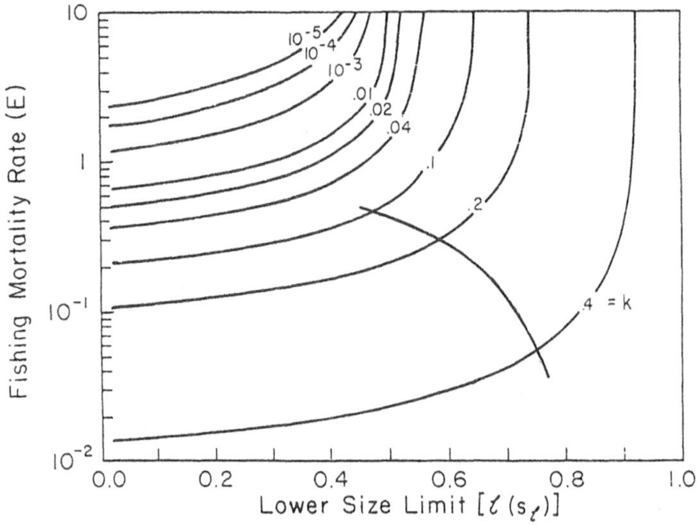

Figure 7. Combinations of values of lower size limit and fishing mortality rate that lead to constant egg production at various levels (k is proportional to total egg production). The combination of size limit and fishing mortality rate that maximizes yield is where the slanted line crosses the value of constant total egg production desired (Botsford & Hobbs 1986).

be used to better manage fisheries. One possible improvement in management, suggested here, is the use of information on age dependence of fecundity in deciding where to set size limits and fishing effort. Application of this would imply, for example, different size limits for the various American lobster fisheries along the eastern United States and Canada. Management policy would thus be more closely matched to the different fecundity schedules.

A second way of improving management of crustacean populations is through increased awareness of the potential effects of egg production being too high. There is some concern that higher recruitment could be obtained from lower egg production in the male-only fishery for snow crab (*Chionoecetes opilio*) in eastern Canada (Ennis et al. 1988). Also, the dramatic fluctuations in catch of Dungeness crab along the west coast of North America could, in part, be due to the sex selective fishery (Botsford 1986b).

Although much of the analysis of the impact of various management tactics has been accomplished using models borrowed from work with fish, in the future there will likely be increasing use of models that incorporate characteristics specific to crustaceans. For example, future models should include growth by discrete molts (e.g. Caddy 1979, Ennis 1985) as well as variability in that process [see Botsford (1985) for examples]. These more realistic models of growth should then include more detailed description of the phase relationships between molting and reproductive events (e.g. Ennis this volume, Hartnoll 1985, Nelson et al. 1988, Nelson this volume, Waddy & Aiken this volume). In addition to these, there are other characteristics specific to crustaceans that are only beginning to be included in management. One example is the existence of a terminal molt, either in females (as in blue crab, *Callinectes sapidus*) or males (as may occur in snow crab, *Chionoecetes opilio*, Conan & Comeau 1986, Jamieson 1986).

Another possibility is that female crustaceans can mate only with males of a specific size range as may be the case for red king crab (*Paralithodes camtschatica*), but is apparently not a problem in harvest of snow crab (*Chionoecetes opilio*) in eastern Canada (Ennis et al. 1988). Another characteristic specific to crustaceans is the potential for infectious egg predators (Wickham 1986). Consideration of these characteristics leads readily to the conclusion that management of crustacean harvest may become increasingly different from management of fish harvest, and perhaps more complex.

ACKNOWLEDGEMENTS

I would like to thank N. Caputi, G. Conan, G.P. Ennis and I.G. Somers for their careful reviews of an earlier version of this manuscript.

REFERENCES

Bannister, R.C.A. & J.T. Addison 1986. Effect of assumptions about the stock-recruitment relationship on a lobster (*Homarus gammarus*) stock assessment. *Can. J. Fish. Aquat. Sci.* 43: 2353-2359.

Bennet, D.B. & E. Edwards 1981. Should we ban the berried lobster? *I.C.E.S.C.M. 1981/K:11*.

Beverton, R.J.H. & S. Holt 1957. On the dynamics of exploited fish populations. *Fish. Invest. London* (2) 19: 533 pp.

Botsford, L.W. 1981. The effects of increased individual growth rates on depressed population equilibria. *Amer. Nat.* 117: 38-63.

Botsford, L.W. 1985. Models of growth. In A.M. Wenner (ed.), *Factors in Adult Growth, Crustacean Issues* 3: 171-188. Rotterdam: A.A. Balkema.

Botsford, L.W. 1986a. Population dynamics of Dungeness crab. *Can. Special Publ. Fish. Aquat. Sci.* 92: 140-153.

Botsford, L.W. 1986b. Effects of environmental forcing on age-structured populations: northern California Dungeness crab (*Cancer magister*) as an example. *Can. J. Fish. Aquat. Sci.* 43: 2345-2352.

Botsford, L.W. & R.C. Hobbs 1986. Static optimization of yield per recruit with reproduction and fishing costs. *Fish. Res.* 4: 181-189.

Botsford, L.W., R.D. Methot & J.E. Wilen 1982. Cyclic covariation in the California king salmon, *Oncorhynchus tshawytscha*, silver salmon, *O. kisutch*, and Dungeness crab, *Cancer magister*, fisheries. *Fish. Bull. US* 80: 791-801.

Botsford, L.W., J.E. Wilen & E.J. Richardson 1986. *Biological and economic analysis of lobster fishery policy in Maine*. Report to committee on Marine Resources, State of Maine, 112th Legislature, 98 pp.

Caddy, J. 1979. Notes on a more generalized yield per recruit analysis for crustaceans using size-specific inputs. Manuscript Report 1525, Fisheries and Marine Science, Halifax, Nova Scotia, Canada.

Caddy, J.F. 1986. Modeling stock-recruitment processes in Crustacea: Some practical and theoretical perspectives. *Can. J. Fish. Aquat. Sci.* 43: 2330-2344.

Campbell, A. 1985. Application of a yield and egg-per-recruit model to the lobster fishery in the Bay of Fundy. *North Amer. J. Fish. Management* 5: 91-104.

Campbell, A. 1986. Implications of size and sex regulations for the lobster fishery of the Bay of Fundy and southwestern Nova Scotia. *Can. Spec. Publ. Fish. Aquat. Sci.* 92: 126-132.

Campbell, A. & D.G. Robinson 1983. Reproductive potential of three American lobster (*Homarus americanus*) stocks in the Canadian Maritimes. *Can. J. Fish. Aquat. Sci.* 40: 1958-1967.

Caputi, N. 1988. Factors affecting the time series bias in stock-recruitment relationships and the interaction between time series and measurement error bias. *Can. J. Fish. Aquat. Sci.* 45: 178-184.

Chittleborough, R.G. & B.F. Phillips 1979. Recruitment to the catchable stock of the western rock lobster, *Panulirus longipes cygnus* George. *Rapp. P.-v. Reun. cons. int. Explor. Mer.*, 175: 139-142.

Conan, G.Y. 1985. Periodicity and phasing of molting. In A.M. Wenner (ed.), *Factors in Adult Growth. Crustacean Issues* 3: 73-99. Rotterdam: A.A. Balkema.

Conan, G.Y. & M. Comeau 1986. Functional maturity and terminal molt of male snow crab, *Chionoecetes opilio*. *Can. J. Fish. Aquat. Sci.* 43: 1710-1719.

Crocos, P.J. 1987. Reproductive dynamics of the grooved tiger prawn, *Penaeus semisulcatus*, in the north--western Gulf of Carpentaria, Australia. *Aust. J. Mar. Freshw. Res.* 38: 79-90.

Dawirs, R.R. 1984. Influence of starvation on larval development of *Carcinus maenas* L. (Decapoda: Portunidae). *J. Exp. Mar. Biol. Ecol.* 80: 47-66.

Ennis, G.P. 1985. An assessment of the impact of size limit and exploitation rate changes on egg production in a Newfoundland lobster population. *North Amer. J. Fish. Management* 5: 86-90.

Ennis, G.P. 1986. Stock definition, recruitment variability, and larval recruitment processes in the American lobster, *Homarus americanus:* A review. *Can. J. Fish. Aquat. Sci.* 43: 2072-2084.

Ennis, G.P. 1990. Annual variation in egg production in a Newfoundland population of the American lobster, *Homarus americanus* (this volume).

Ennis, G.P., R.G. Hooper & D.M. Taylor 1988. Functional maturity in small male snow crabs (*Chionoecetes opilio*). *Can. J. fish. Aquat. Sci.* 45: 2106-2109.

Evans, G.T. & J.C. Rice 1988. Predicting recruitment from stock size without the mediation of a functional relation. *J. cons. int. Explor. Mer.* 44: 111-122.

Fogarty, M.J. & J.S. Idoine 1986. Recruitment dynamics in an American lobster (*Homarus americanus*) population. *Can. J. Fish. Aquat. Sci.* 43: 2368-2376.

Garcia, S. 1983. The stock-recruitment relationship in shrimps: reality or artefacts and misinterpretations? *Oceanogr. Trop.* 18: 25-48.

Getz, W.M. & G.L. Swartzmann 1981. A probability transition model for yield estimation in fisheries with highly variable recruitment. *Can. J. Fish. Aquat. Sci.* 38: 847-855.

Hancock, D.A. 1973. The relationship between stock and recruitment in exploited invertebrates. *Rapp. P.-V. Reun. cons. Int. Explor. Mer.* 164: 113-131.

Harding, G.C., K.F. Drinkwater & W.P. Vass 1983. Factors influencing the size of American lobster (*Homarus americanus*) stocks along the Atlantic coast of Nova Scotia, Gulf of St. Lawrence, and Gulf of Marine: a new synthesis. *Can. J. Fish. Aquat. Sci.* 40: 168-184.

Harding, G.C. & R.W. Trites 1988. Dispersal of *Homarus americanus* larvae in the Gulf of Maine from Browns Bank. *Can. J. Fish. Aquat. Sci.* 45: 416-425.

Harding, G.H., J.D. Pringle, W.P. Vass, S. Pearre Jr & S.J. Smith 1987. Vertical distribution and daily movements of larval lobsters *Homarus americanus* over Browns Bank, Nova Scotia. *Mar. Ecol. Prog. Ser.* 41: 29-41.

Hartnoll, R.G. 1985. Growth, sexual maturity and reproductive output. In A.M. Wenner (ed.), *Factors in Adult Growth. Crustacean Issues* 3: 101-128. Rotterdam: A.A. Balkema.

Incze, L.S., D.A. Armstrong & S.L. Smith 1987. Abundance of larval tanner crabs (*Chionoecetes* spp.) in relation to adult female stocks and regional oceanography of the southeastern Bering Sea. *Can. J. Fish. Aquat. Sci.* 44: 1143-1156.

Ivanova, M.B. & S.V. Vassilenko 1987. Relationship between number of eggs, brood weight, and female body weight in Crustacea. *Int. Revue ges Hydrobiol.* 72: 147-169.

Jamieson, G.S. 1986. Implications of fluctuations in recruitment in selected crab populations. *Can. J. Fish. Aquat. Sci.* 43: 2085-2098.

Jamieson, G.S. & W.D. McKane 1988. Proceedings of the International Workshop on Snow Crab Biology, December 8-10, 1987, Montreal, Quebec. Canadian Manuscript Report of Fisheries and Aquatic Sciences No. 2005, 145 pp.

Jamieson, G.S., A.C. Phillips & W.S. Huggett 1988. Effects of ocean variability on the abundance of Dungeness crab larvae. *Can. Spec. Publications Fish. Aquat. Sci.* (in press).

Johnson, D.F. 1985. The distribution of Brachyuran crustacean megalopae in the waters of the York River, lower Chesapeake Bay and adjacent shelf: implications for recruitment. *Estuarine, Coastal and Shelf Science* 20: 693-705.

Johnson, D.F., L.W. Botsford, R.E. Methot Jr & T.C. Wainwright 1986. Wind stress and cycles in Dungeness crab (*Cancer magister*) catch off California, Oregon and Washington. *Can. J. Fish. Aquat. Sci.* 43: 838-845.

Kope, R.G. 1988. Effects of initial conditions on Monte Carlo estimates of bias in estimating functional relationships. *Can. J. Fish. Aquat. Sci.* 45: 185-187.

Lipcius, R. 1985. Size-dependent reproduction and molting in spiny lobsters and other long-lives decapods. In A.M. Wenner (ed.), *Factors in Adult growth. Crustacean Issues* 3: 129-148. Rotterdam: Balkema.

McMullen, J.C. & H.T. Yoshihara 1969. Fate of unfertilized eggs in king crabs *Paralithodes camtschatica* (Tilesius). Informational Leaflet 127, State of Alaska, Department of Fish and Game.

Methot Jr, R.D. & L.W. Botsford 1982. Estimated preseason abundance in the California Dungeness crab (*Cancer magister*) fisheries. *Can. J. Fish. Aquat. Sci.* 39: 1077-1083.

Morgan, G.R., B.F. Phillips & L.M. Joll 1982. Stock and recruitment relationships in *Panulirus cygnus*, the commercial rock (spiny) lobster of Western Australia. *Fish. Bull.* 80: 475-486.

Neal, R.A. 1975. The Gulf of Mexico research and fishery on penaeid prawns. In 'First Australian National Prawn Seminar, Maroochydore, Queensland' (Australian Fisheries Council, ed.) pp. 1-8 *Australian Government Publ. Service, Canberra.*

Neal, R.A. & R.C. Maris 1985. Fisheries Biology of shrimps and shrimplike animals. In A.J. Provenzano (ed.), *The Biology of Crustacea* 10: 1-110. New York: Academic Press.

Nelson, K., D. Hedgecock & W. Borgeson 1988. Effects of reproduction upon molting and growth in female American lobsters (*Homarus americanus*). *Can. J. Fish. Aquat. Sci.* 45: 805-821.

Nichols, J.H., B.M. Thompson & M. Cryer 1982. Production drift and mortality of the planktonic larvae of the edible crab (*Cancer pagurus*) off the northeast coast of England. *Neth. J. Sea Res.* 16: 173-184.

Orensanz, J.M. & V.F. Gallucci 1988. Comparative study of postlarval life-history schedules in four sympatric species of *Cancer* (Decapoda: Brachyura: Cancridae). *J. Crustacean Biol.* 8: 187-220.

Paul, A.J. & J.M. Paul 1980. The effect of early starvation on later feeding success of king crab zoeae. *J. Exp. Mar. Biol. Ecol.* 44: 247-251.

Penn, J.W. & N. Caputi 1985. Stock recruitment for the tiger prawn (*Penaeus esculentus*) fishery in Exmouth Gulf and their implications for management. In P.C. Rothlisberg, B.J. Hill & D.J. Staples (eds), *Second Australian Prawn Seminar*, NPS2, pp. 165-173. Cleveland, Australia.

Penn, J.W. & N. Caputi 1986. Spawning stock-recruitment relationships and environmental influences on the tiger prawn (*Penaeus esculentus*) fishery in Exmouth Gulf, Western Australia. *Aust. J. Mar. Freshw. Res.* 37: 491-505.

Phillips, B.F. 1986. Prediction of commercial catches of the Western rock lobster *Panulirus cygnus*. *Can. J. Fish. Aquat. Sci.* 43: 2126-2130.

Phillips, B.F. & P.S. McWilliam 1986. The pelagic phase of spiny lobster development. *Can. J. Fish. Aquat. Sci.* 43: 2153-2163.

Pollock, D.E. 1987. Simulation models of rock-lobster populations from areas of widely divergent yields on the Cape West coast. *S. Afr. J. Mar. Sci.* 5: 531-545.

Pollock, D.E. & L.V. Shannon 1987. Response of rock-lobster populations in the Benguela ecosystem to environmental change – a hypothesis. *S. Afr. J. Mar. Sci.* 5: 887-899.

Polovina, J.J. 1989. Density dependency in spiny lobster, *Panulirus marginatus* in the Northwestern Hawaiian Islands. *Can. J. Fish. Aquat. Sci. 46: 660-665.*

Reeves, J.E. 1985. Expected stock status and catch for Bristol Bay red king crab in 1985. Report to North Pacific Fish. Management Council. Natl. Mar. Fish. Serv., NW Alaska Fish. Center, 7600 Sand Pt. way NE, Bldg 4, Bin C 15700, Seattle, WA. 9 pp. (mimeo).

Ricker, W.E. 1954. Stock and Recruitment. *J. Fish. Res. Board Can.* 11: 559-623.

Robinson, D.G. 1979. Consideration of the lobster (*Homarus americanus*) recruitment overfishing hypothesis: with special reference to the Canso Causeway. *Fish. Mar. Serv. Tech. Rep.* 834: 77-99.

Rothlisberg, P.C. 1982. Vertical migration and its effect on dispersal of penaeid shrimp larvae in the Gulf of Carpentaria, Australia. *Fish. Bull. US* 80: 541-554.

Rothlisberg, P.C. & C.B. Miller 1983. Factors affecting the distribution, abundance and survival of *Pandalus jordani* (Decapoda, Pandalidae) larvae off the Oregon coast. *Fish. Bull. US* 81: 455-472.

Saila, S.G. & J.M. Flowers 1965. A simulation study of sex ratios and regulation effects with the American lobster, *Homarus americanus*. *Proc. Gulf and Carribean Fish. Inst.* 18: 16-78.

Scarratt, D.J. 1973. Abundance, survival, vertical and diurnal distribution of lobster larvae in Northumberland Strait 1962-73, and their relationships with commercial stocks. *J. Fish. Res. Board Can.* 30: 1819-1824.

Stasko, A.B. 1980. Possible causes for decline of lobster landings in Northumberland Strait. *CAFSAC Res. Doc.* 80/3: 7 pp.

Sutcliffe Jr, W.H. 1973. Correlations between seasonal river discharge and local landings of American lobster (*Homarus americanus*) and Atlantic halibut (*Hippoglossus hippoglossus*) in the Gulf of St. Lawrence. *J. Fish. Res. Board Can.* 30: 856-859.

Walters, C.J. 1985. Bias in the estimation of functional relationships from time series data. *Can. J. Fish. Aquat. Sci.* 42: 147-149.

Walters, C.J. & D. Ludwig 1981. Effects of measurement errors on the assessment of stock-recruitment relationships. *Can. J. Fish. Aquat. Sci.* 38: 704-710.

Wenner, A.M., H.M. Page & P.R. Siegel 1985. Variation in size at onset of egg production. In A.M. Wenner (ed.), *Factors in Adult Growth. Crustacean Issues* 3: 149-164. Rotterdam: Balkema.

Wickham, D.E. 1979. Predation by the nemertean, *Carcinonemertes errans*, on eggs of the Dungeness crab, *Cancer magister. Mar. Biol.* 55: 45-53.

Wickham, D.E. 1986. Epizootic infestations by nemertean brood parasites on commercially important crustaceans. *Can. J. Fish. Aquat. Sci.* 43: 2295-2302.

Winnor, R.A. 1966. Population fluctuations of the market crab (*Cancer magister* Dana) in the San Francisco area. *Calif. Dep. Fish. Game, MRO Ref.* No. 66-29: 14 pp.

List of authors

D.E. Aiken, Biological Station, St. Andrews, New Brunswick E0G 2X0, Canada
J.H. Annala, Fisheries Research Center, Box 297, Wellington, New Zealand
R.T. Bauer, Center for Crustacean Research, University of Southwestern Louisiana, Lafayette, LA 70504, USA
L.W. Botsford, Department of Wildlife & Fisheries, University of California, Davis, CA 95616, USA
M.S. Brody, US Environmental Protection Agency, Washington, DC 20460, USA
W.H. Clark, Jr., Bodega Marine Laboratory, Bodega Bay, CA 94923, USA
S. Corey, Department of Zoology, University of Guelph, Guelph, Ontario N1G 2W1, Canada
P.J. Crocos, CSIRO Division of Fisheries, Cleveland, Queensland 4163, Australia
J. Dugan, Department of Biology, University of California, Santa Barbara, CA 93106, USA
G.P. Ennis, Department of Fisheries and Oceans, St. John's, Newfoundland A1C 5X1, Canada
E.M. Hill, School of Animal Biology, University College, Bangor, Gwynedd LL59 5EY, UK
D.L. Holland, School of Animal Biology, University College, Bangor, Gwynedd LL59 5EY, UK
J.V. Huner, Center for Crustacean Research, University of Southwestern Louisiana, Lafayette, LA 70504, USA
A.M. Kuris, Department of Biology, University of California, Santa Barbara, CA 93106, USA
O.V. Lindqvist, University of Kuopio, Kuopio, Finland
K. Nelson, Bodega Marine Laboratory, Bodega Bay, CA 94923, USA
T. Onbé, Faculty of Applied Biology, Hiroshima University, Saijo, Higashi-Hiroshima 724, Japan
M.C. Pillai, Bodega Marine Laboratory, Bodega Bay, CA 94923, USA
D.E. Pollock, Sea Fisheries Research Institute, Private Bag X2, Rogge Bay 8012, Cape Town, South Africa
N.N. Rabalais, Louisiana Universities Marine Consortium, Chauvin, LA 70344, USA
J.D. Shields, Department of Parasitology, University of Queensland, St. Lucia, Brisbane, Queensland 4067, Australia

K.M. Somers, Department of Zoology, University of Toronto, Toronto, Ontario, M5S 1A1, Canada
D.H. Steele & V.J. Steele, Department of Biology, Memorial University of Newfoundland, St. John's, Newfoundland A1B 3X9, Canada
T. Subramoniam, Department of Zoology, University of Madras, Guindy Campus, Madras 600 025, India
P. Talbot, Department of Biology, University of California, Riverside, CA 92521, USA
N.B. Terwilliger, Department of Biology, University of Oregon, Eugene, OR 97403, USA
C.L. Van Dover, Woods Hole Oceanographic Institute, Woods Hole, MS 02543, USA
S.L. Waddy, Biological Station, St. Andrews, New Brunswick E0G 2X0, Canada
H. Wells, Department of Biology, Tulsa University, Tulsa, OK 74104, USA
A.M. Wenner, Department of Biology, University of California, Santa Barbara, CA 93106, USA
A.B. Williams, National Marine Fisheries, Systematics Laboratory, National Museum of Natural History, Washington, DC 20560, USA

Index

For Product Safety Concerns and Information please contact our EU representative GPSR@taylorandfrancis.com Taylor & Francis Verlag GmbH, Kaufingerstraße 24, 80331 München, Germany

Batch number: 10731069

Printed by Printforce, the Netherlands